RUSSIA AND THE NEAR

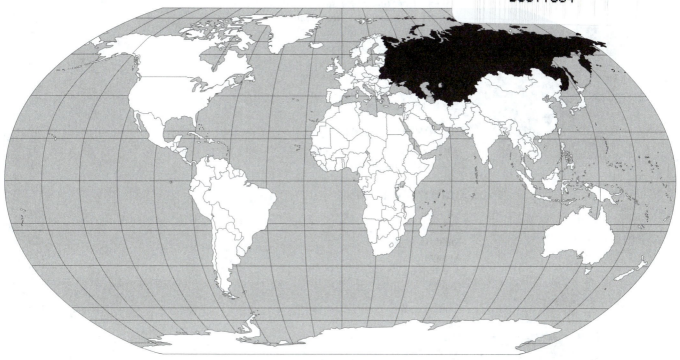

AUTHOR/EDITOR

Dr. Grigory Ioffe

Radford University

Grigory Ioffe, the author/editor of *Global Studies: Russia and the Near Abroad,* is professor of geography at Radford University in Radford, Virginia. He is the author of *Understanding Belarus and How Western Foreign Policy Misses the Mark* (Rowman and Littlefield 2008) and a coauthor of *The End of Peasantry? The Disintegration of Rural Russia* (University of Pittsburgh Press 2006), *Continuity and Change in Rural Russia* (Westview Press 1997) and two other books published in the United States. He has published articles and book reviews in the *Eurasian Geography and Economics, Europe—Asia Studies, East European Politics and Societies,* and other journals and has received research grant awards from the National Council for Eurasian and East European Research, the National Science Foundation, and NASA. A native of Moscow, Russia, Grigory Ioffe is well travelled in the former Soviet Union.

Contents

Articles from the World Press

Russia

Near Abroad

Using *Global Studies*: Russia and the Near Abroad

THE GLOBAL STUDIES SERIES

The Global Studies series was created to help readers acquire a basic knowledge and understanding of the regions and countries in the world. Each volume provides a foundation of information—geographic, cultural, economic, political, historical, artistic, and religious—that will allow readers to better assess the current and future problems within these countries and regions and to comprehend how events there might affect their own well-being. In short, these volumes present the background information necessary to respond to the realities of our global age.

Each of the volumes in the Global Studies series is crafted under the careful direction of an author/editor—an expert in the area under study. The author/editors teach and conduct research and have traveled extensively through the regions about which they are writing.

MAJOR FEATURES OF THE GLOBAL STUDIES SERIES

The Global Studies volumes are organized to provide concise information on the regions and countries within those areas under study. The major sections and features of the books are described here.

Regional Essay

For *Global Studies: Russia and the Near Abroad,* the author/editor has written two indepth essays discussing the peculiarities of the former Soviet Union at large and of its largest former republics, the Russian Federation.

Country Reports

Concise reports are written for each of the former Soviet republics, except Lithuania, Latvia, and Estonia discussed in the Global Studies: Europe.

The country reports are composed of several standard elements. Each report contains a detailed map visually positioning the country among its neighboring states; a summary of statistical information; a current essay providing important historical, geographical, political, cultural, and economic information; and "graphic indicators," with summary statements about the country in terms of development, freedom, and health/welfare.

A Note on the Statistical Reports

The statistical information provided for each country has been drawn from a wide range of sources. Every effort has been made to provide the most current and accurate information available. However, sometimes the information cited by these sources differs to some extent; and, all too often, the most current information available for some countries is somewhat dated. Aside from these occasional difficulties, the statistical summary of each country is generally quite complete and up-to-date. Care should be taken, however, in using these statistics (or, for that matter, any published statistics) in making hard comparisons among countries. We have also provided comparable statistics for the United States and Canada, which can be found on pages x and xi.

World Press Articles

Within each Global Studies volume is reprinted a number of articles carefully selected by our editorial staff and the author/editor from a broad range of international periodicals and newspapers. The articles have been chosen for currency, interest, and their differing perspectives on the subject countries. There are 23 articles in *Global Studies:* Russia and the Near Abroad.

The annotated table of contents, found on pages iv–v, offers a brief summary of each article, while the topic guide indicates the main theme(s) of each article. Thus, readers desiring to focus on articles dealing with a particular theme, say, the environment, may refer to the topic guide to find those articles.

Glossary, Bibliography, Index

At the back of each Global Studies volume, readers will find a glossary of terms and abbreviations, which provides a quick reference to the specialized vocabulary of the area under study and to the standard abbreviations used throughout the volume.

Following the glossary is a bibliography, which lists general works and national histories covering the region.

The index at the end of the volume is an accurate reference to the contents of the volume. Readers seeking specific information and citations should consult this standard index.

Currency and Usefulness

Global Studies: Russia and the Near Abroad like the other Global Studies volumes, is intended to provide the most current and useful information available necessary to understand the events that are shaping the cultures of the region today.

This volume is revised on a regular basis. The statistics are updated, regional essays and country reports revised, and world press articles replaced. In order to accomplish this task, we turn to our author/editor, our advisory boards, and—hopefully—to you, the users of this volume. Your comments are more than welcome. If you have an idea that you think will make the next edition more useful, an article or bit of information that will make it more current, or a general comment on its organization, content, or features that you would like to share with us, please send it in for serious consideration.

A Word from the Book's Author

I was born and raised in the Soviet Union, specifically in Moscow, Russia. Twenty years ago my family and I left our home country. At that time, in order to leave Russia for permanent residency abroad one had to give up one's citizenship; and we were even required to pay to do that. In 1995 I became a naturalized citizen of the United States, of which I am proud. See, if you were born American, you cannot count your citizenship as a personal achievement, but I can. No first-generation immigrant, however, can fully acculturate unless he or she made the move at a very early age. There is a noticeable difference even between my son, who was nine when we left Russia, and my daughter, who was one at the time. But there is an even bigger difference between both of them and me, for I left my native country at the age of thirty-eight. During the next twenty years, I vastly improved my English, and I authored and coauthored five English-language books, not counting this one. Several Americans and Britishers have published substantive reviews of my books, whereby "substantive" is no less important than "positive" (most were positive as well) because I was not sure I could strike a chord with readers raised in the West. Resonating, after all, is not just a function of grammar and vocabulary. People whose formative experiences occurred in far-flung countries are informed by different patterns of perception and explanation as well as by different stereotypes. But while the substantive reviews boosted my self-esteem, it was still obvious to me that I had a long way to go. It is likely that I am forever stuck between Russia and America, and if so I may be well positioned to write a book like this one, as it requires a combination of insider knowledge and some ability to render explanations tailored to outsiders' backgrounds.

My story, however, is more intricate, which is upsetting because I crave simplicity. Many Russians, for example, would maintain that I was never Russian to begin with, as my last name is unmistakably Jewish, and because it usually takes some negative feelings toward the country or its government to leave one's home forever. While both of these are certainly true in my case, I feel that my outlook is now more Russian than it ever was—an ironic outcome of my 20 years in America. I see two reasons for that. First, unlike in my native country, in America I am indeed referred to as a Russian because I came from Russia, my native language is Russian, and because Americans are not well versed in identity nuances in faraway countries. Second, as all immigrants, I am not immune to the us-versus-them mentality whereby "them" are Americans-by-birthright. For example, as befits a true Russian, I have miserably failed to adopt political correctness as a reasoning tool. I found it disheartening that a respected American academic, a one-time president of the Association of American Geographers, wrote days after 9/11 that "bin Laden is the Samuel Huntington [a prominent American political scientist (1927–2008)] of the Arab world. Like the Western pundits, bin Laden collapses ontology into geography; the key move of the modern geopolitical imagination."[1] Because bin Laden was referred to as someone who used a "mirror-image security mapping of the world to that offered by Huntington and other prophets"[2] in describing the contemporary conflict of civilizations, it follows that Huntington is the Osama bin Laden of the Judeo-Christian world, calling the West to battle. I find that absurd. Pointing out a cause of crucial contrast between large groups of people (as Huntington did) scarcely equals inciting a culture war. Besides, I used to live and travel extensively in the Soviet Union, Tajikistan and Kyrgyzstan being the only two union republics I never visited. In the Soviet Union, the straightjacket of the same rigid political system was imposed across a culturally diverse country, but even that did not make Soviet Turkmenistan the twin of Soviet Estonia. Now that the straightjacket is gone, arguing that the contrast between them has a pronounced cultural dimension is too easy, like pushing against an open door [a Russian saying, by the way]. To any Soviet insider, however, it was obvious all along. Living in Russia, in between the two culture realms those two countries represent, one could easily feel and reflect on the contrast between "East" and "West" and on the fact that Russia lay between them not only geographically but culturally as well—Russia is the "middle world of Eurasia," as the Slavophile Vladimir Lamansky once defined it. Thus, "collapsing ontology into geography" makes a whole lot of sense, and Russia's geo-cultural niche is just one example of that. But

there are many others. For instance, the Ukrainian language is closer to Russian and Polish than Russian and Polish are to each other—precisely because Ukraine is located between Russia and Poland, and the same applies to the Belarusian language and Belarus. As a discipline, geography would have a hard time justifying its existence if nothing indeed depended upon location.

Just as I am not friends with political correctness, likewise, I could never bring myself to share the native born American's enthusiasm for promoting democracy all over the world, although perhaps this evangelical passion is beginning to fade. In fact, speaking to the UN General Assembly on September 22, 2009, the president of the United States recognized that "democracy cannot be imposed on any nation from the outside. Each society must search for its own path, and no path is perfect. Each country will pursue a path rooted in the culture of its people and in its past traditions. And I admit that America has too often been selective in its promotion of democracy."[3] I cheered when I read this, as I thought that very little separates the statement of Barack Obama from Samuel Huntington's assertion that "Western belief in the universality of Western culture suffers three problems. It is false; it is immoral; and it is dangerous. . . . Imperialism is the necessary logical consequence of universalism."[4]

Just a couple of years ago, I wrote an article about Belarus in which I stated that "most people in the non-Western world want order, not democracy," and President Lukashenka of Belarus epitomizes Belarusians' longing for order. When I received page proofs from the editor, the above-mentioned words were rendered in the following way: "Most people in the non-Western world want order even more than democracy." I realized that I was in the midst of a fascinating cross-cultural experience and decided that I should strike a compromise. I deleted the word "even" from the edited version, and that was how the sentence was published.[5] More than anything else, this experience taught me to treat cultural dissimilarities with respect, staying away from politicizing and caricaturing them.

This is a commitment that I hope I have not betrayed in this book, which consists of three parts: (1) an essay about the Soviet Union focusing on the seeds of destruction harbored by that country since its birth in the November 1917 Bolshevik Revolution in Russia and on the actual factors, processes, and personalities instrumental in the demise of the Soviet Union in December 1991; (2) an essay on Russia's post-communist transformation from 1992 to 2009 emphasizing privatization and governance against the backdrop of Russia's authoritarian tradition and singling out some history-making personalities; and (3) essays on 11 other former Soviet republics (Latvia, Lithuania, and Estonia, whose 1940 incorporation in the Soviet Union was never recognized by Western governments, are discussed in a different Global Studies volume—that devoted to Europe). These essays have three important foci: ethnicity and ethno-national identity; accomplishments of the Soviet period; and post-Soviet developments. The focus on ethnicity is important as nation-building on the basis of a titular ethnic group became a driving force of post-communist development, and it had to either revisit or reinforce elements of popular mythology that took shape in the pre-Soviet past. In some cases, however, national identities were first constructed during the early Soviet period, as was the case in Central Asia with its history of national delimitation. The focus on what was accomplished during the Soviet period is important for the simple reason that national economies in the post-Soviet realm are path-dependent. That is, to a very significant extent they fall back on fixed assets, infrastructure, and resources created or developed during the Soviet period. In some cases, the Soviet legacy is an impediment or something to overcome, as for example, the layout of pipelines affixing Central Asia to Russia, but in many other cases it is the only available foundation for current livelihoods and future development. Analyses of post-Soviet events, processes, and personalities (the third of the three foci) do not need justification. I have to admit that keeping track of all of them is getting more and more difficult because, despite a legacy of belonging to one polity, the former Soviet republics are on increasingly divergent paths. Yet at least from a Russian perspective, they are still prodigal children, an idea reflected in the Russian cliché *blizhneye zarubezhye,* whose literal translation (a calque) is "near abroad." To wit, it is *kind of* abroad, whereas *dal'neye zarubezhye* or farther abroad *really* is abroad. I therefore extend an apology to whoever finds the title of this book offensive. I did not intend it to be mean.

Finally, I extend my gratitude to my colleague Susan Woodward, geography professor at Radford University, for bettering my English and to Minton E. Goldman of Northeastern University, the author of *Russia, the Eurasian Republics, and Central/Eastern Europe,* the title of earlier editions of this Global Studies book. Some of his material (on the Soviet Union and Russia) has been retained in this edition.

NOTES

1. John Agnew "Not The Wretched of the Earth: Osama bin Laden and the 'Clash of Civilizations'"; http://users.fmg.uva.nl/vmamadouh/awg/forum2/agnew.html.
2. Ibid.
3. http://www.nytimes.com/2009/09/24/us/politics/24prexy.text.html?pagewanted=7&_r=1
4. Samuel Huntington, *The Clash of Civilizations and the Remaking of World Order* (1996), p. 310.
5. Grigory Ioffe, "Nation-Building in Belarus: A Rebuttal," *Eurasian Geography and Economics,* 2007, No. 1: 68.

Internet References for *Global Studies: Russia and the Near Abroad*

**(Some websites continually change their structure and content, so the information
listed here may not always be available.)**

GENERAL SITES

Russia Today
www.europeaninternet.com/russia/

This website provides current headlines and commentary on topics of interest
from Russia.

Library of Congress Country Studies
www.lcweb2.loc.gov/frd/cs/cshome.html

An invaluable resource for facts and analysis of 100 countries' political, economic, social, and national-security systems and installations.

REESWeb: Russia and East European Studies
Internet Resources
www.ucis.pitt.edu/reesweb

Here is a comprehensive index of electronic resources on the Balkans, the
Baltic states, the Caucasus, Central Asia, Central Europe, the C.I.S., Eastern
Europe, the NIS, the Russian Federation, and the former Soviet Union. Direc-
tories are by discipline and by type.

United Nations System
www.unsystem.org

Official Web site for the UN. Everything is listed alphabetically, and data on
UNICC and the Food and Agriculture Organization are available.

UN Environmental Programme (UNEP)
www.unep.org/

Official site of UNEP with information on UN environmental programs, products, services, events, and a search engine.

Central Intelligence Agency
www.cia.gov/index.html

This site includes publications of the CIA, CIA maps, and more.

U.S. Department of State Home Page
www.state.gov/www/ind.html

Country Reports, Human Rights, International Organizations, and more are
organized alphabetically.

RUSSIAN

The Russia Journal
www.russiajournal.com/

The Russia Journal is Russia's top read English-language business weekly.
Published since 1998 in Moscow, The Russia Journal is acknowledged as the
world's leading newsmagazine on Russian affairs and read for its opinion,
insight, and analysis. With contributions from distinguished in-house writers
and leading Russian journalists, it has become firmly established as the most
authoritative and influential publication on and from Russia.

Russia Home Page
*www.csudh.edu/global_options/375Students-Sp96/Russia/Default
.htm*

A comprehensive Russian site is found here that includes history, physical and
energy resources, demographics, technologies, overall development plans (both
political and economic), cultural characteristics, and links to other sources.

Russia on the Web
www.valley.net/~transnat/

This project of the Transnational Institute is an interactive Internet access to
everything Russian. Its offerings include Culture and Art, Language and Literature, Society and Politics in Russia, plus information and news on Business and
Economy, History, Religion, Education, and Philosophy.

EURASIAN REPUBLICS

Russia and Eurasian Republics
www.magenheimer.com/peter/russia/

This is a Web page for teachers covering the Eurasian republics, including
geography and a brief fact list for each country.

Interactive Central Asia Resource Project (ICARP)
www.icarp.org/

ICARP leads to a wide-ranging Central Asia group of institutes and organizations that includes the American Association for the Advancement of Slavic
Studies, Amnesty International Online, the Association for Asian Studies, and
the Center for the Study of the Eurasian Nomads, to name just a few.

The United States (United States of America)

GEOGRAPHY

Area in Square Miles (Kilometers): 3,794,085 (9,826,630) (about 1/2 the size of Russia)

Capital (Population): Washington, DC (563,400)

Environmental Concerns: air and water pollution; limited freshwater resources, desertification; loss of habitat; waste disposal; acid rain

Geographical Features: vast central plain, mountains in the west, hills and low mountains in the east; rugged mountains and broad river valleys in Alaska; volcanic topography in Hawaii

Climate: mostly temperate, but ranging from tropical to arctic

PEOPLE

Population

Total: 301,139,947

Annual Growth Rate: 0.89%

Rural/Urban Population Ratio: 19/81

Major Languages: predominantly English; a sizable Spanish-speaking minority; many others

Ethnic Makeup: 82% white; 13% black; 4% Asian; 1% Amerindian and others

Religions: 52% Protestant; 24% Roman Catholic; 1% Jewish; 13% others; 10% none or unaffiliated

Health

Life Expectancy at Birth: 75 years (male); 81 years (female)

Infant Mortality: 6.37/1,000 live births

Physicians Available: 2.3/1000 people

HIV/AIDS Rate in Adults: 0.6%

Education

Adult Literacy Rate: 97% (official)

Compulsory (Ages): 7–16

COMMUNICATION

Telephones: 177,900,000 main lines

Daily Newspaper Circulation: 196.3/1,000 people

Televisions: 844/1,000 people

Internet Users: 208,000,000 (2006)

TRANSPORTATION

Highways in Miles (Kilometers): 3,986,827 (6,430,366)

Railroads in Miles (Kilometers): 140,499 (226,612)

Usable Airfields: 14,947

Motor Vehicles in Use: 229,620,000

GOVERNMENT

Type: federal republic

Independence Date: July 4, 1776

Head of State/Government: President Barak H. Obama is both head of state and head of government

Political Parties: Democratic Party; Republican Party; others of relatively minor political significance

Suffrage: universal at 18

MILITARY

Military Expenditures (% of GDP): 4.06%

Current Disputes: various boundary and territorial disputes; Iraq and Afghanistan; "war on terrorism"

ECONOMY

Per Capita Income/GDP: $43,800/$13.06 trillion

GDP Growth Rate: 2.9% (2006)

Inflation Rate: 3.2%

Unemployment Rate: 4.8%

Population Below Poverty Line: 12%

Natural Resources: many minerals and metals; petroleum; natural gas; timber; arable land

Agriculture: food grains; feed crops; fruits and vegetables; oil-bearing crops; livestock; dairy products

Industry: diversified in both capital and consumer-goods industries

Exports: $1.023 trillion (primary partners Canada, Mexico, Japan, China, U.K.)

Imports: $1.861 trillion (primary partners Canada, Mexico, Japan, China, Germany)

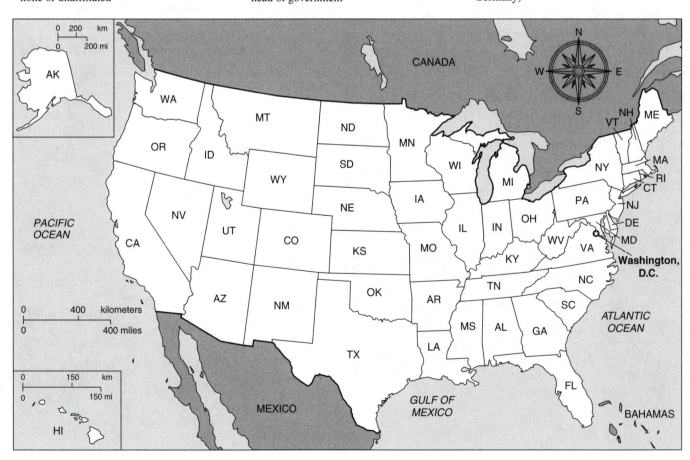

Canada

GEOGRAPHY

Area in Square Miles (Kilometers):
3,855,103 (9,984,670) (slightly larger than the United States)
Capital (Population): Ottawa (1,560,000)
Environmental Concerns: air and water pollution; acid rain; industrial damage to agriculture and forest productivity
Geographical Features: permafrost in the north; mountains in the west; central plains; lowlands in the southeast
Climate: varies from temperate to arctic

PEOPLE

Population
Total: 33,390,141(2007)
Annual Growth Rate: 0.87%
Rural/Urban Population Ratio: 20/80
Major Languages: both English and French are official
Ethnic Makeup: 28% British Isles origin; 23% French origin; 15% other European; 6% others; 2% indigenous; 26% mixed
Religions: 42.6% Roman Catholic; 27.7% Protestant; 12.7% others; 16% none.

Health
Life Expectancy at Birth: 77 years (male); 84 years (female)
Infant Mortality: 4.63/1,000 live births
Physicians Available: 2.1/1,000 people
HIV/AIDS Rate in Adults: 0

Education
Adult Literacy Rate: 97%
Compulsory (Ages): 6–16

COMMUNICATION
Telephones: 20,780,000 main lines
Daily Newspaper Circulation: 167.9/1,000 people
Televisions: 709/1,000 people
Internet Users: 22,000,000 (2006)

TRANSPORTATION
Highways in Miles (Kilometers): 646,226 (1,042,300)
Railroads in Miles (Kilometers): 29,802 (48,068)
Usable Airfields: 1,343
Motor Vehicles in Use: 18,360,000

GOVERNMENT
Type: federation with parliamentary democracy
Independence Date: July 1, 1867
Head of State/Government: Queen Elizabeth II; Prime Minister Stephen Harper
Political Parties: Conservative Party of Canada; Liberal Party; New Democratic Party; Bloc Québécois; Green Party
Suffrage: universal at 18

MILITARY
Military Expenditures (% of GDP): 1.1%
Current Disputes: maritime boundary disputes with the United States and Denmark (Greenland)

ECONOMY
Currency ($U.S. equivalent): 0.97 Canadian dollars = $1 (Oct. 2007)
Per Capita Income/GDP: $35,700/$1.181 trillion
GDP Growth Rate: 2.8%
Inflation Rate: 2%
Unemployment Rate: 6.4% (2006)
Labor Force by Occupation: 75% services; 14% manufacturing; 2% agriculture; and 8% others
Natural Resources: petroleum; natural gas; fish; minerals; cement; forestry products; wildlife; hydropower
Agriculture: grains; livestock; dairy products; potatoes; hogs; poultry and eggs; tobacco; fruits and vegetables
Industry: oil production and refining; natural-gas development; fish products; wood and paper products; chemicals; transportation equipment
Exports: $401.7 billion (primary partners United States, Japan, United Kingdom)
Imports: $356.5 billion (primary partners United States, China, Japan)

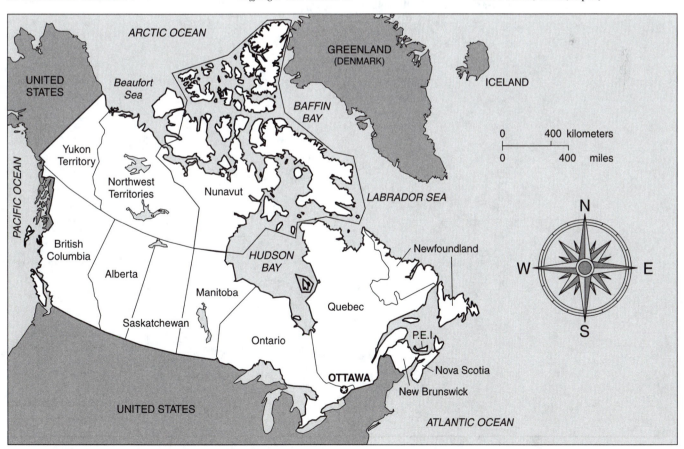

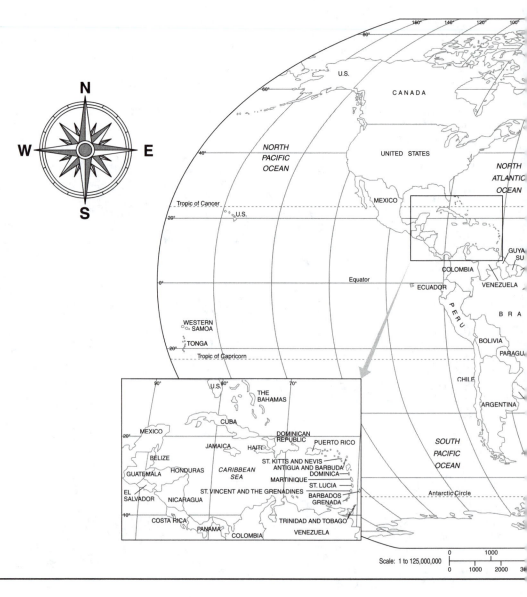

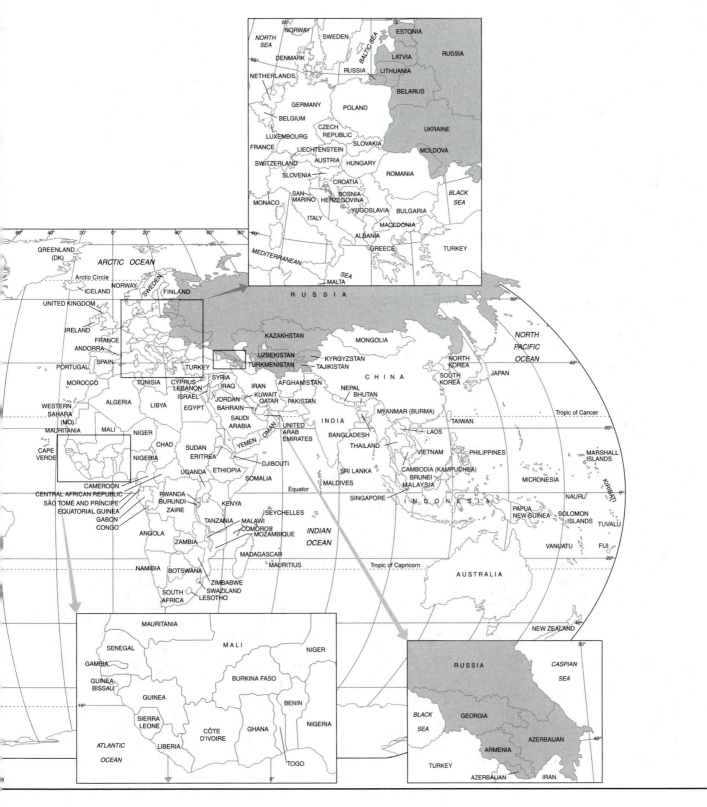

Russia and the Near Abroad

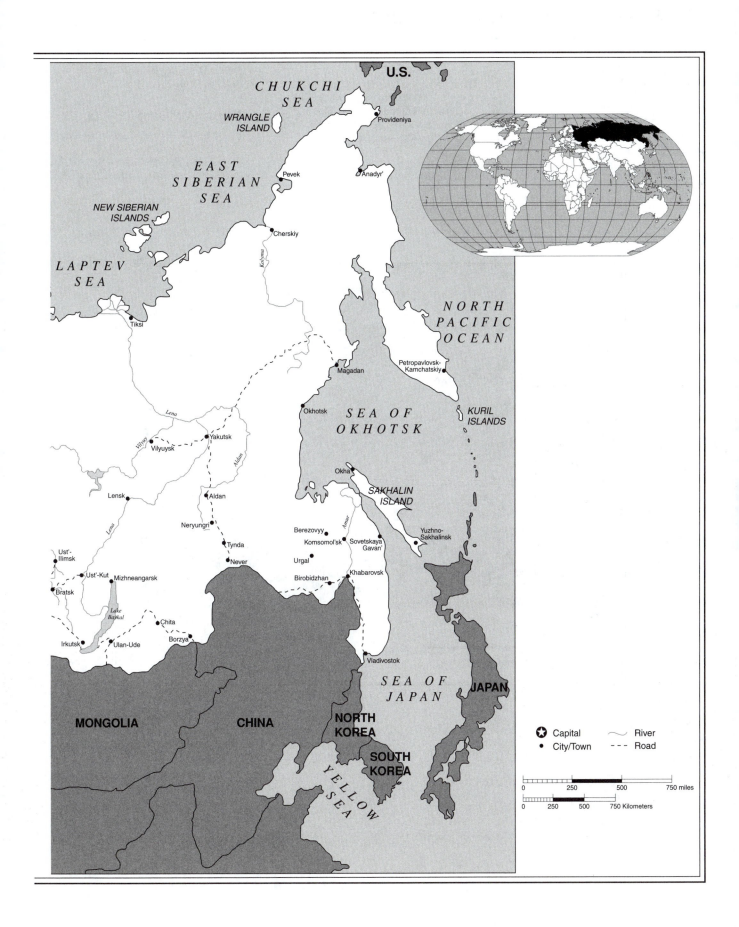

Russia (Russian Federation)

Area in Square Miles (Kilometers):
17,075,200 (6,591,027)

Capital (Population): Moscow
(10,524,400)

Population

Total (millions): 141.8
Per Square km: 8

Percent Urban: 73
Rate of Natural Increase (%): −0.3
Ethnic Makeup: Russian 79.8%; Tatar
3.8%; Ukrainian 2%; Bashkir 1.2%;
Chuvash 1.1%; other or unspecified
12.1% (2002 census)

Health

Life Expectancy at Birth: 61 (male);
74 (female)

Infant Mortality Rate: 9/1,000
Total Fertility Rate: 1.5

Economy/Well-Being

*Currency ($US equivalent on 14 October
2009):* 29.6 rubles = $1
2008 GNI PPP Per Capita: $15,630
*2005 Percent Living on Less than US$2/
Day:* <2

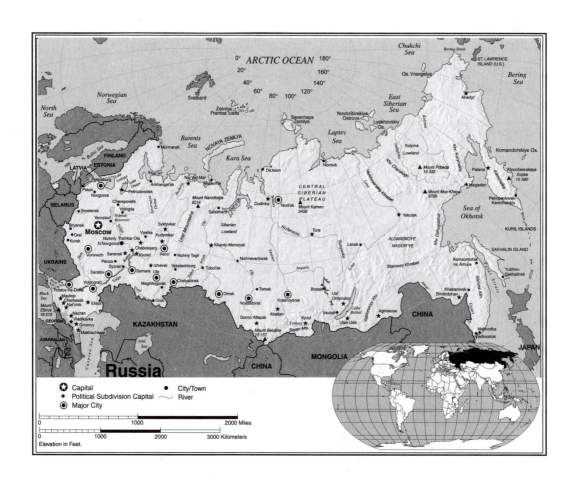

The Eurasian Republics

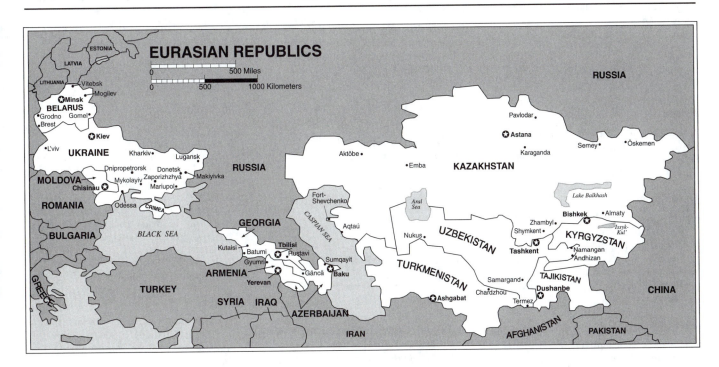

Armenia

Area in Square Miles (Kilometers):
11,503 (29,800) (about the size of Maryland)
Capital (Population): Yerevan (1,356,000)

Population

Total: 3,100,000
Per Square Km: 104
Percent Urban: 64
Rate of Natural Increase (%): 0.6
Major Languages: Armenian; Russian
Ethnic Makeup: 97.7% Armenian; 0.9% Russian; 0.4% others
Religions: 94.7% Armenian Orthodox; 4% others

Health

Life Expectancy at Birth: 68.25 years (male); 76.02 years (female)
Infant Mortality Rate (Ratio): 25/1,000 22.47 deaths/1,000 live births
Total Fertility Rate: 1.7

Government

Political Parties: Pan-Armenian National Movement; National Democratic Union; Armenian Revolutionary Federation; Christian Democratic Party; others

Economy

Currency ($ U.S. Equivalent):
385.4 drams = $1
Per Capita Income: $6,310
Percent Living on Less than US$2/Day: 43
Exports: $800 million
Imports: $1.5 billion f.o.b.

Azerbaijan

Area in Square Miles (Kilometers):
33,428 (86,600) (about the size of Maine)
Capital (Population): Baku (1,917,000)

Population

Total: 8,800,000
Per Square Km: 101
Percent Urban: 52
Rate of Natural Increase (%): 1.2
Major Languages: Azeri 89%; Russian 3%; Armenian 2%; other 6%
Ethnic Makeup: 90.6% Azerbaijani (Azeri); 2.2% Daghestani; 1.8% Russian; 5.4% Armenian, and others
Religions: 93% Muslim; 7% Russian Orthodox, Armenian Orthodox, and others

Health

Life Expectancy at Birth: 59.78 years (male); 68.13 years (female)
Infant Mortality Rate (Ratio): 11/1,000
Total Fertility Rate: 2.3

Government

Political Parties: New Azerbaijan Party; Musavat Party; National Independence Party; Social Democratic Party; Azerbaijan Popular Front; Communist Party; others

Economy

Currency ($ U.S. Equivalent):
.80 manats = $1
Per Capita Income: $7,770
Percent Living on Less than US$2/Day: <2
Exports: $6.117 billion
Imports: $4.656 billion

Belarus

Area in Square Miles (Kilometers): 80,134 (207,600) (about the size of Kansas)
Capital (Population): Minsk (1,830,000)

Population

Total: 9,700,000

Per Square Km: 47
Percent Urban: 74
Rate of Natural Increase (%): −0.3
Major Languages: Belarussian;
   Russian
Ethnic Makeup: 81.2% Belarussian;
   11.4% Russian; 3.9% Polish;
   3% Ukrainian; 1.1% others
Religions: 80% Eastern Orthodox; 20%
   others

Health

Life Expectancy at Birth: 65 years
   (male); 76 years (female)
Infant Mortality Rate (Ratio): 5/1,000
Total Fertility Rate: 1.4

Government

Political Parties: Belarussian Popular
   Front; United Democratic Party;
   Social Democratic Party; Party of
   Popular Accord; Belarussian Party of
   Communists; Belarussian Socialist
   Party; Women's Party Nadezhda;
   others

Economy

Currency ($ U.S. Equivalent): 2,738
   Belarussian rubles = $1
Per Capita Income: $12,150
Percent Living on Less than US$2/Day:
   <2
Exports: $16.14 billion
Imports: $16.94 billion

Georgia

Area in Square Miles (Kilometers):
   26,904 (69,700) (slightly smaller
   than South Carolina)
Capital (Population): Tbilisi (1,480,000)

Population

Total: 4,600,000
Per Square Km: 66
Percent Urban: 53
Rate of Natural Increase (%): 0.2
Major Languages: Georgian; Russian;
   Armenian; Azeri; Abkhaz
Ethnic Makeup: 83.8% Georgian; 5.7%
   Armenian; 1.5% Russian; 6.5% Azeri;
   2.5% other
Religions: 83.9% Georgian Orthodox;
   9.9% Muslim; 10% Russian Orthodox;
   3.9% Armenian; 0.8% Catholic; 0.8%
   other; 0.7% none

Health

Life Expectancy at Birth: 71 years (male);
   79 years (female)

Infant Mortality Rate (Ratio): 13/1,000
Total Fertility Rate: 1.4

Government

Political Parties: Citizens Union of
   Georgia; United Republican Party;
   National Democratic Party; Socialist
   Party; others

Economy

Currency ($ U.S. Equivalent):
   1.67 laris = $1
Per Capita Income: $4,850
Percent Living on Less than US$2/Day:
   30
Exports: $1.4 billion
Imports: $2.5 billion

Kazakhstan

Area in Square Miles (Kilometers):
   1,048,878 (2,717,300) (slightly 4 times
   the size of Texas)
Capital (Population): Astana (750,630)

Population

Total: 15,900,000
Per Square Km: 6
Percent Urban: 53
Rate of Natural Increases (%): 1.3
Major Languages: Kazalj; Russian
Ethnic Makeup: 53.4% Kazakh; 30%
   Russian; 3.7% Ukrainian; 2.4%
   German; 2.5% Uzbek; 1.7% Tatar;
   4.9% others
Religions: 47% Muslim; 1.4% Uygur; 44%
   Russian Orthodox; 2% Protestant; 4.9%
   others

Health

Life Expectancy at Birth: 62 years (male);
   72 years (female)
Infant Mortality Rate (Ratio):
   32/1,000
Total Fertility Rate: 2.7

Government

Political Parties: People's Unity Party;
   Communist Party; Republican People's
   Party; others

Economy

Currency ($ U.S. Equivalent):
   151 tenges = $1
Per Capita Income: $9,690
Percent Living on Less than US$2/Day:
   17
Exports: $30.09 billion
Imports: $17.51 billion

Kyrgyzstan

Area in Square Miles (Kilometers):
   76,621 (198,500) (about the size
   of South Dakota)
Capital (Population): Bishkek
   (1,250,000)

Population

Total: 5,300,000
Per Square Km: 27
Percent Urban: 35
Rate of Natural Increases (%): 1.6
Major Languages: Kirghiz; Russian;
   Uzbek
Ethnic Makeup: 64.9% Kirghiz; 12.5%
   Russian; 13.8% Uzbek; 1% Ukrainian;
   1% Uygur; 1.1% Dungan; 5.7%
   others
Religions: 75% Muslim; 20% Russian
   Orthodox; 5% others

Health

Life Expectancy at Birth: 64 years (male);
   72 years (female)
Infant Mortality Rate (Ratio): 31/1,000
Total Fertility Rate: 2.8

Government

Political Parties: Social Democratic
   Party; Kyrgyz Democratic Movement;
   National Unity; Communist Party;
   others

Economy

Currency ($ U.S. Equivalent):
   43 soms = $1
Per Capita Income: $2,130
Percent Living on Less than US$2/Day:
   52
Exports: $759 million
Imports: $937.4 million

Moldova

Area in Square Miles (Kilometers):
   13,008 (33,843) (slightly larger than
   Maryland)
Capital (Population): Chisinau (formerly
   Kishinev) (610,150)

Population

Total: 4,100,000
Per Square Km: 122
Percent Urban: 41
Rate of Natural Increases (%): −0.1
Major Languages: Moldovan; Romanian;
   Russian; Gagauz (a Turkish dialect)

Ethnic Makeup: 78.2% Moldovan/
  Romanian; 8.4% Ukrainian; 5.8%
  Russian; 4.4% Gagauz, 1.9%
  Bulgarian, and 1.3% others
Religions: 98% Eastern Orthodox;
  1.5% Jewish; 0.5% others

Health

Life Expectancy at Birth: 65 years (male);
  73 years (female)
Infant Mortality Rate (Ratio):
  12/1,000
Total Fertility Rate: 1.3

Government

Political Parties: Christian Democratic
  Popular Front; Yedinstvo Movement;
  Social Democratic Party; Agrarian
  Democratic Party; Democratic Party;
  Democratic Labor Party; others

Economy

Currency ($ U.S. Equivalent):
  11.1 lei = $1
Per Capita Income: $3,210
Percent Living on Less than US$2/Day:
  29
Exports: $1.04 billion
Imports: $2.23 billion

Tajikistan

Area in Square Miles (Kilometers):
  55,237 (143,100) (slightly smaller than
  Wisconsin)
Capital (Population): Dushanbe
  (679,400)

Population

Total: 7,500,000
Per Square Km: 52
Percent Urban: 26
Rate of Natural Increases (%): 2.3
Major Languages: Tajik; Russian
Ethnic Makeup: 79.9% Tajik; 15.3%
  Uzbek; 1.1% Russian; 1.1% Kyrgyz;
  2.6% others
Religions: 85% Sunni Muslim; 5% Shia
  Muslim; 10% others

Health

Life Expectancy at Birth: 64 years (male);
  69 years (female)
Infant Mortality Rate (Ratio): 65/1,000
Total Fertility Rate: 3.4

Government

Political Parties: People's Democratic
  Party; Tajik Socialist Party; Communist
  Party; Islamic Renaissance Party;
  others

Economy

Currency ($ U.S. Equivalent): 4.4
  Tajikistani somoni = $1
Per Capita Income: $1,860
Percent Living on Less than US$2/Day:
  51
Exports: $950 million
Imports: $1.25 billion

Turkmenistan

Area in Square Miles (Kilometers):
  188,407 (488,100) (about the size of
  California)
Capital (Population): Ashkhabad
  (Ashgabat) (1,000,000)

Population

Total: 5,100,000
Per Square Km: 10
Percent Urban: 47
Rate of Natural Increases (%): 1.4
Major Languages: Turkmen; Russian;
  Uzbek
Ethnic Makeup: 85% Turkmen; 4%
  Russian; 5% Uzbek; 6% others
Religions: 89% Muslim; 9% Eastern
  Orthodox; 2% unknown

Health

Life Expectancy at Birth: 61 years (male);
  69 years (female)
Infant Mortality Rate (Ratio): 51/1,000
Total Fertility Rate: 2.5

Government

Political Parties: Democratic Party of
  Turkmenistan; opposition parties are
  outlawed

Economy

Currency ($ U.S. Equivalent): 14,232
  Turkman manats = $1
Per Capita Income: $6,210
Percent Living on Less than US$2/Day:
  50
Exports: $4.7 billion
Imports: $4.175 billion

Ukraine

Area in Square Miles (Kilometers):
  233,028 (603,700) (Slightly smaller
  than Texas)
Capital (Population): Kiev (2,800,000)

Population

Total: 46,000,000
Per Square Km: 76
Percent Urban: 68
Rate of Natural Increases (%): −0.5
Major Languages: Ukrainian; Russian;
  Romanian; Polish
Ethnic Makeup: 77.8% Ukrainian; 17.3%
  Russian; 0.6% Belarusian; 0.5%
  Moldovan; 0.5% Crimean Tatar; 0.4%
  Bulgarian; 0.3% Hungarian; 0.3%
  Romanian; 0.3% Polish; 0.2% Jewish;
  1.8% other
Religions: Ukrainian Orthodox;
  Kiev Patriarchate; Ukrainian
  Autocephalous Orthodox; Ukrainian
  Catholic; Protestant; Jewish
  (percentages unknown) Moscow
  Patriarchate

Health

Life Expectancy at Birth: 63 years (male);
  74 years (female)
Infant Mortality Rate (Ratio): 10/1,000
Total Fertility Rate: 1.4

Government

Political Parties: Green Party; Liberal
  Party; Liberal Democratic Party;
  Democratic Party; People's Party;
  Social Democratic Party; Ukrainian
  Christian Democratic Party;
  Republican Party; others

Economy

Currency ($ U.S. Equivalent):
  8.2805 hryvnia = $1
Per Capita Income: $7,210
Percent Living on Less than US$2/Day:
  <2
Exports: $38.22 billion
Imports: $37.18 billion

Uzbekistan

Area in Square Miles (Kilometers):
  172,696 (447,440) (about the size of
  California)
Capital (Population): Tashkent
  (2,180,000)

Population

Total: 27,600,000
Per Square Km: 10
Percent Urban: 36
Rate of Natural Increases (%): 1.8
Major Languages: 74.3 Uzbek;
  14.2 Russian; 4.4 Tajik; 7.1%
  other

Ethnic Makeup: Uzbek 80%; Russian 5.5%; Tajik 5%; Kazakh 3%; Karakalpak 2.5%; Tatar 1.5%; other 2.5%

Religions: 88% Muslim (mostly Sunni); 9% Eastern Orthodox; 3% others

Health

Life Expectancy at Birth: 65 years (male); 71 years (female)

Infant Mortality Rate (Ratio): 48/1,000

Total Fertility Rate: 2.6

Government

Political Parties: People's Democratic Party (formerly the Communist Party); Fatherland Progress Party; Social Democratic Party; others

Economy

Currency ($ U.S. Equivalent): 1,498 soms = $1

Per Capita Income: $2,660

Percent Living on Less than US$2/Day: 50

Exports: $5 billion

Imports: $3.8 billion

The Soviet Union   Regional Essay

For generations, the Union of the Soviet Socialist Republics, founded by Russian Marxists in the early twentieth century and led by them until its collapse in 1991, survived despite its many flaws. Many people abroad as well as within that communist state believed it would endure a long time. The Soviet Union, however, collapsed and disappeared from maps at the end of 1991, having harbored the seeds of its own destruction since its birth in the November 1917 Bolshevik Revolution in Russia. The Soviet Union was the largest country in the world, encompassing one-sixth of the world's total landmass and hosting about 150 different ethnic groups. Russians made up the largest of those groups, but by the time of the Soviet Union's breakup they accounted for a little less than one-half of that country's 289-million population. Along with Russians, fourteen other major ethnic groups were titular nationalities of the so-called union republics, the major political subdivisions of the Soviet Union; that is, Russians were in majority in Russia, Ukrainians in Ukraine, Armenians in Armenia, and so on.

After the demise of the Soviet Union, starting in 1992, the fifteen republics into which it had been divided administratively, became independent and sovereign states and slowly scrapped most communist institutions in favor of a new political order. For lack of a more precise term, this order has been labeled post-Communist, although this description obscures more than it reveals. By now each successor state of the Soviet Union has followed its own unique path; and it is getting more and more difficult to corral them into the same epithet, so disparate and at times divergent have these paths become. If there is any conceptual model that might render this mindboggling variety more comprehensible, it is probably a controversial concept of civilizations to which the political scientist Samuel Huntington contributed. Huntington believed that with the end of the Cold War, cultural identities, antagonisms, and affiliations, not ideologies, would play the major role in establishing the world order and relations between states. Huntington regarded belief in the universalism of Western values unjustified, arrogant, and dangerous. He warned that, outside Western civilization, the implantation and promotion of market economy and Western-style democracy would be difficult to achieve; and so far this prediction has proved valid.

Within the broad expanse of the Soviet Union, three major civilizations co-existed: Western, Islamic, and Byzantine-Orthodox. Today, the Baltic States of Latvia, Lithuania, and Estonia which fall within the Western civilization sphere are the most successful of all post-Soviet countries in terms of market economy and democratic forms of governance. The Baltic States are also integrated into Western structures, such as the European Union and NATO. It may be of some importance that the Baltic States were independent sovereign countries between World War I and World War II and were incorporated into the Soviet Union in 1940, so their Soviet experiences were shorter than those of other republics.

At the other extreme, the Central Asian *Stans,* part and parcel of Muslim civilization, have achieved least in establishing market infrastructure and democracy, and, if anything, have even experienced setbacks in such areas as living standards, employment, and gender roles. Yet all of the Central Asian republics outperform by far neighboring Afghanistan, home to some of the same Central Asian ethnicities (Uzbek, Turkmen, and particularly Tadjk)—a testimony to the fact that Central Asia may have actually gained much from its Soviet experiences, an observation legitimately at odds with the evaluation of communist legacies in the Baltic States and most countries of East Central Europe except, arguably, Bulgaria, Romania, and parts of former Yugoslavia.

Russia, Ukraine, Belarus, Armenia, and Georgia, all with Byzntine-Orthodox heritages, should be placed in between these two extremes (that is, between the Western-integrated Baltic States and Muslim Central Asia); whereas Muslim Azerbaijan, though geographically contiguous with the aforementioned countries, is closer developmentally to Central Asia, its fast economic growth fueled by oil extraction notwithstanding. Apparently the 70-year-long Soviet experience did not quell the cultural ferment of three long established civilizations. Even when the Soviet Union existed it was quite obvious to a well-travelled Muscovite that, say, Estonia and Turkmenistan belonged in different worlds, and Russia lay somewhere in between, and not just geographically. The intermediate niche of Russia has long been evidenced by the parallel and at times combative existence of two intellectual trends of Russian thought, those held respectively by Westernizers and Slavophiles.

Slavophiles and *Westernizers* are designations for two groups of intellectuals in mid-nineteenth-century Russia that represented opposing schools of thought concerning ways to develop Russia. The Slavophiles held that Russia represents a self-styled civilization, unique and superior to Western culture because it was based on such institutions as the Eastern Orthodox Church, the village community, or *mir,* and the ancient popular assembly, the *sobor.* Both mir and sobor are forms of direct democracy, where "direct" means not mediated by such elected institutions as courts and parliaments. The Slavophiles supported autocracy and opposed political participation. Slavophiles became increasingly nationalistic, which actually made the very term Slavophile (a person that loves Slavs, that is, not only Russians but also, say, Poles, who are Slavs as well) a misnomer. The Westernizers believed that Russia's development depended on the adoption of Western technology and liberal government. In their approach they were rationalistic and often agnostic rather than emotional and mystical, mindsets that tended to characterize Slavophiles. Some remained moderate liberals while others became socialists and political radicals.

Similar rivalries between Western- and Eastern-leaning national projects characterize the nation-building processes in Belarus and Ukraine, countries dissected by Huntington's "civilizational fault line" (see Map on next page) in a way that their smaller parts belong in the Western civilization culture region and larger parts in the Byzantine-Orthodox world.

Grey line adapted from Samuel Huntington, *The Clash of Civilizations* (New York: Simon and Schuster, 1996). Black line (partially dashed) adapted from Piotr Eberhardt, "The Concept of Boundary between Latin and Byzantine Civilization in Europe," *Przegląd Geograficzny* 76, no. 2 (2004):169–88.

There is no question that this tri-part Western, Islamic, and Byzantine-Orthodox typology is an oversimplification. The real world is incomparably more nuanced; and the stereotypical perception of an outsider who is a consumer of world news from not always unbiased sources must also be reckoned with. For example, since their respective "colored revolutions"— orange in Ukraine and rose in Georgia—these countries have been lauded in the West as true democracies, whereas Russia and particularly Belarus have been castigated for the undemocratic behaviors of their political regimes. But if one considers the everyday life of rank-and-file citizens of those countries and not Western reactions to their leaders (reactions that are arguably more grounded in geopolitics than in the actual democratization of the respective countries), then the civilizational perspective appears to be largely valid. Relatively few

Ukrainians have benefited from their fledgling democracy, the country's political leadership is ravaged by an incessant power struggle, and seemingly democratic Ukraine is now lagging behind autocratic Belarus on such socio-economic indicators as GDP per capita, real personal income, and infant mortality. Georgia not only suffered a serious blow to its economy during the 2008 five-day war with Russia, but also the democratic credentials of its America-friendly president have been called into question on more than one occasion.

Perhaps challenging the civilizational concept, Muslim Kazakhstan has succeeded economically so much more than other *stans,* but it has a significant (34%) and reasonably well-integrated Slavic population[1] as do the Baltic countries of Latvia and Estonia. And the Central Asian country of Kyrgyzstan, although home to the third colored revolution (the tulip one),

A WORD ON TERMS

The terms *Russia* and *Soviet Union*. Prior to December 1991, the terms "Russia" and "the Soviet Union" were frequently used interchangeably. This was not quite accurate. *Russia* was the largest, most populous, and most politically influential of the 15 constituent union-republics that comprised the Union of Soviet Socialist Republics, commonly known as the Soviet Union, or the U.S.S.R. Thus, Russia was part of the Soviet Union, not synonymous with it. However, because the Soviet Union actually encompassed the same geographic space that before the Communist Revolution of 1917 was called Russia (more specifically, the Russian Empire), equating the Soviet Union with Russia cannot be disqualified as something entirely groundless. Moreover, even after the Communist Revolution, the country was still named Russia for some time (Soviet Russia to be exact). Then, in 1922, some non-Russian ethnicities acquired their demarcated homelands within the former Russian Empire. These semi-autonomous homelands became known as union republics, and thus the Soviet Union was born. Initially there were just four republics, but their proliferation throughout the 1920s and '30s drove their overall number to 15. The Soviet Union was thus a federation that was formally established by the Soviet Communist Party in 1922. The Russian word *Soviet* has a double meaning: advice and council. The second meaning is essential for interpreting the name of the country, whose life span turned out to be nearly 70 years (1922–1991). The word *Soviet* came into popular usage in the Russian Revolutions of 1905 and 1917, when insurgent workers, peasants, and soldiers formed soviets, or councils, in St. Petersburg, Moscow, and other Russian cities to oppose and create an alternative to czarist authority. "Soviet" thus referred to the formal parliamentary type of government of the federation and its constituent union-republics. The Soviet type of government prevailed in all levels of administration in the Soviet Union, down to rural districts and towns.

has experienced more of a change in ruling clans than a far-reaching democratic transformation.

Our coverage of the post-Soviet space will begin by outlining its shared Soviet experiences and the processes that led to the breakup of the Soviet Union. We will then focus on the Russian Federation, the bulwark of the former U.S.S.R., its largest republic, and the only true successor state, since it acquired what used to be the Soviet seat on the Security Council of the United Nations. The essay on Russia will be followed by shorter essays devoted to each of the other former Soviet republics except the Baltic States of Latvia, Lithuania, and Estonia.

PHASES OF SOVIET/COMMUNIST RULE 1922–1991

In the 69-year period of its existence from 1922, when it was officially set up and called the Union of Soviet Socialist Republics, until its collapse at the end of 1991, the Soviet political system went through phases of change and development that coincided with phases of leadership starting with Lenin in 1917, following the Bolshevik Revolution, and ending with Mikhail Gorbachev's resignation as Soviet president in December, 1991.

The Lenin Phase (1917–1922)

The Lenin phase was harsh and brutal as the Bolsheviks, now calling themselves communists, sought to consolidate their hold on power, make it permanent, and use it to start the reforms leading to socialism and, ultimately, communism. The changes accomplished during the Lenin phase were monumental: victorious conclusion of a long civil war against military remnants of the czarist system and the foreign intervention on its behalf; establishment of institutions of government and affirmation of the Communist Party's supremacy in Soviet politics; suppression of criticism and dissent both inside and outside the Party throughout the country by censorship and the subordination of interest groups, including trade unions; and creation by the so-called "New Economic Policy" or NEP of a mixed economy in which private ownership of some means of production was allowed, especially in agriculture, light industry (e.g., textiles, footwear, and food processing), while other sectors of the economy, notably heavy industry, production of energy, and mining—which Lenin called the commanding heights—were nationalized.

The "Interregnum" (1922–1927)

When in 1922 Lenin's health deteriorated (Lenin died two years later, in 1924), the change of leadership took place with virtually no popular involvement. There was competition among prominent political figures like Lev Trotsky, Lev Kamenev, Nikolai Bukharin, and Grigory Zinoviev. At the time, it seemed to all those leaders of the Communist Party's warring factions that Joseph Stalin would be the best choice for the position of the Party's General Secretary. This was because Stalin was not a member of any faction and did not show too much political ambition. However, once Stalin became Lenin's successor, he began to isolate and discredit those who opposed him. In the economic sphere, Stalin was prepared to get rid of NEP and to advance collectivization of agriculture and full-scale nationalization of industry.

The Stalin Phase (1927–1953)

The Stalin phase witnessed monumental changes in the life of the new Soviet state, notably an intensification of political repression and the exercise of state power over all aspects of public life. Under Stalin, the Soviet system came under the rule of a single, omnipotent leader in control of all the major instruments of political power, especially the Communist Party and the secret police. Stalin became that object of intense fear and veneration often referred to as a "personality cult."

One extremely draconian move of Stalin's was the "Great Purge" of the 1930s of critics and would-be opponents of his policies. In the Great Purge, secret police used unprecedented powers of investigation and interrogation and a politically subservient court system. Victims of the Great Purge ran into the tens of millions and its targets were people as diverse as top political figures with whom Stalin had worked in earlier years, high-ranking military officers, lower-level Communist Party members, government workers, prominent men and women of the arts and sciences, and ordinary citizens.

(Public Domain)
Joseph (Iosif) Stalin (Djugashvili) ruled the Soviet Union from 1922 to 1953.

Stalin's brutal and repressive rule had other dimensions, including a highly centralized bureaucracy with the virtual extinction of representative institutions. Parliaments on both national and local levels of government were deprived of influence—they were subordinated to executive bodies. At the same time, the Soviet judiciary became a pliant tool of the government and of the Communist Party.

Stalin also presided over a profound socioeconomic transformation of the Soviet Union involving the destruction of capitalism in both industry and agriculture and expansive industrialization. Under Stalin's first Five-Year Plan, begun in 1928, peasant household farms were consolidated into cooperatively owned collective farms and state controlled farms. Those who resisted collectivization, notably well-off peasant farmers called *kulaks,* who had benefited from the reforms of Lenin's NEP, were arrested and exiled either to newly colonized areas of Siberia or to labor camps. Stalin also undertook a crash program to industrialize the predominantly rural and agrarian country, not only because it was a necessary component of socialist development, a prerequisite for achieving the ideal utopian communist society, but also because it would strengthen Soviet military power. Up until World War II, the Soviet Union enjoyed double-digit industrial growth, and by 1939 had become a strong industrial power, second only to Germany (in Europe). Importantly, under central planning, priority

was given to the production of so-called producer goods (otherwise known as heavy industry), such as machine tools, tractors, pipes, turbines, heavy trucks, etc., rather than to consumer goods, such as clothing and other necessities.

In agriculture, the Soviet economy fared much worse than in industry. First, unwilling to surrender livestock to collective farms, peasants slaughtered many times more animals than what they needed for personal consumption. For decades thereafter, the Soviet Union could not restore the pre-1930s number of cattle. Second, because in areas with the most fertile soils more peasants resisted collectivization, the government forcibly expropriated grain stored by rural households, including seed grain set aside for planting the next crop. Coupled with the 1932–33 drought in Ukraine and adjacent regions of Russia, this expropriation led to starvation on a grand scale. Altogether about seven million peasants died from this cause, half of them in Ukraine. Third, unlike heavy industry, agriculture proved to be the economic activity particularly ill-suited for central planning. This is because agriculture must adjust to local variations of the physical environment, whereas under central planning most managerial and agronomic decisions were made by bureaucrats detached from farms and ensconced in their urban offices. But despite low yields and low labor productivity, the sheer number of peasants (now called collective farmers) managed to meet the demand for food generated by the largest cities of the Soviet Union; whereas small-town dwellers and country folks produced much of their own food themselves. It took quite some time and precipitous rural out-migration for the implications of collectivization to seriously undermine food production in the Soviet Union. But it is important to point out that agriculture came to be the first stumbling block of the centrally planned (communist) economy, not only in the Soviet Union but also in other communist countries. The Soviet Union simply led the way.

Stalin led the Soviet state through World War II (1939–1945), which the Soviet Union entered on June 22, 1941, when Nazi Germany invaded it in violation of the non-aggression pact that Stalin's People's Commissar for Foreign Affairs (Molotov) signed with Adolph Hitler's Minister of Foreign Affairs (Von Ribbentrop) in August 1939. Throughout the first two years of the war, the Soviet Army for the most part retreated, leaving an enormous area (all of the Baltic republics, Ukraine, Belarus, and a sizable part of European Russia) under German military occupation. But eventually, in the summer of 1943, the Soviets succeeded in turning the tide of war. Following the Kursk battle, the largest tank battle of all wars, the Germans retreated more or less steadily. Russian historians make a lot out of the fact that the idea to open what they call the second front of World War II occurred to the Western allies only after it became clear to them that the Soviets might actually lead their advancement far west of their own western frontier, potentially sweeping across all of Western Europe. Be that as it may, the Soviet Union lost more than 25 million people in World War II, more than any other country participating in that war. Although the war was a terrible drain on Soviet society, it resulted in strengthening the Soviet dictatorship. The war generated patriotism and national unity. In the minds of many ordinary citizens of the U.S.S.R., defending one's own country from brutal aggression and defending communism merged into a single grand idea.

The war also greatly strengthened the Soviet military machine. The Soviet Union emerged from the war as second in power only to the United States. Stalin's death a few years after the war, in March 1953, brought his rule to an abrupt end.

The Khrushchev Phase (1953–1964)

After Stalin died, several senior party officials shared power in what came to be known as "collective leadership," which contrasted with the single person leadership of Stalin. Within that leadership, however, several personalities jockeyed for absolute power. It is in the context of that struggle for power that one ought to interpret the sudden fall of Lavrenty Beria, one of the major perpetrators of Stalinist purges and the boss of the secret police. Beria was accused of being a British spy and got a death sentence for treason. Soon thereafter, occurred the "unmasking" of the so-called anti-Party group composed of such staunch Stalinists as Viacheslav Molotov, Georgy Malenkov, Nikolai Bulgarin, and Lazar Kaganovich, who were expelled from the Communist Party's Central Committee. Eventually Nikita Khrushchev emerged as the Party leader.

Himself one of Stalin's yesmen, Khrushchev now recognized that many of Stalin's ruthless actions had been unnecessary, perhaps even counterproductive and, in his view, "un-Leninist" in the sense that had Lenin been in Stalin's place he never would have pursued some of the more extreme policies of Stalin such as the Grand Purge. Khrushchev openly discredited Stalin's policies in order to change or abandon them, notably condemning the personality cult. Importantly, he placed the secret police under party control and allowed limited opportunity for criticism of government policy in the press. It was under Khrushchev that the surviving prisoners of the labor camps were liberated and for the most part absolved of accusations. Upon returning home, these labor camp survivors shared their stories—which hardly conformed to official history textbooks. One of those returnees was Alexander Solzhenitsyn, whose short novel "One Day in the Life of Ivan Denisovich," published in the Moscow literary magazine *Novyi Mir* in 1962, proved to be particularly influential in shattering the mindset produced and fed by the communist propaganda machine. Thus Khrushchev presided over what became known as a (political) thaw. Although officially aimed at unmasking Stalinism for the sake of an allegedly pristine legacy of Lenin, the thaw initiated a never-ending quest for truth about all aspects of Soviet society, something that eventually brought the Soviet political and economic system to its end.

In an effort to rationalize and legitimize the socialist system in the post-Stalinist era Khrushchev introduced some reforms in the working of the government apparatus, especially the legislative branch, which had been all but ignored by Stalin for most of his rule. Nevertheless, the essence of the dictatorship remained. For example, Khrushchev called for the reinvigoration of legislative bodies at all levels of administration. These popularly elected bodies were supposed to more stringently monitor the economy, especially the performance of farms and factories. Modest changes in the judicial system provided individuals with more opportunity to defend themselves in trials.

Khrushchev initiated economic change as well, expressing a concern for the ordinary citizen by trying to increase

(Gerald R. Ford Presidential Library)

Gerald Ford and Leonid Brezhnev meeting in Vladivostok, November, 1974.

food supplies and to make available more manufactured consumer goods such as home appliances and automobiles. Under Khrushchev a serious boost was given to housing construction, and Soviet citizens were allowed to visit satellite countries of East Central Europe on organized tours. In the largely optimistic environment created by Khrushchev's thaw, the new Program of the Communist Party was adopted in 1961 which predicted that full-fledged communism—understood as a virtual cornucopia of consumer goods (fitting the principle "from everybody according to his/her ability, to everybody according to his/her needs)—would be established in the Soviet Union by 1980. In the 1980s a joke made the rounds in Moscow: a stadium radio announcer says that "some changes in the program have been introduced and should be brought to public attention; namely, instead of communism there will be Olympic Games." In the summer of 1980 Moscow hosted the summer Olympics, so the joke fit the occasion.

The Khrushchev phase ended in October 1964 with his sudden and unexpected ouster by the Soviet Communist Party's Central Committee, which had become dissatisfied not only with Khrushchev's domestic and foreign policies but also with his political style. For example, despite his condemnation of the personality cult, Khrushchev had begun to re-create it around himself, taking control of the offices of both Premier and First Secretary in 1958, thereby ending the shared authority in top-level decision making that he had agreed to accept when he first took power. Other factors contributed to Khrushchev's ouster. For example, he was weakened by acute shortages of grain and dairy products in the early 1960s—the delayed but inevitable consequences of collectivization. His efforts to streamline party organization produced chaos and conflict among party administrators. He was blamed for alienating China, letting the United States get the better of the Soviet Union in the Cuban Missile Crisis, and accomplishing nothing toward the reunification of Berlin under East German rule.

The Brezhnev Phase (1964–1982)

Leonid Brezhnev succeeded Khrushchev as Communist Party General Secretary (the term *First Secretary* was dropped) in October 1964. Under Brezhnev, there was a reappraisal and alteration of Khrushchev's policies.

While Brezhnev did not openly "rehabilitate" Stalin and his personality cult, criticism of Stalinist policies initiated under

Khrushchev and the attendant quest for truth about Soviet history were now deemed excessive, and attempts were made to muzzle both. Stalin was now praised as a war hero; and some intellectuals who had published works critical of the Soviet past in the West were arrested and given prison terms. Repressions never resumed on the scale they had been conducted under Stalin; yet they resumed, and so each potential dissident was aware of the price he or she might pay for disseminating ideas not conforming to the Communist Party blueprint. Some of the more fearless dissidents were put in psychiatric hospitals; and those who were not only personally fearless but had acquired some name recognition in the West were forced to emigrate from the Soviet Union into permanent exile. One of those forced into exile was Alexander Solzhenitsyn.

But Brezhnev also preserved important aspects of the Soviet political system developed by Khrushchev. The top party leadership remained collegial. Brezhnev never became more than "first among equals" in the party Politburo. Brezhnev also tolerated some degree of *institutional pluralism*—namely, discussion and debate in newspapers, periodicals, and books; regime sensitivity to bureaucratic interests on different levels of the party organization; and an increasing articulation of special interests seeking to influence policy making.

The era of Brezhnev's rule of the Soviet Union is now called the "period of stagnation" because of a conspicuous slowdown of economic growth and lack of large-scale technological modernization. While housing construction in major Soviet cities, launched under Khrushchev, expanded and living standards rose, agriculture continued to deteriorate, and a growing share of quality consumer goods and food was purchased in the West in exchange for oil, the export of which significantly increased under Brezhnev. To a significant extent the somewhat decent quality of life in several major Soviet cities was a product of high international oil prices after 1973. However, many people in the Soviet Union endured living standards on a par with the Third World. At the same time, the Soviet education system was of high quality, particularly in science and engineering, and so a new generation of Soviet citizens—well educated, articulate, and to some extent knowledgeable about the better conditions in the West—was disaffected by the regime's failure to deliver the material well-being promised in the ideology of communism.

Another "Interregnum" (1982–1985)

During the Brezhnev years, the Communist Party's Politburo consisted of people in their 70s and 80s. So, talk of a gerontocracy and jokes about the Politburo's old geezers were wildly popular. After Brezhnev died in 1982, his two successors at the helm of power, Yuri Andropov (1982–1984) and Konstantin Chernenko (1984–1985), each only spent 1–1.5 years as general secretary before passing away. Both did little to address the country's major socioeconomic problems, but Andropov, being the more reflective of the two and having had more significant leadership experience (including as the boss of the KGB), had some ideas which were later invoked and praised by another ex-boss of the KGB, Vladimir Putin. However, all that ailing Andropov managed to accomplish was to launch a discipline-tightening campaign with police raiding Moscow's communal bathhouses (a signature of urban life in Russia, these bathhouses

are places where people also drink, eat, and negotiate business deals) in search of public officials who used to spend their working hours in such establishments. As for Chernenko, a lackluster bureaucrat, all that is left in people's memory about him is a joke playing on his first name (Konstantin) and his patronymic (Ustinovich). According to this joke, Moscow was supposed to be renamed Ust-Constantinople. Jokes aside, by the mid-1980s, many people in the Soviet Union—definitely most Muscovites and residents of Saint Petersburg—shared the overall perception of a deep crisis in the entire political and economic system. In order to deal with the crisis, some cohesive political organization was required to take the lead. But in the Soviet Union, there was only one well-organized political force, the Communist Party. Numbering about 18 million members, the Party had cells in every Soviet institution be that a factory, farm, college, middle school, military unit, or retail outlet. Although a retrograde political force, the Communist Party was not insulated from public opinion, so a pro-reform leader was eventually appointed (by the members of the Politburo) General Secretary.

The Gorbachev Phase and "Perestroika" (1985–1991)

In March 1985, at the time of his promotion within the Communist Party Politburo to the position of General Secretary, Mikhail Gorbachev was unusually young. Born in 1931, he was "only" 54 at the time of appointment, much younger than his predecessors; and he took cues from Alexander Yakovlev, former Soviet ambassador to Canada now promoted to Secretary of the Central Committee of the Communist Party responsible for propaganda. Yakovlev became the ideological father of *perestroika,* or restructuring, a program of economic and political reforms over which Gorbachev officially presided. By no means was *perestroika* intended to dismantle the communist system as such—indeed it was supposed to salvage it by way of making it more flexible and able to regenerate economic growth. But Gorbachev and Yakovlev believed that economic improvement depended on the introduction of other changes—notably, a more tolerant and open political environment, greater

(Courtesy of Grigory Ioffe)

The Red Square in Moscow

14

ANDREI SAKHAROV

Andrei Sakharov was born in Moscow on May 21, 1921, the son of a physics teacher. A brilliant student, he studied at Moscow University under Igor Tamm, winner of the Nobel Prize for theoretical physics. During World War II, Sakharov served as an engineer in a military factory. In 1945, he entered the Lebedev Institute in Physics and soon joined the Soviet research group working on atomic weapons. Author of numerous scientific articles during this period, his achievements were broadly recognized both within and outside Soviet Russia. In 1953, at the age of 32, he became the youngest person ever elected as a full member of the Soviet Academy of Sciences.

Between 1950 and 1968 Sakharov conducted top secret research on thermonuclear weapons in a secret location. He also developed an acute awareness of the dangers of nuclear testing activity and the irreversible consequences of nuclear war. His activities as a dissident can be dated from the period of relative intellectual freedom under Nikita Khrushchev in the late 1950s, when Sakharov began to send letters to Soviet leaders urging a halt to nuclear testing. In November 1958, *Pravda* allowed him to publish a lengthy article criticizing a plan to send children talented in mathematics and physics to the countryside for farm work. He also published several prominent articles in *Atomnaia Energiia* and other Soviet journals arguing against continued nuclear testing and the arms race. His views apparently carried weight with Khrushchev and others, with whom Sakharov communicated directly, and influenced the Soviet decision to sign the first test ban treaty in 1963.

The freedoms Sakharov and others enjoyed in these relatively liberal years had enormous effect. The ability to think and write openly about critical social issues was not easily repressed, despite the concerted efforts of Khrushchev's conservative successor, Leonid Brezhnev. In 1966 and 1967, Sakharov openly warned against efforts to rehabilitate Stalin and pressed for civil liberties. With the Soviet invasion of Czechoslovakia in 1968 and the brutal repression of the Prague Spring, Sakharov and others became more militant, expressing their criticism more openly and sometimes standing vigil at trials of those arrested for protest activities. It was at this time that Sakharov published his most prominent and eloquent political essay, *Reflections on Progress, Peaceful Coexistence and Intellectual Freedom*, urging cooperation between East and West, civil liberties, and an end to the arms race.

It was while standing vigil at one such trial in 1970 that Sakharov, a widower, met Elena Bonner, who soon became his second wife and strongest supporter. The publication of *Reflections in the West* resulted in Sakharov's removal from most of his scientific projects and his dismissal as principal consultant to the Soviet Atomic Energy Commission. It soon became difficult for him to publish scientific works as well, although he continued his research and writing. In these difficult circumstances, Sakharov, assisted by Bonner, rapidly assumed a leading role in the Soviet dissident movement.

His writings and protests throughout the 1970s generally touched four themes: the treatment of individuals, particularly other dissidents arrested or otherwise harassed for their political views; the suppression of civil liberties in the U.S.S.R. and elsewhere; attacks on Soviet "totalitarianism," as he described it, and demands for political freedom in Russia; and the grave dangers of the arms race and nuclear development and testing, plus the likely consequences of nuclear war. Sakharov's great international prestige as a nuclear physicist (and his particular knowledge of the Soviet Union's nuclear weapons program) gave special significance to his views and also for a time helped protect him from arrest and expulsion.

Toward the end of the 1970s Sakharov became increasingly alarmed about the Soviet arms buildup. A strong advocate of East-West parity in nuclear weapons, he saw the development of new Soviet missiles as a reflection of aggressive and expansionist designs. He frequently expressed his views to foreign reporters, and much of his samizdat (a clandestine publishing system within the Soviet Union, by which forbidden literature was reproduced and circulated privately) writing appeared in the West. His outspoken criticism of the Soviet invasion of Afghanistan in late 1979 reflected these concerns and led, finally, to his detainment and expulsion from Moscow. In a celebrated incident, Sakharov was banished by administrative order to Gorky, (now Nizhny Novgorod), a large city 250 miles east of Moscow, which was off-limits for foreigners. Thus began a period of almost total isolation and constant harassment by the KGB (secret police).

Sakharov's plight became, in the 1980s, a constant sore in Soviet-American relations. In 1983 he reportedly considered emigration, but was refused because of his knowledge of Soviet state secrets. Continued protests against Soviet militarism resulted in new threats and warnings to him and to family members. On several occasions Sakharov engaged in hunger strikes to call attention to these threats and to gain the right of family members to go abroad. In 1983, President Reagan proclaimed May 21 "National Sakharov Day" in recognition of his courage and his contribution to humanity.

Sakharov was detained in Gorky for almost seven years, released at last by Mikhail Gorbachev in 1986. The remaining three years of his life were spent traveling abroad—something he had never previously done, despite his international fame. He died of a heart attack on December 14, 1989, in Moscow.

Three times named "Hero of Socialist Labor" (1953, 1956, 1962), winner of the Order of Lenin, the Stalin Prize, and the Lenin Prize, Sakharov also received the Nobel Peace Prize in 1975 for his tireless work for nuclear disarmament and his outspoken criticism of human rights violations everywhere, especially in his homeland. He was for many, inside the Soviet Union and out, a noble symbol of courage, intelligence, and humanity.

---

popular influence over the civil and military institutions of government and over the organization and behavior of the Communist Party, and a foreign policy less ideological and more responsive to domestic needs and priorities.

Because the quintessential features of national existence (such as freedom of political debates, elections, kinds of property allowed and promoted in the economy, the nature of economic incentives, and the relationships with the outside world) were to be affected, perestroika was viewed as a long-term program to transform the Soviet Union into a different kind of polity. By all accounts, Gorbachev was guided by noble instincts. But, because he was apparently unable to foresee the implications of the changes set in motion by perestroika, his leadership phase turned out to be the last phase of Soviet political development. Perestroika essentially weakened the Soviet system, creating the conditions and circumstances that led inevitably to its disintegration and collapse by the end of 1991.

COLLAPSE OF THE SOVIET UNION

Out of the two groups of factors, economic and political, that were instrumental in the collapse of the Soviet Union in 1991, economic factors had primacy. Political factors were added to the equation only after the reins of totalitarian, Communist Party-led control over Soviet society were weakened because of endemic economic problems that had come to a head in the late 1980s. The primacy of economic factors may seem counterintuitive to some. It is, after all, common knowledge that the Soviet system was oppressive; so why would popular discontent not be the most decisive factor? A brief outline of Russian political culture provided on **pp. 51–55** may at least partially address this question. At this point it is important to stress that, with the exception of the Baltic republics and perhaps the westernmost part of Ukraine—that had democratic experiences in their past, however limited—most of the Soviet Union had not inherited any democratic tradition whatsoever from preceding polities. In other words, if the Soviet Union was autocratic, its immediate predecessor, the Russian Empire, was hardly any better; and in terms of discriminating against people on ethnic and religious grounds, it was arguably worse. No, or indeed very little, domestic democratic tradition explains people's susceptibility to authoritarian rule. With relatively few exceptions, people were willing to abide by undemocratic rules because they did not know any differently and because they were willing to exchange democracy for an acceptable quality of everyday life. To be sure, that quality was decidedly lower than in Western Europe, Japan, Australia, Canada, the United States, and in some communist states of East Central Europe (notably East Germany, Czechoslovakia, and Hungary), but it was higher than in much of the rest of the world.

For all the endemic flaws of the communist system of governance, it presided over a cradle-to-grave welfare state with a no-nonsense provision of public good to all citizens. This included free healthcare, free college education, free summer camps for children, decent retirement pensions, heavily subsidized food prices, and, of course, lack of mass unemployment and bankruptcies. It is another matter that the communist system did not meet inflated expectations that it itself generated when it claimed that it would "catch up with and outstrip" the West. Even more importantly, maintaining delivery of all free services embedded in the system became more and more expensive, which is what eventually undermined the Soviet system.

The single most important factor leading to the collapse of the Soviet Union was the drastic decline in world prices for oil in 1986. Since the early 1970s, the Soviet Union had exported oil to the West and spent the revenues generated on importing consumer goods, animal feed (mostly grain), and industrial equipment. Oil prices dropping from more than $70 to about $20 per barrel (in 2000 U.S. dollars) sapped the Soviet Union of its buying power. A more long-term factor that led to Soviet collapse was its inability to sustain the arms race on terms dictated by U.S.-Soviet geopolitical rivalry. But if such factors could undermine a country with superpower status, the economy of that country had to have had some profound systemic flaws.

Systemic Flaws in the Soviet-Communist State

The pivotal systemic flaw of a centrally-planned economy— beside which any other pales—is the lack of incentives to modernize and to work better and more efficiently. Under a market economy, competition decides the fate of a business, which will succeed if its revenue exceeds production costs and profit can be invested into modernization and/or expansion. However, the business will go under if revenue falls short of production costs. When under a market economy, some businesses get preferential treatment by the state (for example, they enjoy a reduced tax rate or get bailed out by taxpayers), this is decried as corporate welfare. But under the Soviet system, all businesses were under corporate welfare. If and when some production units were better managed than others—a normal situation—then profitable units bailed out the unprofitable ones by way of an all-encompassing profit redistribution.

This redistribution-based economy kept all businesses afloat, but sent unrelentingly damaging messages to both successful and unsuccessful businesses. The message that the management team of an unprofitable factory would receive is, why should we work better if we are going to be bailed out by the state anyway; whereas a profitable and efficient management team would wonder, why on earth should we strain ourselves if some of our profit and possibly much of it is going to be withheld from us. Even if profits were not taken away, the successful managers would have no say in their possible modernization and/or expansion plans, in part because the mainstay of the economy was profit redistribution and in part because all production units were owned and managed by the state. Prices were set by the state, not derived from a supply-demand trade-off. This created situations unthinkable under a market economy wherein prices for scarce goods invariably go up whereas abundant goods become less expensive. In the Soviet Union, prices on most consumer goods and food items were set low. Not only did this undermine the profitability of the businesses producing them, but it also deprived central planners of accurate signals about the economy. Meanwhile, consumers would stand in long lines in order to buy something (butter, meat, or toilet paper) that was cheap.

Under a market economy, scarcity drives prices up, sending a message to investors that they may make a profit by pouring money into producing pricey products. As an alternative or interim process, foreign trade may patch the hole in supply. Under central planning, this whole switchboard of economic signals is non-existent, because all decisions about how to assign resources are made by specially designated bureaucrats who are supposedly bestowed with excellent foresight and knowledge of consumer demand. Such a system opens incomparably more opportunities for malfunction, breakdown, or failure than a market economy. Under a market economy, the state sets the rules of the economic game, but then it may fail to enforce those rules, which can generate economic crisis. But under central planning the state not only sets the rules and enforces them, it is also the only player in the game. This makes the entire economy supremely inflexible. Not only do consumers line up to buy cheap goods, but also a perennial shortage of first one of those goods and then another occurs, since no group of experts can accomplish what the profit-maximizing decisions of members of an entrepreneurial class can. The environment becomes polluted because, when the state owns and maintains all businesses, imposing fines for poisoning, say, the atmosphere or ground water works much like robbing Peter to pay Paul or rather taking money from one pocket and putting it into another pocket of the same pair of pants. The economy lacks accurate signals and, most importantly, it does not dispense penalties and rewards in proportion to the quality of managerial decisions and the overall effort at meeting consumer demand. Under such conditions, one can perhaps ensure that a certain number of manufactured products of strategic significance be produced by a narrowly circumscribed group of well supplied elite production units. But sustaining a progressively more diversified economic activity dependent on thousands of tradeoffs and links between economic sectors within an area spanning eleven time zones becomes a nightmare.

Realizing the pitfalls of planned economies, one should perhaps wonder why it took the Soviet economy so long to collapse rather than question why it did so. Indeed, rooted in the benign and even admirable longing for overall equality, the communist system issued slogans that long swayed many people within the Soviet Union and beyond. Perhaps if people are kept in relative isolation from the rest of the world such "bliss" can be artificially extended indefinitely. Driven by idealism, some people will agree to exchange material benefits for mere praise and exhortation. But this can last only up to a point. The breaking point was drawing closer and closer in the late 1980s.

Economic Mismanagement and Environmental Degradation

By the mid 1980s it had become apparent to the Soviet leadership that the Soviet economy was suffering from a pervasive condition of economic underperformance. Underperformance meant that what came out of the Soviet economy was far too little given the enormity of investment of land, labor, and capital. This phenomenon was evident in chronic shortages of agricultural and manufactured goods.

Agriculture

Contrary to a widespread stereotype, the Soviet Union and its largest republic of Russia had a sufficient amount of reasonably fertile agricultural land. While its percentage share in the overall land area is small (roughly 10%), the Soviet Union was the world's largest country, spanning eleven time zones. The area with favorable conditions for agriculture is shaped like a wedge whose broader part is located in the European (west of the Urals) section of the former Soviet Union and encompasses a large space between Saint Petersburg in the north and the Caucasus in the south. Indeed, much of Ukraine and some adjacent regions of Russia have some of the world's most fertile and productive soils. As one proceeds east, the wedge progressively narrows so in eastern Siberia, west of Lake Baykal, it becomes a narrow strip along the Trans-Siberian railway line. However, at midpoint, in Western Siberia and in northern Kazakhstan, there are again millions of hectares of arable land of decent and at times supreme quality. East of Baykal, the wedge stops narrowing and widens a bit (particularly north of the city of Blagoveshchensk) but not nearly to the point of its north-south dimension in the European section of the country. Overall, no fewer than 150 million hectares were suitable for agriculture in the former Soviet Union, including close to 40 million hectares of supreme quality, largely in Ukraine and the adjacent regions of Russia down to the foothills of the Caucasus. The wedge-shaped expanse of agriculturally suitable land in the former

Soviet Union is conditioned by physical geography, notably by the dominance of westerly airflow and the attendant spatial trend in climate known as continentality: as one proceeds east along the same parallel of latitude (in northern Eurasia) winters become harsher and longer. The same climatic conditions actually explain the geographic distribution of population in the former Soviet Union and in Russia proper, wherein fewer than 30 million people out of Russia's current population of roughly 140 million live in the Asian part (east of the Urals), known as Siberia, the part of Russia which accounts for 76 percent of its overall area. The aforementioned supply of quality agricultural land was more than enough to meet the demand for food generated by 289 million people (the 1989 population of the U.S.S.R., two years prior to its breakup). And so if agriculture malfunctioned, which it did, it was entirely for socioeconomic reasons.

Soviet agriculture malfunctioned first and foremost because (as was mentioned earlier) by its very nature it is the economic activity least compatible with tenets of centrally planned economy. By the end of the Soviet Union's existence, 25 percent of foodstuffs was produced on just 2 percent of the agricultural land. This 2 percent encompassed land on which country folks had their household farms. Usually attached to one's residence (typically a wooden house with a firewood stove for cooking and heating and without plumbing), a small parcel could host a vegetable garden and a cowshed. Aside from cows, rural folks used to own pigs and chickens, or in Muslim regions of North Caucasus and Central Asia, sheep. Apple and cherry trees were (and still are) common on household farms in the Ukrainian and southern Russian countryside, and tangerines were produced by households in the Trans-Caucasus region (mostly in the republic of Georgia). That one-quarter of the food is produced on 2 percent of the land implies that if one owns the land, one tends to work hard in order to meet one's own demand and perhaps sell surpluses in a nearby or even remote city; but if the land belongs to a collective or state farm, the worker is not going to strain all that much.

A related and no less potent reason for low agricultural productivity in the Soviet Union was precipitous rural out-migration well outpacing the potential of the Soviet agricultural management and technology to replace manual labor. What is more, the spontaneous selection of movers versus stayers was essentially a negative selection for the economic prospects of the Soviet countryside. The most industrious people left while the most passive, resigned, and prone to drinking stayed. This had to do with the fact that the spatial pattern of quality of life in much of the Soviet Union was a function of settlement size: the larger, the better. In other words, large cities not only provided multiple job opportunities as cities do all across the world, but they also offered amenities (such as plumbing and modern heating installations) that in rural villages were largely unavailable. Because of this, productivity of Soviet collective farms depended, as a rule, not only upon natural fertility of the soil but also on proximity to major cities. But in the Soviet Union, particularly in Russia, much agricultural land was located far from cities; and, with the inferior quality of roads, this created a sense of abandonment not conducive to hard and efficient work on the land.

As early as 1963, just two years after the Program of the CPSU informed the Soviet people that communism was going to be achieved by 1980, the Soviet Union resorted to large-scale import of grain primarily for animal feed. From that time forward, a country that once exported food was importing it in ever-increasing quantities; and much of it was being purchased from those same capitalist countries (such as the United States and Canada) that the Program of the CPSU wanted the Soviet Union to "catch up with and outstrip." Since the early 1960s, the quality and variety of food offered by state-run grocery stores in Soviet cities progressively deteriorated. Moscow, Saint Petersburg, and the capital cities of all union republics were, for a long time, exempt from that trend; but by the 1980s, in virtually all Soviet cities with the exception of Moscow, Saint Petersburg, and the capital cities of the Baltic republics, meat and sausage were hard to find in state-run grocery stores, whereas free market prices at urban-based farmer's markets (so-called *kolkhoznye rynki*) where rural folks used to sell the surpluses from their household farms, had become much higher than the heavily subsidized prices at which state-run stores had previously sold food.

Industry

The state of affairs in Soviet industry was even more at odds with resource potential than in agriculture. At least in agriculture environmental constraints to cultivation of most crops applied to much of the country, although the overall supply of high-quality land was adequate. But in industry, no such constraints existed. The Soviet Union was arguably the most self-sufficient country in the world as far as industrial resource endowments are concerned. Metallic ores, fossil fuels, and mineral salts were (and still are) overabundant. It is true that many deposits are located in thinly populated areas with harsh environments that made extraction costly, but virtually any known resource was available within the country, so there was no need to import it. Moreover, the Soviet education system (definitely a success story, particularly in math and sciences) supplied industry with high quality personnel and researchers.

Indeed, a combination of brain power, resource endowments, investment, devotion, and close supervision at times produced remarkable results. In 1957 the Soviet Union launched the world's first artificial satellite, and the Russian word *Sputnik* (meaning satellite) became known across the world. On April 12, 1961, the world's first astronaut (or cosmonaut, as he and his successors were called in the Soviet Union), Yuri Gagarin, orbited the planet. This was a striking success for what, just a couple of generations earlier, had been an economically backward peasant country. The United States later recaptured world leadership in spacecraft industry, but that does not diminish the Soviet effort as well as the idea that, in certain areas and under certain circumstances, Soviet industry could in fact deliver.

However, on closer inspection, those achievements in space had to do with national defense; rarely did Soviet industry succeed in any industrial innovation unrelated to the military-industrial complex. For example, fairly advanced for the late 1970s when they showed up, air-cushioned boats for river passenger traffic were an offshoot of their cousins originally designed for the Soviet navy; and tractors, whose late 1980s' output in the Soviet Union exceeded that in the United States by a factor of six (in terms of sheer number of vehicles), were cousins of tanks. The case of tractors is particularly revealing as Soviet agriculture, their main consumer, performed poorly,

which made a mockery out of the Soviet lead in tractor production. This was in line with the output of steel, in which the Soviet Union in the late 1980s exceeded the United States by 80 percent. However, this dubious superiority was achieved some twenty years after advances in the iron and steel industry stopped being indicative of the scale of a nation's industrial prowess. By the late 1980s, in much of the industrialized world, steel had been partially replaced by other materials; but not in the Soviet Union, which wasted much steel on a poorly utilized fleet of tractors and many other useless or inefficient products.

Soviet industry suffered from underperformance with poor quality, high costs, and economic waste for some of the same reasons as agriculture. Chief among these reasons was a centralized decision-making apparatus. Managerial personnel had little authority to make decisions. Central planners, who had the bulk of decision-making authority, could not substitute free market instruments of monitoring and meeting consumer demand. Central planners set targets for output based upon arbitrary assessments of need. They also determined output goals in many industries in terms of number of tons or in ruble value and did not pay adequate attention to the quality of what was to be produced. The Soviet industrial machine also lacked labor-saving automation. In general, it simply took too many people too long to make decisions at the center and to implement them in the production unit.

A disproportionate growth in heavy industries (that is, in production for the sake of further production) and lack of success in consumer-oriented industries exacerbated industrial underperformance. Furthermore, the regime continued to invest in inefficient units of production for the sake of full employment. This unbalanced investment policy encouraged poor quality, to say nothing of high production costs.

Black Market and Corruption

Because some food items (quality meat, sausage, smoked fish—a staple of the diet in Slavic and Baltic republics; butter, and cheese; citrus fruits and bananas to name just a few of the most widespread deficits) as well as plenty of other consumer goods were in short supply and/or of poor quality, a rampant black market for domestic and particularly foreign-made products developed in the Soviet Union. It worked as a pervasive re-distributor of income, as people who had access to goods in limited supply illicitly sold them to the general public and/or used them as a quasi-currency to buy high-quality and nominally free services. A popular medical doctor, for example, would receive his or her modest salary but would be able to supplement it by tapping into black market sources. And the "most important" people, part of the communist *nomenklatura,* would be assigned to special retail outlets to which the general public had no access. Also, people who were cleared by the Communist Party to make business trips abroad and had access to hard currency (e.g., U.S. dollars) were mandated to surrender it upon coming back home, but in exchange they received special certificates accepted by a network of *Beriozka* stores. Such stores were opened in the late 1970s, particularly in Moscow; they offered domestic and imported foods, clothing, and cosmetics in short supply elsewhere. However, access to these stores was limited; only those with "certificates" were allowed in. The ability to travel abroad, thus constituted a major perk: not only could those minions of fortune use Western consumer goods while abroad, but they also could buy surplus quantities of some goods (tape recorders, for example, or pieces of clothing) and re-sell them at home; and in addition to that, they could access the exquisite network of *Beriozka* retail outlets at home.

By far the biggest material benefit all too often associated with political loyalty and with having access to scarce goods was improvement of living conditions. For decades, most Soviet urbanites lived in so-called communal apartments. These were the former apartments of the pre-1917 nobelemen and bourgeoisie. Most of their owners had either emigrated or perished in the 1917–1921 civil war. With multiple bedrooms but one kitchen, one toilet, and one shower room, these apartments, after the Bolshevik Revolution, were assigned to working class people, one family per room, so that all of them shared the apartments' facilities. These communal apartments hosted the overwhelming majority of Soviet urbanites until housing construction was launched under Khrushchev on an unprecedented scale. The new apartment blocks offered one small apartment (typically with one or two bedrooms) per family, a significant improvement over communal apartments. In the absence of a housing market (or any recognized form of market except the already mentioned retail outlets where peasants used to sell surpluses from their household farms), new apartments were distributed through places of work. Anyone who could document that living space in his or her household was below a certain threshold (say 5 square meters per person) could register a need in dwelling improvement. That usually meant putting a person on a waiting list. As soon as new apartment blocks were commissioned, people who were first on the waiting list would be offered new apartments for free. The waiting time, however, was often in excess of ten years, during which time one's family size could change several times due to births and deaths.

During the last two decades of the Soviet Union's existence, an alternative way of improving one's housing conditions became possible through housing cooperatives wherein one was expected to furnish a relatively large down payment for a new apartment and pay a higher rent thereafter. That way, one would avoid a waiting list. Curiously, even joining a cooperative required registering a housing need, but that could be sidestepped if and when one bribed a certain official. In both cases, however, with or without a waiting list, the best and better located apartments ended up in the hands of the most politically loyal people as well as people who had accumulated significant wealth through black market channels. Just as with the exquisite retail outlets for the privileged few, this way of rewarding political loyalty or illicit wealth flagrantly contradicted the much-vaunted egalitarian foundations of Soviet society. Just like business trips abroad, a privilege of relatively few people cleared by the Communist Party, this was essentially legalized corruption.

Environment

A major aspect of economic mismanagement concerned the Soviet Union's polluted physical environment, which began to arouse both official and public concern only in the late 1980s. Because the Soviet Union was, for the most part, sparsely settled—which ensured significant physical distance between sources of industrial pollution—many areas remained pristine. But pollution affected most urban clusters, where the concentration of toxic elements in the atmosphere and at times in drinking water was by far exceeding thresholds set by law. Also, because data about pollution were classified, public attention was not necessarily drawn to the most hazardous cases. Thus, although attention to the ongoing pollution of the world's deepest lake—Baykal—by a pulp-and-paper plant was justified, much worse cases, such as radioactive waste burial sites in the Urals (not far from the city of Chelyabinsk) or chemical contamination in and around the city of Dzerzhinsk (not far from Nizhny Novgorod), were long unknown to the broad public.

Two area-specific cases of environmental pollution developed into full-blown environmental disasters. One of them was the ravaging of the Aral Sea by pesticides, defoliants, and fertilizers. In order to produce cotton in sunny and arid Central Asia, massive irrigation was required. The only sources of water for irrigation were two rivers, Syr-Darya and Amu-Darya. With sources in mountain glaciers in Central Asia's extreme south (Syr-Darya) and in Afghanistan (Amu-Darya), both rivers flow to the Aral Sea, a salt lake located in the middle of Central Asia. The irrigation canals built in the 1950s and 1960s diverted so much water from these rivers that they stopped reaching the Aral Sea. Located in an area where evaporation greatly exceeds precipitation (which is what makes a salt lake in the first place), the sea began to shrink; and its natural salts, combined with chemicals used on cotton plantations, transformed a large natural reservoir into a stinking marsh. Not only the Aral Sea itself but also most sources of ground water in a large surrounding area became contaminated. The locals, who are mostly poor, could not afford bottled water, so infection and disease spread. The case of the Aral Sea illustrates a classic contradiction between economic and environmental goals. Stopping pollution would require cutting back on the output of cotton, and cotton was one of the few commercial staples of the entire region.

Even more publicized is the second environmental disaster—that associated with the meltdown of a nuclear reactor at the Chernobyl power plant in the northernmost part of Ukraine, just 7 km south of its border with Belarus. This meltdown brought home to Soviet authorities the dangers of its commitment to double the Soviet Union's nuclear power-generating capacity by 1990. The Chernobyl accident on April 26, 1986 still ranks as the world's worst civil nuclear disaster. It took ten days to extinguish the fire in the reactor. A cloud of radioactive fallout drifted hundreds of miles northward and westward beyond Ukraine into other parts of the Soviet Union, Scandinavia, and Central/Eastern Europe, ultimately affecting areas as far away as Italy, France, and Britain, although the most hazardous impact was confined to parts of Belarus (where 70 percent of the radionuclides were deposited), Ukraine, and Russia. The most widespread health consequence attributed to exposure to radiation appears to be thyroid cancer caused by a radioactive isotope of iodine. This isotope has a half-life of only eight days. In contrast, the half-life of radioactive strontium is twenty-nine years. Much of the delayed effect from Chernobyl is caused by accumulation of such isotopes in soil, vegetation, and groundwater.

Soviet investigators accused managers at the plant of negligence, incompetence, complacency, irresponsibility, and lack of discipline—in sum, a carelessness born of a conviction that what did happen could not happen. Western scientists concluded that there had been serious flaws in the construction of the emergency shut-off component of the Chernobyl-style reactor.

Evacuation of residents from highly contaminated areas presented a problem. Not since World War II had Soviet authorities had to deal with the movement of so many people. In the month after the meltdown, it was necessary to move 92,000 people and provide them with food, clothing, and shelter. Subsequently, about 300,000 were relocated. Hundreds of thousands of children in Kiev and nearby cities were sent to contamination-free areas in summer camps located in the south. But the region where the Chernobyl plant was located really has never become safe to enter, and remedial activity continues today.

As has been shown, the most critical existential problems that the Soviet Union faced were of an economic nature; environmental disasters only served to exacerbate the perception of crisis. But because growing domestic and external criticism of the Soviet economy was integral to the mainstream ideological struggle of the Cold War, in which the West was getting the upper hand (and the most well-informed Soviet citizens were aware of that), it was the political criticism of the Soviet system that sounded louder than criticism of Soviet economic arrangements. Moreover, for many people political goals (e.g., freedom to debate, to set up political parties, and to emigrate; competitive elections) are more readily understandable than debates about political economy. Whereas with hindsight (derived from the experience of still communist China and such culturally dissimilar but formerly equally autocratic countries as Chile and South Korea) we know that profound economic reform can predate political reform and that such a sequence may actually prove to be more assured and successful for both facets of change—economic and political—the Soviets chose the simultaneous option. However inconsistent and ambivalent, reforms were undertaken in both the economy and politics. These reforms changed the nature of the Soviet Union so profoundly and within such a short period of time that key domestic political actors (the Communist Party, army, secret police, central planning authorities, and their representatives bodies within and across major ethnic groups) were overwhelmed, and they failed to renegotiate some sort of a social contract that would sustain the seventy-year-old polity that emerged in 1917–1922 out of the wreckage of the Russian Empire. Perestroika unleashed irreconcilable conflicts that precipitated the end of the Soviet Union.

CATALYSTS OF COLLAPSE IN THE LATE 1980s AND EARLY 1990s

Perestroika and the Economy

Gorbachev's restructuring of the Soviet economy had several dimensions: 1) curtailment of the power of central administrators to control agriculture, industry, prices, and foreign trade; 2) use of profitability as an economic indicator in the functioning of economic enterprises; 3) expansion of worker participation in the management of enterprises; 4) tolerance, even encouragement, of a limited private entrepreneurialism in areas of the economy having to do with consumer services and food production; 5) the introduction of major agricultural reform, involving

a gradual movement away from collectivization toward private control and ultimately private ownership of land; and 6) the pursuit of joint ventures with foreign enterprises to facilitate Soviet acquisition of investment capital and technology.

Perestroika and Society: Glasnost

Policies known as *glasnost* (the Russian word for openness) called for full, frank, and public discussion of what Soviet citizens thought was wrong with their society. This openness was a stunning departure from the rigid and repressive control over thought and speech practiced by all previous Soviet leaders. Glasnost called upon party and state officials to respond to public inquiries and public criticisms about policy making in different sectors of the administration, from economic enterprises to the KGB. Glasnost thus envisaged abandoning censorship and secrecy practiced by all Soviet political officials as a matter of course throughout the history of the Soviet state. Censorship meant that certain officials cleared every manuscript and every TV and radio broadcast to make sure it did not include anything that was ideologically incompatible with the policy of the Communist Party. Every publication used to be subject to censorship, even fashion magazines and cook books. This practice effectively eliminated debate from public life and created a semblance of public unanimity in regard to major issues facing the Soviet society.

In the late 1980s, glasnost undermined censorship and thus led to a substantial expansion of public debate on national problems. For example, the Soviet press was allowed to publish criticisms of economic slackness and some of its causes, such as alcoholism, drug abuse, bureaucratic corruption, and economic waste. There was more media coverage of major accidents and injustices of the court system.

Glasnost also included historical revisionism—the "rewriting" of history. Gorbachev condemned Stalinism anew; cast the Khrushchev era in a much more favorable light than Brezhnev had; and referred in non-condemnatory ways to early Soviet revolutionary figures reviled by Stalin, like Leon Trotsky and Nicolai Bukharin, anticipating the official "rehabilitation" of these antiheroes and their ideas and policies.

Glasnost and the Non-Russian Nationalities

With glasnost came a veritable explosion of nationalist-inspired protest and dissent among some of the non-Russian populations of the Soviet Union. Nowhere was nationalist sentiment stronger than in the Baltic republics. Thousands of people in Estonia, Latvia, and Lithuania publicly protested the fiftieth anniversary of the 1939 Hitler-Stalin pact in which Germany gave the Soviets a green light to annex the Baltic States, which had received their independence from Russian rule at the end of World War I. These demonstrations affirmed the intensity of anti-Russian feelings among the indigenous Baltic populations (i.e., ethnic Estonians, Latvians, and Lithuanians), despite Moscow's efforts to integrate them into Soviet society. Baltic nationalists were especially critical of the influx of Russians into their republics. In Latvia, where ethnic Latvians at the end of the 1980s, comprised just a little more than 50 percent of the population (whereas on the eve of Latvia's incorporation in the Soviet Union, in 1940, they accounted for 77 percent), they sought both to curtail Russian immigration and to limit the political influence of Russians on local government. In 1990, the Supreme Soviet (parliament) of Lithuania voted for secession from the Soviet Union.

In Ukraine, anti-Russian nationalism was particularly strong in the western part of the republic. In 1990 nationalists, many of whom belonged to a popular-front organization known as Rukh and a new anti-Communist organization calling itself the Union of Independent Ukrainian Youth, admired and supported the movement toward independence in the Baltic republics. They called for secession of Ukraine from the Soviet Union. Ukrainian nationalists were very sensitive to the separate and special cultural identity of Ukrainians and emphasized contrasts with Russians.

Glasnost gave rise to nationalism also in Moldavia, now Moldova but also called Bessarabia when the area was under Romanian control before World War II. Moldavians yearned for independence and began to campaign for greater popular control over their local economic and political affairs. Moldavian nationalism, like Baltic nationalism, had a strong anti-Soviet and anti-Russian undercurrent because of the Kremlin's not so subtle efforts to undermine Moldavia's Romanian cultural origins, notably through forced use of the Cyrillic alphabet.

Armenian nationalism, also encouraged by glasnost came from a strong sense of community based on a shared memory of the 1915 genocide, the influence of the Armenian Church, the view of Armenians living abroad of Soviet Armenia as a homeland, and the fact that Soviet Armenia was the most ethnically homogeneous of the 15 constituent republics, with more than 90 percent of its population ethnically Armenian. Armenian nationalists annoyed the Kremlin with their demand for the transfer to Armenian control of Nagorno-Karabakh, an ethnic Armenian enclave inside of the republic of Azerbaijan, despite the fact that its population had suffered discrimination by the Azeri government.

Non-Russian peoples predominantly of the Muslim faith in the Soviet Union's Central Asian Republics (Kazakhstan, Turkmenistan, Uzbekistan, Tajikistan and Kyrgyzstan) and in Azerbaijan, were far less nationalistic and anti-Russian than other ethnicities in the Soviet state. The Central Asian economies and societies were less developed than those in the European part of the Soviet Union. Hence, they benefited more from and were far less critical of Soviet policies and of their membership in the highly centralized Soviet state. Likewise, no secessionist movement whatsoever developed in Belarus.

Perestroika and the Communist Party

The Communist Party of the Soviet Union underwent profound change under perestroika because Gorbachev believed that the party, more than any other institution, was responsible for the pre-crisis situation inherited from the Brezhnev phase. In his view, the Communist Party had to liberalize its highly authoritarian structure; loosen its stranglehold over the Soviet economy and society; and rid itself of a self-interested, unresponsive, and corrupt managerial elite. Perestroika ultimately weakened the Soviet Communist Party's grip on power by allowing the existence of new political organizations, called "popular fronts," offering an alternative to the Communist Party and eventually opened the way to political pluralism and

the emergence of a multiparty system. Indeed, as the influence of democratic forces increased between 1985 and 1991, that of the Communist Party diminished with non-Communists and anti-Communists replacing party officials on the local levels of Soviet administration.

Perestroika and the Government: "Democratization"

"Democratization" of governmental agencies on all levels of Soviet administration, especially on the national and republic levels, was another aspect of perestroika. Gorbachev spoke of the need for everybody to participate in shaping society. "Democratization" was intended to give a sense of common purpose to the Soviet people and to engage them in a process of self-monitoring of their political and economic activity.

Democratization initially affected legislative institutions on the national and local levels: the Soviet presidency; the relationship between the republics and the central (federal) government in Moscow; the criminal justice system, including the KGB; the civilian bureaucracy; and the military. In these instances, democratization involved either increasing popular influence over state agencies or assuring their responsiveness to popular interests and needs.

In December 1988, the Supreme Soviet in Moscow approved the introduction of multiple candidacies in the nomination and election of members of all legislative bodies—an attempt to reinvigorate the local soviets; and the replacement of the old Supreme Soviet with two new bodies, the Congress of Peoples' Deputies (CPD) and a new, small, bicameral Supreme Soviet which would convene yearly to discuss constitutional, political, social, and economic issues.

The most critical task of the CPD was to select from its membership a new Supreme Soviet—smaller and more influential than the old one—that would have real power to pass on or veto legislative proposals as well as monitor the policy-making agencies, in particular the selection of a prime minister previously done by the top leadership of the Soviet Communist Party. Elections for membership in the new CPD were held on March 26, 1989. They were open and hectic, with much campaigning by candidates of non-party–controlled groups and individuals as well as by party-sponsored people. In that respect, they were unprecedented in recent Soviet political history. Soviet voters themselves were exhilarated—or befuddled—by the enormous freedom of choice.

Perestroika and the Soviet Military

Gorbachev undertook major reforms of the military involving almost all aspects of its operation. For example, as a result of a new emphasis on improving the quality of life of Soviet consumers, it was necessary to persuade the army to get used to the fact that it would no longer have first claim on national resources and that it would no longer get whatever it said it needed to perform its strategic and security-related functions. Gorbachev told the military that it had to pay more attention to managerial skills and, in particular, to improve the harsh and unpleasant living conditions typically experienced by recruits. In the new atmosphere of openness, multiple instances of hazing of new recruits were reported by the Soviet media. In many cases, hazing resulted from inter-ethnic tensions. Thus, the situation in the army served as an advance warning about one of the most important factors instrumental in the eventual collapse of the Soviet Union (see the following). It stands to reason that, with the exception of a few large cities and regions where people of different ethnicities lived and worked side by side, in much of the Soviet Union opportunities for inter-ethnic contacts were relatively limited, as people of different backgrounds lived in far-flung corners of a vast country. In the military barracks, however, all these different people were brought into close contact with one another, and the percentage share of non-Russian recruits grew quickly because of a significant and lasting disparity in birth rate between Russians and peoples from the Muslim republics.

A fairly significant influence upon the morale of the military was exerted by the war in Afghanistan (1979–1989), in which more than 13,000 Soviet troops perished. Although the Soviets were, generally speaking, fighting the same enemy the Americans were to face just twelve years and seven months after the Soviet withdrawal from Afghanistan (February 1989), the Western cold war geopolitics legitimately treated the Soviet invasion of that country as an attempt to extend the Soviet sphere of influence. Protesting the Soviet invasion of Afghanistan, Americans even declined their participation in the Moscow Olympic Games of 1980. This was a strong message, and many Soviet intellectuals, particularly those taking cues from the West, were relating Soviet victims in that war to its actual goal of propping up a puppet regime in Kabul. So, criticism of the Soviet military doctrine was mounted not only abroad but also within the country. In that atmosphere, the military was obliged to accept a new strategic doctrine that emphasized defense rather than offense. Gorbachev reduced the traditionally heavy role of the military in foreign policy making and excluded its participation, except in an advisory capacity, in decision making on arms reduction. The army was required to cooperate with glasnost and to accept a new degree of public scrutiny. Needless to say, the military leadership was uneasy about perestroika, even though it welcomed economic restructuring, which it believed would ultimately benefit not only the country at large but also the armed forces. And, to maintain civilian control over the army, Gorbachev tried to diminish the military's exaggerated stature in Soviet society.

Perestroika Assessed

If perestroika was intended to assure the longevity of Soviet socialism, it must be considered a failure, along with Mikhail Gorbachev, who presided over it. It disrupted the country's economic life and disappointed people's expectations of an improvement in living conditions; it profoundly compromised Soviet administrative and political unity, dividing Soviet society and greatly weakening the Communist Party's capacity to run the country. Perestroika unleashed three major conflicts which ultimately did away with the Soviet Union. The first of these conflicts was within the Russian society. Specifically, the old ideological argument between the Westernizers and Slavophiles, or rather "national patriots" as the intellectual heirs to Slavophiles were now called, was revisited. Although not a struggle between institutionalized groups, the arguments about how to develop Russia, the bulwark of the Soviet Union,

were vigorous and at times acrimonious. Should Russia imitate Western forms of government and create a market economy or should it draw upon its long tradition of autocracy, fusion of state and business, and opposition to the West? Conducted in the media and within legislative bodies, these arguments created a new division line in Russian society while simultaneously blurring or even erasing a division line between members and non-members of the Communist Party. This was because both intellectual camps enlisted both people who were and people who were not party members. In all fairness, though, heated debates between Westernizers and Russian nationalists played out in Moscow and Saint Petersburg for the most part. But in the heavily centralized polity and society of Russia, it is these two cities whose intellectual elites typically sway the rest of the country.

The second conflict was between the federal power base in Moscow and power bases in the union republics. While the republics were given more power under Gorbachev, some of them, notably the Baltic republics, wanted still more say over their domestic affairs. With this in mind, Gorbachev attempted to develop a new union treaty in which prerogatives of the federal center and the republics would be carefully separated and the authority of republics enhanced in exchange for their upholding and respecting the somewhat diminished but crucial authority of the federal government. Gorbachev's effort was seemingly close to success when it was thwarted by an attempted coup d'etat. Only occasionally did the conflict in question assume the nature of open anti-Russian sentiment. In the 1980s, the CPSU's Politburo had representatives of at least seven non-Russian ethnicities (Ukrainian, Belarusian, Georgian, Latvian, Azeri, Uzbek, and Kazakh) out of the total of fifteen members. From the early 1960s, the republics were overwhelmingly run by a communist cadre of titular nationalities, although the position of Second Secretary of the local (i.e., republic-based) Communist Party was usually assigned to an ethnic Russian. The languages of the republics' titular nationalities were used in local media alongside Russian; and, while in some republics, the degree of Russification (i.e., switch to Russian as the language of everyday communication) of local intelligentsia and public life was significant (notably in Belarus, Ukraine, and in Turkic-speaking republics of Central Asia and Azerbaijan), it did not necessarily derive from some perfidious ploys of Russian imperialists. In Turkic-speaking republics, the Russification process never extended beyond their capital cities. In the Baltic republics, however, the secessionist attitude had been always imbued with anti-Russian sentiment among the indigenous populations. This is because Estonians, Latvians, and Lithuanians believed, not without grounds, that Russians wrested them from the Western civilization to which those indigenous Baltic ethnicities had belonged. It is no accident that only in the Baltic republics did the new pattern of elections into legislatures, introduced under Gorbachev, bring secessionists to power in numbers sufficient to win the parliamentary vote for breaking away from the Soviet Union. However, when in violation of already established tradition, in the republic of Kazakhstan, the post of the Communist Party's First Secretary was in 1986 assigned to an ethnic Russian, Gennady Kolbin (following the death of Dinmukhamed Kunayev, an ethnic Kazakh and close associate of Brezhnev), anti-Russian rallies took place in the much-Russified capital city of Kazakhstan, Almaty.

The third conflict, or rather the third type of conflict, arose between numerous ethnic groups living side by side in such multi-ethnic regions as the Trans-Caucasus and Central Asia. In the late 1980s, none of these conflicts involved ethnic Russians, although Russian communities outside the Russian Federation and within ethnic homelands of the Russian Federation itself were feeling less welcome than before. In the new atmosphere formed by glasnost, true elections, and lack of censorship, people of all ethnicities were prone to learn more about their past, attend traditional places of worship and become outwardly more religious, and they were free to interpret various events from their past and speak publicly and write about them. That historical legacy, however, abounded in inter-ethnic rivalries for land and water, at times with sectarian violence. Under such conditions, the top-down imposition of democracy was fraught with consequences similar to those that the introduction of elections (often under Western pressure) brought about in countries like Algeria, Lebanon, and Palestine from the 1990s on, when destructive forces were able to gain power as a direct result of the newly obtained freedom to vote. The most acute conflicts during the latest phase of the Soviet Union's existence broke out between Muslim Azeri and Christian Armenians, between Uzbeks and Meskhetian Turks (who had been relocated to Uzbekistan from Georgia in the 1930s), as well as between Uzbeks and Kyrgyz in the Kyrgyz city of Osh. In all cases, the minority ethnicity (Armenians in the Azeri city of Sumgait, Turks in Ferghana, Uzbekistan; and Uzbeks in Osh) were subjected to pogroms, and the Armenian-Azeri rivalry over the Nagorno-Karabakh enclave developed into an armed conflict.

All three conflicts (Westernizers vs. Slavophiles, Federal power center vs. union republics, and inter-ethnic clashes) cracked the foundation and threatened the continued existence of the Soviet polity. So, the most retrograde forces within the Communist Party decided to act. Ironically, then, perestroika helped to lay the groundwork for the collapse of the Soviet Union, which it was intended to strengthen and invigorate. Not surprisingly, the man in whom so many Russians had placed their hope for a better future and who tried so hard to vindicate that popular faith in him became more popular abroad, particularly in Germany and in the United States (where in 1988 he was proclaimed the Man of the Year by *Time* magazine) than in the country over which he presided—not exactly a comfortable position for any politician. In much of his *own* country, however, by the end of 1991, he had become, for all too many, an antihero, hastening rather than averting the collapse of the Soviet Union.

THE AUGUST 1991 ANTI-GORBACHEV COUP ATTEMPT

The August 1991 coup attempt against Gorbachev accelerated the collapse of the Soviet Union. The coup further weakened the already debilitated political system, encouraging its critics to continue their assaults. It certainly weakened Gorbachev, who survived the coup but lost most of the respect and the political authority he had accumulated since 1985. Most importantly, the coup confirmed that the most radical communists would not hesitate to use violence to undo reforms conducted under perestroika.

In some ways, Gorbachev had himself to blame for the coup. In an attempt to maintain his grip on power he constantly maneuvered between pro-reform forces in the Party and conservatives, forging alliances of convenience with the latter for the simple reason that conservatives outnumbered reformers in the upper echelons of power. For example, in December 1990, Gorbachev replaced his liberal interior minister, Vadim Bakatin, with former KGB general Boris Pugo and replaced Nikolai Ryzhkov as prime minister (premier) with Valentin Pavlov, a command administrative economist hostile to reforms. Alienated hard-liners in the civilian and military leadership agencies in Moscow gained again at a session of the Congress of Peoples' Deputies, also in December 1990, when Gorbachev named a colorless career Communist *apparatchik,* Gennady Yanayev, to be his vice president, a move that stunned reformers, including some of his closest supporters, such as Eduard Shevardnadze and Alexander N. Yakovlev.

In early 1991, the hard-liners turned on Gorbachev, first on the issue of Lithuanian separatism. They wanted Gorbachev to remove from power in Vilnius the nationalist and anti-Soviet chairman of the Supreme Soviet of Lithuania, Vytautas Lansbergis, but Gorbachev refused. But, what provoked their wrath and a conspiracy to force Gorbachev to reverse perestroika reforms was his sympathy for a radical decentralization of the Soviet state that could—and eventually did—lead to its demise. In late July, Russian Republic president Boris Yeltsin, eight other republic presidents, and Gorbachev accepted a draft union treaty that provided for transforming the 1922 Soviet Union into a confederation of near-sovereign constituent republics. To communist party conservatives this treaty presented a real threat of national disintegration. They decided to force Gorbachev out of power through a coup.

The plotters failed primarily because at the time of the coup the Russian secret police and the military were as divided as the society at large. Specifically, the plotters failed to ensure that the members of a special KGB unit would follow their orders to arrest Russian Republic president Boris Yeltsin, who was loyal to—even though he deeply disliked and distrusted—Gorbachev. Indeed, local KGB commanders had mixed feelings about the coup, which was why they refused to seize Yeltsin. The Soviet military also was not completely behind the coup against Gorbachev. In addition, the plotters botched mobilization of the party for the removal of Gorbachev.

In sharp contrast to the plotters, who had no blueprint for what they would do if they succeeded in gaining control of the state apparatus in Moscow, the reformers loyal to the Gorbachev leadership knew exactly how they should act. Yeltsin and his supporters did not take immediate advantage of Gorbachev's obvious weakness and move to take power themselves. Rather, they called for the restoration of Gorbachev as the legitimate president of the Soviet Union, both to ensure continuity of state power and to give constitutional legitimacy to their own de facto exercise of that power, a gesture that was useful domestically and internationally during and immediately after the crisis.

The beginning of the end of the coup attempt occurred when, on August 21, a mission of loyalists went to the Crimea, where Gorbachev was being held hostage for the third straight day by the hard-liners. The rescuers brought Gorbachev and his family back to Moscow. Shortly after Gorbachev's arrival, the Supreme Soviet formally reinstated him as president.

THE COMMONWEALTH OF INDEPENDENT STATES (CIS) AND GORBACHEV'S RESIGNATION

In the following weeks, Yeltsin, in his own quest for power, eventually took advantage of the Soviet president's evident political weakness. Yeltsin appeared on national television reiterating a series of decrees he had issued in the Russian Federation during the coup, including one suspending the Soviet Communist Party's branch in Russia and its major daily newspaper, *Pravda.* Yeltsin systematically rid the Russian government (and ultimately those parts of the Soviet government under Russian control) of the old *nomenklatura,* replacing them with young, liberal, and modernizing professionals.

Moreover, on September 2, 1991, the Congress of Peoples' Deputies approved a plan of administrative decentralization, although by this time Yeltsin had lost interest in preserving the 1922 Union Treaty (which converted Soviet Russia into the Soviet Union), however loose it might be. He believed, at least for the moment, that the Russian Republic would benefit if relieved of the burden of bankrolling other Soviet republics. By the early 1990s, conflicts unleashed by perestroika produced an atmosphere in which almost every major ethnic group of the Soviet Union thought that it was being taken advantage of. Russians wanted to "stop aiding others for free," while other groups thought that they had been exploited by Russians and occasionally by other ethnicities due to some Soviet arrangements. In this view, the Soviet central government had been the biggest obstacle to the economic reforms needed to improve living standards. A principal mouthpiece for the Russian rendition of this secessionist outlook, Yeltsin was now more a Russian nationalist than the Soviet reformer he once had been. At least some other republic leaders shared his quest to find an alternative to the 1922 Soviet Union.

Yeltsin and the Founding of the Commonwealth of Independent States

On December 8, 1991, the leaders of Russia (Boris Yeltsin), Ukraine (Leonid Kravchuk), and Belarus (Stanislav Shushkevich)—the three states which together controlled 73 percent of the Soviet population and 80 percent of the territory of the Soviet Union—convened in a government retreat in the westernmost part of Belarus and signed an agreement that declared the Soviet Union at an end. This agreement was made quite independently of Gorbachev, whom they did not even bother to consult, effectively blocking him from interfering to prevent the inevitable demise of the Soviet state and with it, of course, his presidency.

Recognizing the vital significance of inter-republic ties for the economy of each republic, the three leaders developed a proposition for an alliance of former Soviet Republics, the so-called Commonwealth of Independent States (CIS) and brought this proposition for ratification by their parliaments. The CIS came into being on December 10, 1991. Most other republics subsequently sought entry into the CIS, and, on December 20, 1991, a formal agreement was signed in Kazakhstan by leaders of eight republics. The signatories recognized one another's sovereign independence and agreed to determine shortly the disposition of the Soviet military. By December 22, Armenia, Azerbaijan, and Moldova had joined the Commonwealth to bring the total membership to 11 former Soviet republics. Georgia joined the CIS in 1993. Only the three Baltic States declined joining the CIS.

Gorbachev's Resignation

With Yeltsin as president of the Russian republic (Russian Federation), the heart of the Soviet state steadily weakened. For example, on December 20, the government of the Russian Federation assumed formal jurisdiction over the Soviet Foreign Ministry, the KGB, the Supreme Soviet, and even Gorbachev's presidential office. With a disquieting aggressiveness, the Russians took over the Soviet money supply and Soviet trade in oil, gold, diamonds, and foreign currency, fueling suspicion, jealousy, and animosity among the other republics.

With no support among the republics for his leadership, Gorbachev resigned as president of the Soviet state. On December 26, the Supreme Soviet passed a resolution acknowledging the demise of the Soviet Union. Russia subsequently was assigned the Soviet Union's seat in the Security Council of the United Nations, and the former Soviet Union's embassies abroad replaced the Soviet flag with that of the Russian Federation.

NOTE

1. In fact, in Kazakhstan, "integrated" does not mean fluent in Kazakh or accepting some traditional life style of the indigenous population. Rather, it refers to urban Kazaks who continue to use Russian as their chosen language for everyday communication.

Russian Federation: Economic and Political Change 1992–2009

In the years after the collapse of the Soviet Union, the Russian Federation made strides away from the dictatorship of the communist and czarist eras toward a kind of democratic political system modeled in spirit, and to a substantial degree in practice, after the developed democracies of the West. While at present Russia is criticized in the West for the resurgence of authoritarianism, even the harshest critics admit that it is far from what it was under communist rule.

As was mentioned in our account of perestroika, the Soviet leadership under Gorbachev opted for simultaneous economic and political reform despite the fact that the Soviet Union was about to collapse under the burden of economic problems. The same simultaneous reform option was now pursued in the Russian Federation, only in a much more straightforward and radical way. The idea of fighting, as it were, on two fronts at the same time is immensely important to take into account when assessing virtually all the post-1991 developments in Russia and other post-Soviet countries. A critical problem associated with *concurrently* reforming the economic *and* the political fabric of an egalitarian society with a strong ethos of dependency on the state and, consequently, utmost dedication to its entitlements, is that a market economy implies a cardinal departure from equality and breeds social stratification. As a result, large groups of people become disenfranchised, and those are precisely the people whose support and participation are needed for democracy to take root. In Charles Tilly's terms, "democratization becomes possible when trust networks integrate significantly into [ruling] regimes, and thus motivate their members to engage in mutually binding consultation—the contingent consent of citizens to programs proposed and enacted by the state" (Tilly 2007: 74).

Russia did not have a tradition of such integration to begin with, and whatever was achieved under perestroika along the lines of boosting public trust in the government was quickly sacrificed on the altar of economic privatization, from which only a privileged few benefited. It is primarily because of this that many intellectual and politically active Russians now view China's reform path as the one that Russia should have pursued if only Western, primarily American, advisers had not pointed them in a different direction. In China, the development of market economy occurred well before democracy ever took root, and due to that (or so it is believed), the Chinese economy has lifted millions from poverty, and thus created preconditions for democracy (yet to come to fruition). The veracity of this way of thinking is debatable, to say the least. But not just China's example matters when assessing Russian reform.

What matters even more is a dominant paradigm, a kind of lens through which Westerners view the rest of the world. The most revealing fact about paradigms is that they change. Prior to the late 1970s, the idea that no democracy can exist without a sizable middle class was a matter of course, a truism hardly challenged by anybody in the West; and because the middle class does not fall from the sky, democracy was viewed as an organic outgrowth of *economic* development. This idea was dealt a blow with the advent of neoliberalism heralded by the administrations of Margaret Thatcher in Great Britain and Ronald Reagan in the United States. Since the early 1980s, many things have been thought achievable in a top-down fashion or even by way of foreign military intervention: Market economy, for example, became—at least in the perception of many intellectuals—a miraculous remedy producing democracy almost by way of immaculate conception. Neoliberalism is "a theory of political economic practices that proposes that human well-being can be best advanced by liberating individual entrepreneurial freedoms and skills within an institutional framework characterized by strong property rights, free markets and free trade. . . . Furthermore, if markets do not exist (in areas such as land, water, education, health care, social security, and environmental pollution), then they must be created by state action if necessary. But beyond these tasks the state should not venture" (Harvey 2005: 2). Although the sway neoliberalism held over foreign policy thinking did not last more than two decades, this ideology has definitely left its mark on the American perceptions of the post-Soviet space.

Since the late 1990s we have been witnessing a profound and pervasive dissatisfaction with neoliberalism, a dissatisfaction further exacerbated by the advent of the global economic crisis in late 2008. After the U.S. government effectively nationalized several major banks and insurance conglomerates (due to their insolvency caused by activity of financial speculators), the number of people still willing to glorify unfettered capitalism has predictably dwindled. And while the idea of forceful democracy promotion appears to be more tenacious than market fundamentalism, it is now in retreat as well following huge but predictable disappointments with implanting democracy in the Middle East as well as in Russia. This ongoing change in the dominant paradigm is important to keep in mind when assessing developments in Russia and elsewhere outside the Western world.

DEVELOPMENT OF A MARKET ECONOMY 1992–2009

President Boris Yeltsin and his successor, Vladimir Putin, had the mammoth task of transforming the communist centrally planned economy in Russia into a free market-based economy in the post-Communist era of the 1990s and the early years of the new century. Both Yeltsin and Putin were convinced that the only way to expand productivity, increase growth, and raise living standards was to replace state control of the country's economic life with free enterprise. But there was no recipe for how to achieve that. A joke that conveys the nature of the predicament is that everybody knows how to get toothpaste out of a tube, but nobody knows how to insert it back. To wit: it is understandable how they nationalized the economy more than seventy years ago, but where is the entrepreneurial class to whom this economy, piece by piece, should now be entrusted? And if that class is not there, how is one supposed to be created from scratch? Because few if any Russians had any assured response to these questions, Western, particularly American, advisors were, at least initially, looked at as demigods. One of these advisors, possibly the most influential one, was Harvard's Jeffrey Sachs, who had already gained experience helping to successfully put the Polish economy on a market base albeit cushioning that sort of transformation with a hefty $15 billion aid package from American donors. Based on that experience, Russia, a country more than three times as big as Poland (population-wise), would require proportionally greater aid. However, no such aid was received, and yet privatization was given the green light. Incidentally, when it was in full swing, two American advisors, Harvard economics professor Andrei Shleifer and his associate Jonathan Hay from the Harvard Institute for International Development, were discovered to have been directly profiting from the market they were busily creating in Russia. The U.S. District Court in Boston partially upheld the accusation after a seven-year investigation and required that Shleifer and Hay pay millions of dollars to the U.S. Treasury. In her well-publicized book, *The Shock Doctrine*, Naomi Klein sees poignant irony in the fact that moneys were to be paid to the United States, not to ordinary Russians, who were never consulted about the entire privatization affair. The actual or perceived role of American advisors at the initial stage of market reform in Russia subsequently fed into a strong wave of anti-Americanism under President Putin. Very few Russians ever approved of what was done to the Russian economy in the early 1990s. One of the very first moves was to decontrol prices except for bread; and as a result, prices skyrocketed and most Russians instantly lost their life savings. On the benefit side, grocery stores that had stood almost empty since the late 1990s immediately filled with all sorts of foods which, however, most people with income from the public sector could not afford to buy. Another early move was to purge the state monopoly on foreign trade. Soon, some people with access to management of oil, gas, and non-ferrous metal extraction made fortunes capitalizing on the gigantic gap in labor costs between Russia and the West. Yet another early move was to issue and distribute the so-called vouchers with a nominal value of 10,000 rubles so they could be used to privatize industry. Because most people suddenly found themselves living in poverty they sold their vouchers to speculators. Nonetheless, as early as 1992–93, the movement of the economy away from communism toward capitalism was well under way.

Privatization of Industry

The key to developing a free market capitalist economy was privatization, that is, the process of turning state-run businesses over to private hands. The privatization of consumer-service enterprises such as retail stores proceeded smoothly and quickly, as people who accumulated vouchers or profited from reselling scarce goods obtained ownership of those outlets. But privatization of large-scale industry was problematic because, at least initially, only the ex-Communist enterprise

managers had both the skills and the resources needed to buy and run firms once owned and operated by the state. In some cases, workers did pool their meager savings to join with management to purchase and control the companies they had worked for to ensure job security. These transactions, however, occurred in only a small part of the state-controlled economy. For the most part, people with access to political power came to own the most lucrative businesses. However, buying enterprises with vouchers could not possibly turn much of Russian industry private because even when underestimated, the price of the largest industrial firms was in the billions of dollars. So the biggest booster to privatization of industry was the loans-for-shares scheme. In 1995, facing severe fiscal deficit and in desperate need of funds for the 1996 presidential elections, the government adopted a loans-for-share scheme proposed by banker Vladimir Potanin and endorsed by Anatoly Chubais (then a deputy prime minister) whereby some of the largest state-owned industrial assets (including state-owned shares in Norilsk Nickel, Yukos, Lukoil, Sibneft, Surgutneftegaz, Novolipetsk Steel, and Mechel) were leased through auctions for money lent by commercial banks to the government. The auctions, however, lacked competition, as they were largely controlled by favored insiders. Because neither the loans nor the leased enterprises were returned in time, this effectively became a form of selling for a very low price. Economically, this turned out to be a short-term success, as government managed to cease subsidizing the then-inefficient enterprises, and their performance improved under new ownership, contributing to the Russian economic growth in the 2000s. However, the scheme was perceived by many as unfair, and it is the loans-for-shares scheme that gave rise to the class of Russian *business oligarchs,* who have concentrated enormous assets, further increasing the wealth gap in Russia and contributing to political instability. The prices at which some industrial firms were sold to the oligarchs were significantly below their market value. For example, Norilsk Nickel (a group of nickel and copper refineries) was sold for $170.1 million, when just one year's profit for that firm in the 1990s was close to $1.5 billion and the firm's market value in 2007 was $22 billion. Another company, Yukos (an oil-extraction business), was sold to the subsequently jailed tycoon Mikhail Khodorkovsky for $159 million; this is a company that in 2005 extracted 24.5 million tons of crude oil.

Privatization significantly boosted social stratification in Russia. In 2008, there were 86 billionaires—in a country with an average monthly salary of about $700. Note that the Russian egalitarian mentality has been long nurtured by the redistributive peasant commune (see the following) and subsequently by 70-odd years of state socialism. The October 2007 issue of *Forbes* magazine (Russian version) published a list of the 200 largest privately owned firms. The ten leading positions in that list are occupied by Russian Aluminum; UGMTK (producing copper and headquartered in Sverdlovsk Oblast); GK TAIF (Tatar petrochemical company headquartered in Kazan); Metalloinvest (ferrous metallurgy); GAZ (automobiles; Nizhny Novgorod); Russneft (oil); Mefalon (mobile phone systems); Eldorado (electronics and hardware stores); Euroset (stores selling mobile phones); and United Metallurgic Company (pipes). Firms whose headquarters are not indicated are centered in Moscow as are 140 of the 200 leading private firms.

Considering that by 1997 close to 70 percent of Russian industry (by output) had been already transferred to private hands, the pace of privatization was considered swift. How was such a swift privatization possible? One of the answers is contained in the question. That is, it was possible *because* it was swift; because it did not involve any consultation with the broad masses of the Russian public or their representative bodies such as any democratic system would require. In other words, privatization benefited from the state of shock and from the absence of democracy. In fact, the entire market transformation of Russia went down in history as "shock therapy," in accord with Milton Friedman's school of economics at the University of Chicago. A usual Russian comment to that label is that, indeed, there was a lot of shock but precious little therapy. Another reason for "success" was the availability of a robust Westernizing minority which took over in Russia during the late 1980s and early 1990s. Although Boris Yeltsin presided over the entire country of Russia, his *de facto* constituency that "called the tune" consisted of the cities of Moscow and Saint Petersburg, where most politically active people belonged to the Westernizers' camp (to an incomparably higher degree than now, incidentally). For the entire Soviet Union minus the Baltic States, that was a truly unique constituency. Muscovites subsequently came to possess the lion's share of personal fortunes created in the early 1990s. In no other part of Russia or the CIS at large were people willing to endure hardships in exchange for something as obscure as democracy and market economy.

By now much blistering critique of Russian privatization is available in print and there are plenty of accounts of how several self-made, brazen entrepreneurs "privatized" Russia in the early 1990s. In an unusually candid interview with Piotr Aven, now a billionaire but who used to be a minister in the Gaidar-led 1991–92 government, some light is cast on the human side of the whole privatization affair in Russia. Aven portrays a group of young and committed conspirators who managed to fill the political vacuum that ensued after Gorbachev was sidelined by Yeltsin and did their best to irretrievably transform the economic system of Russia. To the reader versed in Russian classic literature, Aven's description would remind him or her of episodes and characters from Fyodor Dostoyevsky's novel *The Devils,* only those devils of the 1860s were preoccupied with *installing* the regime that the Gaidar-and-Chubais-led group set out to *dismantle* in the early 1990s. But the two groups match each other in arrogance, zealotry, and secretive ways. Firm believers in the ideas of Milton Friedman's school, the Gaidar-led "Chicago boys" knew that they should stay away from any home-grown economic managers because contacts with them might compromise the tough stand of the reformers and mitigate their zeal. Throughout the two years preceding his 1991 ministerial appointment, Aven did not live in Russia and was stationed in Laxenburg, Austria, at the International Institute of Applied System Analysis; prior to that he was a PhD student. In his interview, Aven acknowledges that the group he became a part of did not engage in any dialogue with Russian society but did transform the entire fabric of Russia's economy. Aven recognizes that he and his comrades-in-arms will never be appreciated by their fellow-countrymen, most of whom believe (unduly as far as Aven is concerned) that their lives worsened as a result. Indeed, according to the January 2007 national survey by the Russian polling firm VTSIOM, 27 percent of Russians have a decidedly

negative opinion about the accomplishments of the Gaidar-led government, and 31 percent evaluate those accomplishments as "rather negative." Only 6 percent credit that government with staving off Russia's ultimate ruin (Aven's view). Considering the direction taken by the Gaidar government, only 18 percent believe in retrospect that it was generally right but nothing much was achieved, 28 percent said that the government led the country in the wrong direction, and 25 percent subscribed to the view that the reformers consciously destroyed the economy. One half of Russians are sure that "shock therapy" was not the only way to go and that a less painful alternative was possible; only 19 percent are of the opinion that there was no other way. In its obituary for Boris Yeltsin, the Associated Press pointedly stated that many Russians "were outraged . . . by his sale of the nation's industrial might and natural resources in shadowy auctions, by the disintegration of the public health care system and by pensions that turned to cinders in the fires of raging inflation."

The above account of popular Russian attitudes toward Russian reform is not intended to convey a personal value judgment. Rather, the point is that popular attitudes are significant in and of themselves, regardless of what anybody thinks about them. Now, 17 years after the start of reform, these attitudes allow one to label the economically successful Muscovites and Peterburgers (not only those who struck it as rich as Aven) as the market-dominated minority which succeeded in imposing its ways on the silent majority. "Market-dominated minority" is a term Amy Chua uses in her book *World on Fire.* While the author of this text is not inclined to read as much into the ethnic makeup of the minority in question as Chua did, the percentage share of ethnic Russians within it is indeed below that of the general population, and many Russians see this as a problem.

Yet another argument put forward in the aforementioned book by Chua, an argument which is worth taking issue with, is that the "United States has been promoting throughout the non-Western world raw, laissez-faire capitalism—a form of markets that the West abandoned long ago." It is *likely* that the role of American advisors is hugely overstated. Scapegoating, after all, is a mainstream tool of domestic political debates in Russia. It is extremely *un*likely that the capitalism that emerged in Russia could possibly be anything but raw with or without American stewardship. For that there are two major reasons. First, Western capitalism today is an outgrowth of a long period of development, whereas Russian capitalism is in its early stage. Today's Russian oligarchs might not think or act altogether differently than the American robber barons of the nineteenth century. Second, Russia—as well as most other post-Soviet countries—does not have any enduring democratic tradition, and the civic attitudes dominant in Russian/Soviet society were, and still are, far from refined.

In the above-mentioned account of Russian privatization by Aven, there is a sobering confession triggered by the following question from his interviewer: "By now, has your worldview changed compared to the one you embraced at the start of [Russian] reform?" "I began to think more about integrity and fairness," responded Aven, "this is number one. Secondly, I began to think more about protecting people who do not understand what is going on, and about the factor of trust and attitude toward those reforms . . . [n]ot just about the mechanism of reform itself but about its perception and trust—these are fundamental parts of reform. This is what I would emphasize today. Next in importance is everything that pertains to the fight

(© Tatyana Nefedova)

A rural village in Kostroma Oblast (in the northeast of Russia's heartland).

against corruption. The problem of corruption and its psychological and institutional causes have been entirely neglected by us but they turned out to be the most important."

A side effect of swift privatization was the equally swift deindustrialization of Russia because quite a few factories and plants halted operations due to inadequate demand for their output. Thus, in 1996–97, Russia's industrial output amounted to about 60 percent of that in 1991. Only by 2008 had industrial output reached its 1991 level. But that reversal is largely ephemeral as it was fueled by super-high international oil prices. Between 1992 and 2003, industrial employment declined from 40 percent to 22 percent of the total employment.

Development of Agriculture

The advent of market reform in the early 1990s has been the second—or by some accounts third—coming of market economy to Russia. Russia's first encounter with capitalism dates back to the last quarter of the nineteenth century; it was facilitated by the 1861 emancipation of serfs. This encounter was inconsistent and relatively short, as it was interrupted by World War I and the 1917 Communist revolution. During the years of the so-called New Economic Policy, 1921–1927, Russian communist leaders themselves loosened the recently acquired reins of centralized control and flirted with market forces, yet they made a political decision to squelch those forces. Only in the early 1990s did capitalism make a comeback following the 80-year-long ostensible deviation from the "European" path of development.

In 1897, 87.4 percent of Russia's population lived in the countryside; and with a minor discount for the gentry and the clergy, this rural sea of people consisted of peasants. Since then, Russia had been significantly modernized. Today, 73 percent of Russia's population is urban, and agriculture officially accounts for 11 percent of Russia's labor force. While there is no denying the seriousness of that change, the elements of continuity deserve more attention than they are usually given. Surprisingly few Russia watchers contemplate the idea that Russia, in fact, continues to be an exceedingly rural country. First, it is rural in terms of mentality: because urbanization was delayed, most of

Russia's urbanites were born and raised in a traditional Russian village. According to Anatoly Vishnevsky, in 1990, when urbanites made up 66 percent of the entire Soviet population, only 15–17 percent of those who were 60+ had been born in urban areas; of those in their forties, about 40 percent had. Only among those 22 and younger did the percentage of urban-born exceed 50 percent.

Second, Russia is also exceedingly rural in a more immediate sense—that is, in terms of rank-and-file Russians' continuing involvement in the working land. In today's Russia, 39 percent of the rural population of working age is employed by agricultural enterprises (down from 48 percent in 1990), and approximately 25 percent of it is engaged in subsistence agriculture, that is, working only on one's own household farm (up from 10 percent in 1990). Altogether, there are 16 million rural household farms in Russia. Additionally, 14 million urbanites own a plot in the so-called "collective orchards," and 5 million urbanites own a plot in "collective vegetable gardens." There are also 2 million dacha-owners, most of whom grow food as well. The total is 37 million families, or more than 100 million people. Additionally, over 20 million statistical urbanites (small and medium-town dwellers) permanently live in traditional rural cottages with vegetable gardens attached to them. Altogether, this means that no fewer than 120 million Russians (out of the total population of 144 million) are working the land one way or another! Consequently, Russia is a more agricultural country than the official statistics would lead you to believe. Officially, only 11 percent of its labor force works the land, and 73 percent of its population are urbanites. But by some accounts, the total labor input in agriculture amounts to 665 million person-hours, which is 60 million person-hours more than labor input in industry. Ninety-two percent of all rural villagers and 44 percent of urbanites depend on their own vegetable gardens and/or livestock. Altogether 38 million households (or about 100 million people) are engaged in highly diverse agricultural operations. To our knowledge, the far-reaching implications of this mass involvement of Russians in farming are rarely given appropriate attention, the attention that would be commensurate with the actual significance of this fact for understanding the Russian world view and the Russian psyche.

When the Soviet Union existed, the stereotypical Western view of Russia's collective-and-state farm system was that Russian peasants craved to become independent farmers. One only had to liberate them from the shackles of collectivization. During the early stage of Russian reform (in 1993), a book titled *The Farmer's Threat* was published in the United States. In the opinion of that book's editor, the independent private farmer represented an existential threat to centrally planned agriculture, so the ex-Soviet *nomenklatura* would resist the disbandment of collective farms. The reasoning behind this view was twofold. First, collectivization, in many instances, had been forced because it ran against the economic self-interest of the Russian peasant. Second and most important, family farming was the mainstay of agriculture in the West, so why should that not be the same in Russia? While the answer to a question like that could be the subject of an entire book (two such books are actually available: see Ioffe and Nefedova 1997; Ioffe, Nefedova, and Zaslavsky 2006), in a nutshell, the problems that Russian agriculture faced turned out to be far different from stereotypical Western views of them.

Russian agriculture did malfunction primarily because agriculture does not lend itself to central planning. But production was not the only function of Soviet collective farms, and by the end of the Soviet Union's existence, not even the most important function for many of them. After Stalinist repressions subsided, collective farms developed into peculiar vehicles for the collective survival of the weakest. They maintained social services, enabled household farming operations (as members of collective farms routinely used fertilizers, tractors, animal feed, and often the collective farm's land for their own purposes), and ensured some tolerable (or rather, tolerated) quality of life to millions of people who did not have to work very hard as collective farmers. In many outlying rural districts, collective farms developed into almshouses, as retirees accounted for up to one-third of the entire local population. Rural out-migration has long been depriving the countryside of the younger and more enterprising people, as well as people with stronger labor ethics. Nowhere in the Soviet Union was the social erosion of agricultural communities as profound and the spread of such impairments as alcoholism as pervasive as in Russia.

One must understand, though, that the low quality of rural life that had long fueled the out-migration of the best and brightest was not a result of the Soviet system of economic management alone. At least in part, it was also a result of the enormity of Russia's space. Even in European Russia, that is, west of the Urals, the average size of a civil division of the oblast level is 57,000 sq km; oblasts are divided into largely rural districts with 2850 sq km per district on average; and the average former collective or state farm had 3,000 hectares of farmland. Even after investment was redistributed in favor of the countryside, which came to account for as much as 28 percent of all investment, it was difficult to create and maintain an adequate network of paved roads to all 27,000 central settlements of collective farms—much less to second-order rural villages—and to ensure that other amenities and utilities available to them would be of decent quality as well. So, only villages in proximity to major urban centers benefited from modernization.

Not all collective and state farms in Russia, however, were in a miserable state as production units by the end of perestroika. About one-fourth of these farms—those located in proximity to oblast capitals, in some regions with high natural fertility of the soil (particularly in Krasnodar and Stavropol regions), and in the republic of Tatarstan—maintained a fairly high level of productivity (e.g., grain yields of about 20 hectograms per hectare and milk yields in excess of 3,000 kg per cow), usually lower than under comparable natural conditions in the West, but higher than in most of Russia.

In the 1980s, the majority of collective farms were kept afloat by huge subsidies as the low retail prices on food maintained in the Soviet Union would not have ensured a reasonably high profit margin even on a well-run Soviet collective farm. In the early 1990s, subsidies suddenly dried up, and the stratification of collective farms by location (i.e., near large cities) and quality of soil became even sharper than before. The process of land abandonment, under way since the 1960s, intensified as well. By the end of the twentieth century, about 20 million hectares of arable land had been abandoned in the European section of Russia. Almost

exclusively that land was located in the peripheral parts of Russian regions and/or under the least favorable conditions of physical environment (poor soil, steep slopes, short growing season, etc.).

Not only did the stratification of collective farms deepen in the 1990s, but the breakdown of the agricultural output by mode of production and the makeup of Russia's foreign trade in food also changed. Under the Soviets, there were two modes of production, large collective-and-state farms and tiny household farms with a symbiotic relationship between them and their parent collective farms. As was mentioned previously, by the end of the Soviet Union's existence, 25 percent of foodstuffs was produced on the 2 percent of the land assigned to household farm operations. In the Russian Federation, the share was 26 percent in 1990. While in truth household farms did use land well beyond vegetable gardens and barns attached to rural residences (for example, when grazing privately owned cows on collective farm land), formally speaking, household farms held only 2 percent of all cropland in 1990.

The changes of the 1990s included conferring the status of joint-stock firms to most collective-and-state farms whereby each member was allotted a nominal but documented land share. If an employee decided to withdraw from a former collective farm in order to set up his or her own private business, he or she would be assigned a separate parcel. Very few people volunteered, however, even though initially the prospects looked bright.

The first registered family businesses in Russian agriculture had emerged in the late 1980s, during Gorbachev's Perestroika, the manifestation of the emergence of a new mode of farming. In 1991, there were only about 4,000 registered private farmers in Russia. By 1995 there were 279,000. Such growth was enabled by a combination of factors: significant tax credits and discounted loans, opportunities to buy farm implements at discounted prices, relative ease in obtaining land, a generally supportive political climate, and personal enthusiasm. At that time, private farmers were mostly self-recruited from the rural elite—agronomists, animal technicians, engineers, and even the leaders of collective and state farms, which thus lost their most enterprising cadres.

In the mid-1990s, the preferences and discounts were cancelled, and private farmers found themselves in the same conditions as former collective farms. Yet, private farms could not compete with larger producers. On average, a private farm in Russia has 50 hectares of land; in European Russia, the most widespread private farm landholding is only from 20–30 hectares in size. Using 7 percent of Russia's farmland, private farmers account for just 3 percent of the total agricultural output in monetary terms. The most widespread specializations of private farms are sunflower (14% of Russia's total output) and grain (8%), that is, products that are relatively cheap, hence such a small share of output in monetary terms. Around 2000, more than one half of private farms were suffering losses, and every fourth such farm had more than half of its land lying idle. According to Bashmachnikov, the President of the Russian Association of Independent Farmers, "the 10-year experience of registered private farms in Russia shows that few of them could build or purchase machines and livestock themselves. Out of 270,000 registered farmers, only about 30,000 have not been crushed by the press of market and have been able to set up a viable commercial farm. Another third just feed themselves. And the remaining third have quit."[1]

In the beginning of the new century (2001) only 4 percent of Russia's agricultural output was produced by registered commercial family farm businesses, which accounted for 9 percent of cropland. At the same time, as much as 52 percent of output was produced by household farms, which are not registered private businesses but represent an outgrowth of the traditional subsistence or natural economy. If anything, the elevation of an archaic mode of farming (more in line with nineteenth century than with twenty-first) to the leading position in Russian agriculture is not what any Russian reformer had envisaged. Some experts believe, though, that 52 percent of output is an overstatement. Not only do household farms use more of collective farms' land and other resources than ever before; but there is also a tendency to under-report the output of large centralized farms and over-report the output of household farms since the latter is relieved from taxes.

Nevertheless, by all accounts, commercial family farming has failed to become the principal mode of farming operation in Russia despite the fact that all legal obstacles have been removed. No serious agricultural expert in Russia believes that this is going to change in the foreseeable future. Reluctance to become a private farmer may be attributed to the lack of, or only a fledgling, market infrastructure, including the reluctance of retailers and processors to buy output from small producers, as well as to inadequate financial support. But the much depleted and eroded human capital and apathy of the traditional Russian village and its deeply entrenched spirit of dependency is one of the most important reasons.

Agriculture has been the most troublesome sector of the Russian economy for as long as anyone can remember. Russian agriculture went through some crucial turning points in the past, such as the abolition of serfdom and its re-imposition in the guise of collectivization. But the magnitude of the current travails seems to exceed what has befallen Russian agriculture in the past. The unrelenting demographic trends do not leave hope for long-term retention of rural labor in much of the Russian agrarian ecumene. As one member of Russia's Council of the Federation (the upper chamber of the parliament), put it, "The hand of Adam Smith, which has already clutched the Russian peasant by his throat, will soon squeeze the life out of him." Note that the gist of this complaint is not too little but too much market that the Russian peasant faces!

By the mid-1990s, the Russian state treasury, swollen by petrodollars, resumed agricultural subsidies, although on a much smaller scale than under the Soviets. Regional subsidies, particularly in ethnic homelands such as Tatarstan and Chuvash Republic, began to prop up collective farms. Also, vertically integrated agribusiness emerged in Russia, composed of entities controlling each level of production, from farming to processing to marketing. Successful food processors initiated these operations. The idea of contractual links between large farms and food processors stemmed from the general success of the latter and their ultimate dependency on the former to supply perishables such as milk. Growth in agriculture commenced in 1999, but most collective farms are losing money. In contrast, many domestic food processors, including dairy, juice, and sometimes sweetshop producers, have been profitable since the mid-1990s, but their further development has been stymied by a deficit of high-quality farm produce. The 1998 default and ruble devaluations made agricultural imports costlier. So the idea that industry, both domestic and foreign, would pull Russian farms out of the quagmire became popular. The contractual links between farms and processors that have emerged since then range from long-term agreements (stipulating exchange of agricultural products for agricultural investment) to wholesale purchase of entire farms that become incorporated into agroindustrial holdings.

According to Dmitry Rylko of the Moscow-based Institute of Agrarian Markets, vertically integrated companies are "currently picking up the last unaffiliated farms of Moscow Oblast and continue to buy up the best farms in other regions." However, according to Gennady Frolov, the manager of Cherkizovsky, "No more than 10 percent of all Russian collective farms may be of interest to those buyers. The rest are irremediable."

Significant changes in the structure of Russia's agricultural output have occurred since the collapse of the Soviet Union. Throughout the last three to four decades of the Soviet Union's existence, a perennial shortage of animal feed forced the Soviet government to import grain feed. However, the introduction of a market system prompted a drastic decrease in livestock. For example, where in 1993 there were 20 million cows in Russia, by 2006 their number had decreased to 9.5 million; as a result, milk output declined by 44 percent and meat output by 51 percent. So meat and dairy products are increasingly imported by Russia now, but there is no longer a need to import grain. Conversely, there is a grain surplus in Russia, occasionally up to 18 million tons; and so Russia has become an important exporter of grain. Needless to say, grain is much less expensive than products of animal husbandry, so it is unlikely that Russia has gained anything from the market-driven changes in the makeup of its agricultural output.

Because it has long become abundantly clear that the market transformation of Russian agriculture has produced but a few positive results, not much Western advice has been solicited, in marked contrast to large-scale industry. A positive side effect of this is that blaming the West for what has transpired in the extensive Russian countryside since the fall of communism is not as widespread in Russia as for other areas of life.

PROBLEMS WITH THE MARKET ECONOMY IN THE 1990s

Private enterprise was, and continues to be, plagued by a lack of effective regulation, especially safeguards for investors, restrictions on profit making, and difficulties of tax collection from big business. A very serious—indeed, socially dangerous—side effect of expanding private enterprise has been the appearance of an alarming gap between a small, aggressive, and ostentatious class of hyper-wealthy people and the rest of the population, much of which is stuck in low-paying jobs and barely able to get by. In Russia as a whole, the income ratio of the wealthiest 10 percent of the people to the poorest 10 percent is 15.3 to 1; in the city of Moscow, it is 41 to 1, according to the official reports which usually underestimate the income of the richest stratum. This gap was, and remains, the case, especially with pensioners and the elderly, some of whom have even taken to begging in cities just for food money—something they never had to do under the Soviets. New private enterprise also brought with it serious social ills such as corruption among high- and low-level government officials responsible for regulating business; crime, especially among young people; and unemployment in some company towns.

Explosion of Crime

Since the early 1990s, the extraordinary accumulation of wealth and at the same time relative decline in official remuneration of all kinds of bureaucrats gave rise to organized crime often referred to as the *mafia* (following the well-known Sicilian term). Members of the Russian mafia are conventional crooks demanding in different ways a share of the new opulence. Although mafia gangs are involved in every conceivable kind of criminal activity, from money laundering and trafficking in narcotics to the disposal of expensive automobiles stolen in the West, a favorite and familiar activity is "protection." As private enterprise gets started, crooks seek a cut for themselves in the profits of new businesses by offering them "protection" against destruction of property. If individuals refuse to buy protection, the mafia uses violence against them and their property. The phenomenon has driven Russian investors to great lengths to conceal their wealth, but in doing so, they forfeit whatever real protection legitimate authorities can provide. And the authorities in some instances had little idea of the extent of private business and virtually no control over its behavior. Another key weakness of Russian law enforcement is corruption, with many officials in league with the criminals they are supposed to control. The above-quoted billionaire, Piotr Aven, was quite on target as far as corruption is concerned.

Corrupt Business-Government Ties

Private enterprise has fostered a corrupt relationship in the form of a coziness between newly privatized business and administrators in Moscow and provincial capitals. Many foreign investors as well as Russian economists see a web of official corruption so great as to compromise the economy's well-being.

According to an estimate published in a Russian government newspaper, the overall sum of bribes paid in Russia exceeds $240 billion a year. According to estimates by the *Indem Foundation,* the total sum of bribes skyrocketed from $33 billion in 2001 to $316 billion in 2005 (close to 20% of Russia's GDP); and the average bribe that a businessman pays to a bureaucrat during the same period grew from $10,000 to $136,000. More than half of all Russian adults have personal experience with bribing somebody; and 80 percent of all businesses pay bribes. According to the 2007 Transparency International's ranking of 180 countries on corruption perception index, Russia is 143rd in that ascending order of corruption. For comparison, Turkey is 58th, China is 72nd, and India is 85th.

Corruption has become so endemic that it is perceived as normal. Here are some of the bribe-infested areas and reasons behind bribing and extortion.

1. Customs: for permission to accept prohibited goods; for return of confiscated goods and currency; and for imposing lower-than-requisite tariffs
2. Health care institutions: for purchasing overpriced equipment and medications; for furnishing false medical diagnoses (for example to avoid a draft); and for serving some people ahead of and at the expense of others
3. Divisions of motor vehicles: for issuing driver's licenses (e.g., without a test) and inspection certificates; and for falsifying data about accidents
4. Judges: for biased verdicts; for violating procedural norms; and for using the court as an instrument for taking over someone's business
5. Tax inspections: for partial tax evasion and for terminating a competitor's business
6. Law-enforcement agencies: for initiation and cancellation of cases and for subjecting them to additional investigation; and for avoiding legal penalties
7. Institutions of higher learning: for faked diplomas; for inflated grades; and for accepting people without requisite tests.

These are only some of the areas involving bribes; there are many others. And while it is certainly true that bribing was not uncommon in Russia under the Soviets and in earlier historical periods, never has it risen to the level it has during the post-Soviet time. Needless to say, the consequences of burgeoning corruption are negative, if not disastrous, for Russia. Apart from undermining respect for the rule of law—a prime prerequisite for the survival of democracy—corruption of the government provides all sorts of political extremists with an attractive platform on which to campaign.

The August 1998 Financial Crash

In August of 1998, the finances of the Yeltsin government collapsed. The roots of the crisis lay in the fact that the Yeltsin government had few resources to support the country's faltering economy. Russia's tax system was—and to a great extent remains—complex, confusing, and occasionally mean-spirited, for both new, private businesses and individual wage-earners. Tax evasion became widespread. Business enterprises unable to pay all that they owe might bribe tax collectors to go easy on them—or, in worst-case scenarios, simply have the tax collectors murdered! In financial terms, the government's increasing difficulty with collecting revenue brought it to the threshold of bankruptcy by early 1998.

Compounding these difficulties, the government had to meet its obligation for government bonds at extremely high rates of return. Investors looking for a quick return on accumulated capital had bought these bonds as an alternative to buying shares in enterprises that the government was auctioning off. The government had compromised its privatization effort to meet its short-term need for cash.

But the long-term consequence was far worse. The Yeltsin government fell into an ever-worsening trap of uncontrollable debt. To pay off the interest on the loans that it had taken, it raised cash through foreign borrowing and by selling off industries to the private sector, issuing high-interest-bearing domestic bonds. But, as lenders began to recognize heightened risk, they demanded ever-higher returns on these government instruments. A financial bubble was in the making, and like all such bubbles, it finally burst. In August 1998, the Russian government confessed that it could not honor its obligations, and it declared a moratorium on repayment. The government let the value of the ruble against the U.S. dollar decline to one-third of what it originally was, making its bonds almost worthless. Not surprisingly, Russia's credit disappeared, and an economic collapse ensued. Importers went out of business, and the standard of living for most Russian consumers sharply deteriorated. Popular faith in capitalism reached a nadir, and the Russian Communist Party seemed on the threshold of a revival in its opposition to market reforms.

The cloud of financial collapse did have a silver lining for the Russian economy that became evident in the Putin era. Domestic producers had benefited from the ruble devaluation because the sharp hike in prices of imports caused by the ruble devaluation forced consumers to turn to domestically manufactured, reasonably priced goods and it also boosted exports and made some domestic producers (e.g., food processors) turn away from imports. This had a positive impact on the economy, stimulating productivity and expanding growth as the 1990s drew to a close. Vladimir Putin came to power in the wake of this economic rejuvenation and drew substantial political benefit from it as he prepared for the March 2000 presidential election. From 1999 to 2008, Russia enjoyed uninterrupted and robust economic growth.

The revenue-collecting situation also improved under Putin. Taxes were reduced, and some significant tax reforms to rationalize revenue collection and assure its compatibility with sustained economic growth were put through. In essence, the country's progressive but largely dysfunctional personal income tax became replaced by a flat corporate tax rate of 13 percent. This reform has encouraged Russians, especially the wealthy who had been holding out, to pay taxes and ease the government's job of tracking down tax scofflaws.

RECENT ECONOMIC GROWTH AND DEVELOPMENT

In the early years of the new century, the Russian economy experienced robust growth. The economy made real gains of an average 7 percent per year (2000: 10%, 2001: 5.7%, 2002: 4.9%, 2003: 7.3%, 2004: 7.1%, 2005: 6.5%, 2006: 6.7%, 2007: 8.1%), making it the eighth largest economy in the world in purchasing power parity. Russia's nominal Gross Domestic Product (GDP) increased sixfold, climbing from 22nd to 10th largest in the world. In 2008, Russia's GDP exceeded that of the Russian Federation in 1990, meaning it has overcome the devastating consequences of the 1998 financial crisis and preceding recession of the 1990s.

During Putin's eight years as president (2000–2008), industry grew by 76 percent and investments increased by 125 percent. Real incomes more than doubled and the average monthly salary increased sevenfold from $80 to $540. From 2000 to 2006 the volume of consumer credit increased 45 times. The number of people living below the poverty line decreased from 30 percent in 2000 to 14 percent in 2008. The infusion of petrodollars boosted Russia's economy and disguised economic

problems. The share of oil and gas in Russia's GDP has more than doubled since 1999 and in 2008 it exceeded 30 percent. That means that the Russian economy failed to diversify. Oil and gas account for 50 percent of Russia's budget revenues and 65 percent of its exports. Some oil revenue went to the stabilization fund established in 2004. As the fund accumulated proceeds from oil export, Russia was able to repay all of the Soviet Union's debts. In early 2008, the fund was split into the Reserve Fund (designed to protect Russia from possible global financial shocks) and the National Welfare Fund, whose revenues will supposedly be used for pension reform.

In December 2007, Vladimir Putin was named *Time* magazine's "Person of the Year" for bringing stability and renewed status to his country. Justifying the magazine's controversial decision, Charles Stengel wrote that "*Time*'s Person of the Year is not and never has been an honor. It is not an endorsement. It is not a popularity contest. At its best, it is a clear-eyed recognition of the world as it is and of the most powerful individuals and forces shaping that world—for better or for worse. It is ultimately about leadership—bold, earth-changing leadership. Putin is not a boy scout. He is not a democrat in any way that the West would define it. He is not a paragon of free speech. He stands, above all, for stability—stability before freedom, stability before choice, stability in a country that has hardly seen it for a hundred years."

To be sure, the tone of international assessments of Russia's transition to market economy and democracy began to change from avowedly pessimistic to mixed around the middle of Putin's tenure as president. A 2004 article by Andrei Shleifer and Daniel Treisman in the influential *Foreign Affairs* magazine made the point that Russia is gradually becoming a "normal," middle-income country, and the doom and gloom of previous pronouncements were overstated. But pessimism got another boost with the advent of global recession, when it became clear that (a) Russia's dependency on fuel exports is excessive for a developed country; and (b) during the years of economic success, Russia failed to utilize its petrodollar bonanza to create preconditions for steady and more broad-based economic growth. Signs of impending crisis multiplied in late 2008, that is, after the changing of the guard in the Kremlin (see the following account of that change), a change to which Russians apply the chess game term, castling (interchanging the positions of the king and a rook), with the king (Mr. Putin) becoming the prime minister and a rook (Mr. Medvedev) president.

Russian Economy and the World Crisis

The commencement of a worldwide financial crisis led to a collapse in demand for the oil, natural gas, and nonferrous metals that form the basis of the Russian economy. As of mid-July 2008, the price for a barrel of oil peaked at $147, but already in December the price had plunged below $40 (It was $37 at this writing in January 2009). As a result, share prices on Russian stock markets have drastically declined. Shares of Gazprom, the leader of Russia's oil and gas industry, dropped more than 70 percent in 2008. On Sept. 16, 2008, the government suspended trading on Russia's stock exchanges for an hour after the worst one-day fall since 1998. Despite words from the finance minister assuring investors that there was no systemic crisis, trading had to be suspended again September 17 and 18, 2008.

The Russian stock market fell 19 percent on Oct. 6 and an additional 14 percent on Oct. 9. Banks that had made huge loans with shares of energy and mining companies as collateral teetered on the brink of insolvency, as they had in 1998.

However, in the 1998 crisis, the Russian government lacked the financial resources to act. As 2009 began, the country's coffers were full after a decade of climbing energy prices. Russia's reserves stood at $600 billion in August 2008, so the government was able to move aggressively. In September 2008, the government lent the country's three biggest banks $44 billion. That same month, the government injected $20 billion into the financial markets and then lent an additional $110 billion to the country's banks. October brought $36 billion more in loans to banks. To head off runs on those and smaller banks, the government raised the limit for deposit insurance to $25,000. As in the U.S. crisis, even after an injection of liquidity from the government, Russia's banks refused to lend to each other. Interest rates on interbank loans climbed to 23 percent. Nine banks had failed by the end of November 2008. The government allowed others to keep operating, even though they were insolvent. But keeping the domestic banking system from insolvency is only one front in the war. At the start of the crisis, Russian companies owed $450 billion to overseas investors. With the Russian stock market falling and the ruble down 20 percent since August, many of those overseas investors demanded their money back. The government has authorized state loans of $50 billion to make up for overseas loans that had come due and where overseas investors refuse to roll them over. And troubles befell the ruble. On January 16, 2009, the Russian currency traded at 43.1 rubles to the euro: That's the lowest value ever for the ruble against the euro. Against the U.S. dollar, the Russian currency is doing better: At 32.1 rubles to the dollar, the ruble is cheaper against the dollar than it's been since 2005. But the ruble was down more than 40 percent in 2008 against its official dual currency basket of 45 percent euros and 55 percent dollars.

The big worry is that Russia won't have enough in the bank to win a war being fought on these three fronts. Reserves that stood at $600 billion in August 2008 were down to $450 billion as of December 19, 2008. Spending to control the economic, financial, and currency crisis is accelerating. In the week that ended December 19, 2008, for example, the central bank spent $7 billion to buy rubles to slow the currency's decline. Reflecting those fears, Standard & Poor's has downgraded the country's long-term sovereign credit rating to negative from stable. Reflecting the perception of lowered returns on investment, Russian banks and companies began to transfer liquidity abroad. Whereas in 2007 Russian bank accounts enjoyed the net inflow of $83.1 billion, in 2008 they were drained of almost $130 billion, which invoked comparisons with a precursor of the 1998 Asian financial crisis, wherein, in 1997, close to $100 billion fled Southeast Asia's banks.

What could Russians possibly do to soften the impact of the world crisis on their economy? In the opinion of Yuri Mamchur, the founder and editor of the popular Russia Blog, when faced with a glut of petrodollars the Russian government failed to direct requisite attention to four areas: (a) science and innovation; (b) entrepreneurship and leadership; (c) small business loans; and (d) infrastructure. While it is debatable whether or not investment in these areas could have possibly led to groundbreaking changes during nine years of economic growth, Mamchur's diagnosis of Russia's crucial vulnerabilities is on target. Indeed, the prestige of science is well below that in Soviet years, and most scientists cannot make ends meet unless they receive grants from the West. The entrepreneurial culture remains low, and small businesses are harassed by corrupt bureaucrats. Perhaps worst of all, "[t]oday Russia is probably the only developed country that isn't inner-connected by a freeway system. Never mind having a system; there isn't even a freeway. The Russian government modernized the railroads and built the highways in major cities. But the cities across the country remain disconnected and the villages forgotten. Railroads are the remainder and reminder of the centralized economy. Rails ignore the interests of small businessmen and bypass small towns." (Yuri Mamchur, "Russia's economic crisis could have been avoided," *The Seattle Times,* December 30, 2008.) Addressing the anti-American hysteria in Russian media, Mamchur admits that "the United States can be blamed for many things, including long-term neglect of Russian relations after the collapse of the U.S.S.R. But America can't really be blamed for Russia's current economic problems. Seven hundred-fifty billion greenbacks in the Russian vaults as of early 2009 was enough to hire professors, build laboratories, give loans to small businesses, and pave freeways. The Russian economy would have been diversified and not nearly as dependent on the price of oil.

In early 2009, it became clear that Russia's super-rich are also super-losers in the financial crisis. According to the Russian business magazine *Finans,* the top 10 wealthiest Russians lost about two-thirds of their fortunes since the beginning of the crisis. The magazine's annual list of Russia's richest shows them suffering breathtaking losses as the country faces its worst financial crisis in a decade. Oleg Deripaska, who had topped *Finans'* list in the previous two years, fell to eighth place after losing some 85 percent of his wealth—down to $4.9 billion from $40 billion, *Finans* estimated. Mikhail Prokhorov, the playboy metals and banking billionaire who sold his stake in mining company Norilsk Nickel early last year, moved up from seventh place to top the list with a fortune of $14.1 billion, down from $21.5 billion a year ago, the magazine said. Roman Abramovich, owner of Britain's Chelsea Soccer Club and a stake in steelmaker Evraz, held on to second place. But his fortune, estimated last year at $23 billion, has shrunk to $13.9 billion.

Deripaska, whose business interests include metals, construction, and energy, was forced in 2008 to cede stakes in Canadian auto components maker Magna and German construction firm Hochtief. He previously has objected to the estimates put on his wealth, saying that they failed to consider the size of his debts. Indeed, Deripaska built his fortune up on the back of aggressive borrowing, and his Rusal aluminum company reportedly owes approximately $17 billion. Other losers include steel magnates Vladimir Lisin, owner of a majority stake in NLMK, and Alexei Mordashov, owner of Severstal. Lisin, who clings to third place, has seen his fortune plunge by 65 percent to $7.7 billion. Mordashov's fortune is down 81 percent to just $4.1 billion.

Finans, which publishes its rich list two months ahead of the better-known U.S. *Forbes* list of billionaires, said the list had halved from 2008, when they counted 101 billionaires. In 2008, *Forbes* confirmed Deripaska as Russia's richest man.

It is the close relationship between Russia's oligarchs and Russian leaders, notably Prime Minister Putin, which is now in the forefront of critical attacks from the small but vociferous liberal (Westernizing) camp. Thus Boris Nemtsov, a vice premier under President Yeltsin accuses Putin of extending a helping hand to his super-rich friends at the expense of rank-and-file Russians. "The official propaganda declares," according to Nemtsov, "that by aiding the most important enterprises, the government prevents their bankruptcy and rescues jobs. But very concrete actions of the government have hardly anything to do with jobs, wages, and with genuine economic interests of Russians." The domestic price of gas in the capital of Russia, the world's second largest producer of oil, is 70 cents per liter, whereas in New York City it is now 50 cents per liter. The above-mentioned Deripaska received $4.5 billion from the National Welfare Fund, whose initial task was to bankroll the Russian pension system. This amount of money was transferred to Deripaska via Vnesheconombank, a Russian bank of which Putin happens to be the chairman of the board of trustees. The money Deripaska received was immediately paid to the Royal Bank of Scotland, one of Deripaska's lenders. In exchange for $4.5 billion, the oligarch transferred 25 percent of shares of Norilsk Nickel (a nickel-producing conglomerate) to Vnesheconombank. Another super-rich man that received generous financial aid ($1.5 billion) from the same fund is Roman Abramovich. His above-mentioned steel-making business, Evraz, is now worth only $3 billion, down from $45 billion just one year ago, due to vastly reduced global demand for steel, whereas the debts of this business amount to $10 billion. Also huge sums were channeled to prop up some of the most recent additions to the circle of Russian oligarchs—that is, to former associates of the KGB (Mr. Sechin and Mr. Bogbanchikov) who are now in charge of the oil producing assets of the jailed Mr. Khodorkovsky. According to Nemtsov, the avowed goal of these and other financial transfers, prevention of bankruptcy, is not a worthwhile goal, as the crisis is an excellent opportunity to get rid of ineffective entrepreneurs; they should have been left to the mercy of their lenders, who would have found better managers for the businesses involved.

Improved Tax Collection

To deliver on the expectations of ordinary Russians and to fund the resurgent military, the government needs more resources and needs to get its tax-collecting capacity in order. In that area, some success was achieved under president Putin. The Russian government's revenue has increased, moving it away from the strapped times of the 1990s, when it could barely afford to pay state salaries to some of its employees. Moscow's failure in that era to collect the taxes due it from corporations that did everything possible to avoid paying their required share of taxes has at last been reversed, as evident in the public prosecution of former Yukos Oil chief Mikhail Khodorkovsky, who has been portrayed as one of Russia's largest and worst tax scofflaws. While the persecution of Khodorkovsky (see box: The Saga of Mikhail Khodorkovsky) has had all the trappings of a political vendetta, few ordinary Russians shed tears about a now jailed oligarch who had been one of the principal beneficiaries of the 1995 loans-for-shares scheme, when lucrative assets were sold for a fraction of their market value to the government's cronies. Moreover, closing loopholes in tax laws that corporations, especially the oil-producing ones, had used to avoid taxes, was lauded by the general public.

By the end of 2003, Putin was well on his way to increasing the tax obligation of enterprises that had been using the most aggressive means to minimize their taxes. And as far as Yukos was concerned, it was given a bill for unpaid taxes at the end of 2003 for $3.4 billion. This bill scared bank creditors of Yukos, who warned its leadership that paying this enormous tax bill could, among other things, prevent the company from servicing its debt. In fact payment of its tax bill encumbered Yukos, increasing its vulnerability to a hostile takeover which occurred in 2005 when its assets were purchased by other energy companies.

Expanding State Control

In addition, the Putin-led Kremlin wants to retain state influence over the production and distribution of Russian energy for both economic and political reasons. Indeed, the Russian oil industry is presently in a state of administrative confusion since it is neither wholly state-owned, as in Saudi Arabia and other countries in the Organization of Petroleum Producing Countries (OPEC), nor fully privatized as it is in the United States. While the Kremlin effectively nationalized Yukos, other oil producing enterprises, notably Lukoil and Rosneft, are privately owned and managed though highly sensitive to what the Kremlin wants them to do.

Gazprom, the largest Russian oil and natural gas producer, is the latest example of this administrative confusion in Russia. In April 2006, Gazprom was ranked the fifth largest corporation in the world in terms of the value of its stock, even larger than Walmart, Toyota, and City-Group. Gazprom is state controlled and extremely sensitive to the wishes of the Kremlin while simultaneously selling shares to private investors interested primarily in profit making. Gazprom has pursued export pricing policies, as in the case of Belarus and Ukraine, inspired by Russian foreign policy interests—Belarus got a discount, while Ukraine got a price hike for reasons having to do with their policies toward Russia. Currently 25 percent of natural gas consumed by the countries of the European Union comes from Russia. The degree of dependency on Russian gas is significantly higher than that of some EU countries. This dependency was tested during the trade war between Russia and Ukraine in the winter of 2008–2009 (December to January) when gas transit through Ukraine was terminated for 19 straight days. Slovakia, Moldova, Bulgaria, and Romania were particularly hard hit, as these countries did not have gas stores that would have allowed them to endure a lengthy stoppage. During the standoff with Ukraine, just as during a similar standoff with Belarus in the previous winter (2007–2008), it was particularly clear that Gazprom is not so much a business as one of the principal vehicles of the Russian state's control over its perceived geopolitical sphere of influence.

From this perspective it becomes understandable why the Putin Kremlin strived to expand its control over the production and marketing of Russian energy. In June 2006, the government decided to limit foreign energy companies to a minority stake in joint ventures. Existing deals were not to be affected but new agreements were to limit foreign investors to the role of junior partners. Several months later the Kremlin suddenly and unexpectedly decided to enforce pollution restrictions against Shell and Exxon Mobil oil drilling operations in the Sakhalin islands, pulling the production license granted to Shell and making clear that henceforth there would be close and careful state monitoring of foreign extractive companies, ending at least a decade of generous Russian concessions to foreign investors. The ostensible reason for this new policy was to keep control of energy and other strategic projects in the government hands and by extension subject to government influence. Similarly, in early July 2006, the State Duma passed a law granting exclusive rights for exporting natural gas to Gazprom giving it a monopoly on the pricing of Russian energy exports that would undoubtedly be responsive to the Kremlin.

Continuing Corruption: The 2006 Motorola Cell Phone Flap

A barely concealed kind of corruption involving the state bureaucracy has become evident in the Putin years. Vladimir A. Ryhzkov, an independent member of the Duma (until 2007), who spoke out on several occasions about what was wrong with Russian government under Putin's leadership, said openly that state bureaucrats brazenly confiscated goods from private companies—they sold what they had stolen for personal gain. Andrew Somers, president of the American Chamber of Commerce in Russia, was even more explicit, reportedly having said that "there does seem to be more avarice, more boldness in trying to extort money" among state officials dealing with economic matters.

The American exporter of cell phones to Russia, Motorola, openly complained about bureaucratic theft of its property which occurred in the Spring of 2006 when Interior Ministry seized 167,000 mobile phones shipped into Sheremetyevo airport on the pretext first of being counterfeits, then contraband, and then a health hazard. Hard-put to provide a convincing justification for its action against Motorola, the Ministry began to destroy about 49,000 of the phones to show it was just doing its job. But Motorola officials smelled a rat when the confiscated phones started showing up in the inventories of street vendors trying to sell them. What seems to have happened was simply a blatant case of ministerial thievery. But, the Interior Ministry stood by its contention that the phones had entered Russia illegally without proper paperwork, hence the seizure of them, and the Ministry's action was protected by a law that allowed the destruction or sale of confiscated property.

Then a Russia phone manufacturer jumped into the fray, saying the confiscated phones constituted a violation of a patent given to it in 2003. It looked like there was no way Motorola was going to come out of the affair as a winner simply because the bureaucracy and the courts held together in denying the American company the justice it deserved. Incidentally, the prosecutor in this case was none other than Vladimir V. Ustinov; he was removed from office by Putin in June, 2006. Although Putin gave no reason for the removal and said he had no complaints about Ustinov, it is quite possible that Putin did not like the latest exposé of government corruption he had said so many times he wanted to eliminate. Eventually, Motorola was reimbursed by the government, which later acknowledged responsibility for the mishap.

The State of the Environment

According to a National Intelligence Council report released in December 1999, "Russia during the next decade will be unable to deal effectively with the formidable environmental challenges posed by decades of Soviet and post-Soviet environmental mismanagement." For example, in 1999 less than half of Russia's population had access to safe drinking water despite a decline in water pollution from industrial resources. Today the main culprit is municipal waste and the way in which nuclear waste in particular is leaching into key water sources. The cost of raising the quality of Russia's drinking water could be as much as $200 billion. Pollution in the air is almost as bad as water pollution, with over 200 cities often exceeding Russian pollution limits. The culprit here is the phenomenal growth of automobile ownership. Auto emissions are likely to offset reductions in industrial air pollution as more and more Russians demand and obtain private cars.

The New York-based Blacksmith Institute, which compiles an annual list of the world's 10 most polluted places, last year included two Russian cities: Norilsk, the Arctic home of the world's largest nickel mine, and Dzerzhinsk, a chemical manufacturing center 400 km (250 miles) east of Moscow.

There is evidence that the Kremlin is sensitive to environmental degradation in Russia. For example, in March 2006, Putin ordered a change in the routing of an oil pipeline to be constructed by the Russian controlled Transneft pipeline construction company to avoid Lake Baykal, the world's largest fresh water lake. The pipeline was originally to run along the northern shore of the lake, endangering it in the event of a leak. The region is seismically active and a minor earthquake or even tremor could break the pipe and cause an oil spill into the lake. Putin ordered the pipeline to be laid 25 miles north of the lake, the route preferred by environmentalists in the Russian Academy of Sciences.

Putin's intercession was significant for politics as well as the Russian environment and Lake Baykal in particular. It demonstrated the president's enormous authority to make things happen and to change things influential people did not want changed. Transneft had preferred the route of its new pipeline to follow the northern shore of Lake Baykal for reasons of cost. The rerouting also strengthened Putin's leadership image by illustrating his concern for the well-being of Russian society.

Medvedev, the new Russian president, has driven the environment to the forefront of the political agenda since assuming the presidency in May 2008. He called an ad hoc meeting of top government officials in June 2008 to discuss clean and efficient energy, while June 5, 2008, was celebrated as the first ever "day of the ecologist."

Russian news agencies quoted President Dmitry Medvedev as saying in June 2008, shortly after taking office that parts of Russia will be uninhabitable within the next three decades if the country does not take better care of the environment. Medvedev, addressing law students in his home city of St. Petersburg, said Russians had been more concerned about survival than the environment in the 1990s. "As a result, there are now many places in Russia on the brink of an adverse ecological situation," agencies quoted Medvedev as saying in response to a question. "If we fail to deal with the ecological situation now, then in 10, 20, 30 years large parts of Russia will be unfit for living," he said.

(Kremlin Presidential Press office)

Dmitry Medvedev and Vladimir Putin: Russia's Ruling Duo

Russia's relative openness in the area of environment is definitely a positive sign. Under the Soviets, data about air, water, and soil pollution were classified and hard to obtain even for researchers, these days detailed annual state reports on the environment are published and available online. For example, a 500-page, 2007 report identifies the 38 cities with the worst air pollution and concrete pollutants are identified for each. Top positions on that list are assigned to Balakovo (a city near one of the largest hydro-electric power stations), Barnaul, Bratsk, Vladimir, Volgograd, Volzhsky, and Ekaterinburg. Two of these cities, Ekaterinburg and Volgograd, are among the largest cities of Russia.

DEMOCRATIZATION RUSSIAN STYLE 1992–2008

The new post-Communist Russian democracy got off to a controversial and conflict-ridden start with the making of a new constitution promulgated in December 1993. Yeltsin tried hard in succeeding years to make Russian democracy work. He was in power for more than eight years, having been elected first president of the Federation in June 1991 when the Soviet Union was in existence. He was re-elected president under the 1993 constitution in early Summer 1996. These were years of intense political bickering that all but paralyzed Russia; and while Yeltsin expended great energy trying to lead the new Russian democracy, its inherent instability eventually wore him out. He resigned in December 1999, designating Vladimir Putin, a former chief of the FSB, the successor to the Soviet era KGB, whom he had appointed prime minister in August, as acting president until presidential elections in March 2000.

Putin won those elections and then was elected to a second term in March 2004. Putin had to deal with a combative political pluralism inherited from the Yeltsin era that at first intimidated him and subsequently alarmed him. He began his first term talking about post-Communist Russia's need for "managed democracy," a term that has become synonymous with curtailment, one might even say, suppression, of criticisms of his leadership. However, by the end of his second term as president, Putin enjoyed genuine popularity with an approval rating hovering near 80 percent. Quite a few ordinary Russians and several public intellectuals wanted Putin to run for president for a third time. This, however, would have violated the Russian

constitution, which limited a president to two *consecutive* 4-year terms. In May 2008, Dmitry Medvedev, Putin's protégé and deputy prime minister, was elected president with Putin occupying the Prime Minister's post. Most analysts believe that, for all intents and purposes, Putin continues to run Russia and that Medvedev's election is a sham that will allow Putin to legitimately run for president again in 2012. While this remains to be seen, especially because the world economic crisis may transform Russia and its political scene in a way that thwarts previous designs, the impression of a sham was reinforced when, shortly after his election, Medvedev initiated the extension of one presidential term to six years. The Russian parliament has adopted a corresponding amendment to the Russian constitution, but this amendment does not apply retroactively to the 2008 election, which means that Medvedev's term in office will still be four years.

The Start-up of Russian Democracy 1992–1996

Yeltsin's determination to proceed with the radical economic reforms nicknamed "shock therapy" brought him into sharp conflict with the Russian parliament. The Congress of Peoples' Deputies was heavily influenced by 300 to 400 conservative deputies and others opposed to an abrupt dismantling of the state-controlled economy. These politicians in the legislature spoke for the managers of the giant Russian military-industrial complex, former Communists, and a large constituency of bureaucrats on the national and local levels of government administration, and also for many ordinary people whose lifelong savings had vanished. Conservatives also called into question Yeltsin's erratic style of leadership. In their view, Yeltsin violated the spirit of the democracy that he said he wanted to establish in Russia.

In December 1992, conservatives in the Congress of Peoples' Deputies attempted to overthrow Yeltsin's government by an unprecedented vote of no-confidence, one that Soviet parliaments could never have taken. But there was not enough support. In the face of such pressures, however, Yeltsin had to sacrifice his reformist prime minister, Yegor Gaidar, an advocate of rapid change to private enterprise. In his place, Yeltsin appointed Viktor Chernomyrdin, a Soviet-era technocrat who was to make a personal fortune during the privatization of the natural gas industry.

The April 25, 1993 Referendum

Faced with an obstructionist attitude among the Supreme Soviet, Yeltsin called for a referendum, to be held at the end of April 1993, for Russian voters to say whom they wished to govern the country: Yeltsin or the Supreme Soviet. After a bitter campaign characterized by accusatory rhetoric and much scandal-mongering focused on President Yeltsin and Vice President Nikolai Rutskoi, almost fifty-nine percent of the voters supported Yeltsin, his reform program, and his wish for early parliamentary elections. Though the strongest support for Yeltsin came from Moscow and St. Petersburg and from young and educated segments of Russian society, Yeltsin received some support everywhere in Russia.

(© RIA Novosti Photo Library)

The Russian White House. The building housed Russia's Supreme Soviet (Parliament) until the Russian constitutional crisis of 1993, when an uprising led to a siege and artillery bombardment on the building that caused a major fire. Today, the White House is home to the executive branch of the Russian parliament.

The September–October 1993 Constitutional Crisis

Unfortunately for Yeltsin and his reformist-minded supporters in the Kremlin, they still had to contend with the conservatives in Parliament. Spearheaded by the Communist Party of the Russian Federation (CPRF), which positioned itself as the defender of the downtrodden, conservatives pressed for the removal of reformers from Yeltsin's government and blocked its efforts to privatize industrial and agricultural enterprises.

The Russian constitutional crisis of 1993 began on September 21, when Boris Yeltsin dissolved the Supreme Soviet, the country's legislature, which had opposed his moves to consolidate power and push forward with economic reforms. Yeltsin's decree of September 21 violated the then-functioning constitution; on October 15, after the end of the crisis, he ordered a referendum on a new constitution. The Congress rejected the decree and voted to impeach Yeltsin. The Vice President, Aleksandr Rutskoy proclaimed himself as

Acting President. On September 28, the lawmakers (i.e., the members of the Supreme Soviet) began rallying their supporters in the streets of Moscow. The legislators found themselves barricaded inside the White House of Russia, the parliament building. Hostilities grew in the next week when the supporters of the Supreme Soviets sparked an armed uprising in Moscow on October 3, seizing the mayor's office and attacking Ostankino—the major television station. In the early morning hours of October 4, troops loyal to the president besieged the White House and urged the members of the Supreme Soviet to surrender. When they refused, the building was shelled by tank-fire. By October 5, 1993, the armed resistance was over. According to government estimates, 187 people had been killed and 437 wounded, while sources close to Russian communists put the death toll as high as 2,000.

During the conflict, President Yeltsin enjoyed the support of the Russian liberal intelligentsia, and the Russian military, as well as U.S. President Bill Clinton, and other Western leaders. Supreme Soviet had the support of the Communist Party and Russian nationalists. For that reason the supporters of the Supreme Soviet were dubbed a "Red-Brown coalition."[2]

Parliamentary Elections of December 1993

In October Yeltsin laid out some of the ground rules for the elections. He said that there would be election of members to both a lower house, the State Duma, which would represent the Russian people by population, and an upper house, the Federation Council, which would represent the population by region. Members in both houses would be elected directly by the voters for a four-year term.

The reformist groups did not do as well as they had hoped. While together they won a plurality (152 seats), they were divided and poorly organized. They could not make a coalition to back Yeltsin and enable him to move forward with his economic reform program. In effect, they had lost the opportunity given them by Yeltsin's disbanding of the old legislative bodies to lead the country. The conservative parties, notably the Communists, however, did better than anyone had expected and won 64 seats. And so did Liberal Democrats, ending up with 24 percent of the popular vote and 78 seats in the State Duma.

The success of the Liberal Democratic Party, led by Vladimir Zhirinovsky, who campaigned on a platform of radical Russian nationalism under the pretext that ethnic Russians, the majority of Russian Federations' population, had been taken advantage of by foreign advisors to Russian government and by a newly emerging cabal of tycoons, most of whom were ethnic minorities, primarily Jews. In 1991, Zhirinovsky (incidentally a half-Jew himself) had been a candidate in the Russian presidential elections but had won only about 7 percent of the popular vote. He subsequently became one of the strongest and most vitriolic critics of Yeltsin's reformist policy at home and conciliatory behavior abroad, referring to Yeltsin on one occasion as a lackey of the Americans.

Zhirinovsky ran an expensive campaign in November and early December 1993, involving a generous and skillful use of television. He said a lot of things that average Russian citizens wanted to hear. For example, he proposed an immediate liquidation of the 5,000 criminal organizations that had sprung up in the wake of the reforms that terminated government control over the economy. He called for the immediate expulsion of non-Russians who had come to Moscow from ex-Soviet republics like Armenia and Georgia to make a killing by selling scarce goods at exorbitant prices. He not only pandered to the popular sense of abandonment and personal humiliation caused by the precipitous decline of living standards (as a result of a shock therapy reform) but also tapped into a xenophobic and messianic nationalism that many Russians had felt for at least a couple of centuries. In hindsight, Zhirinovsky, who subsequently softened his rhetoric but did not give up on his exceedingly telegenic and buffoonish style, seems to have been, at least initially, a decoy pigeon of the late Soviet KGB with the explicit task of attracting and neutralizing radical nationalists, a constituency which grew explosively during the most painful stage of economic reform and which presented no less a danger for the fate of that reform than did the communists. Shared by some Russian analysts, this impression is reinforced when Zhirinovsky's and his faction's vote in the Duma is analyzed. Despite all his rhetoric, Zhirinovsky always cast his Duma vote in support of the policies of Russia's leaders, first Yeltsin and then Putin. Note that Zhirinovsky successfully performed as a kind of nationalist lightning rod well before the Russian presidency itself became the mouthpiece of Russian nationalism, which happened years later with the enthronement of Vladimir Putin.

As for the 1993 election results, they indicated widespread opposition to any "shock therapy" or very rapid enactment of reform. Lacking a cohesive plurality of reformists in the new State Duma and obliged to conciliate moderates and conservatives, Yeltsin appointed Viktor Chernomyrdin as prime minister and the conservative Viktor Gerashchenko as head of the Central Bank because of their cautious approaches to reform.

The Referendum on the New Constitution

On December 12, 1993, in a referendum held concurrently with voting for members of the State Duma, Yeltsin did get voter approval of the constitution he and his advisers had drafted pretty much on their own during the summer of 1993. Prior to that referendum, Russia abided by the 1978 Soviet Constitution of the Russian Federation. The absence of a *post-Soviet* constitution had been viewed as a major factor in the turmoil in executive-legislative relations. The new constitution assigned very significant prerogatives to the President, defined the institutions of government and provided—at least on paper—a credible democratic political system voters liked, hence their approval. Whatever its many shortcomings, the new Russian political system established in the constitution was as democratic as could be expected in a country that had little, if any, experience with real self government in its long history.

THE NEW RUSSIAN POLITICAL SYSTEM

The 1993 constitution provided for the kind of executive-oriented parliamentary democracy that Yeltsin and his advisors preferred. While the constitution makers looked at several Western democratic political systems as a model for post-Communist Russia, the new Russian constitution promulgated at the end of 1993 seemed to resemble in important respects the governmental organization of the French Fifth Republic, in particular its strong presidency.

The Strong Russian President

Under the 1993 constitution, the president of the republic is head of state and commander in chief of Russian military forces, with the responsibility for protecting the Constitution and setting basic domestic- and foreign-policy guidelines. The president can be elected for a maximum of two four-year terms. The president has extensive power to implement policy. The president can nominate and dismiss the prime minister, the cabinet, and the head of the Central Bank, which is no longer under the jurisdiction of the legislature. The president can also dissolve the State Duma; call for the election of a new State Duma after one year has elapsed since the preceding elections and for a referendum; declare martial law and a state of emergency and issue decrees; and propose legislation and veto it. Finally, the president can be impeached by Parliament only on a charge of treason or other grave crime.

The Executive Branch: Prime Minister and Cabinet

The "government" consists of the prime minister and cabinet and is responsible for the day-to-day administration of the country. While the prime minister is nominated by the president, the nominee must receive formal approval of the Duma, which is subject to dissolution if it fails to approve a presidential choice three times in succession. The cabinet is nominated by the prime minister and approved by the president. The cabinet drafts the national budget, which must then be approved by the Duma.

The Parliament

The Parliament is bicameral. The upper house, the Council of the Federation, consists of two representatives from each of the 89 members (oblasts or republics) of the Russian Federation. It confirms changes in internal boundaries, a presidential declaration of martial law or state of emergency, and the president's use of the armed forces abroad. It has the power to remove the president from office in the event the president is found guilty of a criminal act. It has the power to appoint top judges; and, along with the Duma, it can override a presidential veto by a two-thirds majority.

The Council of the Federation, at least theoretically, is much like the U.S. Senate whose members represent each state (two members per state), and in the Russian case each oblast and/or republic. However, in the Russian case, senators were to be elected by regional legislatures and regional governments. Since 2000, they have been actually appointed by the regional leaders. Unlike the U.S. Congress whose members are in fact more influential than most members of the House of Representatives, members of Russia's Council of the Federation are actually less influential than their colleagues in the Duma, who are at least popularly elected.

The State Duma, the lower house, represents the citizens of the Russian Federation by population much like the U.S. House of Representatives. The Duma approves legislation and the state budget. It confirms the president's choice to head the cabinet and the Central Bank and the appointment of senior judges. It can file charges to impeach the president; it can vote no-confidence in the government; and it can override a presidential veto, along with the Council of Federation, by a two-thirds majority of its membership.

The Constitution provided also that the first Duma would serve only two years, with elections for the second Duma in December, 1995, this short term to coincide with the remainder of Yeltsin's term as president. This meant that, at the end of 1995, Russians would have an opportunity to choose a completely new leadership while having a reasonable period of transitional political development to determine exactly what kind of leaders they want.

Federalism

The 1993 Constitution provides for a federal pattern of inter-governmental relations that gives local administrations substantial discretion. Local political leaders demanded this in return for their support of the Constitution and of the unity of the federation. Powers specifically granted to the federal government in the new Constitution include jurisdiction over the continental shelf, establishment of citizenship and federal institutions, the federal budget, establishment of federal financial institutions and currency, taxation policy, the country's energy system, foreign and defense policies, and the court system.

There are 89 administrative subunits (called subjects of the federation) that include 21 republics or ethnic homelands of ethnically non-Russian peoples (see the map above). In some of these republics (notably in the North Caucasus) the titular non-Russian ethnicities are actual majorities, while in some

others (e.g., Buriat, Sakha, Khakas) ethnic Russians outnumber the titular ethnicity due to migration during the Soviet period. Most other subunits are called oblasts. Oblasts and republics have the right to establish their own state institutions. Shared powers concern the maintenance of law and order, the use and protection of natural resources, public health, social policy, and education and economic relations.

A Bill of Rights

The Constitution has an elaborate Bill of Rights that includes freedom of speech and of conscience, freedom from a governing ideology, freedom of movement, freedom of the press, and the right to private property as well as protection of social privileges like free education, housing, medical care, and protection against unemployment. By affirming the existence of basic political freedoms, in particular freedom of expression and freedom of the press, the Constitution guaranteed the new pluralistic environment that gave rise to and encouraged development of political parties. Russians could now freely organize themselves into political groups with a myriad of different programs and policy commitments. In sum, the appearance and consolidation of a political party system was now a permanent feature of Russia's post-Communist polity, which affirmed the country's commitment to a popular based democratic system of government, at least in theory, and to some extent, in practice.

STRAINS ON YELTSIN'S LEADERSHIP 1994–1996

Throughout 1994 and 1995, Yeltsin's leadership of the country was severely strained. One source of strain was his eccentric personal behavior, in particular his alleged periodic bouts

CHECHNYA

Chechnya should not be confused with union republics of which the Soviet Union consisted until its dissolution in December 1991. Unlike those 15 republics, Chechnya was an ethnic homeland *within* the largest former Soviet republic, Russia. This distinction is important in terms of international law. Each Soviet Republic could withdraw from the Soviet Union if at least two-thirds of the republic's legislature voted for secession. It is another matter, of course, that under the Soviet rule such a vote was purely a theoretical possibility—up until this actually happened in Lithuania under Gorbachev's Perestroika. Secession, however, was not even an option for the so-called autonomous republics—a Soviet label for ethnic homelands *within* some union republics, notably Russia. Chechnya was one such homeland. Muslim Chechens had first fought for independence from Russia in the mid-1800s, but their rebellion was crushed militarily. During the first 2.5 decades of Soviet rule, the situation in and around Chechnya was relatively calm, but the memory of resistance to Russian domination was very much alive among Chechens. On February 23, 1944, Chechens and Ingushes, two closely related groups, were collectively deported from their homeland in the North Caucasus to Central Asia. Altogether, more than half a million people were deported. The official accusation leveled at these people was mass desertion, avoiding conscription during war time, and preparation of a rebellion. Such accusations could possibly make sense during the war if applied to individuals; but by no means did they call for collective responsibility and harsh reprisals against the entire people. Although in 1957, Chechens and Ingushes were allowed to return to the North Caucasus and to reestablish their homeland, the horrible experiences of 1944 could not be erased from people's memory. In 1990, as the Soviet Union was about to collapse, a secessionist movement developed in Chechnya. Chechen President Dudayev, a Soviet general and a decorated hero of the war in Afghanistan, became (in 1991) the embodiment of that movement. Yeltsin had no intention of giving up central control over outlying non-Russian populations considered permanent parts of the Russian state. Some experts had warned Yeltsin against a military solution to the problem of Chechen separatism. They warned him not to overestimate the ease with which he apparently thought Russia could bring the Chechens to accept the rule of Moscow, which they had renounced in 1991. Nevertheless, in December 1994, Yeltsin ordered troops into Chechnya, and the first Chechen war was launched, a bloody war which took the lives of 50,000 civilians (mostly ethnic Chechens but also Russians, whose share in the population of Chechen capital was over 50%) and 6,000 Russian military. The Russian army failed to appease Chechen rebels, and in 1996 the war ended with a treaty which granted a very significant level of autonomy to Chechnya, still short of formal independence but very close to it.

Perhaps the most serious strain on Yeltsin was Chechnya. He decided, albeit with some reluctance, to use military force at the end of 1994, to suppress the move to independence of the Chechen Republic, one of the ethnic homelands in the North Caucasus.

Yeltsin's support in the Parliament had virtually disappeared by June 1995, when the Duma voted no-confidence in his government, with 241 in favor of the motion, 72 opposed, and 20 abstentions. The vote was a response to the sloppy handling of a Chechen terrorist attack on a hospital in the southern Russian town of Budyonnovsk. But it also reflected a growing opposition in the Duma to both the style and substance of his leadership. Making the vote particularly ominous for Yeltsin was an accompanying call for his impeachment.

Yeltsin could—and did—ignore this vote. But he was angry and threatened retaliation to discourage the frustrated Duma from proceeding to a second no-confidence vote, which, under the Constitution, his government would not have been able to ignore. Such a vote was eventually taken in early July, but it failed, with only 193 deputies out of 450 willing to support it.

Parliamentary Elections of December 1995

The results of the election were startling but not entirely unexpected. The Communists won the largest number of seats, 157; Our Home Is Russia, the centrist party led by Premier Chernomyrdin won only 54 seats, the Liberal Democrats 51, and the Yabloko Party (of social democratic orientation) won 45 seats. The results were a stunning victory for conservative groups, because by the mid-1990s most Russians had been disillusioned by the economic reform and were in the grips of an identity crisis related to the breakup of the Soviet Union and the loss of superpower status.

Presidential Elections of June–July, 1996

After a hectic campaign that had begun at least six months earlier, a presidential election was held on June 16, 1996. Yeltsin had been elected in June 1991, and his term was up in 1996. According to the electoral rules approved by the Russian Legislature, if none of the official candidates won an absolute majority, a runoff election was to be held within two weeks between the two candidates who had won the largest percentage of popular votes. The candidate with the plurality would then be declared the winner of the runoff. Of 10 candidates in the June 16 balloting, six were prominent political personalities: Yeltsin, Zyuganov, Lebed, Yavlinsky, Zhirinovsky, and Gorbachev. The last decided to run upon learning that the rating of his personal foe, Yeltsin, had declined to single digits.

About 70 percent of eligible Russian voters participated in the June 16 election. Yeltsin and Gennady Zyuganov, head of the Russian Communist Party, ran a very close race, with the president winning about 35 percent of the popular vote and the Communist chief not far behind, with 32 percent. Interestingly, Gorbachev, the last president of the U.S.S.R., won 0.51% of the vote. Whereas Yeltsin did well in Moscow, Saint Petersburg, and other large cities, Zyuganov won the vote of most rural and small-town folks. However, in

with drunkenness, which made him look like a buffoon rather than a national leader. Another was his physical deterioration and occasional incapacitation as a result of a series of heart problems. Yeltsin also alienated his natural allies, liberals who had supported him continuously in preceding years but began criticizing him in 1994 for his dictatorial tendencies and his embarrassing behavior abroad—in particular a drunken stupor that prevented him from keeping an appointment with the Irish prime minister in October 1993.

some provinces of southern Russia, the voting advantage of the communists was overwhelming across the board. Since neither Yeltsin nor Zyuganov received the majority needed to make him a winner, a runoff election was scheduled for early July.

Alongside the hugely unpopular economic reform, the war in Chechnya had been uppermost in the minds of most Russians. Although Zyuganov had tried to make the most of voter anger over Yeltsin's costly and inept effort to bring the Chechens to heel, he was less successful than many had anticipated. Yeltsin tried to assuage a hostile public by his hasty conclusion in early June of a peace treaty with the Chechen leadership and his insistence that the war was finally over, even though sporadic fighting continued.

Yeltsin's margin of victory over Zyuganov on June 16 was slight and fragile, and by no means assured him victory in the runoff. The team of economic reformers under the informal leadership of Anatoly Chubais was working round the clock to avert the electoral defeat of Yeltsin. In order to achieve this, Yeltsin appointed general Lebed, one of his rivals in the first round of elections, to an important government post and fired some unpopular people, like defense minister Grachev, who had bragged that he would capture the capital of Chechnya Grozny with one battalion. However, the most important effort on behalf of Yeltsin (i.e., on behalf of economic reform) was through financing a frenetic media campaign. The major TV channels were in the hands of reformers. Compared with Yeltsin and his supporters, Zyuganov was assigned rather minimal time on TV. Shortly before the second round of elections, two political consultants from the Yeltsin camp were apprehended while exiting the Russian White House with a box containing $500,000 in cash. Some analysts still believe that the real fight for the presidency was behind the scene and had to do with counting votes. But it is also beyond doubt that a significant part of the Russian electorate (educated urbanites for the most part) feared one of the perceived evils—the return of communists to power, even more than the other, market reform, and were prone to vote against Zyuganov and therefore, grudgingly, for Yeltsin.

The runoff occurred without incident on July 3, 1996, and Boris Yeltsin won with 53.70 percent of the popular vote. Zyuganov got 40.41 percent. On July 4, Yeltsin nominated Viktor Chernomyrdin as Prime Minister and directed him to form a government.

YELTSIN'S POLITICAL DECLINE 1996–1999

Yeltsin's leadership after the June–July 1996 presidential elections was unstable, uncertain, and unpredictable. His close advisers and lieutenants struggled with him or among themselves in a byzantine campaign to influence policy making at the expense of his personal authority. At the same time, the Duma tried to weaken and undermine Yeltsin's policies in every way possible.

Continuing Conflict with the Duma

Yeltsin's problems with the Duma continued in the postelection period, reaching a climax in October 1997 over the Kremlin's belt-tightening budget for 1998. The budget included cuts in subsidies for state enterprises and the Russian military. The Communists and their conservative allies, mirroring the fatigue of many Russians in the countryside with five years of uninterrupted hardship, called upon the government to roll back housing reforms, cancel planned changes in currency laws, and share power more equitably and concretely with the Parliament. Both sides eventually backed away from extreme positions. In November, Yeltsin offered to consult with the opposition on ways of adjusting the budget to accommodate both sides, reportedly telling the Communists, "I don't want confrontation or early elections. Don't put me in a difficult position."

"Kaleidoscope" of Prime Ministers

Escalating governmental problems certainly had something to do with Yeltsin's decision on March 23, 1998, to dismiss his entire cabinet, including Prime Minister Chernomyrdin. It was a shock and a surprise, but Yeltsin's abrupt move dispelled any notions that he was too sick to lead the country. Yeltsin demonstrated that he was thoroughly in command and had every intention of remaining so until the expiration of his term in 2000. And while the government had some achievements to its credit, such as the 1997 resumption of economic growth (only to be interrupted by the 1998 financial meltdown) and controlling inflation, it had some equally impressive failures that merited a presidential response. Not the least of these was the inability to collect enough revenue to pay the salaries of the hundreds of thousands of state employees who, after months of receiving no wages, were on the threshold of revolt.

From Chernomyrdin to Kiriyenko

Within days, Yeltsin announced his choice of a successor to Chernomyrdin: Sergei Kiriyenko, who apparently had no dreams of running for president in 2000. Kiriyenko was the former fuel and energy minister, with experience in the oil industry. He had kept a low profile—he was unknown, not only in the West but also among the Russian public. With boyish looks and the demeanor of a PhD student, he was nicknamed kinder-surpryz (a mix of "kindergarten" and "surprise").

Once in power, Kiriyenko had difficulty governing, and there was little Yeltsin could do to help him. The Duma despised Kiriyenko. The new prime minister was too young, too reformist (associated with Chubais and Gaidar, the leading "Chicago boys"), and he was also half-Jewish, something of a stigma in a traditionally anti-Semitic environment. In succeeding months, the Communists fiercely opposed Kiriyenko's austerity program and were angry at his difficulty in paying the back salaries of state employees. Behind the fury of the Communists and other Duma members was a deep-seated suspicion that Kiriyenko was formulating economic policy in response to pressure from the International Monetary Fund (IMF), which was holding up loans until the Russian government's spending and the inflation spirals it engendered were brought under control. Kiriyenko's tenure as Prime Minister was terminated by the 1998 financial default (see previous: The August 1998 Financial Crash) for which he bore hardly any personal responsibility. Nevertheless on August 23, 1998, Kiriyenko handed in his resignation, which Yeltsin immediately accepted.

From Kiriyenko to Primakov

Yeltsin was in a weak position vis-à-vis the Duma. His own popularity in the country was now well below its 1996 election peak and had returned to single digits. Had he stuck by Kiriyenko and defied the Duma, he would eventually have had to dissolve it and call for the election of a new Duma—one in which the Communist opposition might have enjoyed an even stronger position.

Yeltsin now nominated Chernomyrdin, Kiriyenko's predecessor, as prime minister. But the Communists did not think Chernomyrdin would be willing to soften economic reform sufficiently to ease hardships of workers, such as finding money—printing it if necessary—to pay their back salaries. Chernomyrdin's meeting with a high IMF official while he was waiting to be approved by the Duma indicated that he might be inclined to heed IMF advice against policies that would provoke inflation, thus finishing off any chance that the Duma would approve his nomination. Yeltsin then withdrew his nomination of Chernomyrdin in favor of Foreign Minister Yevgeni Primakov. Primakov was in many ways an antipode not just to Kiriyenko but to all Yeltsin's prime ministers. Almost 70 years old at the time of appointment, Primakov was, and still is, a heavyweight with a strong instinct for political survival and a background as an orientalist (a student of Arab language and culture), a journalist, and above all an expert in foreign intelligence who performed secret negotiating missions on behalf of the Soviet government in the Middle East. As foreign minister, he had gained respect at home and abroad as a tough but pragmatic supporter of Russia's interests, and an opponent of NATO's expansion into the former Eastern bloc. By the same token, as Russian Foreign Minister, Primakov was prone to change the foreign policy from largely unconditional support of the United States to a more nationalist defense of Russia's interests. Nevertheless, on May 27, 1997, after 5 months of negotiation with NATO Secretary General Javier Solana, Russia signed the Foundation Act, which for most people marks the end of Cold War hostilities.

The Duma was satisfied and voted overwhelmingly on September 11, 1998, to endorse Primakov as Prime Minister. In the next half year, Primakov restored working relations between the Kremlin and the Duma. He used his well-honed political skills to assuage the Communists and bring an end, at least for the time being, to their campaign to oust Yeltsin, whose physical health increasingly paralyzed him politically. The president seemed to be spending more time away from the job than on it, either on vacation, at home, or in a sanitarium recovering from what the Kremlin reported as bouts of bronchitis or some similar respiratory problem. Primakov brought a measure of stability to the Russian political system that was much appreciated by Russian voters, who seemed to like him for his steady hand on power.

The considerable downside of all this, in the eyes of Russian neo-liberals, was that Primakov was a conservative with very little sympathy for radical economic reform. Moreover, he was clearly unwilling to rock the boat with harsh austerity programs or punitive policies toward enterprises and their owners who were refusing to pay taxes and exporting vast sums of hard currency to safe havens abroad.

In the eyes of Yeltsin and his team of economic reformers, yet another problem was Primakov's developing alliance with

(Courtesy of Grigory Ioffe)

The Headquarters of the Federal Security Service of Russia, the Successor to the Soviet KGB or Secret Police.

Moscow's Mayor Luzhkov, another politician whose prestige was on the rise. Together Primakov and Luzhkov were suspected of plotting to succeed Yeltsin or even unseat him. Although the chances of actual removal of Yeltsin from power were minimal, the fact that the Duma was still contemplating it in the spring of 1999, was enough to convince Yeltsin that Primakov had outlived his usefulness. The incident that the media called the *Primakov loop* may have decided his fate as prime minister. On March 23, 1999, Primakov was heading to Washington for an official visit. Flying over the Atlantic Ocean, he learned that NATO had started to bomb Yugoslavia, a move deeply unpopular with broad masses of Russians. Primakov decided to cancel the visit, ordered the plane to turn around over the ocean and returned to Moscow.

From Primakov to Stepashin

On May 13, 1999, Yeltsin nominated Sergei Stepashin, the minister of the interior, to replace Primakov. Stepashin was a loyalist and had tried to protect Yeltsin from his political enemies. In addition, Stepashin had some credibility with democrats and reformers and, to some extent, ironically, with the Duma. Yeltsin seemed to consider Stepashin a viable future presidential candidate who could use the prime ministership to gain the stature needed to compete with other presidential aspirants.

The Duma quickly approved his appointment as Prime Minister. Opposition to him eroded as it became clear that Yeltsin would fight for his appointee and dissolve the Duma if necessary. This time, Duma deputies were on the defensive, worried about the possibility of defeat and the loss of valuable perquisites during the campaign for parliamentary elections scheduled for December 1999.

From Stepashin to Putin: Russia's Identity Crisis as a Nourishing Environment for Russia's New Leader

Prime Minister Stepashin was in power for a mere 89 days when Yeltsin abruptly replaced him on August 9, 1999, with Vladimir Putin, Federal Security Service chief and Yeltsin loyalist. It is not clear why Yeltsin ousted Stepashin so soon. However, rather than faulting Stepashin (or any of his predecessors) personally, most interpretations circulating in Russia focus on the overall problem of seeking the optimal successor to Yeltsin. This task was taken up by Yeltsin's extended "family," allegedly a mafia-like structure, which usurped much of Yeltsin's own decision-making power at the end of his presidency. The "family" included his actual relatives, particularly Tatyana Dyachenko, one of Yeltsin's daughters, and her partner and subsequently husband, Valentin Yumashev (in 1997–98, Yeltsin's chief of staff). Also part of "the family" was Alexander Voloshin who succeeded Yumashev as Yeltsin's chief of staff; Anatoly Chubais, the father of Russian privatization; as well as the oligarchs Boris Berezovsky and Arkady Abramovich. According to widespread interpretations, "the family" worried about the possibility of legal retribution against its members for the consequences of radical economic reform, deeply unpopular in Russia, and about the safety of their own financial assets and the sources of their replenishment. With this in mind "the family" was looking for someone who would not only run the government but would also succeed Yeltsin as president and put any talk about retribution against him and "the family" to rest. With the broad masses of Russians and most political affiliations dissatisfied with economic reform and with what was exceedingly perceived as the American dictate in foreign and even domestic policy of Russia, it was unrealistic to rely on the next (2000) presidential elections as the way to install Yeltsin's loyalist as president. Hence the idea for Yeltsin to resign before the end of his second term and to appoint such a loyalist who would muster popular support prior to the 2000 election. Clearly this was a risky gambit, a trial-and-error solution tactic, which explains the entire "kaleidoscope" of prime ministers.

Although Stepashin was in fact a Yeltsin loyalist, he did not project the image of a strongman, which most Russians craved. This craving can easily be attributed to Russian political tradition, but, in all fairness, many Russians wanted a strongman simply because, in their perception, the government's capacity to deliver for the general public and to protect ordinary people was in shambles. At the same time, with shock therapy over and economic growth resuming, Russians were receiving a long-expected breather that allowed them to reflect on what had befallen them over the course of the 1990s. The emerging image of a new Russia did not instill much optimism.

First, the disintegration of the Soviet Union was still a bleeding wound. Unlike people in the Baltic States and perhaps in western Ukraine, Russians were used to considering the entire Soviet Union, not just the Russian Federation, their homeland. By the end of the 1990s, tensions between the now independent republics showed up more than once; and other geopolitical centers of power extended their influence on parts of the post-Soviet space, effectively sidelining Russia. Twenty-five million ethnic Russians were left outside the Russian Federation, notably in Ukraine, Kazakhstan, and the Baltic States, and in many cases they were subjected to discrimination. The Russian Federation's borders were a throwback to the Russian Empire of the sixteenth century, with borders between Russia and such countries as Ukraine, Belarus, and Kazakhstan not coinciding with cultural frontiers of any kind. In fact, some frontiers *within* the Russian Federation (notably between ethnic Russians and indigenous

(Kremlin Presidential Press Office)

Boris Yeltsin appearing on TV announcing his resignation on December 31, 1999, six months before the end of his term, and ensuring that Vladimir Putin, his protégé, would be elected the next president.

peoples of the North Caucasus) were more tangible discontinuities than borders between Russia as a whole and those three countries. And yet Belarus, Kazakhstan, and Ukraine showed every sign of becoming full-fledged foreign countries. What is more, they increasingly asserted and identified themselves by distancing from Russia. While it did not feel quite that way in Belarus (before the end of the 1990s), it certainly did in Kazakhstan and Ukraine.

Second, the former satellite countries and the Baltic States were lured into NATO despite promises explicitly made by George H. W. Bush and James Baker (U.S. president and U.S. Secretary of State at the time of the Soviet breakup) to their Soviet counterparts, Gorbachev and Shevardnadze, the Soviet Foreign Minister, not to permit this to happen. Many in Russia believed that NATO had been a vehicle for the deterrence of the Warsaw Pact (the former communist block). But now, with the Warsaw Pact no longer around, NATO not only continued to exist but it had even expanded, thus taking advantage of Russia's weakness. The 1999 bombing of Serbian targets in particular infuriated most Russians, who believed that when the Soviet Union had existed nothing like that could have taken place. While the Serbs did commit more than their share of atrocities and ethnic cleansing, other warring parties in former Yugoslavia (Catholic Croats and Muslim Bosnians) had committed similar acts. Singling out Orthodox Serbs, Russians' co-religionists, as the only villains, was perceived in Russia as outright contempt for Russia's sensibilities. By all accounts, Russia, the former superpower, was relegated to the position of a regional power at best.

Third, within Russia, the advent of the nouveau riche (the new bourgeoisie) with ethnically non-Russians accounting for a disproportionally high share of them, contrasted with the outright misery of retirees and such professionals as teachers and medical doctors. Corruption and felony were up and the sense of economic security was down. Such mainstays of Soviet life as free day care centers, summer camps for children, and free health care were no longer available. Most day care centers and summer camps simply disappeared with their premises being rented out to small businesses. Health care centers remained nominally free but no serious treatment was now available without a bribe and without patients obtaining medications and even cotton wool, gauze pads, bandages, diapers, and toilet paper on their own. The post-Soviet researchers in all disciplines, relatively well-paid during the last three Soviet decades, were now able to make ends meet only if and when they were fortunate enough to be contracted by their Western colleagues, who would dole out money they had received from their grant-making agencies. And the Russian military, in possession of the world's largest arsenal of nuclear arms, was sharing in the overall misery as well. To make matters worse, the communist ideology had given way to a gaping void. While that ideology had not satisfied inquisitive Russians, it still provided moral guidance in the areas of civility, patriotism, family, and interpersonal relationships. In some ways, communist ideology had tried its best to position itself as a state-run and absolutist equivalent of organized religion. And while communist ideology ultimately failed in that capacity, the ensuing moral vacuum was hardly better and was perceived as a sanction for chaos and anarchy.

The plight of the military particularly contrasted with new challenges that Russia was facing on its home turf. The first Chechen war was perceived as a total failure on the Russian federal army's part, and the war's end was humiliating. Effectively self-governed and cleansed of ethnic Russians, Chechnya was still overly dependent on the federal budget and cost it more than any other region of Russia. In the summer of 1999, emboldened by the disarray in the Federal army, Chechen terrorist groups undertook attacks in the neighboring republic of Dagestan, an ethnic hodgepodge with byzantine if informal power-sharing arrangements between the largest ethnic groups and Saudi-educated Muslim clerics, who enjoyed growing influence. On August 7, the most massive incursion of Chechen mujahedeen fighters (Islamic militants) into Dagestan took place under the guidance of Shamil Basayev. Many rank-and-file Russians desperately wanted a strong and decisive national leadership to stamp out these terrorists and bring back domestic order.

The above ingredients made for a political climate instrumental in choosing Vladimir Putin as prime minister and the potential replacement for Yeltsin. With his KGB credentials, Putin projected an image of a strong leader, and his communication skills were to the liking of many Russians. As a Yeltsin loyalist, Putin was apparently committed to protecting Yeltsin himself and "the family" from legal retribution, even as he was ready to internalize and act upon grievances that most Russians had accumulated since December 1991 (when the Soviet Union ceased to exist), which effectively meant parting ways with Yeltsin policies.

When he appointed Putin as temporary head of the government on August 9, 1999, Yeltsin simultaneously declared to the general public that Putin would be his successor. This declaration might not have attracted the requisite attention at the time,

since previously Yeltsin had named other persons as his probable successor, only to see these people mishandle that informally conferred status. This time, however, the trial-and-error process had reached a successful conclusion.

Vladimir Putin was born in Leningrad in 1952. For a long time he resided with his parents in a communal apartment (shared by multiple families). In 1975, he graduated from prestigious Leningrad State University (LSU) where he majored in law. As a student, Putin joined the Communist Party and he remained a member until the Party's temporary ban in 1991. After graduation, Putin was recruited by the local branch of the infamous KGB, the Committee for State Security. However infamous it was for outsiders and those whose lives were touched by Stalinist labor camps, many people growing up in the less ideologically rigid post-Stalin environment, considered working for the KGB their patriotic duty. Putin was one of those people, and he was particularly lured by the aura of foreign intelligence. While working for the KGB, he also studied at the KGB's own institution of higher learning. From 1985 to 1990 he worked in Dresden, East Germany, as a KGB station associate while officially working as director of the Dresden chapter of the Soviet Union—GDR (East Germany) Friendship Society. Upon returning to Russia in 1990, Putin became an assistant to his alma mater's president responsible for international exchange. There, at LSU, Putin got acquainted with Alexander Sobchak, LSU's professor of law. Sobchak was one of Russia's most well-known public intellectuals; he was a devout Westernizer and an ardent proponent of political and economic reform. When in 1991 Sobchak was elected mayor of Leningrad (shortly to be restored to its original name of Saint Petersburg), Putin became chair of the foreign relations department in the city hall. In 1992, Putin officially resigned from KGB service and became a lieutenant colonel in the military reserve. From 1994 to 1996, Putin worked as deputy mayor of Saint Petersburg. By the mid-1990s, in most regions of Russia, people who had taken an active role in promoting reform at the beginning of the decade were losing elections. After Sobchak lost the mayoral race in 1996, Putin moved to Moscow and became a deputy to Pavel Borodin, the person in charge of real estate and physical plant of the Moscow Kremlin. Putin, however, never forgot favors rendered to him by Sobchak, and he would later help his former boss to avoid arrest by organizing his temporary emigration at a time when Sobchak's foes initiated litigation against him. Sobchak returned as soon as the case against him was dropped. Personal loyalty is a signature of Putin, who would promote scores of his Leningrad/Saint Petersburg friends to the highest positions in the federal government. In Moscow, Putin rose steeply in the ranks. In July 1998, he was appointed director of the Federal Security Service, a successor to the KGB.

On his watch, the Service prosecuted environmentalists Alexander Nikitin and Grigory Pasko, naval officers who were accused of espionage for exposing Russia's slipshod handling of nuclear waste. However, far more important for Putin's public image inside Russia was his reaction to the crisis in the Caucasus as well as to the August–September 1999 terrorist attacks in the cities of Moscow, Volgodonsk (Rostov region of the North Caucasus), and Buinaksk (Dagestan). More than three hundred people died as a result of explosions of residential

The Moscow Kremlin, a historic, fortified complex at the heart of Moscow, overlooking the Moscow River. It is the best known of kremlins and includes four palaces, four cathedrals, and the enclosing Kremlin Wall with Kremlin towers. The complex serves as the official residence of the President of Russia.

(Courtesy of Grigory Ioffe)

blocks in those cities, explosions attributed to Chechen terrorists. On September 30, 1999, the second Chechen war began. Unlike the first war, which was not supported by a significant part of the Russian media and in which Russian Westernizers took an anti-war stand and sympathized with Chechens, the second war elicited a different reaction. Now, very few people protested against the war, and most considered it the patriotic duty of the Russian government to defeat the terrorist threat and protect Russia's territorial integrity. Any mention of the Western anti-war pronouncements immediately relegated their authors to the anti-Russian/Russophobe camp. Westernizing views had been dealt a blow anyway by the hugely unpopular economic reforms. Putin effectively rode the wave of the resurgent Russian patriotism and wounded pride.

December 1999 Parliamentary Elections

After much speculation that the parliamentary elections scheduled for December 1999 might be postponed by the Kremlin, which feared a stunning defeat because of a further decline in Yeltsin's popularity, Russian voters in fact went to the polls on Sunday, December 19. The outcome, which showed support for Putin, was a pleasant surprise for Yeltsin and his prime minister. The newly formed pro-Kremlin Unity Party won 76 seats. Other groups sympathetic to the Kremlin also won seats, notably the Union of Right Forces, led by former prime minister Kiriyenko (29 seats); the Yabloko Party, led by Yavlinsky (28 seats); and Zhirinovsky's grouping called the Zhirinovsky Bloc (17 seats). With the help of some of the 125 newly elected independent deputies, the Kremlin now had the chance, for the first time since 1992, to develop a majority of support, however fragile, in the new Duma. With evident popular backing for his chosen successor and citing his poor health, Yeltsin resigned 12 days after the election, asking Russians to forgive him his mistakes and transferring all his power to Putin as acting president. A presidential election was set for late March 2000.

Although the electoral success of reformist groups that might cooperate with the Kremlin was unexpected, the reasons for the outcome became clear once the voting was over. First was the strong expression of voter support for executive leadership and the voters' evident disdain for the legislative paralysis of the Communist-driven Duma. Second, voters expressed support of Putin's tough policy toward Chechnya, viewing the war to restore Moscow's control over the region legitimate and justified; they seem to have admired his apparent decisiveness when compared with Yeltsin. Contrary to the view that Putin suffered from his connection with Yeltsin, Putin had established his own, very positive, political identity.

But some aspects of the elections called into question how truly democratic they really were. There was little grassroots-type campaigning. The competing political groups sought popular support for their candidates indirectly through national television networks. And what they had to say in the media was for the most part mud-slinging, with the pro-Kremlin group condemning the Luzkhov-Primakov group. Discussion of important issues and policies, such as the country's prolonged depression and social decline, was minimal. The Kremlin used its control of the two leading national television stations to smear Luzkhov and his supporters with charges of corruption and murder, claims that were based on half-truths and in some cases outright fabrications. The government's tax authorities harassed media outlets that refused to toe the line. While Luzkhov certainly struck back in kind, his media and financial sources were far fewer than those available to the Kremlin.

PUTIN'S PRESIDENCY 2000–2008

In January 2000, as "acting president," Putin strengthened his own personal grip on power, girding for the March 2000 presidential election. Putin's decisiveness in thwarting the massive incursion of Chechen guerillas into Dagestan, as well as what was perceived as military success in Chechnya itself, aided Putin's popularity, as did new information that Chechen separatists enlisted and used massive international support in terms of both funding of their war effort and mercenaries participating in it. That made the war against separatism and for territorial integrity of Russia even more justified in the eyes of many Russians. While the established links of Chechen fighters led to some Middle Eastern countries, Putin vaguely hinted in some of his speeches that larger and more sinister forces may be interested in the destabilization of Russia. Additionally Putin's popularity was aided by projecting a personal image that could not be more different than his predecessor's. In contrast to flannel-mouthed and ailing Yeltsin, who disappeared for weeks from public view, Putin was young, disciplined, sober, articulate in his native Russian, and even able to speak German. Also, Putin was not implicated in the "shock therapy" of the 1990s or in signing a humiliating treaty with Chechen rebels. And while the very first decree signed by Putin as interim president on December 31, 1999 was a decree absolving the vulnerable Yeltsin and even members of his immediate family from any crimes for life, the idea of distancing himself from his predecessor as soon as possible was not lost on Putin. He quickly moved to replace Yeltsin's advisors with loyalists of his own, for the most part by his former colleagues from Saint Petersburg.

March 2000 Presidential Election

Putin won the presidential election held on March 26, 2000. He obtained the support of 52 percent of the voting public on the first and only ballot, thus winning the presidency without having to go to a runoff. The vote was an impressive personal achievement, providing him with a strong popular mandate to lead the country.

Several conditions and circumstances explain Putin's phenomenal success. He had all the advantages that accrue to an incumbent in any leadership race; for that, he had Yeltsin to thank for having left office in December. As acting president, Putin had opportunities to win popular support—opportunities denied to other would-be candidates, in particular Yevgeni Primakov, Gennady Zyuganov, and Grigory Yavlinsky. Indeed, these opportunities were so overwhelmingly favorable to Putin that Primakov, viewed shortly before the election as the most respected and potentially most popular choice as Yeltsin's successor, dropped out of the race in February, while Zyuganov and Yavlinsky barely campaigned.

Moreover, as Acting President, Putin used the very substantial portion of the mass media controlled by the state to present himself in the most attractive light possible. Importantly, he had the support of Russia's money moguls, such as Boris Berezovsky, who later bragged on more than one occasion that he had actually brought Putin to power.

Voters in many instances followed the advice of their regional leaders, most of whom supported Putin. Putin also went out of his way to cultivate less-well-off voters who had supported Zyuganov in 1996, signaling his sympathy for their economic hardships and a willingness to help them—though in his sphinx-like manner, he refrained from saying exactly what he had in mind to do.

Putin's victory also derived from a number of intangibles, such as his reticence on the domestic policies he would pursue to reinvigorate the dilapidated Russian economy. By speaking in platitudes, he avoided frightening voters and, indeed, seemed to gain their trust—no easy task, given the level of cynicism about government and politics among today's Russians. Putin exuded a measure of self-confidence as well as personal restraint that encouraged voters to think he might succeed in improving living conditions. He helped his image by the articulation of a tough policy to restore Russian control of Chechnya. By election time, it looked as if he had been able to deliver on his promise of returning the rebellious Chechen republic to Russian rule. In sum, Putin was definitely the beneficiary of a popular yearning for strong, confident, successful leadership that could solve problems like Chechnya, unemployment, unpaid state salaries, inadequate pensions, government corruption, and restore respect for Russia in the entire world.

The Communists Take a Beating

Communist Party leader Zyuganov suffered a decisive setback with the loss of a substantial amount of the voter support he had had in the 1996 presidential race. In 1996 Zyuganov had won a majority of voters in about half of Russia's 89 regions. But in March 2000 he won only about 29 percent of the popular vote and carried only five regions. He even lost the Orlovskaya oblast (the Orel region), his home. To some

observers, this result suggested that areas of assumed communist strength no longer existed, because an increasing number of voters had accepted, or had made their peace with, the new market democracy that Putin seemed to represent. Other observers saw Putin's rise to power as vindication of Russia's wounded pride and evidence of the rise of Russian nationalism.

Putin's Authoritarianism

Vladimir Putin established a political regime very different from that of his predecessor, Boris Yeltsin. The mainstream opinion of American foreign policy experts about changes that occurred under Putin and that are associated with his name were best summarized in an article by Michael McFaul and Kathryn Stone-Weiss (*Foreign Affairs,* January/February 2008. Copyright © 2008 by Council on Foreign Relations. Reprinted by permission). While their interpretation of those changes is debatable, the factual part of their narrative appears to be accurate:

Yeltsin was far from a perfect democrat: he used force to crush the Russian parliament in 1993, bulldozed into place a new constitution that increased presidential power, and barred some parties or individuals from competing in a handful of national and regional elections. He also initiated two wars in Chechnya. The system that Yeltsin handed over to Putin lacked many key attributes of a liberal democracy. Still, whatever its warts, the Russian regime under Yeltsin was unquestionably more democratic than the Russian regime today. Although the formal institutional contours of the Russian political system have not changed markedly under Putin, the actual democratic content has eroded considerably.

Putin's rollback of democracy started with independent media outlets. When he came to power, three television networks had the national reach to really count in Russian politics—RTR, ORT, and NTV. Putin tamed all three. RTR was already fully state-owned, so reining it in was easy. He acquired control of ORT, which had the biggest national audience, by running its owner, the billionaire Boris Berezovsky, out of the country. Vladimir Gusinsky, the owner of NTV, tried to fight Putin's effective takeover of his channel, but he ended up losing not only NTV but also the newspaper *Segodnya* and the magazine *Itogi* when prosecutors pressed spurious charges against him. In 2005, Anatoly Chubais, the CEO of RAO UES (Unified Energy Systems of Russia) and a leader in the liberal party SPS (Union of Right Forces), was compelled to hand over another, smaller private television company, REN-TV, to Kremlin-friendly oligarchs. Today, the Kremlin controls all the major national television networks.

More recently, the Kremlin has extended its reach to print and online media, which it had previously left alone. Most major Russian national newspapers have been sold in the last several years to individuals or companies loyal to the Kremlin, leaving the Moscow weekly, *Novaya Gazeta,* the last truly independent national newspaper. On the radio, the station Ekho

Moskvy remains an independent source of news, but even its future is questionable. Meanwhile, Russia now ranks as the third-most-dangerous place in the world to be a journalist, behind only Iraq and Colombia. Reporters Without Borders has counted 21 journalists murdered in Russia since 2000, including Anna Politkovskaya, the country's most courageous investigative journalist, in October 2006.

Putin has also reduced the autonomy of regional governments. He established seven supraregional districts headed primarily by former generals and KGB officers. These seven new super governors were assigned the task of taking control of all the federal agencies in their jurisdictions, many of which had developed affinities with the regional governments during the Yeltsin era. They also began investigating regional leaders as a way of undermining their autonomy and threatening them into subjugation.

Putin emasculated the Federation Council, the upper house of Russia's parliament, by removing elected governors and heads of regional legislatures from the seats they would have automatically taken in this chamber and replacing them with appointed representatives. Regional elections were rigged to punish leaders who resisted Putin's authority. And in September 2004, in a fatal blow to Russian federalism, Putin announced that he would begin appointing governors—with the rationale that this would make them more accountable and effective. There have been no regional elections for executive office since February 2005.

Putin has also made real progress in weakening the autonomy of the parliament. Starting with the December 2003 parliamentary elections, he has taken advantage of his control of other political resources (such as NTV and the regional governorships) to give the Kremlin's party, United Russia, a strong majority in the Duma: United Russia and its allies now control two-thirds of the seats in parliament. Putin's own popularity may be United Russia's greatest electoral asset, but constant positive coverage of United Russia leaders (and negative coverage of Communist Party officials) on Russia's national television stations, overwhelming financial support from Russia's oligarchs, and near-unanimous endorsement by Russia's regional leaders have also helped. After the December 2003 elections, for the first time ever the Organization for Security and Cooperation in Europe issued a critical report on Russia's parliamentary elections, which stressed, "The State Duma elections failed to meet many OSCE and Council of Europe commitments for democratic elections." In 2007, the Russian government refused to allow the OSCE to field an observer mission large enough to monitor the December parliamentary elections effectively.

Political parties not aligned with the Kremlin have also suffered. The independent liberal parties, Yabloko and the SPS, as well as the largest independent party on the left, the Communist Party of the Russian Federation, are all much weaker today and work in a much more constrained political environment than in the 1990s. Other independent parties—including the Republican Party and the Popular Democratic Union, as well as those of the Other Russia coalition—have not even been allowed to register for elections. Several independent parties and candidates have been disqualified from participating in local elections for blatantly political reasons. Potential backers of independent parties have been threatened with sanctions. The imprisonment of Mikhail Khodorkovsky, previously Russia's wealthiest man and owner of the oil company Yukos, sent a powerful message to other businesspeople about the costs of being involved in opposition politics. Meanwhile, pro-Kremlin parties—including United Russia, the largest party in the Duma, and A Just Russia, a Kremlin invention—have enjoyed frequent television coverage and access to generous resources.

In his second term, Putin decided that NGOs could become a threat to his power. He therefore promulgated a law that gives the state numerous means to harass, weaken, and even close down NGOs considered too political. To force independent groups to the margins, the Kremlin has generously funded NGOs either invented by or fully loyal to the state. Perhaps most incredible, public assembly is no longer tolerated. In the spring of 2007, Other Russia, a coalition of civil-society groups and political parties led by the chess champion Garry Kasparov, tried to organize public meetings in Moscow and St. Petersburg. Both meetings were disrupted by thousands of police officers and special forces, and hundreds of demonstrators were arrested—repression on a scale unseen in Russia in 20 years.

In his annual address to the Federation Assembly in April 2007, Putin struck a note of paranoid nationalism when he warned of Western plots to undermine Russian sovereignty. "There is a growing influx of foreign cash used directly to meddle in our domestic affairs," he asserted. "Not everyone likes the stable, gradual rise of our country. Some want to return to the past to rob the people and the state, to plunder natural resources, and deprive our country of its political and economic independence." The Kremlin, accordingly, has tossed out the Peace Corps, closed OSCE missions in Chechnya and then in Moscow, declared persona non grata the AFL-CIO's field representative, raided the offices of the Soros Foundation and the National Democratic Institute, and forced Internews Russia, an NGO dedicated to fostering journalistic professionalism, to close its offices after accusing its director of embezzlement.

While weakening checks on presidential power, Putin and his team have tabled reforms that might have strengthened other branches of the government. The judicial system remains weak, and when major political issues are at stake, the courts serve as yet another tool of presidential power—as happened during NTV's struggle and during the prosecution of Khodorkovsky. There was even an attempt to disbar one of Khodorkovsky's lawyers, Karinna Moskalenko.

Many of Putin's defenders, including some Kremlin officials, have given up the pretense of characterizing Russia as a "managed" or "sovereign" democracy.

Instead, they contend that Russia's democratic retreat has enhanced the state's ability to provide for its citizens. The myth of Putinism is that Russians are safer, more secure, and generally living better than in the 1990s— and that Putin himself deserves the credit. In the 2007 parliamentary elections, the first goal of "Putin's Plan" (the main campaign document of United Russia) was to "provide order."

In fact, although the 1990s was a period of instability, economic collapse, and revolutionary change in political and economic institutions, the state performed roughly as well as it does today, when the country has been relatively "stable" and its economy is growing rapidly. Even in good economic times, autocracy has done no better than democracy at promoting public safety, health, or a secure legal and property-owning environment.

The Russian state under Putin is certainly bigger than it was before. The number of state employees has doubled to roughly 1.5 million. The Russian military has more capacity to fight the war in Chechnya today, and the coercive branches of the government—the police, the tax authorities, the intelligence services—have bigger budgets than they did a decade ago. In some spheres, such as paying pensions and government salaries on time, road building, or educational spending, the state is performing better now than during the 1990s. Yet given the growth in its size and resources, what is striking is how poorly the Russian state still performs. In terms of public safety, health, corruption, and the security of property rights, Russians are actually worse off today than they were a decade ago.

Security, the most basic public good a state can provide for its population, is a central element in the myth of Putinism. In fact, the frequency of terrorist attacks in Russia has increased under Putin. The two biggest terrorist attacks in Russia's history— the Nord-Ost incident at a theater in Moscow in 2002, in which an estimated 300 Russians died, and the Beslan school hostage crisis, in which as many as 500 died—occurred under Putin's autocracy, not Yeltsin's democracy. The number of deaths of both military personnel and civilians in the second Chechen war—now in its eighth year—is substantially higher than during the first Chechen war, which lasted from 1994 to 1996. (Conflict inside Chechnya appears to be subsiding, but conflict in the region is spreading.) The murder rate has also increased under Putin, according to data from Russia's Federal State Statistics Service. In the "anarchic" years of 1995–99, the average annual number of murders was 30,200; in the "orderly" years of 2000–2004, the number was 32,200. The death rate from fires is around 40 a day in Russia, roughly ten times the average rate in Western Europe.

Nor has public health improved in the last eight years. Despite all the money in the Kremlin's coffers, health spending averaged 6 percent of GDP from 2000 to 2005, compared with 6.4 percent from 1996 to 1999. Russia's population has been shrinking since 1990, thanks to decreasing fertility and increasing

mortality rates, but the decline has worsened since 1998. Noncommunicable diseases have become the leading cause of death (cardiovascular disease accounts for 52 percent of deaths, three times the figure for the United States), and alcoholism now accounts for 18 percent of deaths for men between the ages of 25 and 54. At the end of the 1990s, annual alcohol consumption per adult was 10.7 liters (compared with 8.6 liters in the United States and 9.7 in the United Kingdom); in 2004, this figure had increased to 14.5 liters. An estimated 0.9 percent of the Russian population is now infected with HIV, and rates of infection in Russia are now the highest of any country outside Africa, at least partly as a result of inadequate or harmful legal and policy responses and a decrepit health-care system. Life expectancy in Russia rose between 1995 and 1998. Since 1999, however, it has declined to 59 years for Russian men and 72 for Russian women.

At the same time that Russian society has become less secure and less healthy under Putin, Russia's international rankings for economic competitiveness, business friendliness, and transparency and corruption all have fallen. The Russian think tank INDEM estimates that corruption has skyrocketed in the last six years. In 2006, Transparency International ranked Russia at an all-time worst of 121st out of 163 countries on corruption, putting it between the Philippines and Rwanda. Russia ranked 62nd out of 125 on the World Economic Forum's Global Competitiveness Index in 2006, representing a fall of nine places in a year. On the World Bank's 2006 "ease of doing business" index, Russia ranked 96th out of 175, also an all-time worst.

Property rights have also been undermined. Putin and his Kremlin associates have used their unconstrained political powers to redistribute some of Russia's most valuable properties. The seizure and then reselling of Yukos' assets to the state-owned oil company Rosneft was the most egregious case, not only diminishing the value of Russia's most profitable oil company but also slowing investment (both foreign and domestic) and sparking capital flight. State pressure also compelled the owners of the private Russian oil company Sibneft to sell their stakes to the state-owned Gazprom and Royal Dutch/Shell to sell a majority share in its Sakhalin-2 project (in Siberia) to Gazprom. Such transfers have transformed a once private and thriving energy sector into a state-dominated and less efficient part of the Russian economy. The remaining three private oil producers—Lukoil, TNK-BP, and Surgutneftegaz—all face varying degrees of pressure to sell out to Putin loyalists. Under the banner of a program called "National Champions," Putin's regime has done the same in the aerospace, automobile, and heavy-machinery industries. The state has further discouraged investment by arbitrarily enforcing environmental regulations against foreign oil investors, shutting out foreign partners in the development of the Shtokman gas field, and denying a visa to the largest portfolio investor in Russia, the British citizen William Browder. Most World Bank governance indicators, on issues such as the rule of law and control of corruption, have been flat or negative under Putin. Those on which Russia has shown some improvement in the last decade, especially regulatory quality and government effectiveness, started to increase well before the Putin era began.

The second supposed justification for Putin's autocratic ways is that they have paved the way for Russia's spectacular economic growth. As Putin has consolidated his authority, growth has averaged 6.7 percent—especially impressive against the backdrop of the depression in the early 1990s. The last eight years have also seen budget surpluses, the eradication of foreign debt and the accumulation of massive hard-currency reserves, and modest inflation. The stock market is booming, and foreign direct investment, although still low compared to other emerging markets, is growing rapidly. And it is not just the oligarchs who are benefiting from Russia's economic upturn. Since 2000, real disposable income has increased by more than 10 percent a year, consumer spending has skyrocketed, unemployment has fallen from 12 percent in 1999 to 6 percent in 2006, and poverty, according to one measure, has declined from 41 percent in 1999 to 14 percent in 2006. Russians are richer today than ever before.

The correlations between democracy and economic decline in the 1990s and autocracy and economic growth in this decade provide a seemingly powerful excuse for shutting down independent television stations, canceling gubernatorial elections, and eliminating pesky human rights groups. These correlations, however, are mostly spurious.

The 1990s were indeed a time of incredible economic hardship. After Russia's formal independence in December 1991, GDP contracted over seven years. There is some evidence that the formal measures of this contraction overstated the extent of actual economic depression: for instance, purchases of automobiles and household appliances rose dramatically, electricity use increased, and all of Russia's major cities experienced housing booms during this depression. At the same time, however, investment remained flat, unemployment ballooned, disposable incomes dropped, and poverty levels jumped to more than 40 percent after the August 1998 financial meltdown.

Democracy, however, had only a marginal effect on these economic outcomes and may have helped turn the situation around in 1998. For one thing, the economic decline preceded Russian independence. Indeed, it was a key cause of the Soviet collapse. With the Soviet collapse, the drawing of new borders to create 15 new states in 1991 triggered massive trade disruptions. And for several months after independence, Russia did not even control the printing and distribution of its own currency. Neither a more democratic polity nor a robust dictatorship would have altered the negative economic consequences of these structural forces in any appreciable way.

Economic decline after the end of communism was hardly confined to Russia. It followed communism's collapse in every country throughout the region, no matter what the regime type. In the case of Russia, Yeltsin inherited an economy that was already in the worst nonwartime economic depression ever. Given the dreadful economic conditions, every postcommunist government was compelled to pursue some degree of price and trade liberalization, macroeconomic stabilization, and, eventually, privatization. The speed and comprehensiveness of economic reform varied, but even those leaders most resistant to capitalism implemented some market reforms. During this transition, the entire region experienced economic recession and then began to recover several years after the adoption of reforms. Russia's economy followed this same general trajectory—and would have done so under dictatorship or democracy. Russia's economic depression in the 1990s was deeper than the region's average, but that was largely because the socialist economic legacy was worse in Russia than elsewhere.

After the Soviet collapse, Russian leaders did have serious policy choices to make regarding the nature and speed of price and trade liberalization, privatization, and monetary and fiscal reforms. This complex web of policy decisions was subsequently oversimplified as a choice between "shock therapy" (doing all of these things quickly and simultaneously) and "gradual reform" (implementing the same basic menu of policies slowly and in sequence). Between 1992 and 1998, Russian economic policy zigzagged between these two extremes, in large part because Russian elites and Russian society did not share a common view about how to reform the economy.

Because Russia's democratic institutions allowed these ideological debates to play out politically, economic reform was halting, which in turn slowed growth for a time. During Russia's first two years of independence, for example, the constitution gave the Supreme Soviet authority over the Central Bank, an institutional arrangement that produced inflationary monetary policy. The new 1993 constitution fixed this problem by making the bank a more autonomous institution, but the new constitution reaffirmed the parliament's pivotal role in approving the budget, which led to massive budget deficits throughout the 1990s. The Russian government covered these deficits through government bonds and foreign borrowing, which worked while oil prices were high. But when oil prices collapsed in 1997–98, so, too, did Russia's financial system. In August 1998, the government essentially went bankrupt. It first radically devalued the ruble as a way to reduce domestic debt and then simply defaulted on billions of outstanding loans to both domestic and foreign lenders.

This financial meltdown finally put an end to major debate over economic policy in Russia. Because democratic institutions still mattered, the liberal government responsible for the financial crash had to resign, and the parliament compelled Yeltsin to appoint a left-of-center government headed by Prime Minister Yevgeny Primakov. The deputy prime minister in charge of the economy in Primakov's government was a Communist

Party leader. Now that they were in power, Primakov and his government had to pursue fiscally responsible policies, especially as no one would lend to the Russian government. So these "socialists" slashed government spending and reduced the state's role in the economy. In combination with currency devaluation, which reduced imports and spurred Russian exports, Russia's new fiscal austerity created the permissive conditions for real economic growth starting in 1999. And so began Russia's economic turnaround—before Putin came to power and well before autocracy began to take root.

First as prime minister and then as president, Putin stuck to the sound fiscal policies that Primakov had put in place. After competitive elections in December 1999, pro-reform forces in parliament even managed to pass the first balanced budget in post-Soviet Russian history. In cooperation with parliament, Putin's first government dusted off and put into place several liberal reforms drafted years earlier under Yeltsin, including a flat income tax of 13 percent, a new land code (making it possible to own commercial and residential land), a new legal code, a new regime to prevent money laundering, a new regime for currency liberalization, and a reduced tax on profits (from 35 percent to 24 percent).

Putin's real stroke of luck came in the form of rising world oil prices. Worldwide, prices began to climb in 1998, dipped again slightly from 2000 to 2002, and have continued to increase ever since, approaching $100 a barrel. [*In July 2008, oil prices peaked at $146 per barrel; but they plummeted to less than $40 in November–December 2008 after the commencement of the world financial crisis—G.I.*] Economists debate what fraction of Russia's economic growth is directly attributable to rising commodity prices, but all agree that the effect is extremely large. Growing autocracy inside Russia obviously did not cause the rise in oil and gas prices. If anything, the causality runs in the opposite direction: increased energy revenues allowed for the return to autocracy. With so much money from oil windfalls in the Kremlin's coffers, Putin could crack down on or co-opt independent sources of political power; the Kremlin had less reason to fear the negative economic consequences of seizing a company like Yukos and had ample resources to buy off or repress opponents in the media and civil society.

Mentioned in passing in the above narrative, the Nord-Ost and Beslan incidents deserve to be elaborated upon as do events in Chechnya and around it. All of these have contributed to a significant rise in Russian nationalism and its morbid offshoot, a besieged fortress mentality, which Putin capitalized upon.

The March 2003 Chechen Constitution

While inclined in the early months of 2003 to escalate the Russian military effort in Chechnya to suppress the insurgents after their October 2002 terrorist attack in Moscow, Putin decided to try a new conciliatory political tactic, influenced by parliamentary elections scheduled in December 2003 and presidential elections to be held in March 2004. He did not want to increase the risk that Russian voters would blame him for the unrelenting war in Chechnya he had promised to end.

In March 2003, Putin turned to the Chechen population to open the way to peace. He asked Chechens to vote in favor of a constitution that provided for the legal recognition of Chechnya as a permanent part of Russia in return for self rule and money to rebuild the country and its destroyed capital of Grozny, a city now almost without housing and other kinds of infrastructure. Driven by a longing for peace, security, and normality, Chechen voters overwhelmingly supported the new constitution. But, while the voting looked like a major success for Putin, fighting in Chechnya did not stop in 2003. Chechen separatists who opposed the constitution, denouncing it as a Russian trick, had told Chechen voters not to approve it and continued their war against the Russian presence in Chechnya, killing about two dozen Russian soldiers a week.

The October 2003 Chechen Presidential Election

That said, it also was true that in 2003 a new phase of the Russian conflict with Chechnya seemed in evidence with the election to the Chechen presidency of the pro-Moscow Akhmat Kadyrov, formerly a Mufti (a Muslim religious leader) of Chechnya and a guerrilla, who had been acting as Moscow's surrogate since 2000. The deal the Kremlin cut with Kadyrov seemed to be that in return for Russian support of almost any actions, no matter how brutal, to preserve his control of Chechnya and restore normality to Chechen life, he in turn would protect and defend Chechnya's administrative links to Moscow as an integral part of the Russian Federation. It was Putin's plan to give at least the appearance to Russians and Chechens that this second Chechen war was finally coming to an end and that, at least from Moscow's vantage point, Russia had defeated the separatist challenge.

But, Kadyrov was reviled by the separatist guerrilla groups fighting the Russians. They saw him as a traitorous stooge willing to sacrifice his country's independence for the sake of running Chechnya with Russian backing, and they refused to recognize him. Some groups continued to support ex-president Maskhadov, and all groups not only continued to fight Russians in Chechnya but also continued to undertake terrorist acts in Russia proper and in neighboring areas such as Ingushetia. Not surprisingly, in December 2003, some separatists tried but failed to assassinate Kadyrov.

Many Chechens in Grozny were willing to acknowledge that he was trying to restore some order to the city and the country, while fostering a very modest improvement in living conditions. The Kremlin thought so well of him that it started to curtail the size of the military forces it had deployed in the republic to provide further evidence of a return to normalcy in Chechnya on Russian terms.

Throughout 2003, Chechen militant separatists continued to resist Russian authority throughout the republic. The Russians responded in kind, fighting both guerrillas and Chechen civilians.

PUTIN'S REGIME AGAINST THE BACKDROP OF RUSSIA'S AUTHORITARIAN TRADITION

While the factual side of political developments in Putin's Russia narrated by McFaul and Stoner-Weiss[1] is accurate, their interpretation of these developments and value judgments shared throughout that interpretation are problematic. After all, when Putin's second term in office was about to expire, his approval rating hovered near 80 percent, and even the most consistent critics of Putin's authoritarianism (McFaul and Stoner-Weiss included) did not doubt that Putin's popularity in Russia was genuine. Thousands of Russians, not to mention prominent public opinion leaders (like Nikita Mikhalkov, the famed Oscar-winning movie director) petitioned that Putin amend the constitution and run for a third term, but he chose not to satisfy these demands, preferring his new role as second-in-command—at least *de jure*. There is a widespread belief in Russia that Putin reestablished order and thus laid the foundation for economic growth. This opinion may belie facts, but it is so tenacious that McFaul and Stoner-Weiss call it a "conventional narrative." To them, this narrative is wrong and is "based on a spurious correlation between autocracy and growth." Consequently, one of the main tasks of their article in *Foreign Affairs* was to showcase the spurious nature of this correlation. Moreover, the authors make the point that "whatever the apparent gains of Russia under Putin, the gains would have been greater if democracy had survived." This is a strong and loaded claim, as first and foremost it implies that under Yeltsin there was in fact democracy in Russia—a point which stands in strong contrast with Russian public opinion, which is overwhelmingly negative in regard to Yeltsin's years at the helm of power. One can even go further and express doubt as to whether Russians in fact value democracy at all. Also, the claim in question involves conditional tense (what would have been *if*) which has limited currency in a rational analysis. Apparently cognizant of how shaky their reasoning is, McFaul and Stoner-Weiss take a step backward, conceding that internationally a form of government and economic growth are not correlated ("On average, autocracies and democracies in the developing world have grown at the same rate for the last several decades."). From this line of defense they furnish criticism of Kremlin officials who evoke China as a model to emulate: a modernizing autocracy that has delivered an annual growth rate of over ten percent for three decades. "For every China," write McFaul and Stoner-Weiss, "there is an autocratic development disaster such as the Democratic Republic of Congo; for every authoritarian success such as Singapore, there is a resounding failure such as Myanmar. . . ." Apparently McFaul got so enamored with this pronouncement that in his December 10, 2008, lecture at the Washington, DC-based Hoover Institute he delivered its slightly modified version: "For every China there is a Zimbabwe." But even as a rhetorical flourish, this pronouncement is vulnerable in the extreme. Moreover, it exhibits disregard for culture and any available knowledge of historical and/or geographical specificity of the world countries, and for area studies as a source of that knowledge. For there is no such thing as "every China!" China is the world's largest country with a long history of self-reflection. "When we try to tell China how to organize its internal affairs," wrote Henry Kissinger, "it is easy to forget that China has had 14 imperial dynasties, 10 of which had individual histories longer than the entire history of the United States."

Among other things, the Chinese are used to high state capacity, that is, to the omnipotent role of the state in people's everyday life and are not used to Western-style self-governance. At the same time, Congo, that McFaul compared with China, has tragically not even become a full-fledged country and no central authority has ever spoken on behalf of its diverse ethnic groups. The case of Zimbabwe is different but no less tragic and dysfunctional.

Although Russian civilization is younger than Chinese, it is 1,000 years old and exhibits a history of self-reflection imbued with skepticism about the applicability of Western societal norms in Russia. Suffice it to refer to such authorities as Fyodor Dostoyevsky and Alexander Solzhenitsyn. That means that just as any other country, Russia is path-dependent and it does not seem likely that it would bend its path to the liking of foreign policy consultants that pitch democracy promotion as a selling point on their resume. From that perspective, rather than being carried away by a bold assumption as to what Russia would have achieved if it had not been its own self, it makes more sense to ponder over what sustains autocracy in that country and why it has always returned to habitual authoritarianism after short interludes construed as political thaws (whether that was under Tsar Alexander the Second or Nikita Khrushchev or Gorbachev and Yeltsin). Not only famous Russians weighed in on inapplicability of some Western customs in Russia. Although Huntington's seminal work about civilizations is mostly remembered (and often condemned) in conjunction with the relationship between the West and the Muslim world, Huntington also published a map of the eastern border of Western civilization and made some clairvoyant (for 1993, when his thesis was first published as an article) remarks about the difficulties of implanting market and democracy to the east of that border. It would be instructive to contrast Huntington's not quite detailed but articulate argumentation to that of Anders Aslund, an experienced Washington-based Russia hand. Huntington admits that "the lesson of Russian history is that centralization of power is the prerequisite to social and economic reform." He also quotes the Slavophile Danilevsky who "*in words that we also heard in the 1990s,* denounced Europeanizing efforts as 'distorting people's life and replacing its forms with alien, foreign forms,' 'borrowing foreign institutions and transplanting them to Russian soil,' and 'regarding both domestic and foreign relations and questions of Russian life from a foreign, European viewpoint, viewing them, as it were, through a glass fashioned to a European angle of refraction." In contrast to that, Anders Aslund claims that Russia failed to conduct political reform because Western "economists told Yeltsin what reforms were necessary, but the political scientists did not."[2] However, as a CNN editor, put it, "Russia has been allergic to offers of aid from the West ever since hundreds of overpaid consultants arrived in Moscow after the collapse of Communism, in 1991, and proceeded to hand out an array of advice that proved, at times, useless or dangerous"[3]

But what exactly stands behind the "value gap" between Russia and the West? Although such a gap repeatedly shows in sociological surveys, its recognition runs into peculiar obstacles. One of them was reflected upon by Viacheslav Nikonov, an influential political commentator and a grandson of Viacheslav Molotov, Stalin's minister of foreign affairs (and whose name is also immortalized in *Molotov cocktail*). "A Canadian diplomat, who has lived more than ten years in Moscow, shared an interesting idea," writes Nikonov. "The biggest problem of perceiving Russians through the eyes of Westerners is that Russians are . . . white. If they were green, pink, or of any varied color, the problem would evaporate. The Westerners would simply say that Russians are different, which is what they routinely say about people from Asia, Africa, and other countries, whose political regimes are well harsher than ours. But we look like Westerners, and so they expect us to exhibit the same behavior, reactions, customs, beliefs, and institutes they have in the West."

(continued)

PUTIN'S REGIME AGAINST THE BACKDROP OF RUSSIA'S AUTHORITARIAN TRADITION (continued)

A Torn Country

One additional perception obstacle is that Russia is, in Huntington's terms, a torn country. It has been torn since the reign of Peter the Great (1682–1725). Since then, there have been two intellectual trends in Russia's self-reflection, Slavophiles and Westernizers. Nicolai Petro, an American political scientist, believes that whereas under Yeltsin, Westernizers dominated the political discourse, under Putin, Slavophiles came to dominate it. An excerpt from a 2005 article by Stephen White, Margot Light, and Ian McAllister highlights the issues:

Russian 'Westernisers' had always been a part of the wider Europe—or at least should aspire towards it. It took its law codes from Napoleonic France, its hierarchy of official positions from Sweden and Denmark, and its Academy of Sciences from Prussia (indeed for many years the Academy was 'dominated, if not actually run, by German and German-trained scientists and scholars'). Russian students went to Europe to complete their education, like Boris Pasternak; its noble families spoke in French, other than with their servants; and the royal family itself took summer holidays in Nice, with its extravagant Orthodox cathedral (indeed, many of the royal families of Europe were directly related at this time). This, surely, was the way forward. A rather different view was taken by the Slavophiles, whose views were strongly influenced by those of the Orthodox Church. For the Slavophiles, Russia had its own special destiny, reflecting a society that was based on very different principles from the liberalism and secularism of Western nations. The West, for writers like Ivan Kireevsky and Ivan Aksakov, meant the individualistic civilization of Rome, the Catholic religion and a degenerate commercialism. Russia's own values were rather different: a tradition of collective discussion and agreement that found expression in the peasant commune, a unity of nationality and religious faith (what was known in Russian as sobornost'), and a natural pacifism. Some of the Slavophiles looked still wider, to an alliance of Orthodox and Slavic nations, 'not excluding Greece', on the one hand, and 'on the other— the entire Protestant, Catholic, and even Mohammedan and Jewish Europe put together'. This view of Russians as a 'chosen people' underpinned their leadership of an international communist movement that, rather later, sought to displace Western capitalism.

The relationship between Westernizers and Slavophiles in Russia is not just of lateral or horizontal nature whereby one can find representatives of both trends of thought in each social stratum. One can definitely find both among Russian intellectuals, but one can hardly find a spontaneous or nourished Westernizer among any other Russian group. In 1921, shortly after fleeing Russia, Count Trubetskoi wrote that "the top and the bottom of Russian culture, its grassroots and its upper crust, gravitate to different ethnographic zones. As long as the edifice of Russian culture was being crowned by the Byzantine cupola, the whole construction was stable. However, ever since this cupola began to be replaced with an upper floor of Roman-Germanic descent, any stability and harmony of the edifice's parts were being diminished, the top tilting more and more until it finally collapsed while we Russian intellectuals, who have spent so much time in an effort to prop up the Roman-Germanic roof sliding down Russian walls, are now standing stunned and stiff before this gigantic wreck and still thinking how to build a new roof, yet again of Roman-Germanic style." Clearly Trubetskoi assigns Westernizers exclusively to the elite, which gave the world

the likes of Piotr Tchaikovsky and Fyodor Dostoyevsky without whom European culture is unthinkable. It is important to point out that most Western students of Russia established personal contact with Russian Westernized elite, not with Russian masses. In the words of Anatol Lieven, this created a peculiar "echo chamber between Russian intellectuals and commentators on Russia in the West, which has generated and sustained damaging illusions about Russian political realities, and about how quickly it is possible to reform Russia. This pattern of Western dependence on unrepresentative groups of pro-Western 'informants' . . . helps create an environment in which both sides lose sight of reality."[4]

Irresponsible Elites

Nicholas Riasanovsky referred to the Russian proverb "Sytyi golodnogo ne razumeyet," which he loosely translated as "A satiated man is not a comrade of a hungry one." "This bit of folk wisdom," writes Riasanovsky, "has undeniable relevance to Russian history. In imperial Russia, some of the satiated tended to be satiated to the full, while the hungry ones went very hungry. During the nineteenth century, Russia possessed the most splendid court and perhaps the most palatial and magnificent capital city in Europe, as well as an exquisite literature and a glorious ballet, but most of its people endured a marginal existence with every drought threatening starvation."[5]

"One hardly needs to be reminded," notes Riasanovsky, "that the division between the elite and the masses in Russia paralleled similar divisions in other countries. Still, the Russian split was not quite like the others, or at least it represented a more extreme species of the same genus. In this, as in so many other cases, the evolution of Russia seems to offer a sharper and cruder version of what happened to the West of it."[6] A social contrast of this magnitude is much like a "separation of charge," which maintains calm up to a point but once in a while leads to an unusually mighty release of the accumulated tension through sociopolitical equivalents of lightning and thunder. "God save me from witnessing a Russian rebellion, senseless and unrelenting," wrote Alexander Pushkin in his History of Pugachev's Rebellion. The social discontent that emerged in Russia early in the twentieth century was on a scale unseen anywhere in Europe, where the upper classes had understood their vested interest in contributing to the well-being of the masses through organized charities and negotiations with trade unions. The Russian elite failed to develop similar self-preservation techniques. This failure makes anybody preoccupied with social stability worry when he or she is faced with the restoration of profound social contrasts in the Russia of the 1990s and thereafter.

Communalism

Russians pride themselves on their alleged spiritual superiority over individualistic and intensely materialistic Westerners. The famed Russian collectivism is rooted in a traditional peasant community or mir. Russia is a country of belated urbanization and, as was shown previously, it is still more rural than the official statistics of labor force composition would allow one to admit. What is more, as recently as 1990, most Russian urbanites older than 20 years of age were born rural villagers. Mikhail Voslensky, the author of the famous book Nomenklatura, noted that the overwhelming majority of the Soviet Communist Party apparatchiks were villagers by birth. For these reasons, a Russian peasant commune, though nonexistent in today's Russia is an important formative feature of Russian culture.

In the very beginning of his superb collation of pre-revolutionary Russian writings about a peasant commune, Robinson shrewdly

PUTIN'S REGIME AGAINST THE BACKDROP OF RUSSIA'S AUTHORITARIAN TRADITION

contrasted the two microcosms of agrarian colonization: that in Russia and that in North America: "Ordinarily . . . Russian colonists did not live the lonely life of an American frontier family. In the number and in the close inter-relationship of its inhabitants, the smallest Russian settlement had usually the double character of a village and a family."[7] When a village included several households, each usually had one or more strips of land in each of the several fields. And in each of these fields all the households were obliged to abide by the same cycle of cropping and fallowing. It was the custom to redivide the meadows each year before the mowing, while the pasture-lands and forests were generally used in common. Among the powers of the commune were thus "the distribution of land and the regulation of its use, the apportionment of *tiaglo* (feudal obligations), the election of village authorities (elders) . . . the collection of funds for expenditures of the *mir* (the peasant commune in its administrative capacity), the organization of mutual aid, and the resolution of civil and minor criminal affairs. A Russian peasant commune thus involved many manifestations of collectivism. Through periodical repartition and mutual social control, a commune provided for peasant schooling in group action.

In a peasant commune, the possibility always existed of displacing a household head and appointing another member of the same family in his place. A peasant, therefore, depended on a commune not only in terms of access to land but also in terms of social status. Combined with land repartitioning, such social organization did not facilitate initiative and motivation. Communal leadership, the elders, meddled in the property relationships of peasant families. A head of a household could not arbitrarily bequeath his property to one of his children, let alone transfer it to other hands without communal endorsement. Every son was to inherit a part of his late father's land allotment and a part of his implements in order to fulfill his duties to the commune. However, family members did not have full sovereignty over their interpersonal relationships. Everything they earned, including earnings in a city during their leaves of absence from the commune, was supposed to replenish the family budget. The tradition of mutual material support by family members was very persistent, outliving even instances of a patriarchal family virtually splitting up when some members moved to cities or towns for good. It is still common for Russian urbanites to rely on their rural relatives for food supplies, whereas the latter rely on urbanites during the sowing campaign, harvest, and other land tending chores.

From the perspective of Russia's communal past, it is tempting to view the Soviet period as Russian society's historical exercise at self-regulation: if Russians are indeed prone, as they seem to be, to develop and recklessly reproduce the potentially dangerous level of social inequity, a counteracting force must be called upon to keep this inequity in check. The emergence of a social order whose economic essence is redistribution of profits so that those potentially rich are kept from using their profits as they see fit and those potentially poor are kept from starvation, can be seen as just such a counteracting force. For some analysts, the true antecedents of this order are not so much rooted in Marxism as in the peasant tradition of Russia. A school of thought embracing this view was founded by Nikolai Berdiaev. In his "The Origin of Russian Communism,"[8] Berdiaev attributed communism's most essential features to Russia's oriental affinities (among which he included communal collectivism), not to Marxism. Whereas the latter had come to be the source of communist symbolism, the former provided human capital susceptible to social engineering and made it possible to muster the energy of the entire nation. "In the U.S.S.R., the family is forming, the close-knit circle of relatives, the household teaming with children, with Stalin as the father or the elder brother to all of them,"[9] wrote Andrei Platonov, a superb prose writer and a keen observer of Russian life,

in 1935. The school of thought that promoted this perspective was off to a bumpy start because the ideologically indoctrinated audience was not ready to listen. When Nickolas P. Vakar published his book,[10] in which he showed that Soviet polity was essentially a reincarnation of the Russian peasantry's traditional communal arrangements now extended over the entire society under a smoke screen of Marxist symbols, the Western audience was impressed even less. The idea of Vakar's book was actually not original; in it he restated some earlier thoughts of Berdiaev, ridding them of some elements that would only strike a chord with readers well versed in Russian history, and fleshed them out by the new data. But even rehashing some older concepts from the 1930s was apparently still ahead of its time in 1962. Published in that year, Vakar's book was deemed eccentric and fell into oblivion. The Cold War was in full swing, and it came to a head during the 1962 Cuban missile crisis. At that time, touting some shadowy *cultural* roots for communism was too much for the *politically* agitated reader. Communism not only looked ominous from afar; the Western "agitprop" viewed it entirely as a product of erroneous and misleading *theory*. The fact that Marxism swayed quite a few people in the West aggravated ideological excitement. It was only much later, when communism had retreated from the forefront of international affairs, that Berdiaev's cultural paradigm began to gain ground; in fact, it is now part-and-parcel of the Russian Westernizers' way of thinking.[11] In the meantime, for Russian communists history has come full circle; since the early 1990s and until recently, their most ardent supporters have been peasants from Russia's south, not industrial workers.

A Flip Side of Russian Communalism: Cult of Equality-in-Poverty and Low Level of Mutual Trust

Russian collectivism is not without its vulnerabilities. Some of the most frequently mentioned are a poor labor ethic, begrudging individual success, and a tenacious belief that becoming wealthy is only possible through some dishonest scheme and/or through robbing the poor. These ideas are implanted in sundry Russian proverbs, like *Ot trudov pravednykh ne nazhiviosh palat kamennykh* (You won't earn a brick house by laboring conscientiously), *Rabota ne volk, v les ne ubezhit* (Work is no wolf, it's not going to flee to the woods); *Rabota durakov liubit* (Work loves fools), and many others. It is therefore tempting to suggest that collectivism in Russia is seldom a personal choice; rather, it is a willy-nilly survival pattern spontaneously exercised by and sometimes deliberately marketed for the poor. As long as poverty is almost everybody's lot in a local community, a region, or for that matter the entire country, collectivism becomes the ambient environment. But as soon as any Russian becomes a person of more than average means, the very first thing he/she does is detach his personal space from that of everybody else.

Today, tall and often elaborate fences with hefty gate locks arrest one's attention in the Russian-style suburbia, this playground of the newly rich. Likewise, fences of considerable height surround the enclosures of the more successful rural villagers. Rarely in America, stereotyped as the hotbed of individualism, can one witness such an irresistible urge to physically disassociate from the community. To be sure, fenced landholding is not uncommon in America, but it is not nearly as ubiquitous as in Russia, where personal space finds itself in an uneasy relationship with higher-order spaces, communal, regional, and national.

Russian communities affected by growing income disparity among its members do not normally join in coordinated action to finance beautification and improvement of nearby public spaces, including access roads. The very expensive mansions of the "New

(continued)

PUTIN'S REGIME AGAINST THE BACKDROP OF RUSSIA'S AUTHORITARIAN TRADITION *(continued)*

Russians" and outright poor and often impassable roads leading to them quite often coexist: the homeowners are reluctant to share expenses out of a misgiving that someone who is richer than they are or needs the road more will not contribute his or her fair share. Eventually, the most prosperous neighbors withdraw from any preliminary negotiation to finance an access road linking the community with a municipally, regionally, or federally maintained roadway. They just buy very expensive foreign SUVs and do not care about their neighbors who drive less expensive cars.

Yet another flip side of Russian collectivism is a low level of interpersonal trust. In fact, circles of trust in Russian society are atomized, scarcely overlap, and are almost never integrated into public politics. One of the most well-informed and productive students of Russian society, Tatyana Vorozheikina, believes, for example, that the most important reason why Russian society craved a strong leader—after a period of democratic idealism initiated under perestroika—was that that society became permeated with fear of itself. The public order was shattered as a result of a sudden infusion of freedom, and that was something to fear. To facilitate an understanding of the genuine, not contrived nature of these attitudes, one should perhaps refer to the fact that there is nothing intrinsically Russian about them. Here is what Robert Putnam wrote about southern Italy in the 1970s.

> Public life . . . is organized hierarchically rather than horizontally. The very concept of "citizen" here is stunted. From the point of view of the individual inhabitant, public affairs is the business of somebody else—i notabli, "the bosses," "the politicians"—but not me. Few people aspire to partake in deliberations about the commonweal, and few such opportunities present themselves. Political participation is triggered by personal dependency or private greed, not by collective purpose. Engagement in social and cultural associations is meager. Private piety stands in for public purpose. Corruption is widely regarded as the norm, even by politicians themselves, and they are cynical about democratic principles. "Compromise" has only negative overtones. *Laws (almost everyone agrees) are made to be broken, but fearing others' lawlessness, people demand sterner discipline* (emphasis added).[12]

If one just replaces "piety" with "paternalism," the above quote would match Soviet society most impeccably. Now, after several decades of militant atheism, "piety" has taken back part of its center stage in Russia but has not by any means displaced or shattered paternalism. The last sentence of the above quote is emphasized because it casts some light on what many Americans have a problem understanding, that is, why people in Russia, *and in fact in all non-Western environments,* want order in the first place, not liberty, let alone markets. If only in part, this need for order is due to a well-grounded fear of evil behavior on the part of other members of the same society, a problem so pivotal that everything else pales beside it. Consequently, everybody who shares that fear is in need of protection. When Saddam Hussein was deposed in Iraq, sectarian violence erupted, which Hussein, or anyone else governing with an iron fist, would not have allowed. And when the communist order was removed in Russia, its market dominated minority seized state-owned assets and left most ordinary people out in the cold. Under such conditions, only a lunatic would harbor regrets about lacking democracy. For most people, "democracy" becomes a swear word. In the words of Robert Kaplan, "Democracy loses meaning if both rulers and ruled cease to be part of the community tied to a specific territory." The much-reviled behavior of the so-called New Russians is a case in point. The sundry jokes about them pivot on their combination of social arrogance and uncouth ways. They flaunt their wealth, engage in brazen behavior, breed corruption by bribing bureaucrats, and wholeheartedly despise their less successful compatriots. Just a couple of years ago (in late 2006), one of the news items widely discussed in Russia and beyond was that the Russian oligarch Vladimir Potanin paid George Michael, a British pop singer, $3 million to perform at his New Year's Eve party (on December 31, 2006) twenty miles outside Moscow. Another Russian oligarch, Boris Prokhorov, was arrested at an exquisite ski resort in Courchevel, France, and held for three days by the French police on a suspicion of ties with illicit prostitution networks.

Cult of a Strong Leader and Imperial Nationalism

A low level of mutual trust feeds directly into a cult of a strong leader. This was reflected upon by the above-quoted Russian scholar Viacheslav Nikonov, who wrote that "the personification of political institutions and the great role of leaders . . . *make up for a deficiency in mutual trust.* That is why Russians lean not to institutions but to strong personalities" (emphasis added). To wit, Russians entrust their fate to strong leaders not so much because they trust them but because they trust each other even less. Practically no one in Russia harbors the illusion that the national leader is averse to corruption; rather he is believed to be entitled to it and to being able to quickly enrich himself to the point that further enrichment is simply unnecessary so the leader can now pay his undivided attention to common good and thwart the attempts of others to steal from the public. For the popular mentality, a strong leader is clearly the least evil; while liberalism in both economy and political life is fraught with mischief at best and outright evil at worst. As the greatest Russian standup comedian Mikhail Zhvanetsky pointedly stated, "our liberty is a mess, and our idea is order in that mess." In Russian, this message was spiced by a colloquialism, as the word "mess" was rendered as *bardak,* whose direct meaning is a brothel while the acquired figurative meaning is a mess. Order for many Russians is a sacral condition of life, a recourse against the messiness associated with liberty. Since it is conducive to uncontrollable dealings of multiple persons, whom Russians do not trust (just as they do not have much trust in many people that they happen to know), liberty breeds a dangerous mess.

In a much publicized 2008 survey, sociologists presented Russians with a list of historical figures and asked them to rank those leaders in terms of their all-time greatness. More than 5 million votes by telephone, text-messaging, and the internet were registered in the poll, which named Alexander Nevsky, a medieval warrior prince, as the winner. Stalin had led the poll early on and narrowly missed the top spot. Overall, the brutal dictator took 519,071 votes compared to Nevsky's 524,575 and Stolypin's 523,766. Pyotr Stolypin was, Russia's Prime Minister from 1906 to 1911. What the winners of the contest have in common is that they reinstated order by dealing decisive blows to enemies. While Nevsky whacked external enemies, German knights; Stolypin was particularly effective against enemies inside Russia ("You need great shocks," he famously lectured Marxist revolutionaries, "whereas we need a great Russia."); and, of course, Stalin whacked enemies of both kinds.

Russian social scientists from the Westernizers' strand (a smallish but vociferous minority as of this writing) are divided about the outcome of the poll. Perhaps they are more of a single voice when talking to domestic audiences. However, to a BBC correspondent, Mark Urnov, the dean of political studies at Moscow's Higher School of Economics, had this to say. "This vote is the result of eight years of brainwashing by the mass media [under Putin]. This vote for authoritarianism would never have happened eight years ago." The political commentators Leonid Radzikhovsky and Nikolai Svanidze strongly disagree. While they do not doubt the reality

PUTIN'S REGIME AGAINST THE BACKDROP OF RUSSIA'S AUTHORITARIAN TRADITION

of brainwashing, they point to the fact that under Yeltsin, Russian media had been awash in liberal ideas (pro-democracy, pro-market, and pro-West), but all of them failed miserably to indoctrinate Russians; whereas current preaching in favor of a strong hand and Russian imperial grandeur fits popular mentality and is therefore in high public demand. This is a chicken-and-egg conundrum whereby one cannot state for sure that propaganda is the controlling factor or that popular mentality is solely on the receiving end. Both feed into one another. In 2005, "Vladimir Putin famously stated that the collapse of the Soviet Union was the greatest geopolitical catastrophe of the century." While scores of Western pundits referred to this pronouncement as confirmation of Putin's nostalgia for Soviet times and of his retrograde nationalism, relatively few commentators noted how much in sync that pronouncement was with much of Russian society. The Soviet Union, after all, was the last great territorial empire, much like the Ottoman Empire established by the Turks or the Austro-Hungarian Empire. Russians were the bulwark of their empire, and Russian nationalism bears a distinct imperial imprint; it is very different from the mainstream European nationalism that culminated in the formation of nation states. While some contemporary nationalists do long for a Russian nation state, Russians have never lived in one. Instead, their polity was multi-ethnic for centuries with ethnic Russians accounting for half of the entire population of the Soviet Union on the eve of its dissolution. Ethnic Russians were a collective and benevolent Big Brother to other ethnicities, at least in the Russian mind.

Contrary to some tenacious clichés nurtured by Sovietologists, the Russification of public life in the Soviet Union hardly resembled an ominous state-run conspiracy. In the Soviet Union, there were such powerful overt vehicles of Russification to fall back on that the need to involve anything covert or hidden from the public eye seems questionable. One such overt tool of Russification was service in the Soviet army, which is primarily why command of the Russian language was always higher among non-Russian men than among women. Another reason to adopt Russian was to move up in the ranks, whether in a managerial position (like working for a large, federally controlled enterprise in a supervisory, let alone a top-management, position) or to gain political position (executive power at the federal level or in a Communist Party cell). All these required proficiency in Russian. Migration of ethnic Russians was another powerful engine of Russification outside Russia proper. At the same time, all languages that by the time of the Communist revolution (or incorporation into the Soviet Union, as in the case of the Baltic States in 1940) had become advanced literary languages embraced by national elites thrived in the U.S.S.R.; much effort and investment was directed into introducing alphabets into languages that had had no literary standard and into secondary education in most non-Russian languages. Much effort as well was invested into cultivating folk cultures (like choral singing in the Baltics), obviously under the auspices of communist ideology. In cases where official languages of the union republics were close to Russian, as was the case with Ukrainian and Belarusian, Russification was often spontaneous, not enforced. [When claims to the contrary are made in the Ukrainian and Belarusian national discourse, it is important to understand that the very survival of nationalism in those countries depends on distancing themselves from Russia as the meaningful other. Precisely because Russia is culturally and linguistically so close to Belarus and Ukraine, it has to be portrayed as a separate universe and a source of evil.]

Ethnically non-Russian members of the CPSU's Politburo were quite prominent at all times, including Stalin, an ethnic Georgian, at the helm of power for 31 years. Governmental structures in every non-Russian union republic had been more influential than in Russia. In each union republic but Russia, Communist Parties ensured

not just ideological but often commonsense managerial control over a republics' economic life as well; but there was no Communist Party of Russia that was a branch of the Communist Party of the U.S.S.R. (much like union republics' parties were). In Moscow, where three tiers of political power were concentrated—national (U.S.S.R.), Russia-wide, and municipal, the intermediate level was clearly the least prestigious and least funded. Note that Russia was by far the Soviet Union's largest republic. It should not come as a surprise that the self-effacing nature of Russian imperial nationalism is revealed in the fact that some of the most economically backward and mismanaged regions of the Soviet Union were in Russia. If one excludes Central Asia, a region that did not know even embryonic forms of homegrown capitalism prior to the Communist Revolution of 1917, in no non-Russian republic was there anything analogous to Russia's Non-Black-Earth Zone, a vast region of pronounced rural and agricultural decay, a region that encompasses Muscovy, the nucleus of the first centralized Russian state. Much more investment per capita and particularly per unit of land was directed to union republics than to Russia itself. The habitual complaint of Russian nationalists that ethnic Russians made disproportionate sacrifices for the benefit of other ethnicities (within the Soviet Union) is not without grounds. And with the Soviet Union gone, ethnic Russians, numbering close to 25 million people in non-Russian post-Soviet states, became the largest group of expatriates. These considerations provide the much needed context to Putin's 2005 pronouncement about the breakup of the Soviet Union as the greatest geopolitical catastrophe. By all accounts Russia's trauma from the loss of their "benevolent empire" is still a wound to the national psyche.

NOTES

1. McFaul, Michael and Kathryn Stone-Weiss, "The Myth of the Authoritarian Model," *Foreign Affairs,* January/February 2008.
2. Anders Aslund, "Russia's Capitalist Revolution: Why Market Reform Succeeded and Democracy Failed," Lecture at the Kennan Institute, Washington, DC, April 21, 2008; quoted from the event summary at www.wilsoncenter.org/index.cfm?topic_id=1424&fuseaction=topics.event_summary&event_id=398063
3. Peter Gumbel, "Putin-Dell slapdown at Davos," CNN.com, January 28, 2009; www.money.cnn.com/2009/01/28/news/companies/dell.davos.fortune/
4. *Anatol Lieven,* "A Tragic Feud: Alienation between the Russian State and the Liberal Intelligentsia, Past and Present," Lecture at the Kennan Institute, 4 June 2007; quoted from the event summary at www.wilsoncenter.org/index.cfm?topic_id=1424&fuseaction=topics.publications&doc_id=303419&group_id=7718
5. Nicholas Riasanovsky, "Afterword: The Problem of the Peasant," in *The Peasant in Nineteenth-Century Russia,* ed. Wayne S. Vucinich (Stanford, CA: Stanford University Press, 1968) p. 264.
6. Nicholas Riasanovsky, op. cit, 263.
7. Robinson, G. T. *Rural Russia under the Old Regime,* (New York, 1957), p. 10.
8. Nikolai Berdiaev, *The Origin of Russian Communism,* (London: The Centenary Press, 1937).
9. Andrei Platonov, *Zapisnye Knizhki. Materialy k Biografii,* (Moscow: Naslediye, 2000), p. 157.
10. Nickolas Vakar, *The Taproot of Soviet Society,* (New York: Harper and Brothers, 1962).
11. The most influential books written in a corresponding tradition are Akhiazer, A.S., *Rossiya: Kritika Istoricheskogo Opyta,* Moscow, FO 1991; Vishvevsky, A.G., *Serp i Rubl,* Moscow, OGI 1999; Akhiezer, A.S., Davydov, A.P., et al, *Sotsio-Kul'turnye Osnovaniya i Smysl Bol'shevizma,* Novosibirsk: Sibirskii Khronograf 2002.
12. Robert Putnam, *Making Democracy Work: Civic Traditions in Modern Italy,* (Princeton: Princeton University Press 1993), p. 115.

OCTOBER 2002 CHECHEN HOSTAGE-TAKING IN MOSCOW

During the first two years of the Putin presidency, the fighting between Chechen guerrillas and Russian federal forces continued unabated, and there was little of the promised Russian funding for reconstruction of schools, apartment buildings, and hospitals. This was the context of the Moscow theater hostage crisis, also known as the 2002 Nord-Ost siege. The hostages were seized in a working class area of Moscow about four kilometers (2.7 miles) southeast of the Moscow Kremlin. During Act II of a sold-out performance of *Nord-Ost,* a Russian musical theater production, shortly after 9 P.M., some 42 heavily armed and masked men and women drove a bus to the theater and entered the main hall firing assault rifles in the air. The camouflage-clad attackers took some 850 people hostage, including members of the audience and the cast of performers. Some performers (for example, the popular bard Alexei Ivashchenko) who had been resting backstage fled through an open window and called police; some others were released by hostage-takers on account of urgent medical procedures (e.g., insulin shots) and in exchange for passing their message to the authorities. Also several women with small children were released. Escapees reported the highly unusual fact that many of the terrorists were women. These turned out to be the so-called black widows whose husbands were killed by the Russian army as insurgents. The gunmen—led by Movsar Barayev, nephew of the slain Chechen rebel militia commander Arbi Barayev—threatened to kill the hostages unless Russian forces were withdrawn from Chechnya within one week. Should this not happen, they would start killing the hostages.

After a two-and-a-half day siege, on the early Saturday morning of October 26, 2002, Russian security forces propelled an unknown gas into the building's ventilation system. Some of the Chechens were equipped with gas masks, but members of the audience were obviously not protected. After thirty minutes, when the gas had taken effect, the security forces entered through the roof, the basement, and finally the front door. When the shooting began, the rebels told their hostages to lean forward in the theater seats and cover their heads. As rebels and hostages alike began to fall unconscious because of protracted exposure to the unknown gas, several of the female hostage-takers made a dash for the balcony but passed out before they reached the stairs. In a fierce firefight, the federal forces gunned down those guerrillas who were still awake and methodically executed those who had passed out. Officially, all the terrorists were killed by Russian forces. Among the hostages, 130 people never woke up after inhaling the unknown toxic substance and "falling asleep." Ten of them were children 11 to 16 years of age. Some analysts later suggested that the substance might be phentanylum, a synthetic analgetic. Others suggested that this was ftorotan, another nonflammable anesthetic. But no official statement ever confirmed or disproved these suggestions. President Vladimir Putin defended the scale and violence of the assault in a televised address later on the morning of October 26, stating that the government had "achieved the near impossible, saving hundreds of people" and that the rescue "proved it is impossible to bring Russia to its knees." Putin thanked the security forces as well as Russian citizens for their bravery and the international community for the support given against the "common enemy." He also asked forgiveness for not being able to save more of the hostages, and declared Monday a national day of mourning for those who died. He vowed to continue fighting against what he called international terrorism. This author was closely following the hostage-taking episode, the ensuing storm of the building by security forces, and the aftermath of those events, on two Russian TV channels (courtesy Dishnetwork of America) and by making multiple telephone calls to Moscow. Not only had this author attended the same play in the same theater during a trip to Moscow just a couple of months before the hostage-taking, but it also turned out that he either knew personally or was aware of as many as seven people immediately affected by the tragedy. For an émigré who had left Russia 14 years prior to the event, this was a stunning coincidence. What befell those people? Two of them, Aleksei Ivashchenko and Georgy Vasilyev, who coauthored and staged the musical, were at the time of the incident working in a recording studio, located in the same building. They were preparing for a concert dedicated to the one-year anniversary of the Nord-Ost musical production. While Ivashchenko escaped through the backstage window and broke his leg, Vasilyev stayed throughout the entire ordeal and by some accounts acted courageously by addressing the members of the audience and even entertaining them to the extent possible. Like many other hostages, he passed out and subsequently woke up. Three other members of the audience, Vladimir Shkolnikov, a demographer, his wife Maria Shkolnikov, a well-known Moscow cardiologist, and their son Maxim who suffered from diabetes, were released by the terrorists, and Maria Shkolnikov was asked to pass a letter from the terrorists to the Russian government. Finally, Anna and Sergei Syomin, successful design and construction business owners, lost consciousness and Anna later regained it but her husband did not.

Parliamentary Elections of December 2003

In the atmosphere created by the Nord-Ost siege (see 2002 Nord-Ost Inset), the response to it, and the continuing war in Chechnya, the Russian voting public seemed supportive of Putin's efforts to diminish political pluralism in Russia for the sake of stability. Results of parliamentary elections on December 7, 2003, gave the United Russia Party and other parties aligned with it a solid, reliable majority of seats in the new Duma. Results provided a popular stamp of public approval on Putin's style of leadership. Putin now had the political authority to get most of what he wanted from the Parliament.

Liberal ideas have never been particularly popular with Russians, and they had been additionally discredited in the 1990s by the privatization of Russia's most important economic assets, an exceedingly unpopular action touted as the embodiment of economic liberalism and a path to democracy.

Although the United Russia Party was enlisting support of Russia's most prominent oligarchs who had become rich in the 1990s (and thus owed their new social status to exactly that unpopular affair), it was also a party of national bureaucracy and a party articulating Russia's national interest increasingly on a path diverging with that of America and other Western countries. Thus, allegiance to the United Russia Party was very much a stamp of political loyalty to the Kremlin. As the former Russian premier Chernomyrdin once joked, whatever party we seek to create, it turns out to be the CPSU [i.e., the Communist Party of the Soviet Union]. This joke hits the mark: Russians are not only consistently illiberal, but they also like concentration and personalization of power and do not feel good about divided political loyalties in the ruling class.

To make matters even "worse" (or more traditional for Russian political culture) the Kremlin destroyed any possible threat to United Russia by liberal and leftist parties alike by jailing on charges of corruption their main financial contributor, billionaire Mikhail Khodorkovsky, the CEO of Yukos Oil, a giant, multinational oil-producing and marketing firm. Khodorkovsky's funding was particularly crucial for at least one liberal party, Yabloko; but he also aided communists in an ostensible effort to prop up political pluralism and to prevent the concentration of excessive power in the hands of Putin and his aids. In addition, the Kremlin used every resource available to it to enhance the appeal of the pro-government United Russia Party and blunt the challenges from other opposition parties. These "administrative resources," as they were called by the Kremlin, included use of the state-run media, law enforcement authorities, regional governors, and election officials. For example, in late June 2003, in anticipation of a campaign for the upcoming election in December that would involve some popular criticism of Putin's leadership, the Duma voted overwhelmingly in favor of a bill that prohibited support of extremist activities by the media. The law provided for government closure of radio and television stations and print media, with the government determining what was an extremist activity. Punishment ranged from fines to closure of the offending voice. Editorials critical of Putin and media support of one candidate over another would be fair game. There were complaints, however, that this law was unconstitutional in that it limits free speech and was a form of censorship.

By December 2003, the political stage was set for the bloc of parties committed to Putin to sweep the elections. The bloc collectively won two-thirds of the 450 seats in the Duma, giving Putin effective control of the Duma on the eve of the March 2004 parliamentary elections. Whatever he wanted from the Duma, in the months before the election, he essentially got, making Putin look to Russian voters like a political miracle worker, more than capable not only of running Russia but also actually improving it. Several years of economic growth and the resumption of nationalist rhetoric fed into that way of thinking.

One new opposition party did have modest success in winning seats in the Duma in the December elections, the Motherland Party, led by the hyper-nationalist and anti-Western Dmitri O. Rogozin. Rogozin was more likely to be a nuisance rather than a threat to Putin, because Rogozin's true inspiration was to be co-opted by the ruling executive team. However, when asked after the election what aspects of the Russian constitution he would like to change, Rogozin shot back, "Everything!"

March 2004 Presidential Election

In the presidential elections held on March 14, 2004, Putin won a landslide victory. With a 90 percent turnout, Putin won 70 percent of the popular vote. The outcome was no surprise to Russians or outsiders. Although there were some flaws in the election according to monitors sent to Russia by the Organization for Security and Cooperation (OSCE) in Russia, it was relatively clean. Still, the flaws were serious if a country is claiming to be democratic in the Western sense of that term, which of course is one of the most debatable issues. Indeed, many influential figures in Russia denounce Western universalism and stress that Russian democracy fits Russian political

culture, which is a normal condition. Anyway, according to election observers from the OSCE, "Council of Europe standards for democratic elections, such as a vibrant political discourse and meaningful discourse" reportedly, "were lacking."

While there is no reason to claim that Putin won because he cheated—after all, he was immensely popular, and the results of the election were completely in line with that political reality—the government's behavior in the months preceding the election clearly fixed the campaign environment in ways that would assure, arguably guarantee, Putin's impressive electoral success. Although there were six other candidates, the Communist Party's candidate, Nikolai Kharitonov, won almost 14 percent of the vote, underscoring a pervasive dissatisfaction with Putin's domestic policies to improve living standards. The tallies of the other five presidential candidates were in single digits.

In the campaign, the Kremlin used state-controlled media to aggressively propagandize Putin's candidacy, while his rivals had nowhere near an equivalent opportunity to communicate to the voters. For example, in some places local law enforcement authorities used a variety of pretexts to interrupt and silence rival candidates trying to get their messages out in public assemblies. One trick of the police was to terminate a meeting because of a bomb scare.

Putin changed the top leadership in his government to strengthen his campaign for the presidency. At the end of February 2004, only weeks away from the March election, Putin shrewdly replaced Mikhail Kasyanov and most of the cabinet. In effect, Putin let Kasyanov and his cabinet take the blame for the government's failure to help Russian wage-earning workers. But, Putin had other reasons for removing Kasyanov. The two had different views on how to deal with the case of Yukos Oil head Mikhail Khodorkovsky, who was accused of tax evasion, a crime endemic in the Russian business world and one Putin was ostensibly determined to stop. Kasyanov seemed concerned that the case against Khodorkovsky might arouse the fear of other successful Russian entrepreneurs that their assets could be endangered by an overzealous Kremlin or that foreign investment might be scared off. In addition, Putin wanted an opportunity to appoint more high-level government managers from the security apparatus to further strengthen his grip on power.

Prime Minister Kasyanov's successor in early 2004 was Mikhail Fradkov, a technocrat with little political baggage who could be counted on to loyally do the president's bidding in domestic and foreign policy making. Fradkov also had economic, financial, and intelligence expertise. His appointment signaled Putin's continued determination to stay the course with some Russian nationalist version of market economy. Fradkov pledged to go after tax dodgers in Russian big business. The new cabinet was made up of loyalists, new faces, and technical experts, all of whom were political assets to Putin as Russia went to the polls to elect the president on March 14.

Akhmat Kadyrov's Assassination

In May 2004 the Chechen separatists dealt Putin a blow when they assassinated Chechen President Akhmat Kadyrov, putting the lie to Putin's propaganda that the situation in Chechnya was improving. The assassination occurred just two days after Putin's inauguration as president for a second four-year term—when he had said once again that the situation in Chechnya had improved.

The Khlebnikov Assassination

In early July 2004, Paul Khlebnikov, editor in chief of *Forbes Russia* magazine, a subsidiary of the U.S. journal *Forbes,* which is owned and managed by Steve Forbes, was fatally shot in broad daylight in front of his apartment. Khlebnikov was an American journalist of Russian descent. Observers have suggested Khlebnikov may have made powerful enemies because he investigated corruption and sought to shed light on Russian business. Khlebnikov wrote *Godfather of the Kremlin: The Decline of Russia in the Age of Gangster Capitalism* (Forbes, September 2000), a biography of the Russian tycoon Boris Berezovsky. Berezovsky was openly critical of Khlebnikov's writings, particularly an article published in *Forbes* in 1996 about his alleged criminal activities for which he filed a libel suit and for which *Forbes* was forced to retract the allegations. In 2003, Khlebnikov published his second book *Conversation with a Barbarian: Interviews with a Chechen Field Commander on Banditry and Islam,* in which Klebnikov provides the transcript of his 15 hour conversation with a Chechen crime lord, politician, and guerilla commander, Khozh-Ahmed Noukhaev; the book focused on organized crime in Chechnya.

At the time, the Khlebnikov assassination seemed to be something of a high-water mark in the steady loss of freedom and independence of the Russian electronic and print media in the aftermath of Putin's March 2004 reelection.

In early May 2006, two men accused of the crime were acquitted in a trial closed to the public. The trial judge was suspected of partiality against Khlebnikov, having refused to release records of the trial. The judge also was accused of interfering with the prosecution's efforts to appeal the acquittals in a higher court. The appeal was eventually successful. But in the fall of 2006, an appeals court reversed the decision of the jury and the freed suspects of Khlebnikov's assassination were re-arrested. On December 17, 2007, the retrial, now classified as secret, was restarted and immediately halted again because of the continued disappearance of one of the main suspects. It remains to be seen whether the Kremlin will ever get to the bottom of this crime.

Another Presidential Election in Chechnya in August 2004

In the immediate aftermath of Kadyrov's assassination, the Kremlin made plans for a presidential election in Chechnya in September. Kadyrov's son, Ramzan, seemed to be a possible candidate, but he was only 27 years of age at the time of his father's assassination, three years below the constitutional threshold for Chechen president. For this and perhaps other reasons, the Kremlin's preferred candidate for the Chechen presidency was Major General Ali Alkhanov, the Chechen minister of the interior, who had much professional police experience in Chechnya.

Alkhanov was quite different from Kadyrov, who had had some standing among Chechens because of being a high-level Muslim cleric and his having had involvement at one point with the separatists in Chechnya. By contrast Alkhanov, who was literally paraded about by Putin at the Kremlin and told that in the remaining months prior to the presidential election he would continue to be in charge of funds assigned by the Kremlin for the reconstruction of Chechnya, had a much different background, having fought against the Chechen guerillas in both wars. Although there were at least eight other candidates registered to run for the Chechen presidency, the odds clearly favored Alkhanov. When the election was held on August 29, Alkhanov won, as Chechens expected. The anti-Russian Chechen insurgents were livid, and even before the election, initiated what turned out to be a string of terrorist incidents involving suicide bombers who destroyed two Russian passenger planes, blew up a Moscow subway station, and took women and children as hostages in a school in the town of Beslan in North Ossetia. Russians were traumatized by these events and expressed anger over Putin's evident inability to protect the country from this new wave of terrorism to which Russia was patently vulnerable. In early September the Russian people, especially the citizens of Moscow and other big cities, as well as Putin himself, could not make sense of what had happened and seemed baffled over what to do next.

The September, 2004 Chechen Hostage-taking in Beslan

The Beslan school hostage crisis began when a group of armed terrorists, demanding an end to the Second Chechen War, took more than 1,100 people (including some 777 children) hostage on September 1, 2004, at School Number One in the town of Beslan, North Ossetia-Alania, an Ossetian ethnic homeland in the North Caucasus region of the Russian Federation. Of all North Caucasus ethnicities, Ossetians have been traditionally most friendly to ethnic Russians, and unlike their overwhelmingly Muslim neighbors (like Ingushes and Chechens), most Ossetians are Orthodox Christians. On the third day of the standoff, Russian security forces stormed the building using tanks, thermobaric rockets, and other heavy weapons. A series of explosions shook the school, followed by a fire which engulfed the building and a chaotic gun battle between the hostage-takers and Russian security forces. Ultimately, at least 334 hostages were killed, including 186 children. Hundreds more were wounded or reported missing.

As in other instances of terrorist attacks inside Russia by Chechen insurgents, the Russian government was seen as unprepared either to prevent the attack or to deal with it once it occurred. Aggrieved and angry citizens of Beslan, especially heartbroken parents who had lost children, condemned Moscow for its heavy handed intervention that contributed to the death of many hostages. For the most part, Putin expressed sympathy but had little to say about the possibility of Moscow's culpability.

Federal authorities in Moscow promised investigations that were completed in late 2005. They placed the blame not on themselves but on the local government. But Basayev volunteered publicly his agreement with a view that sheer Russian ineptitude, even stupidity and gullibility, made it possible for his followers to slip into North Ossetia and enter school grounds. It became clear, moreover, that the insurgents who took over the school in Beslan were from Chechnya and Ingushetia, undermining Russian charges that the attackers were foreigners from the Middle East with links to Al Qaeda.

In fact, the Putin charge of linkages between the Chechen insurgents and militant Islamic organizations abroad is open to question. Few Chechens have turned up in other places where Al Qaeda has been present, notably Afghanistan and Pakistan, although at times Chechen presence there is claimed by high-ranking politicians (e.g., by the Pakistani prime minister in his February 2009 *Newsweek* interview). The Kremlin has long insisted on connections between the Chechen insurgents and Islamic extremists, and on more than one occasion hinted at the ominous role of some Western forces facilitating these connections in order to destabilize Russia. According to Russian intelligence, foreign money underwrote not only the siege at Beslan but other recent acts of terrorism in 2005, such as the bombing of two passenger airliners on August 24 and a suicide bombing outside a Moscow subway station on August 31. The Kremlin's insistence on linking the Chechen insurgents to external funding has helped to justify its refusal to negotiate with them.

Ramzan Kadyrov Becomes Chechen Prime Minister in March, 2005

In Chechnya political developments seemed to favor Russia. In early March 2005, pro-Russian President of Chechnya, Alkhanov, on signal from the Kremlin, appointed Ramzan Kadyrov, the son of the assassinated president Akhmat Kadyrov, prime minister of Chechnya. The elevation of the younger Kadyrov—he was only 29 years of age—was, to say the least, bizarre because he was so inexperienced and had questionable loyalty to Russia. But, he was willing to do its biddings in Grozny. Ramzan's real power in Chechnya derived from his leadership of a well armed and violent paramilitary force. He would use this force to preserve Russian authority in Chechnya in place of the Russian army, which Putin wanted to remain on the sidelines to avoid worsening the strong anti-Russian feelings. Like his father, Ramzan had decided to work with the Russians despite his separatist beliefs. Indeed, Ramzan terrified most Chechens because of his readiness to use any kind of brutality to keep the peace and preserve the Russian position in his country. Kadyrov is believed to have amassed a huge fortune from extorting kickbacks from the illegal sale of Chechen oil. He is also accused of propagating his father's and his own personality cults. However, Kadyrov had the support of former Russian President Vladimir Putin; and he was even awarded the Hero of Russia medal, the highest honorary title of Russia. Kadyrov has been credited with finally launching the federally-sponsored renovations of the Chechen capital Grozny, which was nearly obliterated by the fighting.

Ridiculously, Ramzan Kadyrov was named an honorary member of the Russian Academy of Natural Sciences, although no contribution to natural (or any other) sciences by Kadyrov has ever been produced. At this writing (early 2009), Ramzan Kadyrov is still the chief instrument of Russian power in Chechnya. Ironically, unlike pre-war Chechnya, today it is an ethnically cleansed republic with virtually no ethnic Russian presence, except for the military. In 1989,

when Chechnya and Ingushetia were part of a single ethnic homeland called Chechen-Ingush Autonomous Republic, ethnic Russians accounted for 23 percent of the population; together with other groups using Russian in everyday life (e.g., Armenians, Ukrainians, Tatars, etc.), they accounted for 36 percent of the total population and more than half of the population of the city of Grozny. According to the Census of 2002, Russian-speaking people of Chechnya now constitute just 3.2 percent (48,000 out of 1,491,000) of the total population.

The 2005 Chechen Attack on Nalchik

In the aftermath of Beslan there seemed to be an escalation of Chechen desperation. In response to Russian carpet-bombing of Chechnya, the young men of destroyed towns and villages fled to the mountains to join the insurgency led by Basayev. Moreover, women now played an increased role in the insurgency against the Russians. They were called "black widows" and were sisters or wives of dead Chechen fighters. They swore to avenge the death of their men, even if that meant sacrificing their own lives. The residue of the population of these places was primarily old people.

In this environment, in mid October 2005, Basayev led another terrorist attack in Nalchik in the predominantly Muslim south Russian Federation republic of Karabardino-Balkariya. The attack involved hostage-taking and the death of at least 137 people and may have been meant to radicalize the economically depressed Islamic region in the Russian Caucasus to join an anti-Russian campaign to free it from Russian control. The attack also led to the acquisition of weapons belonging to local police authorities.

The attack on Nalchik, like the one on Beslan, seemed to have been carried out under the noses of local police authorities, who seemed uninformed about the activities of militants in areas within their jurisdiction. It did not help the cause of security in Nalchik that the president of Karabardino-Balkariya had been accused of corruption and replaced at the end of September 2005. In any event, the Kremlin sent Russian military forces into Nalchik to restore order.

November 2005 Chechen Parliamentary Elections

In an effort to convince the outside world as well as Chechens that the functioning of government in Grozny continues despite the unrelenting insurgent attacks, the Kremlin managed parliamentary elections in Chechnya in November 2005. The outcome of the elections was predictable. The United Russia Party, loyal to the Kremlin, won about a 60 percent majority in both houses.

While the holding of any elections in a severely disrupted political environment of the kind Chechnya has had for over a decade is a signal event—the European Union called the elections "a step forward for democracy in the rebel region"— there was evidence of a manipulation of the vote counting to make sure the results of the election were what the Kremlin wanted. The heavy Russian military presence in the country to assure security for the vote also assured the Kremlin got the result it wanted, according to Russian human rights activists.

But, the results of the election also reflected a political reality that militant, separatist Chechen nationals have been reluctant to accept. An overwhelming majority of Chechen people are tired of the ceaseless fighting with Russia and are demoralized by the dilapidated environment in Grozny and other centers of Chechnya as a result of the war. Public and private buildings are run down or simply rubble; the modest infrastructure the country had prior to the first Chechen war has been shattered and the standard of living is very low.

Chechens know that the only source of extensive material help to restore normality to Chechen society is Russia, which has refused to undertake anything in the nature of a broad program of reconstruction until the issue of Chechnya's status in the Federation is settled to Moscow's satisfaction. The West has shown little interest in helping Chechens recover from the effects of over a decade of fierce fighting on their soil.

The Death of Basayev

On July 10, 2006, the Russians finally succeeded in locating and killing the notorious Chechen terrorist Shamil Basayev. The FSB[3] had tracked him to Ingushetia, where he had been engaged in weapons collecting activity. A contingent of Russian Special Forces blew up a truck allegedly carrying a large quantity of explosives to be used, according to the Russians, in an act of terrorism to coincide with an upcoming meeting of the G-8 nations hosted that year by Putin in St. Petersburg. Basayev was killed in this action.

After Basayev's death, Valery Kuznetsov, the senior prosecutor in Chechnya, observed that the combat activity of the insurgents had declined. But, Akhmed Zakaev, a Chechen insurgent envoy in Britain thought differently, telling Russia's Ekho Moskvy radio station that the situation in Chechnya would not change, not until there was a new "mutually acceptable" relationship, as he put it, between Russia and Chechnya. And Aleksei V. Maleshenko, a scholar at the Carnegie Moscow Center and a specialist on Chechnya, predicted an "atomized" but continuing anti-Russian insurgent activity in Chechnya. Arguments suggesting that a Russian victory was not imminent, even with Basayev out of the way, seemed plausible in 2006. In later years, however, insurgence within Chechnya subsided, but similar acts of anti-Russian defiance became more frequently recorded in the neighboring Muslim republics, particularly in Ingushetia and Dagestan.

Anti-war Protests, Alleged War Crimes, and Russian Public Opinion

Recent public opinion polls suggest that the majority of Russian voters do not support Chechen independence. There is a sense throughout the country that giving Chechnya its independence would end up encouraging other non-Russian republics, which are constituent parts of the Russian Federation, to seek separation. There seems to be an equally strong undercurrent of public opinion opposed to the war and in favor of bringing it to an end. There have been public anti-war demonstrations in places like Moscow's Pushkin Square. However, anti-war protests were much stronger during the first Chechen war (1994–1996) than during and after the second one, whose active phase took place in 1999 and 2000

(thereafter becoming a low-intensity smoldering conflict). Since 1999, only a few devout opponents of Russia's handling of the conflict in Chechnya have spoken against it, with much of the Russian public being either indifferent or hostile to Chechens and other peoples from the Caucasus. Needless to say, several powerful terrorist attacks outside Chechnya did not endear Chechen fighters to the Russian public, which remains divided even in cases when Chechen war protesters fall victim to heinous crimes such as the assassination of the journalist Anna Politkovskaya (October 2006) and the double killing of the attorney Stanislav Markelov and the journalist Anastasia Baburova (January 2009). Both events took place in Moscow during daytime.

Politkovskaya was widely acclaimed for her reporting from Chechnya and won a number of prestigious awards for her work. She frequently visited hospitals and refugee camps in Chechnya to interview the victims. Her numerous articles, critical of the war in Chechnya, described abuses committed by Russian military forces, Chechen rebels, and the Russian-backed Chechen administration led by Akhmat Kadyrov and his son Ramzan Kadyrov. Politkovskaya chronicled human rights abuses and policy failures in Chechnya and elsewhere in Russia's North Caucasus in several books on the subject, including *A Dirty War: A Russian Reporter in Chechnya* and *A Small Corner of Hell: Dispatches from Chechnya,* which painted a picture of a brutal war in which thousands of innocent citizens have been tortured, abducted, or killed at the hands of Chechen or federal authorities. One of her last investigations was the alleged mass poisoning of hundreds of Chechen school children by an unknown chemical substance of strong and prolonged action, which incapacitated them for many months.

Politkovskaya was shot dead in the elevator of her apartment building on October 7, 2006. At this writing, the investigation into her death has not yet been finished. Shortly after her death, Vladimir Putin, on a state visit in Germany, said that while he considers the assassination of Politkovskaya to be a heinous crime that has to be fully investigated, "her political influence within the country was insignificant and she probably enjoyed more visibility among human rights activists and in Western media. In this regard I think that her assassination caused more damage to Russian authorities and Chechen authorities in particular than her publications." While some Russian commentators regarded these comments as cynical, they appear to be, sadly, on target.

Yet another high-profile (and double) murder in Moscow, in January 2009, is most probably related to a legal attempt to undo the early release of colonel Yuri Budanov, a Russian military officer convicted by a Russian court of war crimes in Chechnya. Budanov is a highly controversial figure in Russia: despite (or as some say owing to) his conviction, he enjoys widespread support among Russians, as some polling results reveal. At the same time, he is broadly hated in Chechnya, even by the pro-Russian Chechens. Budanov's tank regiment had been encamped just outside the Chechen village of Tangi-Chu since February 2000, and Budanov himself had a notorious reputation among villagers. He reportedly searched and looted several homes in Tangi-Chu and threatened to torch several others. From 2001 to 2003, Russian courts tried Colonel Yuri Budanov on charges related to the March 27, 2000 kidnapping, rape (an allegation later withdrawn by the prosecution), and brutal murder of Elza Kungaeva, an 18-year-old Chechen girl. The military has portrayed Budanov's behavior as an exceptional example of wanton criminality by a serviceman. However, the kidnapping and murder (and possibly sexual assault) of Kungaeva allegedly reflect a pattern of violations perpetrated by some members of the federal forces that has been documented by Human Rights Watch and other nongovernmental organizations. A resolution adopted in April 2000 by the United Nations Commission on Human Rights called for Russia, among other things, to establish a national commission of inquiry to investigate such crimes, but Russia has not fulfilled the resolution's requirements.

Budanov was arrested on March 29, 2000. In relation to the case of Kungaeva, he was charged with three crimes: kidnapping resulting in death, abuse of office accompanied by violence with serious consequences, and murder of an abductee. No charges have been brought expressly for the beating and torture Kungaeva endured prior to her death. He was also charged in the beating of a subordinate officer, threatening superior officers with a weapon, and other crimes. Budanov claimed that he detained Kungaeva on suspicion of being a sniper, and that he killed her during interrogation. The investigation, however, reportedly found that no member of the Kungaeva family had in any way been suspected of involvement in anti-Russian activity. The trial began on April 9, 2003, in Rostov-on-Don. Legal proceedings against Budanov, who underwent several retrials, lasted a total of 2 years and 3 months. In a controversial decision, Budanov was initially found not guilty by reason of temporary insanity on December 31, 2002, and committed to a psychiatric hospital for further evaluation; the length of the treatment would have been decided by his doctor. However, in the beginning of March 2003 the Supreme Court invalidated the sentence and ordered a new trial. This took place in the same place but with a new judge. The sentence of 10 years of imprisonment was given on July 25, 2003.

In December 2008, a court granted Budanov an early release and he was released on January 15, 2009. The decision has been protested by Chechnya's human rights ombudsman, Nurdi Nukhazhiyev, who has accused Russian judges of "double standards" with regard to Russians and Chechens.

The lawyer for the Kungaeva family, Stanislav Markelov, who had attempted a last-minute appeal against the release of Budanov, was shot dead in Moscow on January 19, 2009, along with Anastasia Baburova, a 25-year-old journalist for *Novaya Gazeta,* the same independent newspaper that Anna Politkovskaya worked for. Even the pro-government daily *Izvestia* bemoans the fact that many Russians cannot conceal their joy in conjunction with Markelov's and Baburova's assassinations. This joy is well documented in many Russian internet blogs. Some people even uncorked a bottle of champagne at the place of the murders and openly celebrated.

THE 2008 PRESIDENTIAL ELECTION

Putin seemed to have decided by the end of 2005 that he would not support proposals to change the Russian constitution to let him run in 2008 for a third term, convinced that such a

move carried serious political liabilities for him and his United Russia party. Rather, in mid November 2005, some inkling of whom Putin might support for the president was evident in a modest cabinet shake-up that strengthened the influence of Defense Minister Sergei Ivanov. Ivanov was promoted to the position of first Deputy Prime Minister ostensibly to improve a variety of conditions having to do with personnel and weapons inside the Russian military. This promotion turned out to be misleading, as Dmitry Medvedev, another deputy prime minister, later (December 2007) became the presidential candidate of the United Russia Party. Other registered presidential candidates were the communist leader Gennady Zyuganov; the leader of the Liberal Democrats Vladimir Zhirinovsky; and the leader of the virtually unknown Democratic Party of Russia, Andrei Bogdanov. According to the Russian constitution, presidential candidates can appear on the ballot if they are promoted by political parties represented in the Duma; alternatively, a candidate can be registered if he or she is officially proposed at a meeting of a certain political affiliation and presents at least two million signatures in his or her support. Such signatures are subject to verification: if more than 5 percent of all signatures are certified as forged or not representing actual citizens of Russia, the candidate is denied registration. Officially, this is why former Prime Minister Kasyanov, who had become a vocal critic of President Putin, was denied registration. The electoral commission found 13.36 percent of all signatures collected in his support to be forged. In contrast to that, Andrei Bogdanov, who very few Russians had ever heard of, appeared on the ballot. Other high-profile candidates who were denied registration included former Soviet dissident Vladimir Bukovsky (as he did not live in Russia for ten straight years prior to his promotion as a candidate—another provision of the constitution) and a long-time world champion in chess, Garry Kasparov (as no meeting of his political affiliation, *The Other Russia,* took place for the simple reason that in all of Moscow no owner of a public arena like a theater, or stadium, or a large club, would agree to provide premises for such a meeting). Neither of these potential candidates, nor Mr. Kasyanov, could count on any significant public support, as people normally vote for candidates marketed by the media, especially TV. Russians tend to vote for whomever is chosen by the existing political elite, which in the case of a wildly popular sitting president (Vladimir Putin) makes a whole lot of sense. It would then appear that the improbability of success of some candidates at the ballot box casts all designs to disqualify those same candidates early on as irrational. Rationality, however, is not among the main features of Russian politics (or any politics for that matter), whereas vindictiveness to and humiliation of the unwanted candidates by the dominant political machine are. Also, the organizers of the event apparently wanted to avoid any surprises and did not want to create additional reasons for *foreign* media to parade certain personalities. This particularly applied to Kasyanov and Kasparov, although Kasparov was a more frequent face on *Fox News* than on any Moscow-based network anyway. As for Russian TV, the overall breakdown of TV time devoted to each registered presidential candidate by five Moscow channels was 46 percent for Medvedev, 20 percent for Zhirinovsky, 19 percent for Zyuganov, and 16 percent for Bogdanov. Medvedev was marketed to the Russian electorate not so much as a self-styled political

THE SAGA OF MIKHAIL KHODORKOVSKY

Mikhail Khodorkovsky (MK) is currently a jailed Russian tycoon. From 1997 to 2003, he was the CEO of Yukos, the petroleum giant; and by 2004, he was the wealthiest man in Russia and the sixteenth wealthiest man in the world. Many domestic and international observers believe that MK was, and still is, persecuted for political reasons; and often the very same people do not doubt that accusations against him are warranted. Indeed, nothing is straightforward in MK's saga, and in many ways his experience reflects the ambivalent history of post-Soviet Russia.

The Soviet Period: 1963–1991

MK was born in 1963 in Moscow into a mixed Russian-Jewish family. His parents both worked as chemical engineers at a large Moscow industrial plant. MK also showed an interest in chemistry and attended a secondary school with advanced teaching of that subject. From an early age he had a habit of earning additional pocket money through menial jobs, such as garbage collector or helper in a bakery shop. In Soviet Russia, this was not common for a secondary school student. MK also attended a wrestling study group. In 1981, he entered the Moscow Mendeleyev College of Chemical Technology. In college, he became deputy chair of the Young Communist League college-wide cell, a high position and a clear testimony to his leadership skills and career ambitions. One must realize that the Young Pioneer Organization (for children from 9 to 14 years of age) and the Young Communist League (for those from 14 to 27) were the only venues of young public life in the U.S.S.R. Unlike the next step—the Communist Party (which MK also joined while in college)—these two associations were not selective: everybody was expected to enter them and yet only some would land a leadership post. Also, one must understand that, in Moscow, since probably the late 1960s, almost nobody joined these groups or indeed the Communist Party for purely ideological reasons. These institutions were the mainstream vehicles for career advancement, virtually the only game in town. While a college student, MK headed summer construction teams, that is, groups of students who would spend part of their school's summer break building various structures in the countryside (e.g., a barn, a house, a cowshed, etc.), usually under the guidance of licensed construction firms, for which they would earn money. While this practice was more common than gainful employment at pre-college age, it was far from universal. MK played a leadership role in those teams; for example, he would visit some collective farm in Siberia during his winter break in order to enter into contracts and negotiate work assignments and pay for the upcoming summer. In 1986, MK graduated with honors. At that time, Gorbachev's Perestroika was entering its second year, and working as a salaried employee of a state-run firm was no longer the only option, although the overwhelming majority of college graduates did precisely that. Such outlets for entrepreneurial activity as the Centers for Scientific and Technological Creativity of the Young (CSTCY) were set up in Moscow district Young Communist League committees. As a leader of one of these committees, MK organized one such center, a structure which could be contracted by an enterprise or R & D institution willing to outsource some of their projects. Such willingness was motivated by the fact that Soviet institutions were banned from transforming money in their bank accounts into salaries above meticulously prescribed norms (e.g., one's regular salary plus a small bonus for being unusually productive). They actually lacked material incentives to spur creativity, and one of Gorbachev's reform mantras was that this lack was the real reason that the Soviet Union was technologically backward. Setting up special units which would be able to actually transfer spare funds into wages was seen as a step in overcoming this backwardness. The CSTCY charged enterprises and other institutions contracting

with them a fee, which was initially as high as 25 percent of the money transferred by these institutions into CSTCY accounts for the purpose of paying wages. Wages and salaries were paid in cash only; even today, this is a usual but no longer universal practice in most post-Soviet countries, including Russia. Thus, in an over-regulated Soviet economy, a perfectly legal channel was created for money laundering albeit for a noble reason and under the ideological guidance of the Young Communist League. The CSTCY headed by MK was the most successful of all those created, and in 1988 its operations totaled 80 million rubles at the time when a ruble was worth about the same as one U.S. dollar. The next endeavor for MK, who as we have seen had a long history of going the extra step to push the limits of the possible, was to use his CSTCY profits to set up an independent bank as soon as that became legally possible. The Menatep bank was created in 1988 and became one of seven or eight new financial structures entrusted with carrying out the same grand idea of investing in new technologies. Before becoming a full-scale banker, MK had graduated extramurally from one of Moscow's leading economics colleges with a specialty in finance.

The Privatization Era: 1992–1996

On the eve of Yeltsin's privatization initiative, MK was one of several Russian entrepreneurs uniquely positioned to take advantage of it. "Taking advantage" was not seen as something sinister at the time. Quite the contrary, it was part of the long-hoped-for ideological reversal whereby money made by the few would soon have a pervasive trickle-down effect; and Russia would bridge the gap separating it from the West and become a prosperous country. In 1992, MK co-authored a book titled *A Man with a Ruble* (read: *A Man with Money*), in which he shared his pioneering experience in business and speculated about the nature of the relationship between business and power. In 1992, he also signed a collective appeal by Russia's leading entrepreneurs to the government to establish fair rules for doing business. In 1993, without interrupting his duties as the director of the Menatep bank, MK became deputy minister of fuel and energy in the government of Viktor Chernomyrdin, a position he held for a year. In the meantime, his bank announced that it would not make a one-time investment of less than $1,000,000. The areas in which it was active were the petrochemical, food, metallurgy, and construction industries, as well as weapons production. In 1993, Menatep received a no interest, one trillion ruble loan from the government (in January 1993, the exchange rate was 500 rubles for $1). By 1995, the so-called voucher privatization was over, but living standards kept declining. Privatized enterprises and their new owners dodged taxes, so health care and education received incomparably less from the state budget (additionally strained by war in Chechnya) than they had under the Soviets. The 1995 gross domestic product in Russia was just 65.4 percent of that in 1991, and thousands of enterprises closed down, unable to sustain their erstwhile market niche in part because of cheap imports and in part because the state could not finance the production of weaponry at the previous scale. Some important industries—notably oil, gas, ferrous and non-ferrous metallurgy—were still largely under state control yet some striking personal fortunes already had been created. This seeming contradiction—between a lean state budget and the emergent class of New Russians (read: newly rich)—was resolved through the legendary loans-for-shares scheme, of which MK became one of several major benefactors. Simply put, the state solicited loans from super-rich Russians and used some of the state-owned enterprises as collateral. The controlling shares of those enterprises were transferred to the lenders' hands. The condition of a deal was that should the state fail to repay those loans within a year, the enterprises in question would be auctioned, and the proceeds would cover debts.

THE SAGA OF MIKHAIL KHODORKOVSKY

The state indeed failed to repay loans, and the enterprises were auctioned. However, the lenders themselves controlled those auctions and thus were able to obtain enterprises at exceedingly low prices. Thus in 1996, MK was able to obtain 78 percent of the shares of Yukos (at the time Russia's second biggest oil business but sitting on the country's largest share of proven oil reserves) for $309 million through some offshore shell companies. Months later, Yukos traded on the Russian stock exchange at a market capitalization of $6 billion. At the top of the Yukos ownership structure was the holding company, Group Menatep, registered to a Gibraltar post office box. MK owned 28 percent of Menatep; and Menatep owned Yukos Universal Limited, which owned 61 percent of Yukos. Menatep also owned the intricate web of shell companies, in and outside Russia, involved in the Yukos tax evasion scheme.

Russia's privatization from which Khodorkovsky benefited so much was blessed by the IMF. One of its fiercest critics was the future Nobel laureate Joseph Stiglitz. Read his April 2003 article from *The Guardian,* pointedly titled "The Ruin of Russia."

Khodorkovsky as Head of Yukos: 1996

To be sure, MK was not the only benefactor of the loans-for-shares scheme. At least eight other persons emerged from this scheme as leading Russian oligarchs. Together they funded Boris Yeltsin's 1996 presidential reelection campaign; and Yeltsin—whose approval rating had declined to single digits—still managed to defeat his communist rival, Gennady Zyuganov. To persuade voters to cast their ballot for the lesser of two evils required hugely expensive air time on TV, and it was provided.

As the new owner of Yukos, MK resorted to his background as a chemical engineer. He personally toured all the oil extracting units and refineries owned by Yukos and found that the company was in bad shape: theft was pervasive, Yukos's gas stations were in the hands of outright criminals, and the company's bosses had signed some very suspicious oil sales deals. Within a short period of time all leading managers were replaced and payments to local power elites, including law enforcement, were cut. Instead, Yukos developed its own powerful corporate security service. As a result of the 1998 Russian financial crisis, when the ruble lost most of its value, the Menatep bank was unable to repay loans in hard currency and lost its license. "Three banks, the Standard Bank of South Africa, Japanese Daiwa Bank, and German West LB Bank, had lent $266 million to Menatep secured by Yukos shares. MK offered oil instead. They refused and took possession of the shares. They dumped the shares very quickly, (collecting less than half of their loan), prompted in a panic sale by MK public threats of massively diluting their stake with new shares. While lawful in Russia at the time, issuing new shares would not have been so in most of the developed world. Yukos also sold shares in its main production subsidiaries to offshore shell companies believed to be linked to MK. Daiwa and West LB, suspecting they would end up with nothing if they persisted, sold out to Standard Bank in mid-1999, which in turn exited Yukos at the end of 2000. The two deals gave MK, Menatep, and Yukos terrible notoriety in Western financial circles. Only in 2003 did MK regain the confidence to return to Western banks with loan proposals." After the 1998 crisis, Western bankers were reluctant to do business in Russia. This is when MK and some of his colleagues realized that, in order to restore or rather gain their reputations, they would have to make their business empires more transparent. Yukos took the lead in disclosing its allocation of shares. That openness, together with a proven willingness to pay dividends, allowed Yukos to attract Western managers and Western loans. At the same time, Yukos resorted to various corporate tax optimization techniques in order to minimize payments into regional and federal treasuries.

Beginning in 1999, MK became involved in a lobbying campaign in the Russian Duma, providing its members with incentives to adopt business-friendly and particularly Yukos-friendly laws. During the 1999 parliamentary elections, MK rendered generous financial support to left-of-center political parties, including the Communist Party of Russia. In 2000, he funded the electoral campaign of Boris Zolotarev, Yukos's associate, who was a gubernatorial candidate in the Evenk Autonomous District of Eastern Siberia. At that time, regional governors were still subject to election, a rule Putin subsequently outlawed. In fact, the race in the Evenk AD was a fight for administrative control over an area whose oil riches had not yet been tapped. In 2001, together with some other shareholders of Yukos, MK registered his Otkrytaya Rossiya (Open Russia) Foundation. Its function was to issue loans facilitating the (a) spread of information technology, (b) civic education of adolescents, including teaching the basic principles of democracy, (c) spread of objective information about Russia's politics and civil society, (d) monitoring and perfecting lawmaking activities, and (e) research in social sciences and humanities. Some observers saw in the Open Russia Foundation the kernel of a prospective political party. In 2003, MK announced his commitment to finance two democratic parties, while a former Yukos manager opted to use his own funds to finance the Communist Party. With Yukos's capitalization reaching $10.3 billion by December 2003, it was becoming clear that MK, its president and major shareholder, was exceedingly capable of making things happen not only in Russia's economic life but in its politics as well. This put MK (as well as potentially other oligarchs willing to affect political life) on a collision course with another group of power brokers, that consisting of the associates of the formerly all-powerful KGB, the Soviet secret police. Critically weakened during the dissolution of the Soviet Union and sidestepped during the privatization of the 1990s, this group received an unexpected boost when one of their own, Vladimir Putin, succeeded Yeltsin as president. It is unlikely that Putin was handpicked by Yeltsin *because* of his KGB background; it is more likely that the chosen successor was sworn to refrain from ever reconsidering the results of privatization (from which Yeltsin's family benefited), let alone from persecuting Yeltsin. Nevertheless, when Putin began to promote people to positions of power from his own circle of trust, many—in fact most—of the newly appointed turned out to be former KGB associates. These people were now positioning themselves as the true Russian patriots and riding a wave of popular disappointment in the way the post-Communist transformation of Russia had unfolded so far. Most importantly, these people, hitherto devoid of MK's financial power, now controlled something that, in Russia, is even more important than money; they controlled law enforcement and the courts. As early as March 2003, the so-called National Strategy Council published a report titled "An Oligarch Coup Is Brewing in Russia." In that report, oligarchs like MK were named the major enemies of Russia.

2003 Arrest and First Trial

MK was first summoned to the Office of General Prosecutor (GP) on July 4, 2003, to testify as a witness in conjunction with the upcoming trial of Platon Lebedev, MK's business partner accused of financial fraud allegedly committed in 1994. There were other signals, such as a request by some obscure member of Duma to verify whether or not Yukos had paid its 2002 taxes in full. On July 11, 2003, on GP's warrant, Yukos's archive was searched and several people close to MK advised him to leave the country. MK declined. However, together with his family, he visited the United States in the summer of 2003 but promptly returned as planned. On October 25, 2003, MK was arrested by a Federal Security Service

(continued)

THE SAGA OF MIKHAIL KHODORKOVSKY (*continued*)

(successor to the KGB) team in the Novosibirsk airport, where his plane had stopped for refueling on his way to Irkutsk. He was accused of fraud and tax evasion. Much of the incriminating activity either took place or originated in the 1990s, when every would-be oligarch and indeed every business owner in Russia broke the law, if only because it did not yet reflect the new realities of a market economy and because getting rich in a hitherto egalitarian society had suddenly become a virtue. MK's case and that of Platon Lebedev (arrested on July 2, 2003) were consolidated into one trial; and on May 31, 2005, both men were sentenced to nine year prison terms on seven counts, including fraud, failure to carry out earlier court rulings, embezzlement, and tax evasion—all or most by a deliberately criminal group first organized by MK in 1994. On appeal, the sentence was reduced to eight years. MK was kept in a remote penitentiary in East Siberia until December 2006, at which time he was transferred to the provincial capital of Chita. While doing time, MK had to work as a sewing machine operator producing gloves and was several times punished by solitary confinement for violating orders. He also wrote several articles and gave written interviews (see Khodorkovsky's Publications). In April 2006, another inmate assaulted MK, cutting his face with a knife. The same inmate later accused MK of sexual solicitation, but the court threw out the case.

In the meantime (December 2004), the Russian government "auctioned" the major oil producing division of Yukos (accounting for 60% of its output) to the highest bidder, which turned out to be the Baikal Finance Group, for $9.3 billion. This firm, unheard of in Russia as well as on international markets, was a mystery shell company registered at the same address as a grocery store in the provincial Russian town of Tver. The takeover bid left analysts speculating hard over the company's financial backing. Most suggested that the Baikal Finance Group was the front for a Kremlin business. When asked about it at a press conference, Mr. Putin said that, to his knowledge, the Baikal Finance Group is owned by people that have long dealt with energy. However, three days later, the company disappeared, purchased by the Russian state firm Rosneft. Thus Yukos became effectively renationalized and its former owner, MK, deprived of any influence. Nobody doubts that MK broke the law but so did other oligarchs, yet he alone was singled out, which is what is subject to speculation. After all, Russia's weakness has never been the absence or imperfection of laws, but their application. Why indeed was there such a highly selective zeal on the part of the Putin administration to apply the law to MK? Some observers claim that it was MK's alleged desire to sell about one-third of his business to an American firm and Putin's adamant resistance to losing national control over a strategically important industry. Other observers point to MK and Putin bickering over corruption during the February 2003 meeting of the Union of Russian Industrialists and Entrepreneurs. Still others claim that Putin personally demanded that MK stop financing political parties and MK rejected this demand. Indeed, after MK's arrest, two democratic parties (one right- and another left-of-center) could not get any representation in the Duma, and the number of seats that the Communist Party won in 2003 (12.6%) was half of that in 1999 (24.3%).

Khodorkovsky's Second Trial

In February 2007, MK and his partner Lebedev were arraigned and charged with grand larceny and money laundering—a forerunner of the second trial and of the authorities' reluctance to ever let MK out of jail. The new accusations were substantiated in 14 volumes; and in January 2009, both of the accused testified that they had not yet familiarized themselves with them. However, in February 2009, they were transported to Moscow, where on March 3 the new trial began. At this writing (September 2009), it is still going on. In April 2009, MK's attorneys stated that the new accusations are absurd, as MK is charged with stealing all the 350 million tons of oil that his company produced under his leadership. And if all of it was stolen, then how was Yukos able to pay salaries and wages, bore new oil wells, and pay more than $40 billion in corporate taxes? Ironically, for MK's elderly parents and his wife and three children, this new trial gave them the happy opportunity to see him daily in court, which they of course could not when he was imprisoned in Eastern Siberia. The second trial comes across as more of a farce than the first one, and there have been some changes in the general public's attitude toward MK. Whereas before much of that public gloated, because in their view MK "stole from the state" (and so did other oligarchs); today there is an outpouring of compassion, although not to the degree that would force the authorities to stop the farce. There was some hope that Russia's new president, Medvedev, who initially seemed a more liberal statesman than his predecessor, would do precisely that, but most now agree that the difference between the two is in truth slight and the reins of power are still in Putin's hands. As for Putin, the pursuit of MK is a personal vendetta.

Khodorkovsky's Publications

While in jail, MK was able to compose two political manifestos and give two extensive written interviews published by the Russian media. The first of his articles appeared in August 2005 and was titled *The Left Turn*. In that article, MK, who made a fortune out of Russian society by making a "right" turn, (that is, by turning from Marxism-inspired state socialism to wild capitalism), confesses that he has changed his political views and now believes that a return to "left" values such as a strong and well-funded social safety net and other facets of state paternalism is inevitable in Russia, where there is also a need to legitimize the 1990s privatization in the eyes of ordinary people. He writes that Russia's political future is actually in the hands of leftist political parties, including Communist, which would create a broad social democratic coalition. In November 2005, the sequel was published, in which the abovementioned ideas were detailed to the point of budgeting their implementation. In September 2008, in his interview with the English-language newspaper, the *Moscow Times,* MK expressed his support of Russia's actions during the August 2008 war with Georgia and the subsequent recognition by Russia of Abkhazian and South Ossetian statehood. After Barack Obama was elected president of the United States, MK published an article titled "New Socialism. A Left Turn, Part 3: Global Perestroika." In that article, MK avers that the world's response to economic crisis ought to be a global shift to neosocialism. In his responses to questions by the author, Lyudmila Ulitskaya, published in September 2009, MK describes, among other things, the psychological side of his ideological left turn. By his own admission, his attitude changed during the 1998 financial crisis in Russia when oil was worth $8 per barrel while production cost was $12 per barrel and many people working at Yukos-controlled oilfields suffered from malnutrition but still understood the overall situation, including MK's inability to pay them. Prior to that, he had viewed his entrepreneurial activity as pure gamesmanship, gambling; but prolonged contact with Russian people in torment reportedly changed his views.

In summary, MK's saga offers much food for thought about various aspects of the groundbreaking change sustained by Russian society since the breakup of the Soviet Union, including the exceptional role of an entrepreneurial personality in that change, the price of acquired personal wealth, and the price of pride.

personality but as the chosen successor of Vladimir Putin. As a personality, Medvedev came across as young, articulate, and intellectual but not folksy or with outbursts of dark (some say cynical) humor like his predecessor. To be sure, in terms of being articulate and well-informed, Medvedev's predecessor had already become a striking improvement over any Soviet leader since Joseph Stalin. Many observers paid attention to the fact that, though a part of the "Leningrad mafia" (that is, a member of Vladimir Putin's most trusted inner circle from his home base in Saint Petersburg), Medvedev has never worked for the KGB or its successor, the FSB. Along with his government post, Medvedev had been the Chairman of the Board of Directors of Gazprom, Russia's highly successful natural gas monopoly. All public opinion polls predicted Medvedev's victory by a huge margin in the election's first round. In Russia, a second round is conducted if none of candidates exceed the threshold of 50 percent of the vote. On the election day of March 2, 2008, Medvedev won 70.28 percent of the vote; Zyuganov won 17.7 percent, Zhirinovsky, 9.35 percent; and Bogdanov, 1.30 percent. The turnout was fairly high at 69.7 percent of eligible voters. The results were within 1 percent of predictions based on exit polls.

Upon winning the race on March 2, 2008 and becoming Russian president on May 7, 2008, Medvedev quickly suggested that Vladimir Putin be his prime minister, and the Duma endorsed Putin in that capacity. Also Medvedev made a couple of speeches that prompted speculations of a pending political thaw. In particular, Medvedev opted to fight corruption with renewed vigor and what he called the legal nihilism of Russians, that is, their contempt of law and lack of belief in a judiciary system. The thaw, however, was put on hold because of the Russian-Georgian war, also known as the 2008 South Ossetia War, the first-ever military conflagration between Russia and another former Soviet republic.

2008 RUSSIA–GEORGIA CONFLICT

The 2008 Russia–Georgia conflict, also known as the 2008 South Ossetia War, was an armed conflict between Georgia, on the one hand, and Russia together with separatists in South Ossetia and Abkhazia, on the other. It occurred in August 2008 and involved land, air, and sea warfare. The background of this conflict, discussed in more detail in the essay devoted to the Republic of Georgia **(see pp. 150–161 including the map on p. 150),** ought to be attributed, in part to lasting tensions between ethnic Georgians and Ossetians as well as Georgians and Abkhazians. Under the Soviets, Ossetians and Abkhazians lived predominantly in their respective ethnic homelands, known as an autonomous republic (Abkhazia) and an autonomous oblast (South Ossetia) within Georgia and subordinated to the Georgian government in Tbilisi. Both enclaves bordered the Russian Federation. Moreover, ethnic Ossetians had two autonomous units, one to the north of the main Caucasus ridge, in Russia, known as North Ossetia and the other to the south of that ridge, in Georgia, and known as South Ossetia. Shortly after the breakup of the Soviet Union, tensions between Georgians and Ossetians escalated into a military conflict. The 1991–1992 South Ossetia War left most of South Ossetia under the control of an unrecognized government backed by Russia.

Some Georgian-inhabited parts of South Ossetia remained under the control of Georgia. This mirrored the situation in Abkhazia. Already increasing tensions escalated during the summer months of 2008. On the evening of August 7, 2008, Georgia launched a large-scale ground- and air-based military attack on South Ossetia's capital, Tskhinvali. The events of August 7 remain a matter of debate and controversy. Russia responded by sending troops into South Ossetia and launching bombing raids farther into Georgia. On August 8, Russian naval forces blocked Georgia's coast and landed ground forces and paratroopers on the Georgian coast. Russian and Abkhazian forces opened a second front by attacking the Kodori Gorge, held by Georgia, and entered western parts of Georgia's interior. After five days of heavy fighting, Georgian forces were ejected from South Ossetia and Abkhazia. Russian troops entered Georgia proper, occupying the cities of Poti and Gori among others. Following mediation by then EU chairman, French president Nicolas Sarkozy, the parties reached a preliminary ceasefire agreement on August 12, 2008, signed by Georgia and Russia on August 15 in Tbilisi and on August 16 in Moscow. President Medvedev had already ordered a halt to Russian military operations in Georgia on August 12, but fighting did not stop immediately. After the signing of the ceasefire, Russia pulled most of its troops out of Georgia proper. However, Russia established "buffer zones" around Abkhazia and South Ossetia and check points in Georgia's interior (Poti, Senaki). On August 26, 2008, Russia recognized the independence of South Ossetia and Abkhazia. International monitoring personnel were deployed in Georgia on October 1. Following international agreements, Russia completed its withdrawal from Georgia proper on October 8. Russian troops remain stationed in Abkhazia and South Ossetia, including areas under Georgian control before the war, under bilateral agreements with the respective governments. A number of incidents have occurred in both conflict zones since the war ended, and tensions between the belligerents remain high.

The preceding, however, is only the factual side of events, which had a profound effect on Russian society. A significant part of the Russian political elite perceived the war with Georgia as a proxy war with the United States and launched a media campaign which drove the anti-Western, particularly anti-American sentiment of the general public to the highest level since the end of the Cold War. And because *during* the Cold War, anti-American propaganda provoked mixed reactions at the grassroots level, it may be that during, and in the wake of, the 2008 war with Georgia, anti-American feelings were at the highest level ever. This is because Georgia had been receiving American aid, American instructors had been training Georgia military, and a campaign to join NATO had been in full swing during the months and weeks immediately preceding the war. In 2005, George W. Bush proclaimed Georgia a beacon of liberty, which no Russian could take seriously, particularly given that the beacon of liberty status was conferred on precisely the day (May 10, 2005) when the first oil was pumped from the Baku end of the Baku-Ceyhan pipeline passing through Georgia. The 2002–2005 construction of this pipeline was perceived as a thorn in Russia's side, as it offered the first route of Azeri and potentially of Central Asian oil not to pass through Russia. This transit route enhanced the geopolitical significance of Georgia in Western eyes, and that was not lost on Russians. The traditional big-brother condescension in regard to former republics played its role as well, as did a habit of construing their legitimate search for other-than-Russia sponsors as ingratitude. Perhaps a specific role of Georgia in the division of labor within the Soviet Union (or rather public image of that role) facilitated such idiosyncratic views.

A republic with a relatively high quality of life during the Soviet period, Georgia was particularly hard hit by the severing of ties with Russia in the early 1990s. Previously, Georgia extracted benefits from its subtropical climate, supplying Russia, Ukraine, and Belarus with early vegetables and tangerines and hosting hundreds of thousands of summer vacationers who flocked to the Black Sea coast from the north. While some of those vacationers were accommodated in the Soviet style vacation infrastructure (e.g., houses of rest and sanatoria) many more were renting flats and houses from Georgians. This created a situation whereby ordinary people in parts of Georgia could live far better lives than the centrally-planned economy of Georgia alone would allow. The idyll terminated in December 1991. The following year, very few Russians, affected by their shock therapy reform, could afford to go to Georgia. Loss of Russia as the prime financial donor devalued local salaries and pensions, which suddenly became worthless, and with that hundreds of thousands of Georgians began to leave their country in pursuit of employment. There are at least half a million Georgians in Russia alone. Considering that the total population of the republic in 2008 was 4.6 million people, this is a lot. In Europe, a significant Georgian population now exists in Greece.

Sudden pauperization led not only to escalation of old inter-ethnic grievances but also to the search for new financial donors and sponsors. During the reign of Eduard Shevardnadze, Georgia was able to only modestly succeed at that, as the republic was hardly a transition economy showcase. Beset by poverty and pervasive corruption, Georgia languished in the shadow of more prosperous Russia and even of oil-rich Azerbaijan. The situation changed with the Rose Revolution of 2003, when the young American-trained lawyer, Mikheil Saakashvili, came to power. A significant source of funding for the Rose Revolution was the network of foundations and NGOs associated with American billionaire financier George Soros. The Foundation for the Defense of Democracies reports the case of a former Georgian parliamentarian who alleges that in the three months prior to the Rose Revolution, "Soros spent $42 million ramping-up for the overthrow of Shevardnadze." Speaking in Tbilisi in June of 2005, Soros said, "I'm very pleased and proud of the work of the foundation in preparing Georgian society for what became a Rose Revolution, but the role of the foundation and my personal [role have] been greatly exaggerated." But according to Salomé Zourabichvili, Georgian Foreign Minister from March 2004 to October 2005, "these institutions were the cradle of democratization, notably the Soros Foundation. . . . [A]ll the NGOs which gravitate around the Soros Foundation undeniably carried the revolution. However, one cannot end one's analysis with the revolution and one clearly sees that, afterwards, the Soros Foundation and the NGOs were integrated into power." In such a way, the American role in the geopolitical repositioning of Georgia became prominent, and that is what Russians were up against, at least in their minds.

Yet one more factor behind popular Russian reaction to the war with Georgia was that, after humiliating defeats in Chechnya and the habitual image of an ailing and dysfunctional Russian military sustained in public view since the early nineties, the ultimate show of Russia's military superiority came across as a long-desired outcome. It is little wonder then that Russian patriotism tinged with anti-Western sentiment and xenophobia rose to a new height. "We did them in," was the prevalent feeling. "Russia is rising from its knees."

PROBLEMS IN THE RUSSIAN MILITARY

One problem-ridden sector of the Russian military is the continuing reliance on a draft. All Russian males aged 18, except those unfit to serve for medical reasons, are supposed to serve for two years as rank-and-file soldiers. Conditions for Russian draftees remain tough. Draftees are not paid by the state, and they are often subjected to hazing. Second-year draftees indulge in hazing first-year draftees that occasionally reaches such extremes that it drives new recruits to suicide. This subculture of criminality serves the purpose of controlling raw recruits. Junior officers, distracted by the inadequacies of pay, housing, and other necessities of life, do not have the will to deal with this hazing. At the same time, senior officers who know quite well the evils of barracks life for draftees turn a blind eye or say that it is impossible to control.

In 2005, 16 soldiers were officially listed as killed in brutal hazing incidents, but that figure understates the hazing menace which allegedly involved over a thousand Russian soldiers who died as a result of various "crimes and incidents," as the military puts it. The hazing incidents have been referred to as the "rule of the grandfathers," senior soldiers responsible for subjecting recruits to mean spirited activities like conducting menial chores, giving up their food, and allowing themselves to participate in humiliating activities. A major consequence of this unrelenting and brutal hazing is that 18 year olds subject to the draft go to great lengths to avoid induction. Army efforts to address the problem are meager and ineffective. Rarely are guilty soldiers punished and their officers deprived of command. In the case of one exceptionally brutal hazing incident at the end of 2005, in which a young recruit had to have both his legs and genitals amputated to save his life, Putin reportedly called the case "horrible" and said the Ministry of Defense would be ordered to improve discipline.

Whether hazing or the many other problems plaguing Russia's military are resolved remains to be seen. A small but potentially significant step toward such improvement seems to have been taken by the government in the spring of 2006, when the Duma voted overwhelmingly to approve a reduction of the time draftees must serve in uniform from two years to one year. But, that concession was offset by a curtailment of deferments, which no longer will be given to a number of social groups such as rural doctors and teachers, athletes, artists, and other cultural figures, as well as young men with pregnant wives or very young children. The army still remains no place for the country's youth under any circumstances.

It is little wonder that Russian youths particularly dread eventual assignment for duty in areas where there is fighting, such as the Caucasus. Draft dodging is common; only about

30 percent of draftees show up for induction. Many are ready to do anything, with the help of family and friends, to escape the draft. Bribing physicians for false diagnoses that could then "legitimately" save a young man from the draft is a common practice in major urban centers. Such bribes are usually in the thousands of dollars—that is, extremely expensive for most families, and yet families are committed to earning and borrowing money for that purpose. Note that in 2008, the average monthly wage in Russia was only 14,354 rubles or $574 and that year was highly successful for the Russian economy. Subsequent devaluation of the ruble resulted in an average monthly wage now (February 2009) equal to $398. According to a September 2009 statement by General Vassily Smirnov, Deputy Chief of the General Staff of the Russian Army, there are more than 100,000 draft dodgers. This is quite a bit considering that the overall number of the fall 2009 draftees was expected to be slightly below 300,000. More than 10,000 people were dodging the fall 2009 draft abroad.

The Russian army has about 1.5 million people in uniform, with many officers convinced that what is left of the army lacks the strength to defend and protect the Russian state. Some military and political leaders probably believe that the United States no longer takes the Russian military seriously, no longer considers it a mortal threat.

What has most disturbed and demoralized the military, in particular its officers corps, is the deterioration of its status in post-Communist Russian society. In the view of the officers corps, the military has been stripped of its mission; and its history of greatness dating back to the czarist era, when it was the chief instrument of Russian national growth, has been forgotten.

In Chechnya throughout the 1990s and well into the new century, control functions of fighting units have been chaotic and in many instances ineffective, with the left hand unaware of what the right hand was doing in fighting the Chechen separatists, who seemed a good part of the time much better equipped, trained, coordinated, and led than the Russian army and other Russian military personnel deployed in Chechnya. Chechnya was most certainly an eye opener for the Kremlin as well as for the military leadership in terms of the quality and quantity of Russia's military preparedness. One thing that Chechnya showed was that most military units were using equipment that was at least a decade old. In Chechnya, the army could not afford to buy modern Russian-made Akula attack helicopters. Paratrooper units could not practice jump-training because of aviation-fuel shortages. The *kontraktnik* soldiers—the nucleus of a future professional army that would not rely on the draft for recruitment—have yet to prove that they are of much help to the army. Since the mid-1990s, talk about military reform has been an on again and off again public discourse. The ultimate idea is to restore the military's health and efficiency. To this end, Vladimir Putin wanted a reduction of uniformed military personnel by 30 percent. He has been critical of what he called "an unwieldy and extravagant military machine." At first his plans to downsize the military met with fierce opposition from both high-ranking civilians and uniformed military personnel. He soon overcame the opposition, however, as military and law-enforcement expenses were eating up 35 percent of the federal budget even though there was not enough money to provide troops with training in the use of weapons or provide military pilots with sufficient flying practice.

In March 2003, Putin approved the Defense Ministry's plan to develop a volunteer Russian army. But the Russian military leadership is still reluctant to abandon compulsory military service. The opposition is ideological and political. Ideologically, the Russian generals are convinced that the army that Russia needs to defend itself at home and abroad must be a draft army because the country's youth will not come forward in sufficient numbers to serve if only because of the army's dreadful reputation for mistreating its draftees. From a political point of view, the generals believe that the Kremlin will never give them the money needed to develop an effective all-volunteer force—it is simply too expensive at this point in time; simply reducing the term of compulsory military service from two years to one would cost about $4.34 billion.

The conservative generals do acknowledge the need for reform of the army, but they want to achieve this reform by themselves. The Kremlin opposes giving the generals this freedom not only because of their conservatism on the issue of change but also because of rampant corruption among them. As a result, several years after the commitment was made to move forward with the transformation of the Russian army to a volunteer force and do away with the draft, not much has been done beyond the Duma's passing a bill in October 2003—with Putin's approval— restoring the Soviet tradition of mandatory military training in schools.

SOCIAL PROBLEMS AND POLICIES

The post-Soviet reforms have brought with them many social problems that threaten Russia's internal stability and material well-being. These problems include xenophobia, the striking deterioration in the national health care system, population decline, rising crime, and gender issues. Although concurrence in time does not necessarily imply cause-and-effect relationship, the fact that these and some other problems began to be recognized in the early 1990s and seem to have been worsening ever since is not a mere coincidence. If the reins of centralized control get suddenly loosened in what for many centuries was a top-down society, it would be naïve not to expect negative side effects. Norms of civility and mutual trust, in which Russian society had been so profoundly lacking, do not develop overnight. If on top of loosening state control, the delivery of public goods (pensions, socialized health care, schooling, and subsidized transport) lapses and income disparity explodes, then expectations of social harmony and profound joy (expected by some Western observers without adequate exposure to life in Russia) in response to the introduction of capitalism and democracy by fiat are well off the mark. Whoever harbored and nurtured these unrealistic expectations in regard to Russian society is to blame for the lack of their fulfillment, not society itself.

Xenophobia

Xenophobia in Russia appears in the form of negative attitudes and actions towards people who are not considered ethnically Russian. This includes anti-Semitism, a general demeaning attitude to northern indigenous peoples of Russia, and hostility towards various ethnicities of the Caucasus and Central Asia.

In May 2006, Amnesty International (AI) reported that racist killings in Russia were "out of control" and that at least 28 people had been killed in 2005. In 2006, AI registered 252 victims of racist crimes, of whom 21 died. In February 2007, President Vladimir Putin asked the Federal Security Service to combat racism, but hate crimes still increased. From January 1 to July 31, 2007, AI registered 310 victims of neo-Nazi and racist crimes in Russia; 37 of whom died as a result of attacks. According to the Moscow Human Rights Bureau, from January to March, 2008, 49 people were killed in assaults by radical nationalists, 28 of them in Moscow and its suburbs. The number of Russian neo-Nazis is estimated at 50,000 to 70,000, "half of the world's total." The director of the Human Rights Bureau, Alexander Brod, stated that surveys show xenophobia and other racist expressions affect 50 percent of Russians.

Whereas in the past, the most widely-discussed aspect of xenophobia in Russia was anti-Semitism, in recent years hostility to the people from the Caucasus and Central Asia captures at least as much attention.

In Russia, *Caucasian* is a collective term referring to anyone descended from the native ethnicities of the Caucasus. In Russian slang, Caucasian people and Central Asians fall into the category of *black*. This is not associated with skin color but with a non-Slavic facial appearance. Several pogroms directed particularly against Caucasian merchants and migrants have been reported in Moscow and other Russian cities. There was a pogrom on April 21, 2001, in Yasenevo Market in Moscow against merchants from the Caucasus, and well-organized attacks on Caucasian businesses and migrants in Ekaterinburg on September 9, 2004. Racially motivated attacks against Armenians and Azerbaijani have grown so common that the presidents of Armenia and Azerbaijan have raised the issue with high-ranking Russian officials.

Since the collapse of the Soviet Union, the rise of the Muslim population in Russia, and the recently ongoing Chechen war, many Russians have associated Islam and Muslims with terrorism and domestic crimes. In August 2007, a video of two ethnic Russian neo-Nazis beheading two Muslim men, one from Dagestan in the Caucasus and one from Tajikistan appeared on the Internet. In February 2004, a nine-year old Tajik girl was stabbed to death in Saint Petersburg by suspected far-right skinheads. In December 2008 an e-mail containing a picture of the severed head of a man identified as Salekh Azizov was sent to the Moscow Human Rights Bureau. It was sent by a group called Russian Nationalists' Combat Group and has led to protests from the Tajik government.

Unlike hostility to the people from the Caucasus and Central Asia, anti-Semitism is an old phenomenon in Russia, dating back centuries. Under the tsars, Jews were confined to the pale of settlement encompassing the westernmost part of the Russian Empire (mostly in what is now Poland, Belarus, Ukraine, and Moldova) and were not allowed to settle in cities beyond that pale unless they were certified members of the merchant guild, lawyers, or medical doctors. Many Jews fell victim to pogroms, particularly in Ukraine and Moldova. Jews could enter institutions of higher learning in areas outside the pale only within the minuscule percentage share of the total number of college applicants specifically assigned to Jews. In February 1917, all these limitations were abolished, and during the first

two decades under Communist rule, Jews were able to raise their social standing immensely. In fact, they were repeatedly accused by Russian nationalists of being facilitators of Stalinist repressions. In the 1920s and 30s, the share of Jews in positions of power was indeed well above their share in population. This was in part because Jews represented one of the most literate communities in Russia; in 1897, in terms of literacy they were second only to ethnic Germans. Because scores of the most educated ethnic Russians joined the ranks of the Whites (the anti-communist and monarchist forces striving to undo the results of the Communist Revolution) and many educated survivors of the 1918–21 Civil War emigrated from Russia, the acute deficiency of an educated and even functionally literate ethnic Russian cadre conditioned the disproportionate reliance of the early Soviet rule on minorities, notably Jews. However, millions of ethnic Russians, Ukrainians, and other titular ethnicities of the union republics entered colleges in the 1920s and 30s; and upon their graduation, the ethnic imbalance in local, regional, and federal power bases all across the Soviet Union began to be eliminated. Once that change had occurred, the discrimination of Jews resumed. But while the Communist leadership mistreated Soviet Jews by limiting their access to positions of power and to some prestigious schools and jobs (particularly those related to trips abroad), it discouraged public expression of hostility to Jews. In the post-Communist era, however, an increasing number of Russians have publicly expressed their dislike of Jews, blaming them for everything that has gone wrong with the new Russia, especially in the economic sphere. Mikhail Veller, a secular Jew, popular Russian-language author, and frequent guest of talk shows on Moscow TV channels who, however, prefers to live in Estonia, writes that "in the collective unconscious of the Russian people, a Jew personifies the quintessential abstract evil, a reincarnation of Satan on this planet. . . . As charcoal in the stomach of the poisoned, the image of a Jew absorbs all toxins."

There is hardly a day when one cannot find individuals or groups of people speaking disapprovingly about or even demonstrating against Jews—this despite the fact that Jews today are an infinitesimal minority of far less than one half of one percent of the Russian population. According to the Anti-Defamation Committee of the Russian Jewish Congress, there are about 200 newspapers throughout Russia that are openly anti-Semitic. And Communists and nationalists openly complain about the supposed Jewish "control" of banks and the mass media. Many blame Jews for their economic hardship; they are further convinced that Jews are responsible for their problems when they look at the Russian leadership, which, they observe, quite erroneously of course, is made up mostly of Jews. It is true that there are now more people of Jewish extraction in high governmental places than at any time since the end of World War II—and some Russians consider this sinister.

Some positive developments deserve to be mentioned as well. Thus, state-sponsored discrimination of Jews, typical of the Soviet era, has been abolished, opening doors to the highest branches of academia, business, and government. Russian Jews, now estimated to number more than 500,000, have reestablished Judaism as a vibrant religion and culture in Russian life. There are Hebrew schools in Russia, and Jewish charities are free to work as they wish. Moreover, in the Putin years a marked revival of Jewish religious life continued, and Putin was frequently photographed in company of the ultra-Orthodox Rabbi Berl Lazar, who had moved to Moscow from America and is now heading the Lubavitcher community of Russia.

But while anti-Semitism no longer has government endorsement, it persists. In May 1998, a synagogue in the center of Moscow was bombed, injuring two people. Exactly one year later, a similar incident occurred, with the detonation of bombs near two Moscow synagogues. Jewish cemeteries continue to be vandalized. Furthermore, in November 1998, Communist deputies in the Duma, looking for someone to blame for the nation's economic illnesses, launched a bitter tirade against Jews in Russian society, saying that Jews in the government and in the media were aligned with outsiders in a campaign to undermine Russian nationalism and sap the country's strength.

One deputy, Albert Makashov, was particularly vicious in his proclamation that Russia's economic woes were the fault of the *zhidy* (a slur for Jews in Russian). The Duma was in no hurry to reprimand Makashov and rejected a resolution deploring racism and characterizing Makashov's remarks as "sharp and bordering on crudeness." Apparently only one Communist deputy supported this resolution. What was striking about the resolution was the refusal of democrats, in particular the liberal party Our Home Is Russia, to support it; many abstained from the vote.

More recent evidence of the persistence of anti-Semitism, even in some government circles, was an appeal in early 2005 by a group of communist and nationalist legislators to the prosecutor general's office demanding that Jewish organizations around the country be investigated and closed down for antisocial behavior. These lawmakers accused Russian Jews of carrying out ritual killings, controlling Russian and international capital, and staging hate crimes against Jews themselves. This appeal seemed timed to coincide with Putin's trip to Poland to commemorate the 60th anniversary of the liberation of Auschwitz, the infamous Nazi concentration camp in Poland.

Russian human rights activists along with liberal politicians denounced the appeal and demanded its withdrawal, which was promptly done if only because it had no real support in the government, which was embarrassed especially after a diplomatic protest by the Israeli Embassy in Moscow. Indeed, the Russian Foreign Minister said the appeal by the legislators had nothing to do with the official position of the Kremlin. The Israeli Embassy was quite stern in its rebuke, declaring the appeal to be "a classic demonstration of anti-Semitism" requiring a prompt and decisive response from authorities.

The Russian government's formal disavowal of the appeal was punctuated by a vote taken in the Duma in early February 2005 to condemn anti-Semitism. The Duma resolution stated that "there should be no room for anti-Semitism or ethnic and religious hatred" and called for immediate cessation of activities inspired by these sentiments; while 306 Duma members supported it, 58 did not, with Communist party members asserting there is no anti-Semitism in Russia. Nevertheless, Putin's personal condemnation of anti-Semitism reinforced the government's negative reaction to the appeal. Putin said that "even in Russia, which did more than anybody else to crush fascism and liberate the Jewish people, we often see symptoms of this disease today; and we feel ashamed about this."

Health Care: Ambiguity, Unreliability, and Inadequacy

Another problem ridden aspect of Russian society with serious implications for political stability is the dismal state of the country's health care. Diseases that have been virtually eliminated in other advanced industrial societies, notably diphtheria, measles, whooping cough, and typhoid fever, continue to afflict Russians. There have also been major increases of infectious diseases such as salmonella poisoning, hepatitis, mumps, polio, dysentery, malaria, and most recently tuberculosis and AIDS. In the late 1990s, there seemed to be an epidemic of tuberculosis, or, rather, a drug-resistant strain of it, in Russia's crowded prison population. By the end of 1998, according to Deputy Health Minister Gennady Onishchenko, 2.5 million Russians were suffering from the disease.

Tuberculosis and other diseases are spreading throughout Russia primarily because of a shortage of vaccines and other drugs, a sharp fall in the amount of money allocated to prevention and treatment of disease, and the chaotic state of the health care system. To survive in today's Russia, doctors, nurses, and others must find either alternative sources of income in completely different fields, (e.g., in agriculture) or rely on bribes from their patients.

The amount of state investment in the national health care system had been drastically reduced after the Soviet Union's breakup; it then rose again but not to the Soviet level. Russians must fend for themselves as best they can, a disastrous situation for many because of chronic hard times. In the case of a disease like tuberculosis, the cost of treatment can be astronomical not only for the individual but also for the state. For example, while ordinary tuberculosis can be cured with drugs costing about $50, the cure for the new drug-resistant strain going around the Russian prison population is $10,000, obviously a prohibitive figure for most cash-short local and regional governments.

Alcohol, Smoking, and AIDS

Excessive and long-term drinking is as much a problem, or worse, in post-Soviet Russia as during the Soviet period of Russian history. Half of all deaths in Russia in the past decade can be attributed to some degree to alcohol abuse. Russian authorities acknowledge that alcoholism is the "great killer," noting that the life expectancy of an alcoholic is 10 to 15 years lower than that of a nondrinker or social drinker. Contributing to the problem of alcohol abuse is easy access to cheap, low-quality vodka that is smuggled into the country or produced illegally by criminals. The state has tried to make it difficult for Russians to get hold of this vodka by a variety of decrees controlling distribution, but to no avail. A black market in cheap vodka thrives, undermining general health and contributing directly and indirectly to high mortality rates.

The problem has become worse in the past few years with rates of "binge drinking" having soared to new heights. More than 50,000 people die annually from alcohol poisoning compared with about 400 annually in the United States, even though the latter has more than twice the total population. Moreover, Russia under Putin and Medvedev still has few developed programs for alcoholics to curtail their drinking habits, although a similar kind of organization (like America's "Alcoholics Anonymous") has gotten started. The prevailing treatment for the majority of chronic alcoholics is the so-called Dovzhenko method developed in Soviet times. It is more psychological than medical and involves a kind of hypnosis to which patients are subjected and which has the effect of convincing them that if they drink another drop of alcohol they will certainly die. The Dovzhenko method, developed by a Soviet psychiatrist, is still used because it is fast and cheap though it is temporary and increasingly impractical because patients realize it is nothing more than a trick and when they realize that they go back to excessive drinking, defined in Russia as a bottle or a bottle and a half of vodka at one sitting.

Smoking is another serious health problem. About 66 percent of Russian men and 33 percent of Russian women are dedicated smokers, compared to the world average of 47 and 12 percent, respectively. While most Western nations have campaigned against smoking, Russia has done little in this regard; and cigarette production and importation have risen 40 percent since 1986. Russians consume 40 percent more cigarettes each year than Americans. Smokers have a shorter life span than nonsmokers by 7 to 15 years. According to a United Nations report of May 1997, 32 percent of all Russian deaths in a year are attributable to diseases caused or made worse by smoking.

Another health problem is AIDS. In 1996, some 1,500 new cases of HIV infection were reported in Russia, more than the official figure for the first nine years of the epidemic combined. In the first quarter of 1997, according to the Russian Health Ministry, the numbers outstripped those for all of 1996. And Russian health officials at the time suggested that there might be 100,000 people infected with the AIDS virus. By far the major cause of the increase in AIDS is the proliferation of drug addicts who use dirty needles.

By 2003, AIDS seemed to have reached epidemic proportions. As reported by Vadim Pokrovsky of the Russian Center for Aids Prevention and Treatment, at least a half-million Russians—and probably many more—carried the AIDS virus, or more than 1.5 percent of the total Russian population of 147 million people. In a few years, said Pokrovsky, it would be possible that 1 out of 25 Russians could be infected by the virus. In some Russian cities, it was reckoned, more than one percent of the population carried the virus. In addition, according to Pokrovsky, the majority of new cases are young Russians under the age of 30.

The reasons for this recent upsurge are not altogether clear. One seems to be the failure of the Russian government to spend more money on prevention. In 2003, it set aside about $38 million to combat the virus and disease and was slated to receive a World Bank loan of $150 million. At the same time, the Russian government has tried to conceal the new dilemma by underestimating the number of newly infected people, according to Pokrovsky. The United Nations seems to think that Moscow is not doing enough to combat the disease, warning in February 2004 that Russia stands to lose 20 million people to the disease if it is not checked by mid-century.

As of 2008, the HIV epidemic in Russia continues to grow, but at a slower pace than in the late 1990s, according to a report by UNAIDS. At the end of December 2007, the number of registered HIV cases in Russia was 416,113, with 42,770 new registered cases that year. The actual number of people living with HIV in Russia is estimated to be about 940,000. In 2007, 83 percent of HIV infections in Russia were registered among injecting drug users, 6 percent among sex workers, and 5 percent among prisoners. However, there is clear evidence of a significant rise in heterosexual transmission. In 2007, 93.19 percent of adults and children with advanced HIV infection were receiving antiretroviral therapy.

Decline in Population and Its Spatial Redistribution

Between the censuses of 1989 and 2002, Russia's population shrank by 3 million people—a result of deaths exceeding births by 8.6 million and in-migration (almost exclusively from the former Soviet republics) exceeding out-migration by 5.6 million. Having reached a peak of almost 1 million people in 1994, immigration subsided to less than 200,000 in 2003, but "negative natural increase" (excess of deaths over births) continued and is unlikely to be reversed any time soon despite the vigorous pro-natalist (encouraging births) policy of the Russian government. By 2008, the Russian population had declined to 142 million from 148.6 million in 1993. From 1993 to 2005, deaths exceeded births by 850,000 a year.

Such a speedy population decline is acknowledged as one of the most serious problems that Russia is now tackling. Both variables, births and deaths, are to blame. The overall number of births depends on the size of prospective parent cohorts and on fertility (births per one woman). The diminishing number of parents may be attributed to the significant decline in fertility that commenced in the early 1960s when living standards were growing. Thus, this decline is in line with the logic of demographic transition experienced by all countries. However, abysmally low fertility (1.4 for Russia as a whole and close to 1.2 for ethnic Russians) is not and may instead be attributed to postponing childbirth during tough economic times. There is an exorbitantly high frequency of abortions in Russia, exceeding the number of live births. Thus, in 2005, there were 121 abortions per 100 live births, which is actually a change for the better, as in the mid-to-late 1990s abortions exceeded live births by a factor of two and in 1985 by 2.5.

The other side of the coin is high mortality, particularly among Russian men, whose life expectancy at birth is only 60 years of age (and was 59 just recently), whereas women in Russia live 12–13 years longer than men, the largest "female advantage" in the world. Such heightened mortality among Russian men (well above that in many Third World countries) is largely attributed to mass alcoholism.

So far, there has been one publicized turning point in the dynamics of Russia's population: In 1992, total numbers began to decline. The year 2007 saw a new turning point of no smaller significance: For the first time, the depletion of Russia's working-age population (due to retirement and premature deaths) was not compensated by the number of people entering that group—a delayed effect of consistently low birth rate. By 2025, Russia's working-age population will have shrunk by 19 million, which is 28 percent of Russia's overall employment in 2007. This will make labor the most limiting production factor in economically resurgent Russia; and most likely the country will have to compete for immigrants with other labor-deficient economies. The prospect of growing immigrant communities is not yet fully accepted by the Russian public or even by the political class, who pin their hopes on bonuses meted out since

2007 for second, third, etc. births. But even if these bonuses will ultimately prove to be fertility boosters, a couple of decades will pass until the more numerous newborns reach working age. And so the idea that Russia's bright economic future may not be achievable without attracting a large number of immigrants is sinking in, as evidenced in the new immigration law (enacted on January 1, 2007), which simplified registration requirements for foreigners, wrested registration away from the jurisdiction of the endemically corrupt Russian police, and made it much easier than before to obtain employment authorization.

Because Russia is so vast and sparsely settled, the spatial aspect of its downward population spiral is certain to carry weight of its own. Thus far, two spatial consequences have been uncovered: (1) Due to rapid rural and small-town depopulation in the peripheral parts of most Russian regions, a continuous belt of human colonization and settlement is giving way to an archipelago-like pattern; and (2) a reversal of the centuries-long process of eastward expansion of Russia's ecumene (i.e., inhabited space) is in full swing, with populations in Russia's Far East having declined by 12 percent from 1989 to 2002. To these trends, one should add that the role of Moscow and its urban agglomeration in the spatial redistribution of Russia's population has become overwhelming. From 2001–2005, net migration (domestic and international) to the Central Federal Okrug (containing Moscow) exceeded net migration to Russia as a whole, which was possible due to the fact that two out of seven federal okrugs (Siberia and Far East) recorded significant out-migration. The lion's share of migrants to the Central Okrug (85 percent from 2001–2005) headed to Moscow and its oblast. To be sure, the capital city region attracted domestic migrants for decades, but European Russia's south (particularly Krasnodar and Stavropol regions) used to exert an equally powerful pull, at least until the mid-1990s. Moreover, Russia's regional capitals intercepted at least half of the migrants from each region's periphery. Now, instead of relocating to regional capitals and/or to Russia's south, migrants increasingly and overwhelmingly head straight to Moscow and/or its environs. In all likelihood, the Moscow region will be the only area in all of Russia which will be able to meet its demand for labor through domestic migration. Potentially, however, other Russian regions can meet their labor demand through *external* migration, as in this area the leading role of the Moscow region is not as salient as in domestic flows.

Reflecting assumptions about natural increase and migration, three versions of a 2025 population projection have been published by the Rosstat, Russia's equivalent of the U.S. Census Bureau: a pessimistic scenario based on extrapolation of the existing (pre-2005) trends; an optimistic scenario, based on the notion that fertility will have risen significantly by 2025; and an average scenario. Even according to the optimistic scenario, however, the total fertility rate will hover below the replacement level, at 1.85. Apparently the average scenario was deemed the most realistic; it is the only scenario which was disaggregated into regions. The average scenario envisages a more moderate growth in TFR (to 1.65 by 2025); sets life expectancy at birth among men, arguably the most outrageous of Russia's demographic statistics, at 61.9 years (versus 60.3 in 2007); and estimates a total population of 134.4 million in 2025, down from 142 million in 2007. Because the life

expectancy projection is fairly conservative, and some growth in fertility already has been recorded, the "natural increase" component of the average projection can be construed as realistic.

The same, however, cannot be said about the migration component. Net migration to Russia is estimated at 415,000 in 2025 (versus 111,000 in 2005.) According to the Rosstat's projection, throughout the entire 2005–2025 period net migration to Russia will replace only 30 percent of the decrease in the working-age population. It is hard to resist the notion that politics weighed down on the Rosstat's reasoning—reflecting the fact that the idea of large-scale immigration is still frowned upon by many in Russia. One should then wonder how the Russian economy will function under a deficit of labor that the above projection entails. While the labor intensity of the, Russian economy can and perhaps should, be lowered, it is the scale of implied progress along this path that raises doubt.

Trends recorded since the publication of the 2025 projection exacerbate that doubt. For example, in 2007, when the working-age population shrank for the first time and the scope of its recorded decline amounted to 300,000 people, about 2,000,000 employment authorizations were issued to citizens of the CIS countries; additionally about 300,000 authorizations were issued to nationals of other countries. The official (and probably understated) record of temporary foreign migrant employment *exceeding* the decline of working-age population by a factor of seven (!) shows how labor-deficient the Russian economy already is. It is unrealistic to think that deficit will somehow diminish in the foreseeable future when the working-age population shrinks much more than it already has. And, although labor migrants are not relocating to Russia for good, many of them may consider that option given a more welcoming social climate and legal opportunities, and that might be to Russia's benefit.

Gender Issues

By the time Putin was elected president in March 2000, the situation of Russian women had deteriorated from what it had been in the Soviet era, when even then, most Russian women endured a harsh, male-inspired discrimination despite Soviet laws requiring that state agencies reserve employment slots for women. In the post-Communist era women have suffered in several ways. They really did not benefit as a group from the economic transformation to a free market, which favored men, the ones who cut the deals and acquired the wealth in the privatization of state property. When unemployment increased as a result of this transformation, the main impact fell on women: they were the first to lose jobs and the last to regain them as the economy improved at the end of the 1990s and in the early years of the new century.

Limits on Political Activism

There is little Russian women can do—or, in some instances want to do—to curtail discrimination against them in Russia's conservative, male-dominated society. This situation was partly the result of the societal structure, but it was also due to women themselves. A majority of women in Russia are not feminists and are inclined to accept, without much question, the role of homemaker. Russia has a very weak feminist movement largely because feminist activity to improve opportunities for women in the workplace, public and private, is viewed with suspicion and cynicism. This was the case throughout the history of the Soviet Union, when the right to work for women too often meant low-level labor.

Women don't see political participation as the route to the improvement of their status. In many instances, Russian women believe that politics is beyond the reach of ordinary people, asking, why fight when there is little prospect of change? So it seems that Russian women are not yet ready to use protest and dissent, despite its current acceptability, to oppose and reverse sex discrimination against them. The organizations that exist for the benefit of women are social in character—not political—even though it is political action that is needed to change the environment for women.

Change may be coming, however. Russian universities are beginning to teach courses on gender issues. Young Russian women who have grown to maturity in the post-Communist era also seem more imbued with the democratic idea of social equality than women born in the Soviet era. In St. Petersburg, a new women's political organization, called the League of Women's Voters, may well be a straw in the wind of change. The St. Petersburg League has made contact with the League of Women Voters in the United States, a development that suggests that women might be willing to learn from other politically active groups. The Internet also offers opportunities for communication among women's groups and the chance to exchange ideas and experiences. If these trends continue, Russian women may be on the threshold of moving themselves forward in Russian society and gaining political and economic visibility.

Domestic Violence against Women

In a completely different vein is another gender issue that has received publicity in the Russian press in recent years: the growth of domestic violence against women, mostly by a husband in a family setting. Police authorities, when asked for help, are reluctant to get involved or will say the woman is at fault. The police even have a slang term for domestic violence

that is a put-down of it—"sauce-pan brawls." The seriousness of the issue is evident from statistics: police say that one woman is killed by her husband or partner in Russia every forty minutes.

Often the violence involves a man who is drunk; and he is drunk, so the argument goes, to escape his problems with a job or with handling money that is not enough to make ends meet. Moreover, married men who commit violence against their wives expect to do so with impunity. If the police are called, the husband is taken away briefly for being drunk, not for hurting his wife. Violent husbands seem to have the upper hand in most instances, reckoning on the reluctance of the wives they hurt to file a formal complaint or simply leave the house where violence occurs. Wives are financially dependent on the husband and have no other place to go given the absence of shelters for battered women like those that have grown up in other countries, not only in the West but also in post-Communist countries such as Poland.

Women are reluctant to bring complaints to the attention of the authorities because doing so makes them feel as if they have failed in managing family life. Women, lacking political skills, are reluctant to organize for the purpose of lobbying the government for recognition of their predicament, according to Natalia Abubrikova, an activist seeking to call public attention to violence against women.

There is, however, evidence that Russian society is becoming more sensitive to this particular gender issue. A day center for female victims of domestic violence known as ANNA has been set up in Moscow with international money. Victims of abuse can turn to ANNA first, before they do anything else to regain their equilibrium. In addition, Russia now has crisis hotlines and nongovernmental organizations have appeared that run sessions and offer free advice in Moscow and other cities. One agent of ANNA, Larissa Ponarina, says that there is much more recognition of the issue of domestic violence in Russia today than ten years ago but much more care needs to be offered to victims. Ponarina advocates a major education drive to increase the sensitivity of Russians to the issue of domestic violence, which for too long they simply have ignored as a norm instead considering it an exception.

ALEXANDER SOLZHENITSYN

Alexander Solzhenitsyn (1918–2008) and Andrei Sakharov were alter egos of sorts as they embodied two intellectual traditions in Russia, that of Slavophiles and that of Westernizers, respectively. Both were men of unparalleled courage and achievement and both contributed greatly to defeating Soviet communism. But whereas Sakharov was a man of liberal persuasion and a voice for Western values of democracy and tolerance, Solzhenitsyn was a conservative thinker emphasizing Russian spirituality as an antipode of Western materialism and claiming that communism itself was a genuinely Western invention unleashed on Russians in order to enslave and degrade them.

Solzhenitsyn was born in the resort town of Kislovodsk in the North Caucasus and spent much of his adolescence in Rostov-on-Don, the largest Russian city of that multi-ethnic region. There

he entered, and graduated with honors from, the local university where he majored in math. Beginning in February 1943, Solzhenitsyn served as the commander of a sound-ranging battery in the Red army and was involved in some major battles and twice decorated. While serving in East Prussia, he was arrested for writing disapprovingly about Stalin in letters to a friend. Sentenced to an eight-year term in a labor camp, Solzhenitsyn served his sentence in various camps, first in central Russia and then in northern Kazakhstan. One year before (1952) and one year after (1954) his release Solzhenitsyn was diagnosed with cancer and underwent surgery. This experience became the basis of his novel *Cancer Ward*. After his release from a labor camp, Solzhenitsyn worked as a secondary school math teacher, first in Central Asia and then in the Ryazan province of Central Russia. Having dabbled in fiction

ALEXANDER SOLZHENITSYN

writing since adolescence, Solzhenitsyn was now spending many of his nighttime hours writing short stories and a novel. At the age of 42, he approached Alexander Tvardovsky, a poet and the editor-in-chief of *Novyi Mir* magazine, the Soviet Union's most liberal literary magazine whose very existence had been made possible by Khrushchev's political thaw. The manuscript that Solzhenitsyn brought to *Novyi Mir* was of such a controversial nature that Tvardovsky contacted Khrushchev himself for his personal approval before publishing it. Soon (1962) the novella titled *One Day in the Life of Ivan Denisovich* was published by *Novyi Mir* and became an instant success. The story was set in a Soviet labor camp in the 1950s and described a single day of an ordinary prisoner, Ivan Denisovich Shukhov. Its publication was an extraordinary event—never before had an account of Stalinist repression been openly distributed in the Soviet Union. Soon thereafter three more novellas by Solzhenitsyn were published by the same magazine. All of them peered into and reflected upon aspects of everyday life that used to be taboo for Soviet writers. Arguably, these were pieces of higher artistic quality than anything Solzhenitsyn released thereafter, first as a dissident whose works were passed from one Soviet intellectual to another in pale typewritten copies (and at those people's peril) and then as a sworn enemy of the Soviet Union stripped of its citizenship and expelled from the country.

Indeed, the Soviet political elite, which had endorsed Solzhenitsyn's early publications, soon became concerned about their potential to erode the ideological foundations of the world's first communist state. Khrushchev's thaw was coming to an end, and even before it ended, Solzhenitsyn's novel *Cancer Ward* was denied publication. Soon after the enthronement of Leonid Brezhnev, the KGB conducted a search at the apartment of Solzhenitsyn's friend where part of his archive was held. As a consequence, Solzhenitsyn became a dissident. When in 1968 his two novels, *Cancer Ward* and *The First Circle,* were published in the West, the Soviet leadership launched a full scale propaganda campaign against Solzhenitsyn. In 1970, he was awarded the Nobel Prize in literature. Solzhenitsyn did not go to Stockholm to receive his prize because he was afraid he would not be allowed back into the Soviet Union.

On August 23, 1973, the KGB apprehended Yelizaveta Voronyanskaya, Solzhenitsyn's assistant. During the interrogation she betrayed the whereabouts of one copy of Solzhenitsyn's manuscript of a three-volume book about the Soviet prison camp system, *The Gulag Archipelago.* Another copy of that manuscript had already been sent to the United States by special delivery. Upon returning home from the interrogation Ms. Voronyanskaya committed suicide. Having learned about that, Solzhenitsyn asked the YMCA-Press, an American publisher, to release the book. Based on his own experiences, on the testimony of 227 former prisoners, and on his own research into the history of the Soviet penal system, the book shattered the positive image of the world's major communist state, the image the Soviet Union still retained among many people around the world. Within the Soviet Union, the manuscript was circulated in multiple carbon copies (so called samizdat—a Russian term for a clandestine publishing system within the Soviet Union, by which forbidden literature was reproduced and circulated privately) by volunteers and had powerful influence on all who managed to read even an excerpt from it. Not a book of fiction, *The Gulag Archipelago* was based on facts and delivered a harsh verdict, not just on the Soviet penal system but on the entire political system. By doing so, the book facilitated further searches for truth in the minds and hearts of its readers. Shortly after the publication of the book in the United States, Solzhenitsyn was

(© Mikhail Evstafiev. www.evstafiev.com)
Russian writer and Nobel Prize winner Alexander Solzhenitsyn looks out from a train in Vladivostok, summer 1994, before departing on a journey across Russia. Solzhenitsyn returned to Russia after nearly 20 years in exile.

arrested (February 12, 1974), stripped of his Soviet citizenship and placed on a plane bound for West Germany. By the end of March, Solzhenitsyn's immediate family was afforded the opportunity to leave the Soviet Union as well. Solzhenitsyn's archive and his military (World War II) decorations were secretly transported from Moscow by an American diplomat. After spending two years in Europe, Solzhenitsyn moved with his family to the United States, where he settled in Cavendish, Vermont, in a natural setting resembling Central Russia. In 1978, he was given an honorary Literary Degree from Harvard University and on June 8, 1978 he gave his Harvard Commencement Address. Delivering his commencement speech, he called the United States spiritually weak and mired in vulgar materialism. Americans, he said, speaking in Russian through a translator, suffered from a "decline in courage" and a "lack of manliness." Few were willing to die for their ideals, he said. He condemned both the United States government and American society for a "hasty" capitulation in Vietnam. He criticized the country's music as intolerable and attacked its unfettered press, accusing it of violations of privacy. He said that the West erred in measuring other civilizations by its own model. While faulting Soviet society for denying fair legal treatment of people, he also faulted the West for being too legalistic: "A society which is based on the letter of the law and never reaches any higher is taking very scarce advantage of the high level of human possibilities."

In his homeland, attitudes toward Solzhenitsyn had changed by the end of perestroika. In 1990, his Soviet citizenship was reinstated and two programmatic articles of Solzhenitsyn were published by large-circulation Moscow newspapers. Devoted to ways of developing Russia, these articles were not exactly attuned to Russian Westernizers, who were setting the tone in domestic public discourse. Also, in 1990, a literary state award to Solzhenitsyn for *The Gulag Archipelago* was announced. During his first official visit to the United States, President Boris Yeltsin of Russia had a lengthy phone conversation with Solzhenitsyn.

On May 27, 1994, Alexander Solzhenitsyn returned to Russia. He landed in Vladivostok, Russia's far-eastern port city and took a trans-Siberian train to Moscow. In Russia, Solzhenitsyn lived in a country house in the vicinity of Moscow. The piece of land on which Solzhenitsyn and his family built a two-story house had once housed a state dacha and was presented to him as a gift from the Russian government. In 1998, Yeltsin conferred a high-ranking honorary medal on Solzhenitsyn, but he rejected the award saying that he could not possibly accept it from the people "who drove

(continued)

ALEXANDER SOLZHENITSYN *(continued)*

Russia to such a ruinous condition." What he referred to was the whole ideology and practice of privatization which transformed Russia throughout the 1990s. Solzhenitsyn only warmed up to Russian leadership with the enthronement of Vladimir Putin, who paid a much publicized visit to the writer.

Alexander Solzhenitsyn died of heart failure in his home near Moscow on August 3, 2008, at the age of 89. A burial service was held at the Donskoy Monastery in Moscow. He was buried at a place chosen by him in the Donskoy necropolis. Russian and world leaders paid tribute to Solzhenitsyn following his death. Solzhenitsyn's legacy is hotly debated and will continue to elicit wide-ranging opinions, but there is no question that he was a man of great stature who, with great personal courage, contributed to defeating Soviet communism.

PATRIARCH ALEKSY II

Patriarch Aleksy II was elected Patriarch of Moscow in 1990, one and half years prior to the Soviet Union's collapse. The patriarch's secular name was Alexey Ridiger. A person of Baltic German descent, he was born in 1929 in the Republic of Estonia, to which his father had emigrated from Saint Petersburg shortly after the communist revolution of 1917. From his early childhood Alexey Ridiger served in the Orthodox Church.

In 1940, Estonia was incorporated in the Soviet Union, and soon thereafter Germany invaded the Soviet Union, and Estonia was occupied from 1941–44. After Soviet forces returned to Estonia, unlike most of the people with Baltic German roots, the Ridiger family chose to stay put and did not migrate to the West.

Having been closed during war time, after the Soviet annexation of Estonia, the Alexander Nevsky Cathedral in the city of Tallinn was reopened in 1945. Alexey Ridiger, who had become a Soviet citizen, served as an altar boy in the cathedral from May to October 1946. He was made a psalm-reader in St. Simeon's Church later that year; in 1947, he officiated in the same office in the Church of the Kazan Icon of the Mother of God in Tallinn. In 1947, Ridiger entered Leningrad Theological Seminary, and graduated in 1949. He then entered the Leningrad Theological Academy (now Saint Petersburg Theological Seminary) from which he graduated in 1953.

In 1950, he was first ordained a deacon and then a priest and was appointed rector of the Theophany church in the city of Jõhvi, Estonia, in the Tallinn Diocese. In 1957, Aleksy was appointed Rector of the Cathedral of the Dormition in Tallinn and Dean of the Tartu district. In 1958, he was elevated to the rank of Archpriest and in 1959 he was appointed Dean of the united Tartu-Viljandi deanery of the Tallinn diocese. He was chosen to be the Orthodox Church Bishop of Tallinn and Estonia in 1961. In 1964, he was elevated to the rank of archbishop; and, on February 25, 1968, at the age of 39, metropolitan.

After the death of Patriarch Pimen I in 1990, Aleksy was chosen to become the new Patriarch of the Russian Orthodox Church. He had a reputation as a conciliator, a person who could find common ground with various groups in the episcopate. Upon taking on the new role, Patriarch Aleksy became a vocal advocate for the rights of the church, calling for the Soviet government to allow religious education in the state schools and for a "freedom of conscience" law.

Patriarch Aleksy died at his residence in Peredelkino, outside Moscow, on December 5, 2008. There is controversy regarding alleged clandestine work by Aleksy II as a KGB agent since the 1950s. The Moscow Patriarchate has consistently denied he played such a role. Among the sources that cite his alleged KGB service are Gleb Yakunin and Yevgenia Albats, both of whom were given access to the KGB archives. According to Oleg Gordievsky, a Colonel of the KGB and KGB Resident-designate

(© Maxim Massalitin)

Aleksy II, Patriarch of the Russian Orthodox Church

and bureau chief in London, who defected to the United Kingdom, Aleksy had been working for the KGB for forty years, and his case officer was Nikolai Patrushev, the FSBs boss under President Putin.

Whatever the truth, what comes across as a striking exposure and denunciation to some commentators, leaves others largely unruffled. Historically, the Russian Orthodox Church has been the arm of the state and the utmost nationalistic and patriotic force for much of Russian history, so the intensely atheistic period of Russian history which culminated in the closure of many churches and repressions against clerics in the 1930s is almost an aberration.

PATRIARCH ALEKSY II

During World War II, Joseph Stalin, himself a former student at an Orthodox seminary, appealed to the patriotic feelings of Russian believers and reopened many churches. If Aleksy indeed agreed to work for the KGB, as is alleged, it is highly likely that he considered that service to be a patriotic duty instrumental in maintaining Russian influence in Estonia.

What is beyond doubt is that, in his capacity as the Patriarch, Aleksy II was extremely successful in recovering the Church's real estate, art, and relics from the state. Even more importantly, Aleksy II played a significant role in the August 1991 coup by invoking the power of the Church against the coup plotters. At the height of the August 1991 crisis, Aleksy II made a direct appeal to the army: "We call upon the whole of our nation, and particularly our army, to show support, and not to permit shedding of fraternal blood." After this crucial turning point, Aleksy II served as a critical conciliatory figure in Russian history.

NOTES

1. Patsiorkovsky, V.V. *Sel'skaya Rossiya 1991–2001* (Moscow: Finansy i Statistika 2003) p. 83. The quoted statement by Bashmachnikov was made in 2002.

2. Brown is a polemic reference to Nazis. SA men (a paramilitary organization of the Nazi Party, which played a key role in Adolf Hitler's rise to power in the 1920s and 1930s) were often called "brown-shirts" for the color of their uniforms.

3. FSB is a Russian abbreviation of the Federal Security Service, a successor of the KGB.

Russia's Foreign Relations

RUSSIA AND THE BALTIC STATES IN THE 1990s

The three Baltic republics—Estonia, Latvia, and Lithuania—gained independence from Czarist Russia in 1918. The Russian communists tried, but failed, to bring them back under Russian control in the aftermath of the November 1917 revolution. They remained independent until the outbreak of World War II, when, in 1940, they were annexed by the Soviet Union as a result of the implementation of the infamous Ribbentrop-Molotov (or Soviet-German non-aggression) pact of 1939. Each annexed country was given "union-republic" status and remained an integral part of the Soviet state until they separated from it and proclaimed their independence in September 1991.

In the 1990s, Russia was concerned about several large problems in its relations with the newly independent Baltic republics: the persistence of Baltic discrimination against Russian-speaking residents; Baltic efforts to integrate with the West militarily and economically while down-grading their ties to Russia; and the removal of Soviet-era monuments while simultaneously glorifying indigenous Waffen SS legionnaires who fought for Nazi Germany during World War II and participated in numerous Nazi atrocities against Jews and Slavs within and outside the Baltic States. Indeed, the official Russian relationships with each of the three Baltic States have been more acrimonious than with any other foreign state save Georgia. Perhaps too much in these relationships is fueled by Baltic grievances, including the central one—that Russians had wrested Estonians, Latvians, and Lithuanians from the Western civilization to which they legitimately belonged.

Discrimination of Russian Minorities

Inspired by a strong post-Communist undercurrent of anti-Russian sentiment in the societies of all three countries, the Baltic peoples began to discriminate against the Russian communities in the early 1990s. Throughout the Soviet period of their history, the ethnic makeup of all three republics underwent significant changes. Whereas in Lithuania the inflow of ethnic Russians proved insignificant and the major change appeared to be the outflow of ethnic Poles to Poland; in Estonia and Latvia, Russians and representatives of other ethnicities from the Soviet "mainland" began to account for a sizable share of the population. Thus, whereas on the eve of the 1940 incorporation of Latvia into the Soviet Union ethnic Latvians accounted for 77 percent of Latvia's population, after the country reclaimed independence in 1991, their share was only 52 percent. The share of ethnic Estonians in Estonia also declined from 93 percent soon after the war to just 62 percent in 1989. By contrast, in Lithuania, the percentage of ethnic Lithuanians remained relatively high (79.6% in 1989) even at the end of the Soviet era. Because most non-Russian migrants from the rest of the Soviet Union used to adopt Russian as their language of everyday communication, the number of Russian speakers in all three Baltic States significantly exceeds the number of ethnic Russians. Thus, in Latvia, ethnic Russians account for 28 percent of the population whereas Russian speakers account for approximately 40 percent and more than 50 percent in the capital city of Riga. Post-war migrations had been caused by a combination of factors, including large-scale industrialization, for which the local labor force was inadequate; and the Soviet military buildup, as most military retirees that served on Baltic bases chose to remain because in the Baltic republics living standards were higher than elsewhere in the Soviet Union. This quality of life gradient was an important self-styled stimulus to voluntary migration to the Baltic States from Russia, Belarus, and Ukraine.

After independence, all ethnic Russians and other Russian-speaking residents of Lithuania received Lithuanian citizenship. However, in Estonia and particularly in Latvia, where the percentage share of the indigenous population was much lower than in Lithuania, discriminatory naturalization laws were adopted. Citizenship was granted to non-indigenous residents who had lived in Latvia and Estonia for a long time (in Latvia for more than 16 years) and could speak the state language, but many others were denied naturalization and became permanent residents with no electoral rights and some other limitations. One of the naturalization requirements was passing the state language exam. Many Russian speakers had never bothered to learn the local languages. So the Baltic Russians stayed put but complained, providing much grist for the nationalist politicians in Russia, who seek popularity by championing causes of their beleaguered compatriots abroad. The problem has eased somewhat since the mid-1990s when all three Baltic States appeared on the waiting list for accession to the European Union. Since that time many non-indigenous permanent residents have become naturalized. Even so, the Russian-speaking populations of the Baltic States have been under conflicting pressures. On the one hand, any attempt on the part of the authorities to eradicate Russian in the public domain, to rewrite history, to limit Russian-language schooling, and to remove Soviet-era monuments meets with disapproval from many Russian speakers. On the other hand, their disproportionately high participation in the entrepreneurial class of Latvia and Estonia improves their attitude to their host countries, while their disproportionately low participation in government produces the opposite effect. At the same time, an extremely high concentration of Russian speakers in certain compact areas within the Baltic States delays their cultural assimilation. Besides capital cities with significant concentrations of Russian speakers in Soviet-era residential areas, in Estonia the overwhelmingly Russian speaking areas are the industrial cities of Narva (the third-largest city in Estonia) and Kohtla-Järve in the republic's northeast, and in Latvia, it is Daugavpils and other communities in the country's southeast.

Integration with the West

During Yeltsin's years in office, Russia tried to discourage the Baltic republics from seeking NATO membership, offering them instead treaties that would guarantee their independence and territorial integrity. But Baltic leaders did not want security guarantees from the Kremlin, which they deeply distrusted, as an alternative to membership in NATO. Moreover, they were convinced that Yeltsin's assurances were not shared by his nationalist and Communist opponents. In 2002, Estonia, Latvia, and Lithuania, along with four other countries (Bulgaria, Romania, Slovakia, and Slovenia) were invited to join NATO. All three Baltic States became full-fledged NATO members in March 2004, and just one month later they joined the European Union.

RUSSIA AND THE BALTIC STATES AFTER 2000

Problems in Russia's relations with the three Baltic republics evident in the Yeltsin era became even more complicated and more serious for both sides in the years of Putin's presidency of Russia. The most important problems ostensibly remain treatment of the Russian-speaking community in each of the three republics and the titular nationalities reevaluating Soviet history.

Baltic Membership in NATO and the EU

Although the three Baltic republics joined two Western alliances in the spring of 2004, as late as August of the same year the Kremlin was still criticizing Baltic membership in NATO. Defense Minister Sergei Ivanov, in a meeting with U.S. Secretary of Defense Donald Rumsfeld, said that Baltic membership in NATO was a waste of Western taxpayers' money; that the republics were too small to help NATO fight the war against terrorism; and that they were "consumers," not "producers," of collective security in the region. Ivanov reminded Rumsfeld of Baltic sympathies for the Nazi German state on the eve of, and during, World War II and of their "inconsistency" on the matter of human rights. It was clear that Baltic membership in the western alliance, no matter how justified by the Baltic governments, was still a significant bone of contention in Russian relations with them.

In 2005 and 2006, the Kremlin had even more reason to regret Baltic membership in NATO. It seemed to coincide with the buildup of a pro-NATO, pro-American bloc of countries with strong anti-Russian proclivities, notably the three Baltic Republics and Poland. In May 2005, U.S. President George W. Bush went to Riga for a summit conference with the Balkan presidents after publicly declaring U.S. endorsement of the Baltic position that they had been "occupied" by the Soviet Union for 50 years and were entitled to a Russian acknowledgment of this position and some form of Russian repentance for having extinguished their independence in 1940. To the dismay of Moscow the EU went along with the United States, echoing this position.

Other Problems in Russian-Baltic Relations

Another source of annoyance for the Kremlin has been a revival of Baltic demands for reparations for their suffering during Soviet rule after 1940, which they call a "Soviet occupation." For example, on June 17, 2004, the Lithuanian parliament passed a resolution saying that Russia should pay Lithuania compensation of $20 billion for the harm done by the World War II occupation by the Soviet army. Lithuanian moderates knew that this was an incendiary proposal that had originated in 2000 with Vytautas Landsbergis, Lithuania's first non-Communist leader and an ultranationalist, now a member of the Lithuanian parliament.

In 2005, Baltic leaders spoke of the economic and other kinds of deprivation their countries had suffered during their annexation to the Soviet Union from 1940 to 1991. They proposed a Russian payment of a billion dollars for each of the 50 years they were part of the Soviet Union. Washington supported the demand for reparations, raising the specter of a freezing of Russian assets in the United States that could be followed by equivalent action by governments in the EU. The idea of Russian payment of reparations to the Baltic republics infuriates the Kremlin and creates a highly negative atmosphere even as the Baltic republics try to maintain stable relations, which their leaders say they want, with contemporary Russia.

Russia has had additional problems specific to individual countries. Some of these problems have brought Russian relations with individual Baltic republics to a nadir making cooperation with them difficult. In some instances, the West, in particular the United States, has sympathized with or supported individual Baltic republics in their dealings with the Kremlin, much to its annoyance. In its relationships with the Baltic States, Russia can utilize foreign trade instruments. Whereas in Russia's overall exports and imports the role of the Baltic States is negligibly small, the opposite is not true. For example, Russia is the largest trading partner of Lithuania and the third largest trading partner of Latvia.

Russia and Lithuania

There have been several problems in Lithuanian relations with Russia in recent years. Besides Lithuania's membership in NATO, these problems have to do with (a) transit to Kaliningrad; (b) Lithuanian relations with Belarus, Ukraine, and the countries of the Trans-Caucasus, that strike the Kremlin as having an anti-Russian streak; and (c) Russian oil supplies to the sole Lithuanian refinery.

Russia's Kaliningrad exclave (i.e., former East Prussia) can be reached from Russia's "mainland" through Belarus and Lithuania or through Belarus and Poland. Prior to Lithuania's accession to the European Union, rail and automobile transit through Lithuania did not present a problem for Russian citizens. The Russian government wanted to maintain visa-free transit after 2004. However, the Lithuanian authorities now require a visa from regular Russian travelers in transit or a special document qualifying for one-time crossing the Lithuanian border. This creates problems for Russians not only because of visa fees but also because a visa is granted only to those who have a special passport for trips abroad. (In Russia, everybody is required to have a domestic passport as an ID, but the so-called 'foreign' passports, issued for a fee, are not required.) Two fees in addition to a rail ticket become a serious impediment to travel for Russian citizens if they are people of average means.

The tight relationship between Lithuania and Ukraine has long been viewed with suspicion by the Kremlin. Besides a shared antipathy to the former colonial master, these relationships potentially have a geopolitical dimension, as it may lead to lessening these countries' dependency on Russian oil. Some Azerbaijani and Central Asian oil is already transported to Turkey via the pipeline, circumventing Russia. If the capacity of the Baku–Batumi pipeline is increased, then oil tankers can carry oil to the Ukrainian Black Sea port of Odessa and from there it can be transported via the existing Odessa–Brody pipeline with possible extensions to Belarus, Poland, and Lithuania. Given this possibility, the relationship between Lithuania and Belarus is also looked at with suspicion despite the fact that Belarus has so far been Russia's closest ally. But there is significant cross-border traffic between Minsk and Vilnius as well as between the border regions of the two countries, and Lithuania has been lobbying the European Union for a special relationship with Belarus.

Lithuania is home to the only refinery existing in the Baltic States, Mažeikių Nafta, an oil processing plant in the city of Mažeikiai. This Soviet-era refinery used to be supplied with crude oil via a pipeline passing through Belarus and Russia. During the privatization campaign of the early 1990s, the Lithuanian authorities did all they could not to sell the refinery to a Russian business. In fact, the enterprise was sold to a little known American firm. However, having run into problems, the American firm resold it to Yukos, a Russian conglomerate owned by Khodorkovsky, a tycoon with good ties with American business. After Khodorkovsky's arrest, the refinery was again up for grabs; Russian Lukoil, an internationally known oil producer with gas stations as far as America, was well-positioned to buy it. Yet some unknown Polish firm managed to acquire Mažeikių Nafta at the last moment. Soon thereafter, in 2006, Russians terminated the oil supply, referring to technical problems with the pipeline. It is abundantly clear that both Lithuanian efforts to avoid acquisition of an important Lithuanian asset by a Russian business and the termination of oil deliveries by the Russian side are politically motivated. Ironically, the refinery is still using Russian oil, only it is transported by rail, which incurs extra costs absorbed by the buyer.

Russia and Estonia

Neither the president of Lithuania, nor the president of Estonia participated in the Moscow-based May 2005 celebration of the 60th anniversary of the end of World War II, a celebration attended by many world leaders, including President Bush of the United States and the presidents of most European countries. In fact, two out of three Baltic leaders declined an official invitation by the Russian government. Clearly, this was a controversial decision on their part. The official reasoning was that for the Baltic States the end of World War II meant the replacement of one occupying power (Nazi Germany) by another (the Soviet Union), so for Lithuanians, Latvians, and Estonians there was nothing to celebrate. To be sure, all three Baltic States had been assigned to the Soviet sphere of influence by the German-Soviet non-aggression pact of August 1939, an agreement that was subsequently violated by Nazi Germany's invasion of the Soviet Union in June 1941. Moreover, the Western war-time allies of the Soviet Union, the United States and Great Britain,

did not challenge the Soviet acquisition of the Baltic States at the 1943–45 conferences (held in Tehran, Yalta, and Potsdam) that made far-reaching decisions about the post-war borders of Europe. However, the controversial nature of the decision not to participate in the Moscow celebration is rooted not so much in the intricacies of great-power diplomacy and not even so much in what May 1945 meant for the Baltic States as in the attitude of the Baltic political elites toward local Nazi collaborators. Estonia is a case in point. On the one hand, on April 26–27, 2008, the Soviet war memorial, the Bronze Soldier, was dismantled and transferred from a traffic island amidst the central square of Tallinn, a move which met with protest in Russia and in Estonia itself amidst its large Russian-speaking population. On the other hand, the anti-Soviet historical revisionist movement presents the Estonian Waffen FF troops as "freedom fighters," and the Estonian leadership condones that as well as the erection of monuments glorifying and commemorating Estonia's 70,000 strong Waffen SS contingent. The most scandalous of these monuments, in Lihula, near Parnu, was removed in September 2004. When that happened, Mart Laar, the former prime minister of Estonia (1992 to 1994 and from 1999 to 2002) recommended that his successor, Juhan Parts, apologize to "the freedom fighters" for the removal of the monument.

It appears that the two different occupying powers elicit very different attitudes among the Estonian political elite, and that predictably provokes Russian anger. After the removal of the Soviet war memorial from a prominent square in Tallinn, the Russian-speaking minority rallied and some of its younger members vandalized shops and cars in central Tallinn for two straight nights. One of the protesters, Dmitry Ganin, perished in a confrontation with Estonian police. Both Dmitry Ganin and his mother, Vera Ganina, had Russian citizenship but resided permanently in Estonia. The death of Dmitry Ganin was widely publicized in Russia, and Vera Ganina's letter to President Medvedev of Russia and his response to that letter were published and/or broadcast by all major news media outlets. In his letter of February 22, 2009 to Vera Ganina, the Russian president wrote: "Please accept my sincere words of support and deep sympathy on account of your loss. This feeling is shared by millions of Russian citizens. The death of your son amidst dramatic events of April 2007 shocked all of us. Dmitry had an acute sense of justice and openly protested against the blasphemous plan to transfer the monument commemorating the warriors that liberated Europe from fascism. He rightly considered that to be the profanation of their feat. Your son was not only defending the memory of the heroic past, he was vindicating the civil dignity of the people who sacrificed their lives for the peace on this planet." The quote clearly shows that the decision to move the Soviet-era monument to a military cemetery touched a raw nerve and invoked a very different understanding of history by Russians and Estonians. In his essay "To die for Tallinn?" the conservative American columnist Pat Buchanan expounded on that difference and on what he sees as an unreflective American reaction to it:

> To Russians, who lost millions of their grandfathers, fathers and uncles in the Great Patriotic War, the Red Army liberated Europe from Nazism, and their sacrifices ought to be honored. And the Estonians are a pack of ingrates.

To Estonians, the Red Army did not liberate anyone. Having won their independence from the Russian Empire in World War I, they were raped by Russia—to whom they had been ceded as part of the Hitler-Stalin Pact. In June 1940, the Red Army stormed into the three Baltic republics, butchered the elites and shipped scores of thousands off to Stalin's labor camps never to be seen again. While the Soviets were expelled from the Baltic republics by the Germans in 1941, they returned in 1944 and held the Baltic peoples in captivity until the Evil Empire collapsed. It was only then that Estonia regained her independence and freedom.

Why should Estonians honor a Red Army that brutalized them and, after driving out the Germans, re-enslaved them for half a century?

Why should this issue be of interest to America?

If President Putin decides the Estonians need a lesson, and sends troops to teach it, the United States, under NATO, would have to treat Russian intervention in Estonia as an attack upon the United States, and declare war on behalf of Estonia.

So we come face to face with the idiocy of having moved NATO onto Russia's front porch, and having given war guarantees to three little nations with historic animosities toward a nuclear power that has the ability to inflict 1,000 times the destruction upon us as Iran.

The story of the Bronze Soldier is not yet finished, and one of the particularly interesting spinoffs was the January 2009 acquittal by the Estonian court of four people accused of organizing the street protests in Tallinn in late April 2007. One of those acquitted was 63-year-old Dmitry Klionsky, former correspondent of the main communist daily *Pravda* in Tallinn and a citizen of Estonia. Klionsky was implicated in issuing tendentious press releases for the youth movement *Nochnoi Dozor,* which was in the epicenter of the protests turned violent. The acquittal of these people was not lost on a coterie of Russian Westernizers. "The judge of a [regional] court, Violetta Kyvask, acquitted four defenders of the Bronze Soldier . . . ," wrote Victor Shenderovich. "I don't know the details of the lawsuit as such, but I do know political details. Because the lawsuit was in fact political, as those in the dock were and are aliens, that is, people openly hostile to the Estonian state. The office of the prosecutor collected evidence of their guilt. Even the country's prime minister had a chance to weigh in, appealing to the court to apply all the power of the law. . . . The prestige of the state was on the line! Could anyone entertain doubt as to the outcome? What a ridiculous question! Just transfer this situation to our lovely state and recall the face of Judge Kolesnikova [the judge at the process of Mikhail Khodorkovsky – G.I.] or the face of Judge Olikhver [the judge at the process of one of Khodorkovsky's associates], whichever you like better, and imagine the verdict. And having done so, express joy for little Estonia, in which there is room for a free court, as a result of which Estonia has prestige."

Other problems Russia has with Estonia include the failure to sign the 2005 border treaty; the reaction to what Russia deemed their politically incorrect behavior of the Estonian President Toomas Hendrik Ilves in Khanty-Mansiisk (in the deep interior of Russia) and a 2009 spy scandal.

A border treaty was negotiated and signed by Estonian and Russian diplomats in May 2005, which the Estonian government was at least as eager as the Kremlin to consummate because of the requirement that countries cannot belong to the EU while they have border disputes with neighbors. The Estonian parliament dragged its feet regarding ratification, insisting that this treaty be identical to the Treaty of Tartu of 1920, in which the Estonian frontier with Bolshevik Russia was defined to the satisfaction of the Estonians, who had the friendship and support of the Western Powers.

In 2005, the Estonian parliament refused to ratify the treaty because the demarcation of the border in the treaty left a speck of territory historically claimed by Estonia in Russia. The Estonian position was based on another consideration, that is, that the Estonia of 2005 is a continuation of the Estonia of 1919, with no reference to, or recognition of, the Estonia under Soviet communist rule between 1940 and 1991. By the end of the summer of 2005, the Estonian parliament lacked the two-thirds majority needed for ratification.

Not surprisingly, the Kremlin was livid over this perceived anti-Russian stance. On September 1, 2005, Putin announced the withdrawal of Russia's signature of the treaty. To be fair to the Russian side, the Kremlin could not sustain the embarrassment of supporting a treaty which was increasingly questioned, criticized, and opposed by Estonian conservatives and nationalists. The Putin Kremlin had no intention of ceding territory to Estonia claimed by the government in Tallinn and did not appreciate the perceived way in which the United States had a hand in bolstering the Estonian position which, in 2006, was to look to the EU and NATO for help in dealing with Russia.

In June 2008, the Estonian delegation attended the World Congress of the Ugro-Finnic Peoples held in Khanty-Mansiisk, a West Siberian town better known for nearby oilfields than for being the capital of the ethnic homeland of two indigenous Siberian peoples, the Khanty and the Mansi. The point of Estonian as well as Finnish and Hungarian participation in that meeting is that the native languages of all these European peoples are not Indo-European but share common ancestry with the languages of the Khanty, Mansi and several other groups, such as Komi, Mari, and Udmurts, whose ethnic homelands are within Russia. Many centuries ago, the ancestors of Estonians, Finns, and Hungarians left their ancestral lands (now within Russia) and migrated west. During the 2008 Congress, the Estonian president called upon Ugro-Finnic ethnicities' self-determination, which Russian hosts construed as an inappropriate abuse of their hospitality. Further on, in his speech to the Congress a member of the Russian parliament touched upon what he called a Western double standard that he believed revealed itself in differing attitudes toward (a) the beating by unknown thugs of a chairman of the ethno-cultural union of Mari (in Russia) and (b) the beating of Russians protesting the transfer of the Bronze Soldier by the Estonian police. At this the entire Estonian delegation, headed by President Ilves, left the congress.

Finally, it was revealed in February 2009 that one of Russia's highest-placed spies in NATO was Hermann Simm, a former Estonian official. He pleaded guilty to treason and was jailed for 12½ years. The Estonian authorities have released some details of the case that has had the intelligence world buzzing for a year or so. Russia's foreign intelligence service, the SVR, had recruited Mr. Simm during his holiday in Tunisia in 1995. He was a prime catch. He had just finished a stint as a top policeman and was starting a new security job at the defense ministry. The approach was made by Valery Zentsov, once a KGB officer in Soviet-occupied Estonia. Mr. Simm was neither blackmailed nor, at first, bribed; all he wanted was his Soviet-era rank of colonel back. At a third meeting he was put on the payroll and received just over $100,000 in all.

Russia and Latvia

Russia's relations with Latvia have endured some of the same difficulties as the development of Russian relations with Estonia and Lithuania. The issue of discrimination against Latvia's Russian speaking community, which is close to half the country's total population, was especially troublesome to the Kremlin, so much so that during 2005 there was reference in the Russian press to the possibility of economic sanctions to punish the Latvian government for its failure to protect the rights of all Russian-speaking citizens.

Russian speaking residents of Latvia who cannot pass the language test were denied full citizenship rights and a Latvian passport. Until they became full citizens, Russian speaking Latvians could have only a so-called "non-citizen's" passport. When leaving the country with this passport they ran the risk of a denial of permission to return to Latvia.

The Latvian government, to its credit but under pressure from the EU, amended a severe citizenship law. For example, on April 30, 2004, the Secretariat of the Special Assignments Minister for Social Integration in co-operation with the Ministry for Children and Family Affairs sent a letter to parents of the 15,000 children born after 1991 inviting them to register their children as citizens of Latvia and also explaining the procedure. In certain cases, the naturalization examinations have been simplified. Since June 2001, graduates of national minority schools who have passed the Latvian language exam are not required to take the language test for naturalization. Applicants who have reached the age of 65 only need to take the oral part of the language test.

In June 2001, the government reduced the standard naturalization fee by a third and broadened the number of groups eligible for paying a reduced naturalization fee. On average around 40 percent of applicants for citizenship pay either a reduced state fee or are fully exempt from it. As a result of these measures, in the twelve years since the beginning of the naturalization process in 1995 to 2008, the number of non-citizens has decreased from 29 percent to 18 percent (Citizenship in Latvia, Ministry of Foreign Affairs of the Republic of Latvia; 28 March 2008; www.am.gov.lv/en/policy/4641/4642/4651/).

Latvians have been just as determined as their Baltic neighbors to get the Kremlin to agree to recognize Latvia's period of membership in the Soviet Union as a "union-republic" as a "soviet occupation." As long as the Kremlin refuses to accommodate Latvia on the occupation issue, Riga will continue to refuse to conclude a treaty demarcating its border with Russia despite pressure from the EU that members have no border disputes with neighbors. If a country wants membership, it must resolve such disputes before joining. The EU gave the Baltic republics a break in this regard by admitting them before a final resolution of any differences they had over the border with Russia.

Another source of anxiety for the Kremlin and for the vast majority of Russians has been the Latvian government's favoritism toward Latvia's so-called freedom fighters, who fought on Hitler's side during World War II. To be sure, ethnic Latvians fought on both sides of the frontline; and even before that—during and after the 1917 Communist revolution in Russia—a disproportionately high number of Latvians were ardent supporters of communism. Many worked for the forerunner of the KGB. But despite the fact that the Latvian president, Vike Freiberga, became the only leader from the Baltic States who decided to participate in the Moscow-based celebration of the 60th anniversary of the victorious end of World War II, the Latvian political establishment appears to entertain warmer feelings to Latvians who fought on Hitler's side. There were roughly 150,000 of those fighters, many of whom took active part in punitive operations against Soviet partisans in Belarus. "[A]mong the last defenders of Hitler's Reich Chancellery . . . were eighty Latvian soldiers from the fifteenth Battalion of the fifteenth Waffen SS Division. The last commander of this battalion, Lieutenant Neilands, would act as interpreter for the talks on German surrender between the commander of Berlin, General Krebs, and the Soviets. Yet another Latvian, the Soviet colonel Nikolajs Berzarins, would become the first commander of Russian-occupied Berlin." [Modris Eksteins, *Walking Since Daybreak. A Story of Eastern Europe, World War II and the Heart of the Twentieth Century* (Papermac: London, 2000), p. 218.] But today only one tyrant's Latvian soldiers (Hitler's, not Stalin's) are treated as heroes. This was particularly apparent on March 16, 2009, when the procession of surviving veterans of the Waffen SS and their young supporters marched from the Doma Cathedral to the Freedom Monument in Riga and laid flowers. Several people, mostly Russian-speakers, who whistled and shouted "Hitler kaput!" (Hitler is finished!) during that ceremony were apprehended by Latvian police. Russian Ambassador Veshnyakov issued a statement condemning the Latvian authorities' connivance with Nazi collaborators.

It is particularly insulting to Russians that the glorification of the Nazis in Latvia was almost entirely ignored by Western media. Even George Soros's Open Society foundation was mute; and in the past it had even co-sponsored an apologetic film about the Latvian Legion of the Waffen SS.

To be sure, Russian rhetoric in regard to Latvia just as strongly betrays imperial complexes as genuine, and at times justifiable, indignation. This was particularly apparent during the January 2009 violent rallies in Riga against the Latvian government's austerity measures, as Latvia appears to be hardest hit by the recent global financial crisis and recession. "What the Latvian economy hinges (or rather hinged) on," gloated Russian daily *Izvestia*, "is backward agriculture, tips from Brussels, and Russian transit [via Latvian ports]. Can one build something decent on such a foundation? No wonder, the economy collapsed as a house of cards."

RUSSIA AND OTHER FORMER SOVIET REPUBLICS

Russia has obvious and compelling interests in the ex-Soviet Eurasian republics. Belarus, Ukraine, Moldova, Georgia, Armenia, Azerbaijan, Kazakhstan, Turkmenistan, Uzbekistan, Tajikistan, and Kyrgyzstan, as well as, occasionally, the Baltic States are referred to by the Russians as "The Near Abroad." Russia needs close ties to these republics for economic and geostrategic reasons. Their economies are closely linked to the Russian economy. Russia does not want outside countries like Turkey, Iran, and China to strengthen political, economic, and sociocultural ties with the Eurasian republics at Moscow's expense. Most of all Russia is opposed to the growing American influence in the post-Soviet space, culminating in the Baltic States joining NATO and in steps undertaken for Georgia and Ukraine to follow suit. Russia is particularly hostile to the idea of the expansion of NATO within its traditional and lasting sphere of influence. Not only is NATO a chief element of the Western military infrastructure that was set up to contain the Soviet Union's ambitions during the Cold War, but it also, from the Russian point of view, lost its *raison d'etre* when its rival, the Warsaw Pact, was disbanded. When NATO not only continued to exist but actually expanded, it became difficult for Russian authorities to see NATO the way it projects itself, as an instrument for maintaining international peace. To Russian officialdom, NATO is still an instrument for furthering the geostrategic interests of the United States at the expense of Russia.

Russia has certain advantages and disadvantages in dealing with the former Soviet republics. Advantages have to do with using Russian, still a language of inter-ethnic communication in much of the post-Soviet space, and with economic complementarity between Russia and the other former republics. Russia supplies most of the oil, natural gas, and timber needed by other ex-Soviet republics and uses this leverage to make them accommodate to its policies. In varying degrees the republics suffer from underdevelopment, military weakness, and political instability; and therefore they find it difficult (though not impossible) to oppose Russian interests. Moreover, until recently most of the former Soviet republics were outside the protection of the West, which in any case would be hesitant (or so it seemed) to take their side in a conflict with Russia. When the formula of Western non-interference was shattered before and during the 2008 war between Russia and Georgia, relationships between Russia and the West reached their lowest point since the end of the Cold War. The 2004–2005 "Orange Revolution" in Ukraine was also seen in Russia as a result of Western interference. A modest improvement in Russian relationships with the United States in the early phase of the Obama administration was preceded by the United States' backtracking from its enthusiastic support for Georgia's and Ukraine's accession to NATO.

Russia nevertheless faces substantial obstacles to its influence-building policies in the Eurasian republics. The republics are hypersensitive to their newly won independence; none wishes the restoration of a Russia-dominated community. Furthermore, the poverty of the Russian economy and military establishment severely limits the Kremlin's ability to pursue an intrusive policy there. And the Kremlin cannot easily dismiss the opposition of the West to an expansion of Russian influence in the region given other strategic priorities.

The Commonwealth of Independent States (CIS)

The Eurasian republics, along with Russia, make up the Commonwealth of Independent States (CIS), founded in December 1991, by Russia, Belarus, and Ukraine and subsequently expanded to include the other ex-Soviet republics except Latvia, Lithuania, and Estonia. The most important rationale behind the CIS was the enabling of multiple ties that bind together the fragments of what used to be a single national economy. Deprived of those ties, many industrial enterprises would have simply ceased to exist because they depended on fuel, raw materials, and semi-finished products originating from what had suddenly become a foreign country and they often hinged on demand generated on the other side of what became international borders. In many cases ties with Russia were of critical significance, as it had accounted for roughly 70 percent of the Soviet industrial output and supplied many republics with fuel and raw materials. To a significant extent, the breakup of the Soviet Union was caused by strong reluctance on the part of Russia's political establishment to prop up the economies of other Soviet republics "for free." From the viewpoint of the Kremlin, the CIS would enable Russia to use its resources as economic leverage to maintain ties that fit its interests and to discard those that did not. At first, in 1992 and 1993, Russia had hoped that, through its leadership of the CIS, it could expand its influence over the large ex-Soviet space the Russians call the Near Abroad. But the Kremlin soon recognized that the Commonwealth would not easily be an instrument of Russian policy. Most ex-Soviet republics, especially Ukraine, resented and resisted Russia's perceived use of the CIS to promote a Moscow-oriented, Moscow-led union. Increasingly, after 1993, under pressure by nationalist politicians, the Kremlin acted outside the framework of the CIS to influence politics in the Near Abroad. For example, Russia has deployed troops in Georgia and Tajikistan—ostensibly to help local authorities deal with internal unrest. Then, in 1994, Russia succeeded in getting the CIS to acknowledge Russia's leading role as the region's "peacekeeper." Although this development raised concern in the West, there was no official opposition to it.

Erosion of Purpose and Unity

The shadowy, ineffectual character of the CIS was apparent early on when the Baltic republics refused to join or have anything to do with the CIS.

In February 1995, at a meeting in Kazakhstan, twelve CIS states took a decidedly less accommodating position regarding the role of Russia in the region. Taken aback by aggressive and brutal Russian military action in Chechnya, CIS leaders affirmed their reluctance either to allow further integration within the CIS or to assign Russia the leading role it wanted to play in the Near Abroad. At a summit meeting held in the Moldovan capital of Chisinau in October 1997, CIS leaders blamed Russia for the way in which the CIS had become increasingly, in their view, a tool of Russian hegemony.

Former Georgian president Eduard Shevardnadze was especially outspoken, accusing the Russians of mischief making in Abkhazia, where Georgian refugees from a separatist conflict had difficulty returning to their homes because of Abkhazian resistance, believed to be encouraged by the Kremlin. Shevardnadze said that Georgia would consider leaving the CIS if

Russia did not stop impeding the peace process in Abkhazia. He also accused the Russians of playing a role in a recent assassination attempt against him and made no secret of his hostility to both Primakov and Yeltsin.

Shevardnadze and other presidents argued that CIS integration was hindering, not helping, the development of harmonious relations among member states. In this spirit, Uzbekistan President Karimov said that the CIS states had to solve their internal problems and strengthen their individual economies before they could move toward integration. Azerbaijan President Aliyev added that his country could not support integration until the Karabakh conflict was resolved.

After the October 1997 summit, Russia became pessimistic about the future of the CIS, criticizing the phenomenon of "geopolitical pluralism" that meant the steady movement of CIS countries away from Russia. And by the beginning of the new century, CIS countries had concluded a wide variety of unilateral economic and military-political ties among themselves and with outside countries, strengthening their role as actors in world politics without Russia. In November 1999, for example, Georgia and Azerbaijan, backed by the United States and Turkey, agreed to build a pipeline to transport Azeri oil to Western markets on a route that bypassed Russia. Similarly, throughout the 1990s and especially toward the end of the decade, Ukraine and the Central Asian republics have strengthened political and military links to the West, in particular the United States.

On September 19, 2003, Moldovan President Vladimir Voronin called attention to the way in which several CIS members including Russia, Kazakhstan, Ukraine, and Belarus recently had gone outside the framework of the CIS to set up a so-called "common economic space;" though he made clear that, at least at the moment, Moldova had no plans to leave the CIS because of the benefits that the large CIS market provided for interstate trade. And on November 4, Georgia's Defense Minister, Giorgi Baramidze, said that he would not attend an upcoming CIS Defense Council meeting, adding that the CIS is "yesterday's history." And in February 2006, Georgia officially withdrew from the CIS Defense Council, saying that it sought NATO membership and could not be part of two military structures simultaneously.

This Georgian policy obviously weakened still further the unity and purpose of the CIS. The Kremlin retaliated with a ban on the import into Russia of several of Georgia's wine exports. In turn, on May 2, 2006, Georgian president Mikheil Saakashvili did his own bit of anti-CIS and implicitly anti-Russian retaliation by announcing that Georgia was considering withdrawing from the CIS until it could see the benefits of continued membership in an organization which is highly unpopular. Georgia officially withdrew from the CIS following its August 2008 war with Russia.

Furthermore, on April 9, 2005, a high ranking Ukrainian official said at a news conference that the CIS had little hope of evolving into an organization its members could trust and support. He even went so far as to say that the Ukrainian government, despite its close ties to the Kremlin, was considering not making financial contributions to the various CIS organizations. From the very beginning of its membership in the CIS, Ukraine—led by the pro-Russian President Leonid Kuchma—had been concerned about the CIS becoming a tool of Moscow to expand Russian influence. But, when Viktor Yushchenko, who was pro West, won the presidency in 2004, he went even further, questioning the very future of the CIS. On August 26,

2005, Turkmenistan decided to withdraw from the CIS to become what is termed "an associate member." Turkmen President Sepaamurat Niyazov for some time had been boycotting the CIS by not attending meetings of its economic and military bodies such as the Council of Defense. And he did not attend the summit of CIS leaders held in the Russian city of Kazan to commemorate the 1,000th anniversary of the city. He preferred bilateral inter-state diplomacy with his neighbors, including Russia, to address issues and resolve problems.

Reflecting on ongoing tensions within the CIS, President Putin of Russia made a sobering confession at a press conference during his trip to Armenia in March 2005. "Whereas in Europe countries worked collaboratively in order to unite, the CIS was created expressly for the sake of a civilized divorce," said Putin. "There was only one goal: the CIS was created so the process of the Soviet Union's disintegration would unfold in the most civilized manner possible. And this goal was achieved," reiterated Putin. He, at the same time, cautioned against the disbandment of the CIS and called it "a fairly useful club and a platform where national leaders can regularly meet, talk about problems, and either resolve them collectively or make them subject to bilateral agreements." The CIS will therefore continue to exist. That said, however, there are several levels of cooperation within that Russia-centric alliance: a fairly significant political estrangement between Russia and Ukraine against the backdrop of still close economic ties between the two; identical estrangement between Russia and Turkmenistan coupled with dwindling bilateral economic ties; estrangement between Russia and Moldova; a tighter political cooperation between Russia and Uzbekistan (with some apparent ebbs and flows in their relationship); still tighter cooperation between Russia and Armenia, Kazakhstan, Kyrgyzstan, and Tajikistan; and the closest, although still problem-ridden cooperation between Russia and Belarus. Russia's partners from the last two groups participate in two co-extensive Russia-centric alliances that exist side by side with CIS but demonstrate higher levels of cohesiveness compared with the CIS. One of these tighter alliances was forged through the so-called Collective Security Treaty of 1992 and another through the signing of the 2002 agreement about the formation of a custom union, the so-called Eurasian Economic Community.

Relationships between Russia and other post-Soviet countries will be discussed as part of corresponding country essays.

RUSSIA AND THE WEST: AN INTRODUCTION

This is perhaps the most difficult topic to discuss, as simply following the ebbs and flows in mutual relationships between Russia and the West would not do justice to the topic. The newscasts and even speeches of the major players involved appear to be important, but it is easy to get carried away by emphases on specific events whose significance and lasting impact are not self-evident at all. It would be even less meaningful to give vent to stereotypes and clichés which are so tenacious both in Russia (in regard to the West) and in the West (in regard to Russia). Analysis is further complicated by the fact that although most Russians seem to agree on who the major Western actor is—in which capacity most see the United States, Russians do not see the West as a single or cohesive entity. For example, despite all complications in mutual relationships, Germany, France, and

Italy (particularly under Berlusconi) are construed by Russian political elites as all but friends; while the United States and Britain are seen as foes seeking to undermine Russia's stability; and the countries of East Central Europe and the Baltic States (both being integrated into Western structures) are seen as the least friendly and most ungrateful former vassals and telltales fixated on spiting their former boss (Russia) and currying favor with the new one (the United States). But even in this area there are noticeable differences, as Russia's relationships with Poland are incomparably more poisoned by mutual grievances than Russia's relationships with, say, the Czech Republic or Slovakia.

Perhaps the only overarching formula appearing to capture the tenor of Russia's attitude toward the West is that it has reverted to the ambivalence that characterized it before the Communist Revolution. In simplest terms, this is a love-hate relationship. On the one hand, there is a deep-seated inferiority complex and the desire to ultimately imitate Western arrangements in economy and domestic politics. Yet on the other hand, there is resentment over the West's mentoring Russia on democracy and the alleged reluctance of the West to take Russia's interests into account, and even worse, the alleged desire to destabilize and weaken Russia. Perhaps forever destined to oscillate between the opposing views of the West, as reflected in historical arguments between the Slavophiles and Westernizers, Russia's public opinion is currently noticeably closer to the Slavophiles, who claimed that, while Russia may share some traits with the Western civilization and even admire it, Russia is not part of that civilization—hence mutual misunderstanding, disagreements, and geopolitical rivalry. Apparently, despite intense demonization of his views in liberal America (because of his alleged view of Islam), Huntington was justified in stating that "cultural identities, antagonisms and affiliations will not only play a role, but play a major role in relations between states," and that, moreover, "Western belief in the universality of Western culture suffers three problems. It is false; it is immoral; and it is dangerous." To an ideologically unbiased observer of both Russian and, say, American discourse on Russia-and-the-West, Huntington's assertions are impossible to shrug off.

"For three centuries now, our political thinking has been wavering within a false dilemma—whether to blend with Europe or oppose it," writes Andrei Piontkovsky, a Russian liberal political commentator and a darling of some Washington-based think tanks. While he calls the dilemma false, this is a value judgment; and the fact that Russia has been effectively "wavering" for three centuries may, in and of itself, mean more than how we think about it. "This old mixture of attraction and resentment, which is the archetype of Russian political consciousness, has lately become aggravated yet again," writes Piontkovsky. In his essay he exemplifies Russia's vaccilations thusly. "We are part of Europe, but we are being pushed out of Europe. We would like a strategic relationship with the West, but we're being rejected. They didn't believe our striving for peace and friendship, they saw our goodwill as weakness." Dozens of variations of such passages in political prose repeat the motifs of the famous poem written more than 80 years ago:

> Come unto us, from the black ways of war,
> Come to our peaceful arms and rest.
> Comrades, while it is not too late,
> Sheathe the old sword. May brotherhood be blest.

If not, we have not anything to lose.
We also know old perfidies.
By sick descendants you will be
Accursed for centuries and centuries.

The above is an excerpt from the 1918 poem "Scythians" by Alexander Blok. It is provided in full (on p. 000–000) as part of the collection of articles about Russia. As per Piontkovsky, "the 'Scythian Complex' has taken over the whole Russian political class.... All of them are looking at Europe [and on America – G.I.] in a Blok-like way, 'with hatred and with love,' differing perhaps, only in the relative proportions thereof."

RUSSIA AND CENTRAL EUROPE

Today, what is known as Central or East Central Europe is integrated into Western political and economic structures and thus appears to be part of the collective West. Following the collapse of communist rule and Russia's subsequent loss of influence throughout the region in the 1990s, one of the biggest problems facing the Kremlin in the former Soviet bloc countries, was the enlargement of NATO. In 1999, Poland, Hungary, and the Czech Republic were added to the organization. Another expansion came with the accession of seven Northern European and Eastern European countries: Estonia, Latvia, Lithuania, Slovenia, Slovakia, Bulgaria, and Romania. These nations were first invited to start talks of membership during the 2002 Prague summit, and joined NATO on March 29, 2004, shortly before the 2004 Istanbul summit. Most recently, Albania and Croatia joined on April 1, 2009, shortly before the 2009 Strasbourg–Kehl summit. This development seemed to punctuate Russia's loss of influence in a region where it has had extensive interests for centuries and which it had controlled for 45 years after the end of World War II.

Throughout the 1990s the government of President Yeltsin complained bitterly to the NATO allies that admission of these former allies of the defunct Soviet Union was a violation of the pledge given to Soviet president Mikhail Gorbachev, in 1989, that NATO would not take advantage of the new situation in the region. For its part, the Kremlin would refrain from Brezhnev-style interventions to block the ascendancy of democratic and reform-oriented leaders. The betrayal of this pledge, in Russia's view, would usher in a new phase of competition and confrontation between Russia and the West.

The May 1997 Russia-NATO Agreement

To reassure the Kremlin that NATO had no sinister intent as far as Russia was concerned, the Alliance concluded an agreement in May 1997, providing for systematic, institutionalized East-West consultations on strategic and other issues of mutual concern. Called the Founding Act, the agreement established a new body, the 16+1 Council, to be composed of representatives of the 16 NATO members and Russia. It would meet periodically and provide the Kremlin with a formal arena in which to discuss and debate NATO policies. Although Russian president Yeltsin suggested to his government and the Duma that the Founding Act gave Russia a veto over NATO policy making, the Alliance was quick to make clear that nothing of the kind was true and that Russia was not a *member* of NATO as a result of the Founding Act. Russian leaders were angry over this diplomatic setback, with nationalist politicians accusing the Yeltsin government of selling out to the West.

Russia's early misgivings about the integration of East Central Europe into NATO appear to have been justified by the peculiar stratification that developed within NATO, with its new members (former Soviet Union's members and satellites) showing more critical attitudes toward Russia compared with the old members and at the same time following the American leadership more closely than other NATO countries. This division became stereotyped as Old Europe versus New Europe. The term "Old Europe" was first popularized in January 2003 after then-U.S. Secretary of Defense Donald Rumsfeld used it to refer to European countries that did not support the 2003 invasion of Iraq, specifically France and Germany. Since then, the term has been used by pundits and media personalities to describe various combinations of the first-world countries of Europe; comedian Jon Stewart, for example, has applied the term humorously to all Western European Great Powers, most notably Germany, France, and the United Kingdom. Conversely, "New Europe" has come to mean three former Soviet Republics (the Baltic States) plus former members of the Warsaw Pact. But even within the New Europe, however critical it is of Russia, there is no unanimity on this issue, as the Baltic States and Poland are most critical and, in the eyes of the Russian political class, unduly (though predictably) hostile and confrontational. Poland is clearly special in this regard in view of a long history of Polish grievances against Russia.

Russia and Poland

Russian-Polish relations since Putin became president in March 2000, have been plagued by mutual suspicion and distrust arising out of a long history of problems and crises dating back at least to the late-eighteenth century. Since 1999, Russia has been troubled by Polish membership in NATO; by Polish criticism of Russian policy in Chechnya; and by Poland's cultivation of the Baltic republics after they gained independence from Moscow in September 1991—most especially Lithuania, with which Poland has had a relationship dating back a half-dozen centuries. Other irritants in Russian-Polish relationship have been associated with (a) Poland's activism in neighboring Belarus and particularly Ukraine; (b) the July 2005 beating of three children of Russian diplomats in Warsaw and a seemingly masterminded retaliation in Moscow just a couple of weeks later when two Polish diplomats and one journalist were beaten up; (c) Russia's refusal to buy Polish meat on the grounds that it contains forbidden substances and Poland retaliating by torpedoing the resumptions of talks between Russia and the EU to sign a new strategic partnership treaty (the meat conflict lasted for two years and was resolved in 2007); (d) different and seemingly diverging attitudes toward historical events, notably the 1940 slaughter of almost 22,000 interned Polish officers in Katyn in the Smolensk region (Russia initially opened its archives and acknowledged the atrocity as perpetrated by the Soviet security apparatus; but in Poland's view, Russia did not take full responsibility for the crime); and (e) the would-be installation of the American ballistic missile interceptor in Poland (an idea initiated and vigorously pursued by the Bush administration, but reconsidered by the administration of President Obama), which Russia says would alter the strategic balance in NATO's favor.

Of all Central European countries entering NATO, Poland has been the most troubling to the Kremlin because of the known Polish hostility for Russia. Perhaps hostility is not the most accurate term, as Russian popular culture arguably shares more with its Polish counterpart (e.g., aversion to rationalism of the supposedly German type, cuisine, drinking habits; etc.) than with any other Central European culture. But in one way or another, Russia actually dominated Poland from 1717 to 1991, and many events throughout that period left lasting scars embedded in the Polish world view. This was clear when post-Communist Polish leaders eagerly rushed to integrate Poland in every conceivable way with the West and sought guarantees of never again being violated by Russia. Needless to say, Russia resents this anti-Russian streak in the Polish national psyche. There is, however, a noticeable part of Polish national discourse critical and even dismissive of demonizing Russia. Thus Tomasz Zarycki, a leading Polish sociologist, sees Poland's perception of Russia "as a manifestation of a not so unusual phenomenon of interaction between two peripheral regions." According to Zarycki, Poland copes with an inferiority complex vis-à-vis the West; and "Russia's central role in the formation of modern Polish identity" compensates for Poland's perceived "weakness in relations to broadly defined West." Zarycki distinguishes five major functions of a negative image of Russia: the rescaling of Poland's weaknesses or the function of a negative point of reference; the strengthening of Poland's European identity; a unifying threat; the role of oppressor, crucial in Polish victimization-based identity; and finally, as an area of exclusive expertise of Poles. Writing for one of the leading Polish dailies, Piotr Skwecinski makes the point that one of the fundamental elements of Polish thinking about Russia is hypocrisy. "It is simply impossible to imagine such a Russia whose existence many Poles would kindly condone." It seems important to keep such introspective remarks in mind when analyzing Polish pronouncements on Russia.

The Kremlin has not liked Poland's criticism of Moscow's policy toward Chechnya. Poland sees in Chechnya a ruthless use of Russian military power, reminiscent of aggressive Soviet diplomacy after World War II, in particular the Soviet interventions in Hungary in 1956 and Czechoslovakia in 1968, Soviet aggressiveness in Poland in the October 1956 political crisis, and the summer 1980 Solidarity crisis. These parallels of course ignore the fact that Poland, Hungary, and Czechoslovakia were not parts of Russia, whereas Chechnya is precisely that, a part of the Russian Federation with no constitutional right to withdraw from it. Anyway, when in February 2000 the Polish government tolerated—and by implication sympathized with—public demonstrations against Russian policy in Chechnya, Putin, then acting president, made clear his displeasure.

After 2000, Polish governments led by President Alexander Kwasniewski, an ex-communist who sought to balance Poland's pro-West policies with the maintenance of friendly ties to Russia, have toned down criticism of Russia. The Kremlin has not had to remind Warsaw of the strategic reality of Poland's location between Germany and Russia and that good working relations with Russia are not a luxury but a necessity. Putin and Kwasniewski held several summit meetings in Warsaw and Moscow, which demonstrated a strong Russian interest in getting along with Poland.

But despite the mutual understanding that Russian-Polish relations need to be improved, there are periods when the old animosities are clear. Such a time was the summer of 2005. Relations between the two countries had reached a new high of mutual antagonism and hostility. The rhetoric of each toward the other was as mean spirited as ever with a Putin aide saying something like "Poles talk about Russians as anti-Semites talk about Jews" and Polish foreign minister Adam Rotfeld replying in effect: "You [the Russians] are looking for an enemy and you find it in Poland."

What had caused this rift between Russia and Poland were the roles the two countries played in the November–December 2004 presidential elections in Ukraine. Poland made public its strong support of the candidacy of Viktor Yushchenko, the self-described pro-West reformer. Indeed, president Alexander Kwasniewski, and his predecessor, Lech Walesa, both visited Kiev in November 2004, to lobby for Poland's preferred candidate and against his opponent, the conservative, pro-Russian Viktor Yanukovich. The Kremlin was livid over the eventual victory of Yushchenko and of the material support for his campaign from both Poland and the United States. That the Kremlin also had interfered in the Ukrainian election was no reason for it to hold its fire against Warsaw for its interference.

Relations deteriorated further in the summer of 2005 when Putin invited German Chancellor Schroeder to celebrate the 750th anniversary of the Russian city of Kaliningrad, known before the World War II as the East Prussian town of Koenigsburg. Putin did not invite Poland, which borders Kaliningrad. President Kwasniewski was furious over the slight, criticizing it publicly on Polish television along with Berlin's support of a new Russian gas pipeline under the Baltic Sea that would bypass Poland, depriving it of an opportunity to earn money by collecting a toll.

In October 2005, the ultra-nationalist and conservative, Lech Kaczynski, was elected Poland's president, while his brother Jaroslaw became prime minister. Both have turned out to be another source of tension in an already severely strained Polish-Russian relationship; although Kaczynski insists publicly he is committed to good working relations with Russia—if only because Russia has become the chief supplier of energy for Polish industry. Still, he shares the widespread antipathy toward Russia and has said and done little to reassure the suspicious Kremlin that he is sincere in his statements about improving relations. With Jaroslaw Kaczynski being replaced by a more pragmatic and less ideological Donald Tusk as prime minister of Poland (in November 2007), some limited rapprochement between Russia and Poland is currently occurring.

RUSSIA AND SOUTHEASTERN (BALKAN) EUROPE

A major problem for Russia in the Balkans since the end of communist rule in the region, as well as in Russia itself, in the late 1980s and early 1990s was the collapse and disintegration of Yugoslavia. Yugoslavia had been a problem for Russia ever since it became communist under the leadership of Josip Tito after World War II. Tito was independent of the Soviet Union and determined to chart a course of domestic and foreign policy quite independently from, and frequently at odds with, the Soviet Union. And, when Yugoslavia started to break up into separate, ethnic-based republics in the years following Tito's

death in 1980, there were other problems. By the early 1990s, with the outbreak of war among some of the ex-Yugoslav republics, notably Serbia, Croatia, and Bosnia-Herzegovina, the Kremlin found itself on the threshold of an unwanted confrontation with the West, in particular the United States. The confrontation became inevitable as Russia and the West took sides (in the Yugoslav conflict) in perfect compliance with Huntington's clash of civilizations thesis and/or with his map of the eastern border of Western civilization (**see map on page 10**). Moscow sympathized with its co-religionists, Orthodox Serbs (and so did Orthodox Greece) while the West opposed both the Serb insurgency in the Bosnian civil war (1992–1995) and the Serb Republic's brutal crackdown on Albanians in its province of Kosovo (1998–1999). It is worth remembering that the secessionist plans of Catholic Croatia and Slovenia, the implementation of which actually sparked the entire conflict in Yugoslavia back in the early 1990s, had been enthusiastically supported by Germany, Italy, and the United States.

Russia and the End of the Milosevic Era

In the aftermath of the 1999 Kosovo crisis, Russia recognized that the West, led by Washington, meant to isolate, weaken, and defeat Yugoslav president Slobodan Milosevic, who had been indicted by the United Nations War Crimes Tribunal in the Hague. But Russian president Putin sent a message of support to Milosevic. In May 2000, Putin hosted a meeting in Moscow with Milosevic's defense minister, Dragoljub Ojdanovic, another high-ranking Yugoslav official under indictment. Putin made a point of showing Russian support of Milosevic by giving Ojdanovic a conspicuously generous reception, including a loan to Belgrade worth about $102 million and an announcement of a sale of oil to Yugoslavia worth $32 million. Putin's behavior left no doubt about his anger over the NATO military intervention in Serbia and NATO's evident dominance in the south Balkans. It also affirmed the Russian position on the Hague War Crimes Trials—that they were politically motivated and inherently unfair.

But Russia's pro-Serb position soon proved untenable. Yugoslav opposition leader Vojislav Kostunica, a democrat committed to reconciliation and cooperation with the West, defeated Milosevic in the September 2000 presidential election. Milosevic refused to acknowledge the results and tried to reverse them, a move that provoked an explosion of public wrath against his regime. With violent, popular antigovernment demonstrations in Belgrade threatening to plunge the country into civil war, Milosevic capitulated, accepted his defeat, and gave way to Kostunica.

At first the Kremlin stood by Milosevic, implicitly endorsing his efforts to question and undermine the legality of the voting. But as it became clear that his days in power were over, the Kremlin slowly withdrew its support of Milosevic and recognized Kostunica's legitimacy.

Russia and Kosovo Since 2000

Throughout 2001 Russia had problems with the surge of Albanian nationalism and separatism in Kosovo, which threatened Yugoslavia's tenuous hold over the province. Remnants of the Kosovo Liberation Army (KLA) ignored the United Nations authority, terrorizing Serb inhabitants into fleeing Kosovo, to strengthen the Albanian campaign for independence from Belgrade. Worse, armed Kosovar Albanians, many of them members of the KLA now in different guise and garb, stirred up Albanian minorities in southwestern Serbia and northern Macedonia. They wanted the small Albanian minority in Serbia to become part of Kosovo. Kosovar Albanians also made common cause with leaders of the large Albanian minority in Macedonia seeking increased administrative autonomy of Skopje.

In June 2001, Putin expressed his concern over these developments to Serbian President Kostunica in a summit meeting held in Belgrade. Putin and Kostunica both said that the international community should do more to discipline Kosovo or else risk more civil war in the south Balkans, with Macedonia the next center of conflict. Putin called for sealing Kosovo's borders tightly to prevent KLA troublemakers from infiltrating other countries and destabilizing them. Putin also reiterated a Russian commitment to Kosovo's remaining part of Yugoslavia, though he supported Kosovo's autonomy within Yugoslavia, as the West did. For the Kremlin, Kosovo bore a striking resemblance to the Chechnya issue in which Russia opposed independence for the Chechens. In 2006, Russia monitored negotiations between Serbs in Kosovo and the Republic of Serbia and representatives of the Albanian majority in Kosovo over the region's future in which the Albanians remained stalwartly committed to complete independence from Serbia.

By September 2009, 62 countries have recognized Kosovo after its unilateral declaration of independence on February 17, 2008. Russia did not join those countries. According to Russia's political establishment, the West exercised a double standard by supporting Kosovo's independence from Serbia and at the same time denying support to rebellious provinces of Georgia such as Abkhazia and South Ossetia. To be sure, no attempt to explain that these cases of ethnic separatism (Kosovo as a breakaway province of Serbia, on the one hand, and Georgia's breakaway provinces, on the other) were really different in nature and thus warranted different reactions has ever been convincing. During and in the aftermath of Russia's August 2008 military conflict with Georgia, Russian leaders and diplomats repeatedly brought up the issue of a double standard on the part of the West as well as the fact that Russia had repeatedly warned the West about the Kosovo precedent. And while only Russia, Nicaragua, and Venezuela have so far recognized Abkhazia and South Ossetia as independent countries, the West's enthusiastic support for Georgia began to wane by the end of 2008.

RUSSIA AND ITS FORMER COLD WAR RIVALS

By most accounts, Russia's relationships with its former Cold War rivals, the United States of America, U.S.-centered NATO, and "Old Europe" (the bulwark of the European Union) soured during Putin's presidency to such an extent that parallels with the Cold War were invoked by observers and analysts. While most downplayed that comparison, the very fact of invoking the Cold War speaks for itself. The major bone of contention is the roles the United States and multinational Western structures are playing within post-Soviet space, which Russia sees as its inviolable zone of interest.

Russia and NATO

The issue of NATO expansion has had its fair share of American critics. Such critics are available on both sides of the American ideological divide and have Russia's misgivings about NATO at the core of their objections to its expansion. Thus, Pat Buchanan, a conservative pundit, devoted a 2007 op-ed piece to the tussle between Russia and Estonia over the removal of the Bronze Soldier monument. In that piece he writes: "So we come face to face with the idiocy of having moved NATO onto Russia's front porch, and having given war guarantees to three little nations with historic animosities toward a nuclear power that has the ability to inflict 1,000 times the destruction upon us as Iran." In his December 2008 *Newsweek* article, Michael Mandelbaum, a pundit usually advising the Democrats, writes that

> to the Kremlin, the expansion process has also seemed to be based on dishonest premises. U.S. officials advertised it as a way of promoting democracy, of forcing ex-Soviet states to reform. But the democratic commitment of NATO's first ex-communist entrants—Poland, Hungary and the Czech Republic—was never in doubt. And if the Americans truly believed that NATO membership was the best way to guarantee free elections and constitutional rights, why didn't they immediately offer it to the largest ex-communist country of them all, Russia itself? Instead, Moscow was told it would never be able to join. . . . NATO expansion taught Russia another lesson. The process went ahead because Moscow was too weak to stop it. This told the Russians that to have a say in European affairs, they needed to be able to assert themselves militarily. Last summer's war in Georgia was one result.

In his October 17, 2008 speech at the Kennan Institute, Mandelbaum was even more blunt on the issue when he said that "the only coherent rationale for expanding NATO and the reason that the Eastern Europeans, who had no doubts about their own commitment to democratic governance, wanted to belong was to protect them against Russia. The Russians could see this perfectly well, and it did not improve their attitude to the West to have Western, and particularly American, officials tell them, often in patronizing tones, that NATO expansion was not aimed at them, when it was obvious to all that it was."

For its part NATO was not entirely insensitive to Russian concerns and anxieties about the Alliance coming close to its borders. NATO has wanted to demonstrate that if and when this happened, Russia should have no cause for alarm; it is no longer against Russia. Quite the contrary, there was a consensus in the Alliance favoring an improvement of ties with Russia by somehow giving it a role to play with NATO while not admitting it to membership. In May 2002, NATO had devised a solution to the problem in the form of a Russian-NATO pact.

A New Russian-NATO Agreement in 2002

In mid-May 2002, at a Russian and NATO diplomats' meeting in Reykjavik, Iceland, a pact was signed that provided for a new partnership with the Alliance. By the terms of the pact, Russia for the first time becomes an equal partner in discussions among the then 19 members of NATO on a variety of issues, including nuclear nonproliferation, military cooperation, and civilian emergency planning. The pact makes clear and explicit the limitations of the new deal as far as Russia was concerned. Russia is in no way admitted to the Alliance; Alliance decision making on procedural and substantive matters involves only the legal, full members of the organization. Still, what Russia has is quite substantial and extremely significant for Russian relations with the West.

Russia now has a say on issues of direct relevance to its interests abroad. Though separate from the members of the Alliance, Russia is not inferior to them. Its partnership with NATO is of great symbolic importance because it punctuated the end of the Cold War without in any way suggesting a Russian defeat. Moreover, given the fact that NATO was first established to defend its members against the Russian-led Soviet Union, NATO's decision to approach Russia not only as an equal but also as a friend and quasi ally is of historical significance in the history of international relations since the end of World War II.

The pact affirms what had been in evidence since the collapse of the Soviet Union at the end of 1991, namely, Russia's slow but steady movement away from communist ideas, institutions, and processes in domestic and foreign policy toward Western-style democracy—even though it was, and remains, clear that the end of this transitional process still lies in the future.

East-West Tensions in 2005–2008

The Kremlin has a mindset toward the West that is inclined to see Western interests as threatening to Russia. For example, Federal Security Agency (FSB) chief and Putin confidante, Nikolai P. Patrushev, in March 2005, said that Western countries were trying to undermine Russian influence in the ex-Soviet Eurasian republics by supporting parties and politicians opposed to leaders of those republics who were pro-Russian. Patrushev told the Russian Duma that Western inspired anti-Russian activities were going on everywhere in the ex-Soviet space including Russia itself where, he asserted, the West was ready to use the tactics of subversion to influence politics in Russia. These accusations led to legislated restrictions on the behavior of organizations thought to be instruments of, and funded by, the West.

Russia in the G-8

In an effort to counter such views and to bring the interests of Russia and the West into closer alignment, the West invited Russia to be a member of the so-called G-8 club of Western industrial states that includes Britain, France, Germany, Italy, the United States, Canada, and Japan. In 2006, Russia was made chair of the group. Russia's chairmanship, plus the honor of hosting the July 2006 G-8 summit in St. Petersburg, gratified the Kremlin.

As in the case of NATO, there was trouble in the Kremlin's effort to become a credible associate of the Group's other members for not only economic but also political reasons. In February 2006, at a G-8 meeting in Moscow, tensions were evident over a view of some members that Russia did not yet deserve to be a member of the Group. They said Putin was undermining

Russian democracy by maintaining restrictions on political freedom and on the Russian media, much of which had come under state control. They were critical of Putin's attack on, and ultimate state take-over of, the private oil company Yukos and the imprisonment of its chief executive officer, Mikhail Khodorkovsky, for political reasons. Indeed, in the United States before the June summit to be held in St. Petersburg, some members of the U.S. Congress actually called for the ouster of Russia because it did not meet G-8 standards in the political and economic spheres of national life.

At the July 2006 summit Putin responded to G-8 complaints saying in effect that Western critics simply did not understand Russia, and there was no way for Russia to set them straight. Other Russian speakers again rejected Western criticism, saying in effect that Russia had no intention of worrying what outsiders say about its "Europeaness" and, rather, intended to reclaim the role the Soviet Union once had as a prominent and powerful player in a multi-polar international environment.

At least one basis for Russia's assertiveness was its new found power as chief supplier of energy to Western Europe and a potential alternative to unreliable sources of Middle East energy to the United States. Also, Russia had just had success, finally, in killing Shamil Basayev, the head of the Chechen separatists responsible for a seemingly unrelenting series of brutal terrorist acts inside Russia on behalf of the Chechen cause.

But, that said, the St. Petersburg summit was not all that Putin had wanted, notably an American endorsement of Russia's entry into the World Trade Organization and evidence of a continuing Russian-American entente in world affairs, especially regarding Iran and North Korea. The summit, therefore, could not be considered a real success for Putin's diplomacy; and, as has been true in so many other instances, the cause of Russian disappointment was the United States, despite the Bush administration's long standing line that Washington and the Kremlin could work together to resolve big and divisive international issues.

Russia and the United States

The topic of Russian-American relations is exceedingly complex. At a minimum, it includes two relatively independent sub-topics: Russian attitudes toward the United States and; American attitudes toward Russia. These topics are highlighted in the two following sections. Subsequent sections are devoted to Russia's opposition to the American invasion of Iraq, relationships between Russian and American leaders, Putin's 2007 Munich speech, and the contentious issue of Iran.

Russian Attitudes toward the United State

The Russian public's attitude toward the United States was most favorable during the last years of Gorbachev's Perestroika—that is, in 1988–91, when cold war sentiment was predominantly harbored by older generations. While feelings were still warm in the early 1990s, the first setbacks occurred when American advisors, working closely with the Russian government on privatization, did not produce an improvement of living conditions. Quite the contrary, life got markedly worse for the vast majority, while a couple of dozen crooks privatized the most lucrative segments of the Russian economy.

The expectation of rapid improvement due to foreign advisors was unrealistic to begin with, but it did exacerbate public disappointment. The image of being used and duped by perfidious Western, particularly American, advisors (i.e., the image cultivated first by Communists and then by Russian nationalists) became prevalent among ordinary Russians prone to conspiracy theories. The NATO expansion undertaken despite high-profile promises not to do so, and especially the U.S.-orchestrated 1999 NATO bombing of Serbia gave the impression that the United States was taking advantage of Russia's weakness. Since that time Russian attitudes toward the United States have been steadily deteriorating despite some temporary upswings and, most importantly, despite the multitude of institutional and interpersonal ties that emerged between the two countries after the collapse of communism. Some of those ties have been widely publicized, as for example, Harvard University's preserving eighteen bells from the Danilov Monastery in Moscow and returning them to Russia. (In 1930, during the anti-religious crusade, these bells were bought as scrap metal by Richard Crane and delivered to Harvard. In 2008, after almost twenty years of negotiations, the bells were returned to the newly renovated Danilov Monastery in exchange for copies manufactured in Russia.) However, even events like this did not improve the opinions of the masses regarding the United States. Numerous public opinion surveys conducted in Russia reveal that America is seen as a country unfriendly to Russia and constantly striving to undermine its security and natural sphere of influence. To be sure, in Russia, unfavorable views about America are not as pervasive as, say, in the Muslim world. The 2007 Pew Survey, for example, revealed that 48 percent of Russians hold unfavorable views about the United States, while 41 percent hold favorable views. But well over half of all Russians have negative views about NATO, in which the United States is the indisputable leader.

Analysis of Russian attitudes toward America makes obvious that (in the words of *The Economist* magazine) "America looms much larger in Russia's mind than Russia ever did in America's." Indeed, in Russia, America is mentioned in all kinds of speeches, pronouncements, newspaper and journal articles, talk shows, and casual conversations with astounding frequency. At times (perhaps most of the time), references to America are made without any real need, when the context or topics of discussion have no immediate connection to the United States. Perhaps America best epitomizes Russia's love-hate affair with the Western world. Quite frequently the undertone of such references is that America pursues undisguised interests of its own and/or protects its own citizens abroad steadfastly, whereas Russia does not but should. This attitude explains why the dominant Russian emotion after the five-day war with Georgia in August 2008 was that "we won a showdown with America." It is hard to escape the impression that America is both envied and resented because it is successful. America's success is seen against the backdrop of what is construed as Russia's failure—failure to retain its superpower status, failure to secure its sphere of influence, and most importantly failure to inspire awe or outright fear. The last is arguably the biggest of all failures, since possessing crude power to sway the global political order has always been an important part of Russia's self image, much more important than being able to secure a good life for its citizens.

There are different opinions as to what is instrumental in sustaining anti-Americanism in Russia. The most widespread opinion is that anti-American sentiment is cultivated by Russian leaders, who use media to propagate their resentment of America. One of the most shrewd observers of Russian life, Leonid Radzikhovsky, a person of Westernizing and liberal persuasion, disagrees with this. According to Radzikhovsky, the call for anti-Americanism does not come from the top tiers of power; rather, it is generated at the grassroots level. The people at the helm of power simply tell ordinary Russians what they want to hear. "There is such a viewpoint, such a myth," said Radzikhovsky in his extensive interview on *Ekho Moskvy,* a Moscow-based liberal radio channel, "that liberals, to which I have to assign myself, are just crazy about America. I don't know why but I am not. I just don't know America well enough, although I try to understand it. . . . However, my concern is not America as such; rather it is anti-Americanism—and these are totally different things." "You mean anti-Americanism as our home-grown phenomenon?" asked the interviewer? "Yes," replied Radzikhovsky, "it is 100 percent homegrown. It's almost like—let me refer to my favorite subject—Jews and anti-Semites. One can be negatively disposed toward Jews because they have many unpleasant traits [Radzikhovsky is Jewish – GI], both individual and shared by their whole community; but this has nothing to do with anti-Semitism. Anti-Semitism is a kind of psychosis. The case with America is quite the same. Yes, the USA has many unpleasant traits. One of the most unpleasant traits is that it is indeed a very powerful country; and, as is the case with strong countries, it can be too self-confident and too tactless; and America's responsibility is overly great, as its every mistake gets imprinted on the entire world. But while this is all true, anti-Americanism has no relation to it—this is a psychiatric disease of Russian society, a psychosis of Russian consciousness. . . . America is a scapegoat on which you can blame anything. This is a subject that fleshes out the emptiness of one's mind. If you don't know what to do and there is no constructive program of action, you will lull yourself by hating somebody. This is a neurosis entirely harmless for America but very dangerous for the sick mind of our society. It is only because of that that I care about anti-Americanism without even caring all that much about America itself."

Whether or not Radzikhovsky's diagnosis is on target, it was Vladimir Putin's February 2007 speech at the 43rd Munich Conference on Security Policy that vented Russia's frustration with America. The speech elicited diametrically opposite reactions, from displeasure and disappointment in the West to extreme glorification in Russia, where literally everybody (with the exception of a narrow circle of pro-Western liberals) seemed to cherish Putin's speech as the utmost expression of Russia's "getting up off its knees." It is important to remember that the speech was delivered more than 1.5 years before the recession, when petrodollars were still pouring into Russia, creating perhaps a false sense of economic security.

In his Munich speech, Putin gave vent to Russia's grievances against the United States. "We are constantly being taught about democracy," declared Putin. "But for some reason those who teach us do not want to learn themselves." Much of his criticism of the United States had to do with the unacceptable and even impossible idea of the "unipolar model," whereby the sole remaining superpower resorts to unilateral and "frequently illegitimate actions that have not resolved any problems" but instead "caused new human tragedies and created new centers of tension." While this was clearly said in reference to the Iraq war, the spectrum of Putin's remarks extended far beyond criticism of that war. Other rhetorical targets of Vladimir Putin were the (a) American plan to deploy an anti-missile defense system in East Central Europe, which is "the next step in an inevitable arms race;" (b) delay of the ratification of the Adapted Treaty on Conventional Armed Forces in Europe (signed in 1999) by Russia's Western partners; (c) NATO expansion, which "represents a serious provocation that reduces the level of mutual trust," (d) allegedly hostile attitude toward Russian companies buying assets of Western businesses; (e) Western ambivalence in the war against global poverty wherein Western aid funds the donor country's companies and at the same time the West generously subsidizes its own agriculture and undermines farming in poor countries; and (f) activities of the Organization for Security and Cooperation in Europe, which allegedly has become "a vulgar instrument designed to promote the foreign policy interests of a group of countries."

Putin's criticism of NATO expansion was especially full of vitriol:

> We have the right to ask against whom is this expansion intended. And what happened to the assurances our western partners made after the dissolution of the Warsaw Pact? Where are those declarations today? No one even remembers them. But I will allow myself to remind this audience what was said. I would like to quote the speech of NATO General Secretary Mr. Woerner in Brussels on 17 May 1990. He said at the time that, 'the fact that we are ready not to place a NATO army outside of German territory gives the Soviet Union a firm security guarantee.' Where are these guarantees? The stones and concrete blocks of the Berlin Wall have long been distributed as souvenirs. But we should not forget that the fall of the Berlin Wall was possible thanks to a historic choice—one that was also made by our people, the people of Russia—a choice in favor of democracy, freedom, openness and a sincere partnership with all the members of the big European family. And now they are trying to impose new dividing lines and walls on us—these walls may be virtual but they are nevertheless dividing, ones that cut through our continent. And is it possible that we will once again require many years and decades, as well as several generations of politicians, to dissemble and dismantle these new walls?

Apparently some of Russia's grievances, as voiced by Putin, cannot be easily shrugged off; even the staunchest critics of Putin called them "seemingly correct and legitimate" for the most part. To be sure, there are more grotesque forms of anti-Americanism in Russia than those invoked by Russia's former president. In November 2008, the *Izvestia* daily (a media outlet which is particularly close to the Kremlin) published an interview with a professor of the Diplomatic Academy under Russia's Foreign Ministry, Igor Panarin, in which he predicted civil war in the United States as early as 2010 followed by the breakup of the country. Published by *Izvestia*, Panarin's map of imminent disintegration of the United States shows (a) western states that are likely to join China, (b) Midwestern states that are likely to join Canada, (c) northeastern and mid-Atlantic states which are likely to join the European Union, (d) the Texas Republic, encompassing much of the American South and part of its West, which is likely to join Mexico; (e) Alaska, that will join Russia; and (f) Hawaii, which will join either China or Japan.

To the newspaper's delight, Panarin's interview, full of ludicrous and unsubstantiated judgments, was debated by the *Wall Street Journal,* and even the White House spokeswoman, Dana Perino, was asked questions in regard to that interview during her routine meeting with journalists. Moreover, Panarin appeared on CNN, and he was invited to give guest lectures in China and some Arab countries. Leonid Smirnyagin, a Moscow State University professor and a specialist on the United States, averred in his letter to *Izvestia*'s editor that Panarin's alleged popularity in the West, of which the newspaper is so proud, is actually a sad phenomenon, as Panarin is "invited in the capacity of a village idiot but presents his views as if they were in the mainstream of Russia's political thought." In the same letter, however, Smirnyagin conceded that "during Bush's presidency America's moral authority has significantly declined across the entire world, and anti-Americanism acquired rabid forms not only in Russia but also in a majority of countries. . . . So, interest in Panarin in some of those countries is one more indication of mass disgust with the United States, and not only with its rulers but also with the country and its people. This is sad in its own right. But it is especially sad that in our country, anti-Americanism has become an indispensible attribute of Russian patriotism. I am reminded of that all too frequently when I lecture my students about the geography of the United States. And I am trying to explain to them that healthy patriotism ought to be a feature of every citizen, only it should be based on love of one's own country, not hatred of another."

American Attitudes toward Russia

Because this text is not devoted to the United States, it might seem odd to offer an analysis of this issue. However, because virtually all news and commentaries on Russia broadcast or published in the United States are produced by people adhering to certain schools of thought, it is worth providing at least a glimpse of those schools. For quite some time, American foreign policy experts belonged to either an idealist or a realist group when it came to Russia. These groups differed not only in their policy recommendations, but in their analytical frameworks. The idealists emphasized democratization as the leading indicator of development and gave greater weight to the influence of U.S. policy on Russia. The realists emphasized economic development as the leading indicator and argued U.S. policy could influence Russia only on the margins. So, whereas idealists would normally demand domestic policy concessions from Russian leaders in exchange for some foreign policy favors, the realists would not. For example, according to the 1974 Trade Act of the United States, the Jackson-Vanik amendment, named for its major co-sponsors, Sen. Henry "Scoop" Jackson (D-WA) and Rep. Charles Vanik (D-OH), denied most favored nation status to certain countries with non-market economies that restricted emigration rights. Permanent normal trade relations would be extended to a country subject to the law only if the president determined that it complies with the freedom of emigration requirements of the amendment. Although signed into law by the Republican President Ford, the amendment was a legislative initiative of the Democrats. The number one target of the amendment was the Soviet policy denying emigration to Jews. Already, under Gorbachev Perestroika, most restrictions to the emigration of Soviet Jews had been removed; and after the fall of the Soviet Union, anybody could leave Russia for good except for persons under criminal investigation. Russia is now considered a country with a market economy. The irony is, though, that the Jackson-Vanik amendment has never been rescinded, an omission which greatly irritates Russia's leaders. This demonstrates that the idealists' imprint on American foreign policy may be quite enduring.

Most of the time idealists were close to the Democratic Party and advised Democratic administrations, whereas realists were close to the Republican Party. This pattern of party affiliation created a strong opinion among politically active Russians that for Russia's leaders it is always easier to deal with Republicans than with Democrats. For that reason, many, if not most, members of Russia's political establishment sided with Senator McCain as a presidential hopeful despite his harsh criticisms of Russia and his sponsorship of Georgia's President Saakashvili, Russia's perceived enemy. During the Bush years, however, the realist-versus-idealist division was shattered, because a group of neoconservatives began to affect, if not outright determine, American foreign policy. Affiliated with the Republican Party, neoconservatives were actually quite similar to idealists in their pronouncements about Russia. The reference in Putin's Munich speech to someone persistently teaching Russia democracy but not willing to learn himself is actually a reference to American neoconservatives. Perhaps the only difference between them and the idealists was in the degree of desired engagement with Russia. Whereas the idealists have been in favor of such engagement and consultation, neoconservatives were essentially unilateralists or proponents of a policy based on accomplished facts. This difference, though, is barely noticeable at times. Hence, a bitter exchange between the self-proclaimed idealist Michael McFaul (now the principal Russia hand in the Obama administration) and *The Nation* magazine sparked by Richard Dreyfuss's 2008 article "The Rise and McFaul of Obama's Russia Policy." The article claimed that "recent U.S. blunders in regard to Moscow, which have caused a nationalist reaction in that country" were actually engineered by McFaul,who is a neoconservative—a charge McFaul vehemently denied.

Because the emergence of neoconservatives has become integral to the change in the Republican Party's base, whereby fiscal conservatives-and-social-liberals (e.g., Chuck Hagel, Richard Weld, Lyndon Chaffee, or Arlene Specter types) are being squeezed out of the party's leadership, the Republican affiliation with the realist group is no longer a given. Thus, soon after Obama's election it was actually Henry Kissinger who delivered Obama's message to the Russian President Medvedev. Kissinger, of course, is the most prominent realist; he claims that all differences between Russia and the United States pale beside their common responsibility for nuclear non-proliferation. In his February 2009 *Newsweek* article, Kissinger wrote that "the Russian proposal for a joint missile defense toward the Middle East, including radar sites in southern Russia, has always seemed to me a creative political and strategic answer to a common problem." "Always seemed to me" is a strong statement, considering that the Bush administration, saddled by neoconservatives, dismissed Russia's suggestion

and opted for unilateral deployment of missile interceptors and radar in Poland and the Czech Republic, an option that infuriated Russia. Fareed Zakaria, editor-in-chief of *Newsweek International* and an important foreign policy expert, adheres to the same idea as Kissinger. "We cannot deploy missile interceptors along Russia's borders, draw Georgia and Ukraine into NATO, and still expect Russia's cooperation on Iran's nuclear program," wrote Zakaria in his October 20, 2008, *Newsweek* article. The Washington-based think tank the *Nixon Center* also shares this view, as numerous speeches by Dimitri Simes, the *Nixon Center*'s president, show.

Because the neoconservative foreign policy agenda and, to a significant extent, that of the idealists, are now construed as failures, a reappraisal of the American policy toward Russia is in full swing. At least three recent books have contributed to this reappraisal: (1) *Distorted Mirrors: Americans and Their Relations with Russia and China in the Twentieth Century,* by Donald E. Davis and Eugene P. Trani; (2) *American Mission and "The Evil Empire": The crusade for a "Free Russia" since 1881,* by David S. Foglesong; and (3) *America – Russia: Cold War of Cultures,* by Veronika Krasheninnikova (in Russian). The first of these books claims that Americans have looked at Russia and China through different lenses. Russia is seemingly European, and so its differences from the perceived Western ideal are deemed somewhat artificial. The most important of those differences has been Russia's repressive system, including prisons and labor camps. Should one remove the system of oppression, Russia would acquire its genuine European self. As for China, it is exotic to begin with, and so the focus of American perception of that country has been on the legitimate cultural specificity of China, its long history, Confucian traditions, and the specific Chinese view of the world. That China has actually been more, not less repressive than Russia does not interfere with America's fascination with China. Another source of disparity is that throughout history Russia has been more of a geopolitical rival of America than China, which explains why America has actually lauded Russian developments only twice, both times when Russia committed geopolitical suicide: first in February 1917 and again in December 1991.

According to Foglesong,[1] the contemporary vilification of Russia may be less about the rationalization of U.S. interests and policies and more about the affirmation of an American identity. Using visual and rhetorical images of Russia in the United States, the author shows how the political crusade for a free Russia can be seen as part of the broader missionary enterprise that reflects America's national purpose. By tracing American representations of Russia over the last 130 years, Foglesong highlights three of the strongest notions that have informed American attitudes toward Russia: (a) a messianic faith that America could inspire a sweeping overnight transformation from autocracy to democracy; (b) a notion that, despite historic differences, Russia and America are very much akin, so that Russia, more than any other country, is America's "dark double;" and (c) an extreme antipathy to "evil" leaders who Americans blame for thwarting what they believe to be the natural triumph of the American mission. These expectations and emotions continue to affect how American journalists and politicians write and talk about Russia.

The book by Veronika Krasheninnikova[2] draws on similar themes. In fact, the book provided the Russian embassy in Washington, DC with an important publicity stunt. In October 2007, it organized a special meeting to which the most important American pundits weighing in on Russia were invited. Krasheninnikova believes that what America thinks and says about Russia tells more about America than it does about Russia. This is because America looks at Russia through the foreign policy lens of liberal democracy. In Krasheninnikova's view, America does not realize how exceptional it is in its foundation and development. No Old World country was created from a tabula rasa, on virgin terrain, with no previous regime to dismantle and social patterns to reverse. Another underestimated factor that contributed to the viable liberal democratic consensus is the self selection of people who came to settle in the United States—freedom-seeking, risk-assuming, entrepreneurial immigrants who, by conscientious choice, joined the system of values that the New World offered. Values that are self-evident, axiomatic, and non-debatable to Americans appear highly relative to Russians, who in one lifetime could experience three opposing sets of fundamentals. In the twentieth century alone, the political system in Russia changed three times from one extreme to another—from monarchy to autocracy and communism, then to a developing democracy and capitalism. Meanwhile, the American nation was, despite a civil war, far more stable, establishing its political system 230 years ago and working to perfect it, without serious considerations of alternatives. America's very clear and unambiguous definition of democracy starkly contrasts with the Russian usage of the term. Russians have known the Bolshevik "democracy of the people's councils," Stalin's "most democratic constitution in the world," the "tautological people's democracy" of Khrushchev, Brezhnev's "advanced socialist democracy," "expanding and perfecting democracy" under Gorbachev, a Hobbesian state-of-nature oligarchic system of the 1990s that qualified as "democracy" and received much applause from the Clinton administration; and currently a "managed democracy" and a "sovereign democracy" and other creations of the Kremlin political science laboratory. For Russians, basic American values are difficult to comprehend. Startling, for example, is the founding role of God and religious belief in the American state. The American constitution refers to the Creator as the source of unalienable rights—a difficult-to-prove proposition that to Russians is hardly up to the state to decide. Such a nationalistic appropriation of God's will, in combination with other historical peculiarities gives a sacred character to democracy, similar to the past sacredness of the monarch in Russia. It offers a Manichean view of the world, in which all men are divided into Godly and Ungodly, righteous and wicked, and where no shades of gray exist, only a cosmic struggle between darkness and light. It sanctions all actions and absolves all mistakes made in the name of freedom. It supports a powerful messianic spirit and a striving to "civilize" the rest of the world. It creates a naïve expectation that other countries should line up behind America, as if America is a prophet, and follow it in the realization of its mission received exclusively from God. Today, the consensus imposed by America's ideology leads to such conclusions as Russia "rolls back on democracy," "slips into authoritarianism," "constricts the freedom of press," "bullies its smaller neighbors," and "uses oil for political blackmail." Krasheninnikova quotes Stephen Cohen, who noted that when Washington meddles in the politics of Georgia and Ukraine, it is "promoting democracy;" but when the Kremlin does so, it is "neoimperialism." When NATO expands to Russia's front and back doorsteps, it is "fighting terrorism," and "protecting new states;" but when Moscow protests it is "engaging in Cold War thinking."

Despite all disagreements, however, Russia and the United States have tried to address and resolve divisive issues by the diplomacy of cooperation, conciliation, and compromise. At the very least, there has been a willingness of Russian and American leaders to meet and discuss differences and disagreements. Though determined that Russia should go its own way, even when doing so provokes the United States, the Kremlin has vowed to protect and preserve the Russian-American cooperation.

One of the most important positive aspects of the Russian-American relationship has been the Kremlin's support (however qualified) of the Bush administration's war on terrorism.

The Impact of 9/11, 2001

Following the September 11th terrorist attacks, President Putin promptly expressed sympathy for, and support of, the Bush administration's declared "war on terrorism." Putin expressed his readiness to use any and all means to combat the enemy, identified as Osama bin Laden, his Al Qaeda network of terrorist cells throughout the world, and the Taliban rulers in Afghanistan.

One important reason why the Kremlin signed on with the Americans was that Russia was even more vulnerable to acts of terrorism than the United States. It was ill prepared for the difficulties it would have from a terrorist attack on its soil.

Most immediately, Putin needed, as much as the Americans did, to destroy the roots of terrorism in the vast landmass that stretched from the Fertile Crescent to the Hindu Kush. Chechens, in their desperate fight for independence from Russia, seemed ready, willing, and able to practice terror with the help and encouragement of radical Islamist terrorists from the Middle East and Central Asia. Moreover, the Kremlin shared American hostility to the Taliban regime in Afghanistan because of its readiness to spread radical Islamist fundamentalism into Chechnya and Dagestan and for its intimacy with Pakistan, long a rival of Russia for influence in Afghanistan. The Taliban had received economic, financial, and military support from Pakistan in return for its cooperation with Pakistani foreign policy. Moreover, the Taliban regime had all but extinguished the last vestiges of Russian influence in Afghanistan, effectively blocking any effort of Moscow to gain access to the Indian Ocean through Afghanistan and the Baluchistan part of southern Pakistan.

Russian Opposition to the U.S. Invasion of Iraq

But, the Russian-American *entente* after September 11, 2001, was severely damaged by the Bush administration's unilateral invasion of Iraq over the opposition of Russia and some NATO allies, in particular France and Germany. Since the early 1990s, Russia had been at odds with the United States over Iraq, which the Kremlin had been trying for some time to cultivate with a view to restoring the friendly relations that Moscow had had with Baghdad in the Soviet era. In contrast to the U.S. policy since the 1991 Gulf War of using economic sanctions to force the Iraqi government to dismantle its biological, chemical, and nuclear weapons, the Kremlin wanted to ease United Nations sanctions imposed on Iraq following the war. Restoration of Iraqi freedom to export oil would allow it to pay off the enormous debt it had incurred to the former Soviet Union, now held by Russia. The Russians also wanted to discourage punitive American retaliation for Saddam Hussein's interference with, and eventual expulsion from, Iraq of UN inspectors sent in to investigate any clandestine Iraqi production of weapons of mass destruction, forbidden by the United Nations to Iraq under the terms of the treaty ending the conflict over Kuwait. In the American view, throughout 1997–1999, the Russians were a dangerous challenge to U.S. efforts to enforce Iraqi compliance with pledges about disarmament.

The Kremlin rejected the premise on which the U.S. invasion of Iraq and other Bush administration policies had come to rest, namely, the legitimacy of preventive action or the right to move against a country based on perceived threats. Russia called this "Bush doctrine" of preemption illegal, immoral, and just plain dangerous to international peace and stability. Ironically, some Russian commentators likened that doctrine to that of Leonid Brezhnev, who justified the Soviet military intervention in Czechoslovakia in August 1968 and subsequently in Afghanistan in December 1979 in similar terms. The Kremlin faulted Bush for moving against Iraq without UN endorsement. Respect for the UN has become an important pillar of Russian security, a position inspired in part by Gorbachev's foreign-policy principles in his "New Thinking" of the late 1980s.

Moreover, the deployment of a large American military force in Iraq gave the United States, at least in the Kremlin's view, an unacceptable strategic advantage that further strengthened its already solid strategic situation in the Middle East, thanks to close American relations with Israel, Jordan, Saudi Arabia, and Egypt. Indeed, Russia had fought in Afghanistan from 1979 to 1986, partly to oppose American interests in the region. In addition, Russia already had made significant concessions to the United States in the wake of the September 11th terrorist attacks in endorsing the U.S. invasion and occupation of Afghanistan and talking the Central Asian country of Kyrgyzstan into allowing the United States to establish a military base in Central Asia.

In addition, big Russian oil companies, notably Lukoil, had contracts or were promised contracts by the now-defunct Saddam Hussein regime, providing for their involvement in the drilling, refining, and marketing of Iraqi oil. With the invasion of Iraq, those contracts looked to the Kremlin very much like they were lost, never to be retrieved, because U.S. oil companies would be invited to jump in to replace the Russian oil companies. To the dismay of the Kremlin, the Bush administration seems to have paid little or no attention to this Russian concern, leaving the Russians with no incentive to cooperate with Washington on Iraq.

Russian-American Relations after 2002

Once the U.S. invasion of Iraq occurred, Putin's Kremlin made it a point at all times to leave the door slightly ajar for some kind of compromise with the Bush administration's policy. His inclination not to permanently poison Russian-American relations over Iraq was shared by German Chancellor Gerhard Schroeder, who also was concerned about avoiding a serious crisis with Washington.

The Kremlin welcomed American moves in 2004 to transfer administrative power in Baghdad from the U.S. envoy L. Paul Bremer to Iraqi leadership and leave to it the job of formulating

the main lines of Iraq's domestic and foreign policy, including who was to get what in the way of contracts to help in the physical reconstruction of Iraq. Accordingly, Russia supported a joint British-American resolution in the Security Council endorsing the transfer of power at the end of June 2004.

Prior to the 2008 presidential elections in the United States, a working relationship between Russia and the United States was, to some extent, based on what seemed to be a genuine rapport between Putin and George W. Bush. In fact, Bush was repeatedly lampooned by much of the American media for his 2001 comment about Putin. At a press-conference after his first meeting with Putin, held in Slovenia, Bush said "I looked the man in the eye. I found him to be very straightforward and trustworthy and we had a very good dialogue. I was able to get a sense of his soul." These last words in particular were repeatedly invoked as an alleged proof of Bush's deficient foreign policy skills. After their first meeting, Bush and Putin continued to meet regularly and frequently. While the summits sometimes produced little of substance, they always have resulted in an improvement of the diplomatic environment.

The May 2002 Summit in Moscow

Putin and Bush met in Moscow in May 2002 and again in November 2002 in St. Petersburg. At the May summit the two leaders signed an historic agreement to remove stockpiled nuclear arms from operational deployment. The treaty called for the reduction of 6,000 nuclear warheads of each side to 2,000 by 2012, when the agreement expires. But there were a lot of ambiguities and loopholes. For example, the treaty did not set a pace for dismantling. Moreover, decommissioning did not mean destruction, so the treaty in effect allowed both sides to handle the downgrading of the weapons in any way they wished, which meant that both sides could keep the downsized warhead arsenal in ways that could allow reactivation of them without much difficulty. Still, the treaty was important for its symbolic value, which reflected a joint will to cooperate and limit weapons arsenals. In addition, it did give Russia what it wanted in arms reduction given that it was financially strapped and eager to use scarce resources for domestic development.

In addition, Putin and Bush discussed Russia's sale of nuclear technology to Iran, which was deeply troubling to the Americans. Bush said in effect that the clerical dictatorship running Iran today should not receive anything that could enable it to develop a nuclear weapon. Putin's response was not very satisfying to Bush. He insisted that Russia was doing nothing to violate the spirit and rules of nonproliferation because what Russia was selling to Iran involved the enhancement of its nuclear energy capability—not its weapons-making capability. Bush differed, saying that a country like Iran that was oil rich was not in need of increased nuclear energy. On another issue, Putin did agree to join Bush in telling Pakistan to curb anti-Indian militants who were causing trouble in Kashmir, threatening to bring both countries into a confrontation that could escalate in a war. And Bush praised Russian religious tolerance during a visit to Moscow's Grand Choral Synagogue.

The November 2002 Summit in St. Petersburg

At the St. Petersburg summit in November 2002, there was discussion of the terrorist threat, with Putin asking Bush about the connection to terrorist activities of two American allies, Pakistan and Saudi Arabia. The discussion of such a highly sensitive issue for the United States was possible because of the growing personal friendship between Putin and Bush. Both were determined not to allow their contrasting views about these two Islamic countries to complicate in any way their strengthening relationship. While Bush did not answer Putin, neither did he or his advisers at the summit say anything to suggest in any way the possibility of a disagreement, never mind confrontation. There was some discussion of the ongoing Russian war in Chechnya, but all that Bush would say publicly was that the United States hoped for a political solution of the Chechen problem. There was nothing that Putin could fault in that pronouncement. Finally, the two leaders agreed to begin an "energy dialogue" in light of the fact that Russia by now was a major oil producer with plenty to export to any buyer, especially the United States.

The June 2003 Summit in St. Petersburg

In early June 2003, Bush returned briefly to St. Petersburg for another meeting with Putin to try to strengthen ties in the wake of the U.S. invasion of Iraq. Though Bush did not raise the issue of controversial Russian behavior in Chechnya, Putin refused to back away from his opposition to U.S. policy in Iraq. Although this meeting was not as successful as the previous ones, the fact that it was held at all given the sharpness of the disagreement between Russia and the United States over Iraq must be considered in a positive way as a reflection of the determination of the two leaders not to allow their differences to compromise relations between their countries.

The September 2003 Summit at Camp David

Putin and Bush held another meeting in late September 2003, this time in the United States at Camp David, following Putin's address at the annual opening of the UN General Assembly. More for atmosphere than previous meetings, none of the major issues between Russia and the United States, such as Russian sales of nuclear technology to Iran; or American concern about Putin's increasing authoritarian style, as seen in his closing-down of the last independent, national television network on the eve of upcoming parliamentary elections, were addressed at any great length.

Indeed, on the touchy issue of the sale of Russian nuclear technology to Iran, Putin reaffirmed that Russia will not reverse a decision to help Iran build a nuclear reactor, though he did try to soothe the Americans by saying that he would respectfully but firmly tell Teheran that the Russian help to Iran was not intended in any way to make possible the development of a nuclear weapon and that Russia wanted the Iranian government to cooperate with UN-sponsored nuclear inspections. The obvious failure to get Putin to back off on this issue flustered and embarrassed the Bush administration and contributed to a chill in Russian-American relations.

Down-turn in Russian-American Relations 2004–2008

In 2004 there were thick clouds on the diplomatic horizon. The Kremlin was noticeably concerned and perhaps even alarmed by the escalating U.S. military involvement in Central Asia in the name of fighting terrorism in Afghanistan. Russian Defense

Minister Sergei Ivanov reportedly said in early October 2003 that the Russian government expected that as soon as the mission in Afghanistan was completed, the United States would withdraw from bases in Uzbekistan and Kyrgyzstan. Ivanov tried, somewhat obliquely, to tell the Americans that Russia's approval of U.S. use of the bases was solely for the purpose of fighting terrorism in Afghanistan and not for any other reason. He said to NATO that Russia would be obliged to respond in kind if the Alliance remained in an offensive mode. Russia would have to make appropriate adjustments in its own defensive strategy involving, in particular, nuclear weapons, as he put it.

Tensions between Russia and the United States eased a bit with the Kremlin's agreement, after a long period of opposition, to the construction of a privately owned and run pipeline to Murmansk to carry oil being shipped to the United States. The Murmansk pipeline will carry 2.4 million barrels of oil a day. The ultimate advantage to the United States, as Putin well knew, would be to curtail a bit America's dependency on oil from the Middle East. And in May both the Duma and Federation Council ratified the November Moscow arms limitation treaty, despite opposition from Zyuganov's communists and Zhirinovsky's nationalists, who complained about giving too much to the Americans in light of so many differences between the countries.

The Putin Kremlin also held out the prospect of providing a reasonably priced and reliable source of natural gas. The United States was so eager for this deal that it was ready to make a $15 billion investment to develop Russia's giant Shtokman natural gas reserves in the Barents Sea. By the summer of 2004, Russian-American relations seemed on the upswing despite the persistence of a multitude of differences and disagreements involving Iran, Iraq, Georgia, and Central Asia. It remained to be seen how far this tendency would go and how long it would last.

In 2005 and 2006 tensions between Russia and the United States steadily mounted despite the appearance of a kind of personal camaraderie between Putin and Bush, who was still insisting he saw something special in the Russian leader making possible resolution of tough problems. At a summit meeting of Putin and Bush in Bratislava, Slovakia, in late February 2005, although Putin was chastised for policies limiting political freedom in Russia, Bush referred to his Russian counterpart as "my friend," a position the Kremlin had difficulty believing, especially after the comment by Secretary of State Condoleezza Rich in late April that she found "Russian trends worrying."

The Kremlin pointedly blamed Washington for setbacks to its influence building policies in popular presidential elections in Georgia in 2003, Ukraine in 2004, and Kyrgyzstan in 2005. The Kremlin was annoyed by public American criticism of Russia's perceived retreats in democratization, evident in the 2004 Russian presidential election in which the anti-Putin forces were severely undermined in offering the Russian voting public alternative candidates and subsequently with laws restricting organizations seen as critical of, or opposed to, Putin's leadership and policies.

Bush gaffes contributed to the strain in U.S. relations with Russia. For example, on his way to attending celebrations in Moscow in May 2005 to commemorate the 60th anniversary of the end of World War II in Europe, Bush visited the Baltic Republics, now allies in NATO. In a letter to the three Baltic presidents, Bush infuriated the Kremlin by referring to the end of World War II as "the beginning of the unlawful Soviet annexation of Latvia, Lithuania and Estonia." Bush was correct in a way but the Soviet control of the Baltic republics had been made possible by the Nazis (who signed the 1939 non-aggression pact with Stalin and allowed the Soviets to incorporate the three Baltic States), and it was the defeat of the Nazis that was about to be celebrated in Moscow, the capital of the country whose sacrifices in World War II were second to none. Bush poured oil on the diplomatic fire in the Baltic republics by warning his friend in the Kremlin not to interfere with democratization in the Baltic region.

That said, U.S. criticism of Russia was not accompanied by threats and denunciations. Instead, in April 2005, Secretary of State Rice told Putin that the United States was no threat to Russia in the Eurasian region and that the United States had no intention of replacing Russian influence in the region even though it looked like that was exactly what it was doing in its efforts to strengthen political and military ties with the ex-Soviet Central Asian republics.

While in Moscow for the May 2005 "V-E Day" celebrations, Bush did pull off a successful meeting with Putin in terms of friendly atmospherics to diminish the rancor over Bush's prior visit to the Baltic republics that had enraged the Kremlin. Then Bush met with Putin during the G-8 summit in mid July 2006 in St. Petersburg, a mixed success. On the one hand, Bush used the meeting as an opportunity to strengthen personal relations with Putin, but on the other hand he refused to give him what he wanted: American endorsement of Russia's bid to join the World Trade Organization. On the plus side was a decision by the two leaders to cooperate in a global program to track potential nuclear terrorists, detect and intercept bomb making materials, and coordinate responses if terrorists should obtain a nuclear weapon.

An unnecessary strain on Russian-American relations occurred in the Spring of 2006 from highly critical comments in May by Vice President Dick Cheney while attending a conference on democracy in Vilnius. He rebuked Russia "for unfairly and improperly restricting people's rights and using the country's vast resources of gas and oil as 'tools of intimidation or blackmail." The Kremlin was annoyed by Cheney's sympathetic audience that included Polish President Lech Kaczynski and Lithuanian President Valdas Adamkus and was disinclined to collaboration with the United States especially if it ran counter to Russian interests.

Putin's Munich Speech

In February 2007, Putin made his famous (or infamous) Munich address, discussed earlier. After that speech, most commentators agreed that Russian-American relations were experiencing a nadir. In March 2007, the United States announced plans to build an antiballistic missile defense installation in Poland along with a radar station in the Czech Republic. Both nations were former Warsaw Pact members. American officials said that the system was intended to protect the United States and Europe from possible nuclear missile attacks by Iran or North Korea. Russia, however, viewed the new system as a potential threat and, in response, tested a long-range intercontinental ballistic missile, the RS-24, which it claimed could

(White House photo by Eric Draper)
President George W. Bush of the United States and President Vladimir Putin of Russia, exchange handshakes Thursday, June 7, 2007, after their meeting at the G-8 Summit in Heiligendamm, Germany.

defeat any defense system. Russian president Vladimir Putin warned the United States that these new tensions could turn Europe into a "powder keg." On June 3, 2007, Putin warned that if the United States built the missile defense system, Russia would consider targeting missiles at Poland and the Czech Republic. At the June 8–9, 2007 G-8 summit in Germany, however, the eight global leaders, including Putin and Bush, were all smiles. A photo even showed Bush and Putin holding hands, like a bride and a groom about to get married. According to the TruthInTheMedia.org bulletin of June 17, 2007, "the wedding ceremony seemed to be presided by 'Pastor Angel' (German Chancellor Angela Merkel), with Romano Prodi (Italian prime minister) and Nicolas Sarcozy (French president) acting as Bush's bridesmaids. Meanwhile, Tony Blair stood in as Putin's best man. During the actual conference, Putin quickly poured cold water on the overheated coals of Russian-American relations by proposing a joint operation of a military base in Azerbaijan, which would make American missile interceptors and radar to be deployed in the Czech Republic and Poland redundant, as the alleged threat to be monitored was posed by Iran, the next-door neighbor of Azerbaijan. By some accounts, Putin's idea stunned the American delegation, who had come to Heiligendamm, Germany, loaded for bear—the Russian Bear, to be exact—not friendly hugs. With the Russians masterfully diffusing the tension as quickly as they had raised it, the Americans were left speechless. It was the kind of "global warming" that was clearly welcomed by everyone at the conference."

The last meeting of Putin and Bush as heads of state took place at the Russian president's dacha in Sochi on the Black Sea and did not result in any breakthroughs, although most observers noted that seven years after their first meeting their personal chemistry remained intact.

In August 2008, American-Russian relations were strained yet again, when Georgia invaded South Ossetia, resulting in Russia intervening and invading Georgia. Russia claimed that it was a mission to protect the Georgian separatist regions of South Ossetia and Abkhazia from a Georgian military offensive. However, Russian military forces didn't stop in these regions and continued toward the Georgian capital. The United States chose to support Georgia in the conflict, sending humanitarian aid to Georgia and assisting with the withdrawal of Georgian troops from Iraq. On November 5, 2008, Russian President Dmitry Medvedev, in his first annual address to the Federal Assembly of Russia, promised to deploy Iskander short-range missiles to Kaliningrad near the border with American-backed Poland.

The United Stated and Russia during Barack Obama's Tenure

A favorable change in tone of Russian-American relations occurred during the first months of the Obama administration. In February 2009, President Obama sent a secret letter to Russia's President Medvedev suggesting that he would back off deploying a new missile defense system in Eastern Europe if Moscow would help stop Iran from developing long-range weapons. Medvedev's press secretary said Mr. Medvedev perceived the development of Russian-American relations as "exceptionally positive and hoped details could be fleshed out at an early March 2009 meeting in Geneva between Foreign Minister Sergei V. Lavrov and Secretary of State Hillary Rodham Clinton. During that meeting, the key phrase describing the necessity to warm up the bilateral relations was "pushing the reset button" on those relations. Clinton, presenting Russian Foreign Minister with a gift-wrapped "reset" button, called the "little gift" symbolic of what President Barack Obama and Vice President Joe Biden have been saying about the United States and Russia. "We want to reset our relationship," she said. Yet something may have been lost in translation in the gift before closed-door talks in Geneva. The word on the button was meant to say "reset" in Russian. "We worked hard to get the right Russian word," Clinton told Lavrov. "Do you think we got it?" "You got it wrong," Lavrov replied—the word actually means "overcharge." (Indeed, instead of *perezagruzka,* which is Russian for reset, the inscription read *peregruzka,* which is Russian for overcharge). "We won't let you do that to us," Clinton said, with the two laughing. Growing more serious, Clinton told Lavrov: "We mean it and we look forward to it."

On March 31, 2009, in anticipation of the first Obama-Medvedev meeting in London during the G-20 summit devoted to the global financial crisis, Dmitry Medvedev published an article in *The Washington Post.* "It is hard to dispute the pessimistic assessments of the Russian-American relationship that prevailed at the end of last year," wrote Medvedev in that article. "Unfortunately, relations soured because of the previous U.S. administration's plans—specifically, deployment of the U.S. global missile defense system in Eastern Europe, efforts to push NATO's borders eastward and refusal to ratify the Treaty on Conventional Armed Forces in Europe. All of these positions undermined Russia's interests and, if implemented, would inevitably require a response on our part. I believe that removing such obstacles to good relations would be beneficial to our countries—essentially removing "toxic assets" to make good a negative balance sheet—and beneficial to the world."

Speaking at a Moscow news conference, a Russian Foreign Ministry spokesman said the first meeting between Obama and Medvedev (on April 2, 2009), set the basis for practical cooperation between the two countries. Referring to the joint statement issued by the presidents, the spokesman reiterated several areas of cooperation, including strategic arms reduction. Indeed, while setting in motion fast-track negotiations on a replacement for the 1991 Strategic Arms Reduction Treaty, or START, which expires at year's end, the two leaders vowed at the same time to jointly confront other perceived threats. They specifically mentioned the nuclear programs in Iran and North Korea and Al Qaeda militants who have found refuge in Pakistan. They set a nominal July deadline for a substitute treaty for START, a date that coincides with Obama's first presidential visit to Russia. That conceivably would leave time to get the new treaty approved in the U.S. Senate by the December expiration of the current agreement. But arms control experts say December is not a hard deadline so long as there is progress. It remains to be seen whether or not a change in tone of Russian-American relations will be accompanied with a change in substance. Some observers remain skeptical. Nikolai Zlobin, director of the Russia and Eurasia Project at the Washington-based World Security Institute, wrote in the Russian newspaper *Vedomosti* that the true cause of problems in bilateral relations is their "colossal asymmetry," with America having an incomparably greater significance for Russia than Russia for America both in politics and economy.

Russia, the United States, and Iran

The year 2006, saw a worsening of Russian-American differences over Iran's alleged determination to build a nuclear weapon. For some time Russia had been selling Iran huge amounts of weaponry. In December 2004, Russia agreed to sell Iran $1 billion worth of anti-aircraft missiles and to supply technology and equipment to help Iran build a civilian nuclear reactor, which in the view of Western countries could be used also to build a nuclear weapon. Russian help was part of a larger Russian policy forged over years of friendly and cooperative relations between the two countries because the Iranian market was important to Russia and because Iran perceived strengthening ties with Russia as a counterweight to American influence in the region.

In 1995, Russia and Iran signed an agreement according to which Russia would resume building the nuclear power station in Bushehr, not far from the Persian Gulf. Earlier, a German firm had undertaken the project but subsequently abandoned it. Russians re-launched the entire project in 1998. The cost of the first reactor at the Bushehr station was about $1 billion. It was supposed to have been commissioned in 2007, but various delays of logistic and political nature have so far prevented this from happening.

For several years the Kremlin had rejected U.S. requests not to help Iran with nuclear technology. The Kremlin's position was simply that it had the right to sell Iran anything it wanted; Russia needed the business, and Russia needed to keep on the good side of Teheran to gain leverage with it. In addition, the head of the International Atomic Energy Agency, Mohamed El-Baradei, said in June 2004 that Russian sales were not a cause for concern.

Subsequently, it was found that Iran in fact had a nuclear weapons program and that its government was determined to have a nuclear capability. Fearful that a nuclear Iran could destabilize the Middle East and interfere with the region's supply of oil to the West, especially the United States, Western countries invited Russia and China to join in an international consortium to pressure Teheran to abandon its nuclear energy project. Russia, along with China, agreed to cooperate.

From the outset, Russia disagreed sharply with the West, and especially with the United States, over how to influence Teheran: the United States wanted its allies to exert pressure on Iran, even coerce it to abandon nuclear energy development, whereas Russia, also concerned about the development of an Iranian nuclear arsenal, preferred compromise and the avoidance of coercion. The Kremlin was determined not to provoke Iran lest it make trouble for Russia in the Caucasus and Central Asia. In any event it was in no hurry to see the West expand its influence in and around Iran.

In early 2006, Russia, along with China, called upon Iran to freeze its nuclear activity and agreed in principle to authorize the International Atomic Energy Agency to send the Iran case to the United Nations with the understanding from the United States that the case would not be discussed by the UN for a month. During this month of delay, the Kremlin proposed a compromise settlement of the Iranian nuclear issue designed to avoid UN discussion of Iranian policy that the Teheran government furiously opposed. The proposal gave Iran the right to develop nuclear energy on the basis of nuclear fuel from Russia and only for civilian use. But, the Russian proposal said, Russian and international nuclear inspectors should monitor the Iranian work on nuclear energy to make sure it is not diverted for the purpose of producing weapons.

The Russian compromise was accepted by the United States and the Teheran government agreed to discuss it with Russia in February 2006. Eventually, however, the Iranians rejected the Russian compromise, insisting that Iran, as a sovereign state, was free to pursue any kind of nuclear policy it wished and that it would develop a nuclear defense capability come hell or high water.

The give and take between Russia and the United States on this issue continued intermittently. In April 2006, the Bush administration called for an arms embargo on Iran to push it to accommodate Western demands for a nuclear freeze. Although the Kremlin could not persuade Iran to compromise, it opposed an arms embargo. In the meantime, the Iranian case finally reached the Security Council which ended by giving Teheran an ultimatum in July 2006 that said if the government did not back off and cease its nuclear projects, there would be punishment, a subsequence that Iran dreaded and Russia tried to avoid. Much to the chagrin of Washington the ultimatum expired, and so in December 2006 the UN Security Council imposed sanctions on Iran with Russia's grudging support. As noted above, Iran continues to be a thorny issue and possibly a bargaining chip in the Russian-American relations.

RUSSIA AND THE FAR EAST

Throughout the 1990s and into the new century, both Yeltsin and Putin strengthened ties with countries in the Asia-Pacific region, notably China, Japan, and the Republic of Korea (South Korea). Russian initiatives were inspired by trade and security in the first place. In addition, it was axiomatic that when the Kremlin had problems with the West, in particular the European Union, it focused on strengthening ties with Beijing. In addition, the Kremlin was obliged in the new century to pay increasing attention to the policies of North Korea, which Russia as well as South Korea, Japan, and the United States found troubling, even dangerous, to their security.

Russia and China

As Russian relations with the United States became tense in the early years of the new century, Russian relations with China became closer and stronger. The Chinese were eager to expand purchases of Russian weaponry, especially after Israel bowed to American pressure in mid-2000 to cancel the sale of an early warning radar system to China. For their part, the Russians needed the business from China. China purchased two $800 million Russian destroyers, concluded an agreement to share information on the formation of military doctrine and joint training, and conducted joint military exercises with Russia for the first time. By the beginning of 2000, some 2,000 Russian technicians were employed by Chinese research institutes, working on laser technology, the miniaturization of nuclear weapons, cruise missiles, and space-based weapons.

Putin and President Jiang Zemin also made common cause against U.S. foreign policy. At a summit meeting in Beijing in July 2000, they sharply criticized the United States' perceived "unilateralist" approach to foreign affairs and declared their opposition to plans for a space-based missile-defense system that would violate the 1972 ABM treaty. Putin and Jiang emphasized a shared Russian-Chinese concept of global security, rejecting the notion that the United States was the only superpower in the post–Cold War era. Russia and China concluded a number of important economic and military agreements that reinforced their growing friendship and the prospects of long-term cooperation. Russia agreed to build an experimental fast-neutron reactor for China.

Conspicuous by its absence was an agreement, long anticipated, providing for the construction of a Russian oil pipeline from Siberia to China. Eventually, the agreement was signed in April 2009. According to this agreement, Russian companies received a $25 billion loan to be repaid within 20 years. In exchange they will build a pipeline from the East Siberian town of Taishet to the Chinese border plus a 60 km branch on Chinese land. Currently, Taishet receives gas from Russia's largest West Siberian natural gas field, but the pipeline is to cross yet to be commissioned East Siberian natural gas deposits. China is to begin receiving Russian natural gas in 2010. The new friendship with China is difficult to reconcile with the abiding Russian fear of China, especially of the prospects of massive Chinese emigration to adjacent Russian territory in the Far East.

Russia and China: Trans-border Contacts and Migration

The Soviet–Chinese border in the Russian Federation's Far East region was tightly controlled for a good part of the last century, reflecting (despite outward signs of Communist solidarity) a protracted territorial dispute between the U.S.S.R. and China involving islands located in the Amur River, which formed an extended portion of that border. Economic and political reforms preceding and consequent to the disintegration of the U.S.S.R. have made the border between China and the independent Russian Federation somewhat, or considerably, more "permeable" (depending on what is moving across it), although it continues to be distinguished by a special regime reflecting the specific conditions prevailing in the two countries and in the regions adjoining their common border.

It makes sense to view the cross-border contacts in the broader context of Russia's and China's adjoining location and complementarity. The latter has a vivid demographic dimension that fuels concerns on the Russian side. Unlike China, Russia is a country with a shrinking population due to fertility well below the replacement level (a total fertility rate of 1.4 in 2007) and low life expectancy (67 years), particularly among men (60 years). These are the immediate causes of the negative rate of natural increase (−0.3 percent) not offset by positive net migration into the country (2 persons per 1,000 population). In Russia's Far East, the situation is exacerbated by negative net domestic migration. Whereas Russia's overall population declined 1.26 percent between the censuses of 1989 and 2002, the population of the Far East Federal District declined by 16.8 percent. Considering that by 2025 Russia's working-age population will have shrunk by 19 million, a number equivalent to 28 percent of Russia's overall work force in 2007, Russia might be willing to address this deficit by enticing migration from China and other countries. However, the prospect of growing immigrant communities is not yet fully accepted by the Russian public or even by the political class. This is particularly evident in conjunction with actual and potential Chinese migration to Russia.

Unlike Russia, China has explicitly aimed at curbing its population growth and it has succeeded in greatly lowering fertility—from 5.8 in 1970 to just 1.6 in 2007. Nevertheless, China's population is still growing because of momentum— that is, due to the large generation of parents born at a time when fertility was still high. What is more, China's stunning economic success has not been spatially even; many in China's interior provinces and rural areas have not yet shared in the country's economic boom, so that the pool of potential economic migrants remains large. Northeastern China is a case in point. In 1997, three northeastern provinces (Heilongjiang, Jilin, and Liaoning), were home to 40 percent of China's low-paid employees. This situation has most probably affected population redistribution. Whereas in 1975 Liaoning accounted for 6.6 percent of China's overall population, in 2005 it accounted for only 4.3 percent. For Heilongjiang the respective shares are 4.8 percent and 3.0 percent. Of particular relevance is the fact that many people in northeastern China are poor not only by Western standards, but by Russian norms as well. For example, in 2005 the average wage in Heilongjiang was about $150 a month, whereas in Amur Oblast it was $400 a month. In both regions, groceries accounted for 33.5 percent of household budget outlays. However, whereas in neighboring Amur Oblast, the 2005 per capita consumption was 47 kg of meat and processed meat, 153 kg of milk and dairy foods, 307 kg of fruits and vegetables, and 130 kg of grain-based products; in Heilongjiang the respective consumption statistics (available for urban population only) were 18.4 kg, 14.2 kg, 170 kg, and

86.1 kg. During the last thirty years, Heilongjiang households significantly improved their consumption of durables such as refrigerators, washing machines, computers, etc. But ownership rates of all these durables in Heilongjiang are still much lower than in Amur Oblast. This disparity is particularly acute in the countryside.

Furthermore, just as the U.S.–Mexican border marks the most significant drop-off in living standards between any two neighboring countries, likewise the Sino-Russian border marks the world's steepest contrast in population density. According to estimates based on varying area coverage on both sides of the border, the population density on the Chinese side is 15–30 times higher than on the Russian side. In the most populated region of Russia's Far East, Primorskiy Kray, the average population density is only 13.5 people per sq km. Only in the southernmost part of that region does population density reach 30 people per sq km, but along much of the border with China it does not exceed 4–5. In contrast, in northeast China, population density is 130 people per sq km.

Viewed in another way, whereas about 5 million people live in the southern areas of Russia's Far East, the three Chinese provinces directly south of the Amur River have more than 100 million residents, more than three times the entire population of Siberia. The city of Harbin alone boasts more residents than in Vladivostok, Khabarovsk, and Blagoveshchensk (the three largest cities of Russia's Far East) combined.

Despite the fact that Russia is in dire need of working-age migrants, which China can readily furnish, Chinese migration is viewed as a problem and even a threat. Such a view is persistently propagated by Russia's provincial governments, the entities that might in fact be interested in attracting the Chinese labor force to their labor-deficient regional economies. Instead of directly addressing the most obvious solution to the labor shortage and attempting to regulate Chinese migration, Russian authorities have inadvertently forced migration and shuttle-trading (that is, trading with the aid of people crossing the border regularly and carrying the maximum allowed quantity of goods for sale) underground, into illicit channels. Russian provincial newspapers have engaged in fear-mongering with respect to the magnitude of potential Chinese settlement in Russia. According to the rumors frequently circulated by these papers, millions of Chinese have already settled in Russia illegally, more are about to arrive, and the "yellow threat" is real. Two excerpts from Russian periodicals exemplify the letter and spirit of reports about Chinese migration to Russia.

I think there is no need to explain that not all the Chinese reside in Russia legally. This especially applies to Siberian and far-eastern regions, where Chinese communities live in wooded areas into which even extortionists from among militiamen never poke their noses. . . . If the migration trend remains unchanged, one can predict that by 2010 there will be 20 million Chinese in Russia and 40 million in 2020 (Gilbo 2003).

Lately, the number of Chinese citizens illegally coming to Siberia and Far East have been growing steadily; and according to the data from local migration services, "there are already 10–12 million of them at a minimum" (Orlov 2004).

In contrast, Russian migration scholars' estimates of the overall number of Chinese settlers in Russia do not exceed 400,000. Thus, the present character of trans-border exchange between Russia and China is a direct result of contrasting attitudes and approaches to such exchange in those countries. China is explicitly open to exchange: it has instituted visa waivers, actively monitors the shuttle trade, and has established free economic zones in some border towns (e.g., Heihe). In contrast, Russia has not articulated a coherent policy or strategy for trans-border exchange, preferring to manage it through informal and largely illicit channels while occasionally heightening public fears of a "Chinese threat." This approach is both ambivalent and dysfunctional, containing the seeds of a self-fulfilling prophesy. Given the combination of resource endowments, endemically corrupt bureaucrats, and implacable demographic vulnerabilities in the Russian Far East, a more aggressive and widespread form of Chinese expansion will almost certainly commence in the not too distant future.

Upswing in Russian-Chinese Relations after 2002

By the end of 2002, Russia had drawn closer to China in the wake of problems that both countries were having with North Korea, exacerbated by the hard line of the Bush administration toward Pyongyang. In December 2002, Putin went to Beijing to discuss not only North Korea's apparent determination to proceed with the development of a nuclear weapon but also other problems, including a perceived American pursuit of global dominance. Both Moscow and Beijing were troubled by the growing dispute between North Korea and the United States over a perceived North Korean violation of a 1994 treaty providing for American energy aid, including construction of nuclear power plants in North Korea in return for its willingness to refrain from developing nuclear weapons. Putin made clear Russia's interest in working with China to avoid a North Korean confrontation with the United States.

Russia and China also worried about the advance of American military influence in the Central Asian republics to fight the war on terror that, incidentally, both Moscow and Beijing supported because of their own vulnerabilities to terrorist attacks. Both were very sensitive to the Bush administration's preemption doctrine, especially as regarded Iraq, as seen in their jointly stated commitment to a multi-polar world, code words for opposition to the kind of unilateralism practiced by the Bush administration.

The year 2003 got off to a good start for both countries when the new Chinese president, Hu Jintao, visited Moscow at the end of March, telling Putin that Russia was the first place abroad he was to visit in his new capacity because he wanted to show "how important for us is the development of relations with Russia." The meeting produced no substantive agreements beyond a statement to the effect of calling for the United Nations to play a central role in making a political settlement in Iraq. Putin and Hu Jintao also called upon North Korea to give up its quest for nuclear weapons. During Putin's visit to China in March 2006, multiple memoranda, mostly devoted to prospective Chinese purchases of Russian fuel, were signed. Dmitry Medvedev made his first international visit as president (May 2008) to Kazakhstan and China, thus asserting the growing importance of the eastern direction in the Russian foreign policy. Medvedev was accompanied by a group of Russian businessmen. Their goal was to extend Russia's export to China beyond natural resources. At the same time, Russia is interested in Chinese investment.

China, Russia, and ex-Soviet Central Asia

The Kremlin always has to take into account a strong and abiding Chinese interest in the development of post-Communist Central Asia, especially of Kazakhstan, which has a very long, 1,000-mile boundary with China. China wants to expand imports of natural resources from Central Asia and to participate in the economic growth of the Central Asian republics. China would like their pro-Russian governments to strengthen ties with Beijing. From its vantage point, Russia sees China as a potential for influence in ex-Soviet Central Asia and monitors Chinese behavior toward the Central Asian republics carefully if not discreetly—the Kremlin insists on helping the Central Asian republics police their porous frontiers.

On the other hand, Russia and China have strong incentives to avoid rivalry and promote cooperation and mutual understanding. This is especially true with the advent of the terrorist threat to the security and stability of the Central Asian republics and northwestern China where Muslim and Turkic-speaking Uighurs (close relatives of Uzbeks and Kazakhs) occasionally rebel against the Beijing government. Neither Russia nor China welcome, in the name of fighting terrorism, the growing American military, economic, and political presence in Central Asia.

This common resistance to the penetration of other world powers into Central Asia may be the cornerstone of the so-called Shanghai Cooperation Organization (SCO). SCO is an intergovernmental organization founded in Shanghai on June 15, 2001 by six countries: China, Russia, Kazakhstan, Kyrgyzstan, Tajikistan, and Uzbekistan. SCO's working languages are Chinese and Russian. SCO's predecessor, the Shanghai Five mechanism, originated and grew from the endeavor by China, Russia, Kazakhstan, Kyrgyzstan, and Tajikistan to strengthen confidence-building and disarmament in the border regions. In 1996 and 1997, their heads of state met in Shanghai and Moscow, respectively, and signed the Treaty on Deepening Military Trust in Border Regions and the Treaty on Reduction of Military Forces in Border Regions. Thereafter, this annual meeting became a regular practice and has been held alternately in the five member states. The topics of the meeting gradually extended from building up trust in the border regions to mutually beneficial cooperation in the political, security, diplomatic, economic, trade, and other areas among the five states. The president of Uzbekistan was invited to the 2000 Dushanbe Summit as a guest of the host state. As the first meeting of the five heads of state took place in Shanghai, the cooperation mechanism was later known as the "Shanghai Five."

On the fifth anniversary of the Shanghai Five in June 2001, the heads of state of its members and the president of Uzbekistan met in Shanghai, the birthplace of the mechanism. First they signed a joint declaration admitting Uzbekistan as a member of the Shanghai Five mechanism and then jointly issued the Declaration on the Establishment of the Shanghai Cooperation Organization. The document announced that for the purpose of upgrading the level of cooperation to more effectively seize opportunities and deal with new challenges and threats, the six states had decided to establish a Shanghai Cooperation Organization on the basis of the Shanghai Five mechanism.

In June 2002, the heads of SCO member states met in St. Petersburg and signed the SCO Charter, which clearly expounded the SCO purposes and principles, organizational structure, form of operation, cooperation, orientation, and external relations, marking the actual establishment of this new organization in the sense of international law.

On June 17, 2004, the SCO held its annual Summit in Tashkent, Uzbekistan. Much of the pre-summit media attention included what Russian President Putin and Chinese President Hu hoped would facilitate the development of economic relations between the SCO countries. This appeared to have been successful. At the conclusion of the summit, the leaders signed a document titled the Tashkent Declaration. The declaration summarized the outcome of the SCO's work since it was set up, evaluated the activities of the organization's agencies, and set new goals. They also signed agreements on cooperation in fighting drug trafficking and on the protection of secret information in the framework of the SCO anti-terrorist agency, whose headquarters were opened in Tashkent.

Several meetings were conducted on the sidelines of the Summit. Chinese President Hu Jintao met with Afghanistan Transitional President Hamid Karzai, who was a guest of Uzbek President Islam Karimov; and they discussed Afghanistan's attempts to locate and bring to justice the terrorists who attacked Chinese workers there. In other meetings, Kyrgyz President Askar Akayev handed over to Hu Jintao a written document to confirm Kyrgyzstan's stance to recognize China's full market economy status. Hu said the move by Kyrgyzstan will "greatly push forward China-Kyrgyzstan bilateral trade and economic cooperation." Hu made a four-point proposal in meeting with Tajik President Emomali Rakhmonov on strengthening cooperation. Hu said the two sides should support each other on major issues, enhance law-enforcement cooperation to fight against terrorism, separatism, and extremism, as well as improve economic relations and cultural exchanges. Rakhmonov said he agreed with Hu's proposals. In April 2009, Kazakh President Nursultan Nazarbayev visited China, and the two sides signed a number of agreements to cement cooperation primarily in oil and natural gas as well as in other areas. Russia and Uzbekistan signed a strategic-partnership treaty, with Russian President Vladimir Putin and Uzbek President Islam Karimov hailing it as a new stage in long-term relations.

Russia and Japan

Relations between these two nations are hindered primarily by a dispute over the Kuril Islands. On February 10, 1904, a conflict between Imperial Japan and the Russian Empire resulted in the Russo-Japanese war over Manchuria and Korea. Following that war, Japan acquired the Kuril Islands and the southern part of Sakhalin Island from Russia. As part of the Soviet Union, Russia reestablished its sovereignty over these areas as a result of World War II, in which Japan was defeated. Japan believes that the four tiny southernmost islands in the Kuril archipelago had been transferred to Russia illegally and should be returned. Japan calls these islands its Northern Territories.

There is still no post-World War II peace treaty between Russia and Japan.

After the dissolution of the Soviet Union, Moscow took a stand in firm opposition to returning the disputed territories to Japan. Although Japan joined with the Group of Seven industrialized nations in providing some technical and financial assistance to Russia, relations between Tokyo and Moscow remained poor. In September 1992, Russian President Boris Yeltsin postponed a scheduled visit to Japan. The visit took place in October 1993. He made no further concessions on the Northern Territories dispute over the four islands northeast of Hokkaido, a major obstacle to Japanese-Russian relations, but did agree to abide by the 1956 Soviet pledge to return two areas (Shikotan and the Habomai Islands) of the Northern Territories to Japan. Yeltsin also apologized for Soviet mistreatment of Japanese prisoners of war after World War II. In March 1994, then Japanese minister of foreign affairs Hata Tsutomu visited Moscow and met with Russian minister of foreign affairs Andrei Kozyrev and other senior officials. The two sides agreed to seek a resolution over the longstanding Northern Territories dispute, but the resolution of the dispute is not expected in the near future. Despite the territorial dispute, Hata offered some financial support to Russian market-oriented economic reforms.

In August 2006, Russian maritime authorities killed a Japanese fisherman and captured a crab fishing boat in the waters around the disputed Kuril Islands. The Russian foreign ministry has claimed that the death was caused by a "stray bullet."

In September 2006, Russian Foreign Minister Sergei Lavrov said Russia would "continue the dialogue with the new Japanese government. We will build our relations, how the peoples of the two countries want them to be. Foreign Minister Taro Aso remained on his post in the government. We have good, longstanding relations, we will act under the elaborated program."

The dispute over the Southern Kuril Islands deteriorated Russo-Japan relations after July 2008, when the Japanese government published new guidelines for school textbooks to teach Japanese children that their country has sovereignty over the Kuril Islands. The Russian public was outraged by the action and demanded the government counteract it. The Foreign Minister of Russia announced on July 18, 2008 that "[these actions] contribute neither to the development of positive cooperation between the two countries, nor to the settlement of the dispute" and reaffirmed its sovereignty over the islands. Nevertheless, the relations between the two countries remain in reasonably good shape. Since the early 1990s, the Russian Far East has been one of the major markets for used Japanese cars. Despite being produced for left-hand traffic, these vehicles are almost universally used in that region of Russia.

The North Korean Nuclear Issue

As in the recent past, North Korean foreign and defense polices challenged Russian, Chinese, and Japanese interests in the Pacific Rim. In 2003 and 2004, the most difficult and potentially dangerous of these challenges was North Korea's determination to develop nuclear weapons despite international resistance, especially by Japan and the United States. By the late summer of 2003, the Pyongyang regime was still adamant in its policy to continue developing a nuclear weapons arsenal.

Both Russia and China had proceeded cautiously in responding to North Korea's nuclear build-up. They didn't like it; but they both were reluctant to exert punitive pressure on Pyongyang. In 2005–06, Putin displayed a new, more sympathetic perspective on North Korea, inspired by not only a perceived American global expansionism but also opportunities for Russian trade and commerce. Under Putin, Russia has developed energy projects with North Korea and has encouraged Russian investors to get involved in North Korean economic development. Thus, in the fall of 2001, Russia agreed to start supplying the area of North Korea's frontier with Russia with electrical power to get the economic ball rolling. Russia also is building a new rail line into North Korea and in February 2002, agreed to upgrade the commercial infrastructure, such as wharves in Najin, an ice-free port 25 miles away from a railway bridge to Russia. The Russian ministry of transportation reportedly set aside $250 million for construction of a branch of the Trans-Siberian Railway system into North Korea.

But, there is a problem for Russia in drawing closer to North Korea. Pyongyang is fearful, insecure, and, at the least, psychologically not ready to accommodate expansive Russian plans. It is true that Pyongyang wants good relations with Russia to help offset American aggressiveness; but it also still wants to keep contacts with Russia as well as with other countries to a minimum. Strategically, North Korea's main interest in friendship with Russia is to use it for leverage in dealing with China and the United States. With Russia, North Korean strategy for a long time has been to get as much help as possible from the Kremlin while paying as little as possible for what the Russians give.

To some extent, Russian policy toward North Korea is similar to that of China, making possible joint initiatives. For example, cognizant of the security issue, Russia and China proposed their own security agreement with Pyongyang, which was promptly rejected. The North Koreans insisted that the only country that could reassure their security was the United States. But the Bush administration was not willing to offer such guarantees, at least not until North Korea took real and verifiable steps to halt its building of nuclear weapons. Pyongyang continued its insistence that only a security agreement with Washington would be an acceptable quid pro quo.

Both Russia and China continued to oppose sanctions against North Korea to halt its production of nuclear weapons despite Pyongyang's determination to proceed with the development of ballistic missiles and its insistence and its preference for direct talks with the United States on the nuclear issue rather than the recent multilateral discussions involving South Korea, China, Japan, Russia, and the United States. Apart from reservations about the efficacy and risks of "strong-arming" Pyongyang, neither the Kremlin nor Beijing want to be seen as allies of their two rivals, Japan and the United States, no matter how strong the logic for cooperation may seem.

The impact of the crisis in the spring and summer of 2006 (and yet again in 2009) over North Korea's nuclear build-up and missile testing proximate to Japanese territory, however, may complicate Russo-Japanese relations even more. Japan and Russia differed over how to deal with Pyongyang, with Tokyo leaning toward the United States in wanting to threaten North Korea if it did not curtail its nuclear weapons production, while Russia was inclining toward China in proceeding cautiously and avoiding provocation of the hyper-sensitive North Korean leadership. Finally, Russo-Chinese cooperation in dealing with North Korea to the perceived detriment of Japan has been a source of anxiety in Tokyo.

NOTES

1. The following synopsis of David Foglesong's book is based on the meeting report found at www.wilsoncenter.org/index.cfm?topic_id=1424&fuseaction=topics.publications&doc_id=494540&group_id=7718 and devoted to David Foglesong's 9 June 2008 lecture at the Kennan Institute in Washington, DC.

2. The following synopsis of Krasheninnikova's book is based on a handout distributed at the October 15, 2007 meeting at the Russian Embassy in Washington DC devoted to Krasheninnikova's book promotion.

Ukraine

Ukraine Statistics

Area in Square Miles (Kilometers):
   233,000 (603,700)
Capital (Population): Kyiv (2,800,000)

Population

Total (millions): 46.0
Per Square Km: 76
Percent Urban: 68
Rate of Natural Increase: −0.5%

Ethnic Makeup: Ukrainian 77.8%;
   Russian 17.3%; Belarusian 0.6%;
   Moldovan 0.5%; Crimean Tatar 0.5%;
   Bulgarian 0.4%; Hungarian 0.3%;
   Romanian 0.3%; Polish 0.3%; Jewish
   0.2%; other 1.8% (2001 census)

Health

Life Expectancy at Birth: 63 (male); 74
   (female)

Infant Mortality Rate: 10/1,000
Total Fertility Rate: 1.4

Economy

*Currency ($U.S. Equivalent on
   October 14, 2009):* 8.2805
   Hryvnia = $1
2008 GNI PPP Per Capita: $7,210
*2005 Percent Living on Less than US$2/
   Day:* <2

Ukraine Country Report

Ukraine is the largest (603,700 sq km) of all the independent states located entirely within Europe as conventionally defined (from the Atlantic Ocean to the Urals) and the second largest of all European states (considering Russia, the largest of those states). Until 1991, it was habitual in English to use the article "the" with the name of the country (The Ukraine), but the article was subsequently dropped. The reason behind the usage of "the" was that, etymologically, "Ukraine" means located near the edge. So the article implied that Ukraine is really just a region within a greater whole, just as one says "the Midwest" or "the Great Plains." And because in the case of Ukraine that greater whole used to be the Russian Empire, the continuing usage of the article in front of "Ukraine" conveyed, at least to some Ukrainians versed in English, a Russian imperialistic affectation in their use of English. In the imperial Russian discourse and even in the title of the Russian Tsar (*"The Sovereign of all Rus': the Great, the Little, and the White"*), Ukraine was referred to as Malorossiya, which means Little or Lesser Russia. The term Little Russia has become an archaic one, and anachronistic usage in the modern context is considered offensive by Ukrainians, as it often implies the denial of a separate Ukrainian national identity.

Ukraine is a unitary state composed of 24 oblasts (provinces), one autonomous republic (Crimea), and two cities with special status: Kyiv, its capital, and Sevastopol, which houses the Russian Black Sea Fleet under a leasing agreement (set to expire in 2017). When the name of Ukraine's capital is transliterated from its Russian, not Ukrainian name it is usually spelled Kiev. Besides Kyiv, other significant cities of Ukraine include Kharkiv (population 1.4 million), Dnipropetrovsk (1.1 million), Odesa (1.0 million), Zaporizhia (0.8 million), and Lviv (0.7 million), as well as many others.

Ukraine's overall population is 46 million, and Kyiv, its capital city, is home to 2.8 million people. According to the Ukrainian Census of 2001, ethnic Ukrainians make up 77.8 percent of the population. The only other significant (exceeding at least 1% of the population) ethnic group is Russians (17.3%). Other groups, such as Belarusians, Moldovans, Crimean Tatars, etc., are minuscule, although some of them, notably Crimean Tatars and Hungarians, live in compact areas (Crimea and Trans-Carpathian Ukraine, respectively) wherein the concentration of these groups is pronounced.

The official language of Ukraine is Ukrainian, a Slavic language, which is lexically midway between Russian and Polish but somewhat closer to Russian. According to official 2001 census data, about 75 percent of Kyiv's population responded "Ukrainian" to the native language (*ridna mova*) census question and roughly 25 percent responded "Russian." On the other hand, when the question, "What language do you use in everyday life?" was asked in a 2003 sociological survey, Kyivans' answers were distributed as follows: "mostly Russian": 52 percent; "both Russian and Ukrainian in equal measure": 32 percent; "mostly Ukrainian": 14 percent; "exclusively Ukrainian": 4.3 percent. Indeed, in the streets of Kyiv one hears Russian spoken with a noticeably greater frequency than Ukrainian. The preponderance of Russian is even more significant in other major urban areas of eastern (e.g., Kharkiv, Donetsk, Luhansk, etc.) and southern (Odesa, Simferopol, etc.) Ukraine, whereas in western Ukraine Ukrainian is more widely used. Quite a few people in Ukraine, by some accounts up to 18 percent of the population, use a mixture of Russian and Ukrainian, the so-called *surzhyk* (literally a multi-grain bread or dough).

Ukraine possesses rich reserves of iron ore, bituminous and anthracite coal, and manganese-bearing ores located in close proximity to each other. The Donetsk Coal Basin (or Donbas) in east-central Ukraine is the industrial heartland of the country and one of the major heavy-industrial and mining-metallurgical complexes of Europe. Ukraine is blessed with some of the world's most fertile soils, called Chernozem (i.e., black soils). These soils are black and contain a very high percentage of humus.

However, socioeconomically Ukraine falls short of its resource potential. In the most recent (2008) country rankings on the Human Development Index (HDI), a composite measure of development based on GDP per capita, life expectancy at birth, and educational attainment, Ukraine is number 82, which is within the medium human development range (76–153). Of the former Soviet republics, not only are the three Baltic States ahead of Ukraine but so also are Belarus, Russia, and Kazakhstan. If the Caucasus is construed as the southeastern border of Europe, then only one European country, Moldova, falls lower than Ukraine on the HDI.

PRE-SOVIET HISTORY AND UKRAINIAN IDENTITY

Ukrainian historical tradition seeks to anchor itself in the same historical beginnings as Russia, that is, in Rus. That was an Eastern Slavic state which emerged in the ninth century and existed up until it fell victim to the Tatar-Mongolian yoke; around 1240 it disintegrated into separate princedoms. Kyiv was the capital of Rus during its most glorious years (882–1169), which included 988, the year when Christianity became the official religion of the country under Vladimir the Great. Russian and Ukrainian historical traditions focus on the same Kyivan legacy but then part ways. While Russians emphasize the decline of Kyiv (Kiev), which they call the "mother of Russian cities," and the subsequent rise of first Vladimir (city) and then Moscow as centers of the recreated (by the 1400s) Rus-Russia, Ukrainians maintain that Muscovites simply usurped the Kyivan legacy to conceal the fact that they developed from a new ethnic substratum: Slavs merged first with Ugro-Finnic tribes and then with hordes of Asiatic nomads. Hence, to Ukrainian historians, Russia emerged as an Asiatic power valuing autocracy, whereas Ukraine, with its direct lineage from quintessentially European Kyivan Rus, is a consistently European entity valuing individual freedoms. To them, Kyivan Rus was "the first Ukrainian state."

Because Ukraine was not an independent country for much of its existence, a separate Ukrainian identity was late in coming. Andrew Wilson, the author of perhaps the most authoritative historical account of Ukraine in English, notes that for a long time, between the fourteenth and the twentieth century, Ukrainians were carving a niche for themselves between Russia and Poland—the main powers in Eastern Europe after the fall of Rus. In fact, observes Wilson, "Ukraine's entire history could be written in terms of its oscillation between the two sides, with the Russians decisively surpassing the Poles in importance only in the 19th and 20th centuries" (Wilson 2002: 40).

The Ukrainian national identity began to develop in the 1800s. One of the very first documented hallmarks of that process became the 1861 article by Mykola Kostomarov titled "Two Russian Nationalities." However, even before that, the great national poet Taras Shevchenko (1814–1861) had fostered a sense of belonging to Ukraine, to which both Russian and Polish identities were deemed fundamentally hostile. For a long time, indeed virtually until the 1930s, most cities of Ukraine were dominated by Russians, Jews, and Poles, while most people in the largely Ukrainian countryside embraced local and parochial (not all-Ukrainian) identities.

(Courtesy of Grigory Ioffe)

Andriyevska church in Kyiv (Kiev), Ukraine, was designed by Bartolomeo Rastrelli.

Wilson observes that, for the Ukrainians, "identity markers in the west, against Catholic Poland, remained sharper than those in the east, against Orthodox Russia" (Wilson 2002: 70). This is a far-reaching observation, for although Ukrainian and Russian are separate languages, they are both classed as East Slavic (whereas Polish is West Slavic) and are indeed closely related to one each other, while other, non-linguistic cultural affinities between Russians and Ukrainians are fairly strong. This can hardly be otherwise, because from the mid-1600s until the 1991 dissolution of the Soviet Union muchof Ukraine existed within a Russia-dominated polity. Only Galicia (centered in Lviv), west of the Zbruch river, remained outside that polity until as recently as 1939. And neighboring Trans-Carpathian Ukraine (centered in Uzhhorod) was long separate from Russia. In Galicia, strong support of a separate Ukrainian identity has

long been rendered by the Uniate (Greek Catholic) Church, a religious community (established in 1596) which retains the distinctive spiritual, liturgical, and canonical traditions of the Orthodox Church, permits a married clergy, and yet recognizes the supremacy of the Pope. Unlike the western part of Ukraine, no cultural divide developed in eastern Ukraine that would have reliably, if informally, separated it from Russia. Books in the Ukrainian language that would have established a single literary tradition for the entire country were not printed en masse prior to 1905, and no public education system in that language existed in all of Ukraine prior to the early 1920s. Against this backdrop, the largely Ukrainian-speaking countryside was spatially stratified into dialect zones. Within the existing dialect nets, eastern Ukrainian dialects morphed into Russian (as one proceeded east) even more decisively than western Ukrainian dialects morphed into

Polish (as one proceeded west). In addition, there was a religious divide in the west (Catholics versus Greek Catholics and Catholics versus Orthodox), but none whatsoever in the east, where the Orthodox Church dominated communities on both sides of the would-be divide between Ukraine and Russia.

The geo-linguistic situation is important to take into account when reflecting about the dominant features of Ukrainian nationalism. Indeed, its overriding feature is its striving to distance Ukraine from Russia and Russians by all means. And because this is, objectively speaking, not an easy task to accomplish in regard to eastern Ukraine, juxtaposing Ukraine and Russia (in order to contrast them) became the leitmotif, the idée fixe, and the battle cry not only of such radical nationalists as Dmitry Dontsov (1883–1973), whose hatred of Russia was visceral and undisguised, but even of moderate and

Russia-friendly nationalists like Leonid Kuchma, former (1994–2005) Ukrainian president, whose 2003 Moscow-published book was pointedly titled "Ukraine Is No Russia." Needless to say, the juxtaposition in question also derives, in no small part, from a reactive perception by Ukrainians of the imperial thinking shared by many Russians who still (almost 20 years after the dissolution of the U.S.S.R.) cannot quite see Ukraine as a separate country.

Aside from Kostomarov and Shevchenko, the pantheon of pre-Soviet ethnic Ukrainians includes such prominent figures as Bohdan Khmelnytsky, Ivan Mazepa, Mikhailo Drahomanov, Mykola Hohol (Nikolay Gogol), and Mykhailo Hrushevsky. In his own way each of them reflects an uneasy coexistence between Ukraine and Russia.

Bohdan Khmelnytsky (1595–1657) was a hetman (a military commander) of the Zaporizhian Cossacks. "Zaporizhian" literally means "beyond the rapids" on the Dnieper (Dnipro) River—that is, where the city of Zaporizhia currently exists. Historically, the Cossacks (from the Turkic qazaq or free men) were migrants who took advantage of the no-man's land in the open steppe to set up agricultural and raiding communities beyond the reach of the main regional powers—the Polish Lithuanian Commonwealth (Rzeczpospolita), Muscovy (Russia), and the Crimean Tatars and the Ottoman Turks. Ukrainian historical tradition names the Zaporizhian Sich as the Ukrainian Cossack state, the cradle of the Ukrainian identity. However, the Cossacks were at least in part defined by their religion (Orthodox Christian) and their social status (descendants of run-away serfs), rather than their ethnicity. Khmelnytsky led the uprising against the Polish-Lithuanian Commonwealth magnates (1648–1654), and in 1654 he concluded the Treaty of Pereiaslav with the Tsardom of Russia, which led to the eventual loss of Zaporizhian Sich's independence to the Russian Empire. Russian and Soviet historians have always looked at the Pereiaslav treaty as the affirmation of Russian and Ukrainian re-union, meaning that two prongs in the three-pronged but initially monolith Rus-based ethnicity (Russians, Ukrainians, and Belarusians) had eventually merged. Khmelnytsky, therefore, is considered the utmost Ukrainian, and the fact that he was responsible for but a part of the entire land where the Ukrainian identity was yet to form, is disregarded. One of the major streets in downtown Moscow (Maroseika Street) was renamed Khmelnytsky Street on the eve of the 300-hundred year anniversary of the Pereiaslav treaty. Ukrainian historians, however, view that treaty as "simply one choice amongst many, a confederational alliance directed against an external enemy, and as a contract, not an act of fealty" (Wilson 2002: 64).

Ivan Mazepa (1639–1709) was one of Khmelnytsky's successors (since 1687) as hetman of the Zaporizhia Cossacks and a favorite of Peter the Great. Mazepa accumulated great wealth, becoming one of Europe's largest land owners. A multitude of churches were built all over Ukraine during his reign in the Ukrainian Baroque style. In 1702, the Cossacks of what is now right-bank Ukraine (west of the Dnieper River) waged an uprising against Poland, which after early successes was quashed. Mazepa convinced Peter the Great to allow him to intervene, which he successfully did. In the beginning of the eighteenth century, as the Russian Empire suffered setbacks in the Great Northern War, Peter the Great decided to centralize control over his realm. In Mazepa's opinion, the strengthening of Russia's central power could put at risk the broad autonomy granted to the Cossack Hetmanate under the Treaty of Pereiaslav in 1654. The Hetman himself started to feel his post threatened in the face of increasing calls to replace him with one of the abundant generals of the Russian army. The last straw in the souring relations with Tsar Peter was his refusal to commit any significant force to defend Ukraine against the Polish King Stanislaus Leszczynski, an ally of Charles XII of Sweden, who threatened to attack the Cossack Hetmanate in 1708. Peter expected an attack by King Charles of Sweden and decided he could spare no forces. In the opinion of Mazepa, this violated the Treaty of Pereiaslav, because Russia refused to protect the Cossack territory and left it to fare on its own. As the Swedish and Polish armies advanced towards Ukraine, Mazepa allied himself with them on October 28, 1708. However, only 3,000 Cossacks followed their Hetman; the rest remained loyal to the Tsar. Mazepa's call to arms was further weakened by the Orthodox Clergy's allegiance to the Tsar. Learning of Mazepa's treason, the Russian army sacked and razed the Cossack Hetmanate capital of Baturyn, killing the defending garrison and all its population. The Russian army was ordered to tie the dead Cossacks to crosses, and float them down the Dnieper River all the way to the Black Sea in order to intimidate Mazepa loyalists living downstream.

Former President Kuchma of Ukraine explained to Russians the significance of Mazepa in his aforementioned book *Ukraine Is No Russia:* "Whereas Khmelnytsky is a symbol of the Ukrainian statehood's succession from Kyivan Rus to nowadays, Mazepa epitomizes the alternative. In many Ukrainians' eyes he counterbalances Khmelnytsky, that is, he neutralizes his bias. Today treating Mazepa as a traitor is an anachronism which can testify to some psychological immaturity. It is good that our National Bank . . . placed the portraits of both Khmelnytsky and Mazepa on Hryvnia (Ukrainian currency; Mazepa's portrait is on the 10-hryvnia note – G.I.) notes. Some of our people are still waging wars against monuments and portraits. But hardly anybody would tear paper banknotes in protest. Seeing both Khmelnytsky and Mazepa in their wallets, our people will evince more tolerance of our history."

Mykhailo Drahomanov (1841–1895) was an ideologue of Ukrainian nationalism, ethnographer, historian, and public figure, born to a noble family of Cossack descent. In the 1860s, he was involved in the *hromada* (literally meaning "community") movement, briefly worked with peasants in a Sunday school run by members of the Kyiv *hromada* (a Ukrainian national movement) and developed his views on the peasant question. In the early 1870s he was one of the younger members of the leftist wing of the Kyiv *hromada.* He lectured at Kyiv University from 1870 to 1875, but because of the repressions against the Ukrainian national movement peaking up in 1876 was forced to leave the Russian Empire. He emigrated, settling initially in Geneva. As an émigré, he continued his political, scholarly, and publishing activities. In 1885–95, he was a professor at the University of Sofia. Drahomanov wrote the first systematic political program for the Ukrainian national movement. He himself defined his political convictions as "ethical socialism," and acknowledged that he had been deeply impressed by socialist literature as a teenager. He published the periodical *Hromada,* which was financed by members of the Kyiv *hromada.* Mykhailo Drahomanov argued that in the Ukrainian case the national movement had to address the social question. He was against narrow nationalist egocentrism and believed that nationality was just a means to achieve universal human ideals. This viewpoint was criticized by more radical nationalists such as Dmitry Dontsov.

Mykola Hohol (1809–1852), better known as Nikolay Gogol, is one of the most famous Russian writers of all time. His novels such as *Dead Souls* and plays such as *The Government Inspector* are part and parcel of all secondary school curricula in both Russia and Ukraine. These, as well as other texts by Gogol, are not simply dead artifacts covered by schools pro forma; they are among the most read and staged literary works that have been and still are cited and invoked by writers, and the media, including talk shows, on an everyday basis, particularly in conjunction with criticizing some proverbial features of Russian/Ukrainian politics, bureaucracy, and everyday life. Gogol was born in the Cossack village of Sorochyntsi, in Poltava Governorate of the Russian Empire, present-day Ukraine. His mother was a descendant of Polish nobility. His father Vasily Gogol-Yanovsky, a descendant of Ukrainian Cossacks, belonged to the petty gentry and wrote poetry in Russian and Ukrainian. As was typical of the left-bank Ukrainian gentry of the early nineteenth century, the family spoke Russian as well as Ukrainian. As a child, Gogol helped stage Ukrainian-language plays in his uncle's home theater. "I myself do not know whether my soul is Ukrainian [khokhliatskaya] or Russian [russkaya]," wrote Gogol in a letter in 1844, "I know only that on no account would I give priority to the Little Russian before the Russian or to the Russian before the Little Russian. Both natures are too richly endowed by God, as if by design, each of them separately contains within itself what the other lacks—a sure sign that they complement one another."

Here's how Oleh Tiagnibok, a modern Ukrainian nationalist, characterizes Gogol: "Gogol is a split personality torn between the empire and the Ukrainian identity, between glory and the good life in Saint Petersburg and languishing in Dikan'ka [a Ukrainian village where some stories by Gogol were set – G.I]. As a writer, Gogol is undoubtedly great. But his outlook is that of a Little Russian. . . . This was a result of statelessness. If the Ukrainian state had existed at the time, Gogol would have scarcely been a Russian writer. This is a vivid confirmation of the necessity of the state for the normal development of a nation."

Mikhailo Hrushevsky (1866–1934) was Ukraine's greatest modern historian and leader of the Ukrainian national movement. He was head of the Central Rada, Ukraine's 1917–1918 revolutionary parliament and a leading cultural figure in Soviet Ukraine during the 1920s. Hrushevsky was born in Chelm, currently in southeastern Poland, in the family of a reputable teacher of Church Slavonic. Hrushevsky authored the ten-volume *History of Ukraine-Rus,* which is believed to be the first detailed scholarly synthesis of Ukrainian history. According to Hrushevsky, Ukrainians have existed as a separate ethnicity since the early Middle Ages. In Kyivan Rus, Ukrainians constituted the core of the state apart from

the northeastern territorial communities which subsequently became Russians. The immediate heir of Kyivan statehood was not the Vladimir-Suzdal princedom (as claimed by Russian historical tradition) but first the Galich-Volyn princedom (in what is now western Ukraine) and then the Great Duchy of Lithuania. In the spring of 1917 Hrushevsky was elected the head of the Ukrainian Central Rada. He guided the Rada from Ukrainian national autonomy within Russia through to completely independent statehood. Following the German-supported coup of General Pavlo Skoropadsky, Hrushevsky emigrated. While an émigré in Austria, he began to adopt a pro-Bolshevik position. In 1924, Hrushevsky returned to Ukraine, where he concentrated on academic work, becoming a professor at Kyiv State University and a full member of the Ukraine's, and subsequently of the Soviet Union's, Academy of Sciences. Hrushevsky fell victim to the purge of the Ukrainian intelligentsia and, in 1931, he was forced to move from Kyiv to Moscow. In 1934, under the close watch of the Soviet secret police, he died while on vacation in the resort town of Kislovodsk in the Caucasus. While Hrushevsky was buried with full state honors, shortly thereafter all of his works were banned, and many of Hrushevsky's surviving relatives, including his daughter, were arrested and died in the Gulags.

SOVIET HISTORY AND UKRAINIAN IDENTITY

In 1922, alongside the Russian Federation, Belarus, and the Trans-Caucasian Federation Ukraine became a cosigner of the Union Treaty, which ushered in the Union of the Soviet Socialist Republics. The Soviet period of Ukrainian history (1922–1991) was controversial in the extreme. On the one hand, it was during that period that the Ukrainian identity was strengthened and became shared by a vast majority of people residing in Ukraine, due to Ukrainian statehood, however restricted, within the union. It was during the Soviet period that the Ukrainian national territory was united for the first time within one polity: in 1939 western Ukraine, which had been part of Poland since 1921, joined Soviet Ukraine and then the Trans-Carpathian Ukraine joined the rest of Soviet Ukraine in 1945. Later, in 1954, Crimea, a region without any sizable concentration of ethnic Ukrainians, was shifted from Russia to Ukraine. Thus, during the Soviet period, Ukraine enjoyed an expansion and national consolidation without precedent in its earlier history. Last but not least, it was during the same period that the Ukrainian

economy underwent massive industrialization and acquired the potential that, shortly before the Soviet Union's dissolution, was assessed as having no match among the Soviet republics. By most accounts, Ukraine was well-integrated into the Soviet Union, so much so that, for many Ukrainians, Soviet identity became no less and occasionally even more important than the sense of belonging to Ukraine. While this identity shift did not occur in the westernmost part of Ukraine, which bore the indelible imprint of its lasting non-Soviet experiences all throughout the Soviet period, in much of Ukraine this was a widespread phenomenon on a scale unseen elsewhere in the Soviet Union save Russia itself and Belarus. On the other hand, the Soviet experience brought much suffering to Ukrainians such as the extermination of the cream of the crop of the national intelligentsia in the 1930s, hunger in 1932–33, and the Chernobyl disaster of 1986. Obviously, the devastations of World War II, although not (for the most part) attributed to the Soviet side, were also part and parcel of Soviet Ukrainian history.

Arguably, the hallmarks of Soviet history that left lasting impacts on Ukraine are *Korenizatsiya,* purges, *Holodomor,* industrialization, World War II, activities of the Ukrainian Insurgent Army, postwar reconstruction and development, and Chernobyl.

Korenizatsiya (from the word koren' = root) was a largely successful attempt by the Communist Party to introduce the Ukrainian language into all spheres of public life, schooling, and media in the Ukrainian Soviet Socialist Republic. This policy of Ukrainization was conducted from 1923 to 1930. Not only did it result in the increase of Ukrainian language usage in cities, but it also enhanced the status of the Ukrainian literary standard and significantly strengthened the role of ethnic Ukrainians in the political establishment. In 1922, ethnic Ukrainians accounted for just 23.3 percent of the members of the Communist Party of Ukraine, while 53.6 percent were Russians, and 23.1 percent represented other ethnicities, including Jews whose share among Party members was well above their share in total population. By 1927, the proportions were vastly different: 51.9 percent Ukrainians, 30.0 percent Russians, and 18.1 percent others. By 1933, ethnic Ukrainians accounted for 60 percent of the Communist Party members. The effect of the Ukrainization was construed by many as positive, particularly against the backdrop of the policy of forcible Polonization conducted at the same time in western Ukraine under the Polish administration.

However, by the late 1920s, many Ukrainian writers, scholars, and political leaders were being accused of bourgeois nationalism and falling victim to purges. By some accounts, two waves of Stalinist political repression and persecution in the Soviet Union (1929–34 and 1936–38) resulted in the killing of four-fifths of the Ukrainian cultural elite. At that time, however, the peasantry was the most numerous social group in Ukraine, and that group fell victim to forced collectivization. Forced collectivization had a devastating effect on agricultural productivity. Despite this, in 1932 the Soviet government increased Ukraine's production quotas by 44 percent, ensuring that they could not be met. Soviet law required that members of a collective farm receive no grain until government quotas were satisfied. The authorities in many instances exacted such high levels of procurement from collective farms that starvation became widespread. Millions starved to death in a famine known as the *Holodomor.* Available data are insufficient for precise calculations and therefore estimates are wide ranging, from as "low" as 2 million to as high as 10 million deaths. While hardly anybody doubts that *Holodomor* was a crime perpetrated by Soviet authorities, qualifications of that crime vary. Some Ukrainian scholars and politicians have argued that the Soviet policies that caused the famine were designed as an attack on the rise of Ukrainian nationalism and therefore fall under the legal definition of genocide. Indeed in March 2008, the parliament of Ukraine recognized the actions of the Soviet government as an act of genocide against Ukrainians. The alternative interpretation, popular in eastern Ukraine (that is, in the area which fell victim to the *Holodomor* to the utmost degree) and in Russia, is that class warfare was the root cause of that crime, notably, the attitude of the Soviet leadership to the peasantry as an archaic and inherently reactionary (petty bourgeois) group hostile to an urban proletariat. The proponents of this interpretation invoke statistics showing that some ethnic Russian provinces (particularly Kursk and Saratov) were victimized no less and even more than Ukraine. Also, ethnic Ukrainian leaders such as Chubar and Petrovsky were on the side of the perpetrators of *Holodomor* in Ukraine. According to some scholars (e.g., Belarusian historian Yury Shevtsov and Russian historian Roy Medvedev), equating *Holodomor* with the genocide of Ukrainians is part of the identity politics of the Ukrainian political elite which helps them disassociate Ukraine from Russia by nurturing a cult of victimhood amongst Ukrainians. The same idea is expressed in the letter that Russian

President Medvedev sent to President Viktor Yushchenko of Ukraine in November 2008. Interestingly, parallel to November 2008 commemorative events in Kyiv, to which Medvedev was invited but declined to participate, two alternative historical conferences on *Holodomor* were held in Moscow and Kharkiv (the second-largest city of Ukraine). These alternative conferences upheld the official Russian view of *holodomor.*

At just the time when the Ukrainian countryside was undergoing such immense sufferings, the industrialization campaign which transformed Ukraine into a strong industrial power, particularly in the areas of extractive industries (coal, iron ore, and manganese), ferrous and non-ferrous metallurgy, and subsequently mechanical engineering, was in full swing. In fact, out of the total of 1,500 new industrial enterprises under construction in the U.S.S.R. during the first five-year plan of 1928–33, roughly 400 were located in Ukraine. The true symbol of Soviet industrialization was *Dnieproges,* Europe's largest hydroelectric power station (on the Dnieper River), commissioned in 1932. By 1940, industrial output of Ukraine exceeded that of 1913 by a factor of seven.

World War II brought about colossal destruction in Ukraine. Its population declined from 41.6 million in June 1941 to 27.4 million in January 1945, and about one-quarter of the survivors had no dwelling. Out of the estimated 8.7 million Soviet troops who fell in battle against the Nazis, 1.4 million were ethnic Ukrainians. Sixteen thousand industrial enterprises and 714 cities and towns were completely destroyed. The Soviet monetary assessment of war-inflicted property damage was 285 billion rubles in 1941 prices, that is higher than in any other Soviet republic affected by the war. Although the vast majority of Ukrainians fought alongside the Red Army and Soviet resistance, in 1943 some elements of the Ukrainian underground created an anti-Soviet nationalist formation in Galicia, the Ukrainian Insurgent Army (UIA) that at times engaged the Nazi forces and at times collaborated with them. At the same time, another nationalist movement fought alongside the Nazis. In total, the number of ethnic Ukrainians that fought in the ranks of the Soviet Army is estimated from 4.5 million to 7 million. The pro-Soviet partisan guerilla resistance in Ukraine is estimated to have numbered at 47,800 at the start of occupation and peaked at 500,000 in 1944; about 50 percent were ethnic Ukrainians. Generally, the assessments of the Ukrainian Insurgent Army's manpower are wide ranging, from 15,000 to as many as 100,000 fighters. Overall, the scale of

collaboration with the Nazis in Ukraine was more significant than in the neighboring countries such as Poland and Belarus but no more significant than in the Baltic republics. Over half a million Ukrainian Jews perished, not only at the hands of occupiers but also (and by some accounts for the most part) at the hands of local collaborators. In western Ukraine, hostility to both Poles and Russians explained why Germans were sometimes initially received as liberators. However, brutal German rule in the occupied territories eventually turned Nazi supporters against the occupation.

The activity of the UIA is a controversial page of recent Ukrainian history. The UIA was a group of Ukrainian nationalist partisans who engaged in a series of guerrilla conflicts during World War II. The group was the military wing of the Organization of Ukrainian Nationalists—Bandera faction (the OUN-B), originally formed in Volyn (northwestern Ukraine) in the spring and summer of 1943. The OUN's stated immediate goal was to protect ethnic Ukrainians from repression and exploitation by Polish governing authorities; its ultimate goal was an independent and unified Ukrainian state. The organization began as a resistance group and developed into a guerrilla army. The UIA fought a large variety of military forces, including the Nazi German Wehrmacht and Waffen SS, the Polish underground army (Armia Krajowa) and Soviet forces—including Soviet partisans, the Red Army, and Soviet secret police forces. However, on many occasions, the UIA also cooperated with the German Wehrmacht and Waffen SS against the Soviets and Poles. The UIA played a substantial role in the killing and ethnic cleansing of much of Western Ukraine's Polish population and according to some accounts, in Jewish pogroms. After the end of World War II, the UIA remained active and fought against Poland until 1947 and against the Soviet Union until 1949. It was particularly strong in the Carpathian Mountains, the entirety of Galicia, and in Volyn—in modern Western Ukraine. It drew substantial moral support from the Ukrainian Greek Catholic Church. The strength of the UIA was largely a reflection of the popularity it enjoyed among the people of Western Ukraine. Outside of Western Ukraine, support was minimal, and the majority of the Soviet (eastern) Ukrainian population considered the OUN/UIA to have been primarily German collaborators. While UIA was formally disbanded in early September, 1949, some of its units continued operations until 1956. Attitude toward the UIA is still a kind of an ideological barricade in modern Ukraine. While in many towns of western Ukraine monuments to the UIA fighters have been set up, in eastern and southern Ukrainian towns monuments to UIA's victims have been erected in response. In 2007, President Yushchenko of Ukraine posthumously awarded the title "Hero of Ukraine," the country's highest honor, to UIA leader Roman Shukhevych, a move that did not meet the approval of many Ukrainians residing in the eastern and southern parts of the country.

After World War II, the Ukrainian SSR became one of the founding members of the United Nations. Most investment during post-war five year plans was directed to industry, and by 1950, Ukraine managed to produce as much coal and steel as in 1940. Modest amounts of oil and gas were discovered in western Ukraine, and the pipeline linking Dashava (in Lviv Oblast) and Kyiv was built in 1948. Alongside reconstructed old enterprises, scores of new ones were built, including such industrial giants as three iron ore mining and processing plants near Krivyi Rih, a tire factory in Dnipropetrovsk, hydroelectric power stations in Kremenchug and Dnieprodzierzhynsk, and synthetic fiber factories in Cherkassy and Chernihiv. In the 1970s and 80s five nuclear power stations were commissioned in Ukraine. Overall, Ukraine accounted for 17 percent of the Soviet Union's industrial output, becoming an industrial power second only to gigantic Russia.

On April 26, 1986, one of four reactors in the Chernobyl Nuclear Power Plant exploded, resulting in the worst nuclear reactor accident in history (pp. 120). The station was located in the town of Pripyat in Kyiv Oblast, 7 km south of the border with Belarus. At the time of the accident seven million people lived in the contaminated territories, including 2.2 million in Ukraine. After the accident many residents of Kyiv were scared and tried their best to avoid buying food produced in northern Ukraine. Due to direction of airflow, however, most contaminated areas turned out to be located north of the station, in the Republic of Belarus. In the 1990s, a new town, Slavutych, was built outside the exclusion zone to house and support the employees of the plant which was decommissioned in 2000. A report prepared by the International Atomic Energy Agency and World Health Organization attributed 56 direct deaths to the accident and estimated that there may have been 4,000 extra cancer deaths.

INDEPENDENT UKRAINE: 1991–

On July 16, 1990, the Supreme Soviet of Ukraine adopted a declaration that established the principles of self-determination

(Courtesy of Grigory Ioffe)

Kyiv (Kiev): Ukraine Independence Monument.

of the Ukrainian nation, its democracy, political and economic independence, and the priority of Ukrainian law over Soviet law on the Ukrainian territory. On August 24, 1991, after the coup attempt against Gorbachev in Moscow failed (see pp. 21), the Supreme Soviet of Ukraine adopted the Act of Independence. A referendum and the first presidential election took place on December 1, 1991. That day, more than 90 percent of the Ukrainian people expressed their support for the Act of Independence, and they elected the chairman of the parliament, Leonid Kravchuk, to serve as the first President of the country.

After the collapse of the Soviet Union, Ukraine inherited a 780,000-man military force on its territory, equipped with the third-largest nuclear weapons arsenal in the world. In May 1992, Ukraine signed the Strategic Arms Reduction Treaty (START) in which the country agreed to give up all nuclear weapons to Russia for "disposal" and to join the Nuclear Non-Proliferation Treaty as a non-nuclear weapon state. Ukraine ratified the treaty in 1994, and by 1996 the country was free of nuclear weapons.

Ukraine was initially viewed as a republic with favorable economic conditions in comparison to the other regions of the Soviet Union. However, the country experienced a deeper economic slowdown than some of the other former Soviet Republics, including Russia and Belarus. Ukraine lost 60 percent of its GDP between 1991 and 1999 and suffered five-digit inflation rates. After miners in eastern Ukraine went on strike, the Ukrainian Parliament decided to conduct parliamentary and presidential elections before the expiration of the respective constitutional terms. As a result of early presidential elections of July 1994, Leonid Kuchma became President of Ukraine. While his rival, the incumbent president Kravchuk, native of western Ukraine, enjoyed support in the western part of the republic, Kuchma enjoyed support in central and eastern Ukraine. A native of Chernihiv Oblast, Kuchma had established his reputation in the area of nuclear missile and spacecraft R & D. From 1982 he had worked as deputy chief designer and from 1986 to 1992 as director of Yuzhmash, a Dnipropetrovsk-based plant and major component of the Soviet military-industrial complex. From October 1992 to September 1993, Kuchma was Ukraine's Prime Minister. At the time of Kuchma's election to the presidency, two political parties enjoyed most of the backing of the electorate, the Communist Party (primarily in the eastern and southern regions) and Rukh, a party of Ukrainian nationalists emphasizing the need for ultimate detachment from Russia and joining Western multinational structures. Rukh's support was centered in the western and central regions. The hostile relationship between these parties was reflected in the activity of the Ukrainian Parliament throughout 1994–98.

DEVELOPMENT

The Ukrainian economy continued its downward trend throughout the 1990s and gave way to growth only after 1999. A new currency, the *hryvnia,* was introduced in 1996. Mass privatization of the Ukrainian industry was mostly conducted under Kuchma. From 2000 to 2008, the country enjoyed steady economic growth averaging about seven percent annually.

The Ukrainian economy continued its downward trend throughout the 1990s and gave way to growth only after 1999. A new currency, the *hryvnia,* was introduced in 1996. Mass privatization of the Ukrainian industry was mostly conducted under Kuchma. From 2000 to 2008, the country enjoyed steady economic growth averaging about seven percent annually. A new Constitution of Ukraine was adopted in 1996; it turned Ukraine into a semi-presidential republic and established a stable political system. Kuchma was, however, criticized

by opponents for concentrating too much power in his own office, transferring public property into the hands of loyal oligarchs (including his relatives), discouraging free speech, corruption, and electoral fraud. His tenure as president was clouded by scandals, the most publicized of which was the September 2000 disappearance and tragic death of Georgy Gongadze, a journalist of mixed Georgian and Ukrainian origin who was critical of Kuchma and his entourage. In November 2000, opposition politician Oleksandr Moroz publicized secret tape recordings which he claimed implicated President Kuchma in Gongadze's murder. The recordings were said to be of discussions between Kuchma, presidential chief of staff Volodymyr Lytvyn, and Interior Minister Yuriy Kravchenko, and supposedly provided by an unnamed security officer (later named as Major Mykola Mel'nychenko, Kuchma's bodyguard). The conversations included comments expressing annoyance at Gongadze's writings as well as discussions of ways to shut him up, such as deporting him and arranging for him to be kidnapped and taken to Chechnya. In March 2008, three former police officers were sentenced to prison for the actual murder of Gongadze. Mykola Protasov was given a sentence of 13 years, while Valeriy Kostenko and Oleksandr Popovych were each handed 12-year terms. But so far the investigations have failed to show who ordered the murder. A breakthrough in the investigation occurred in June 2009, when General Alexey Pukach, former department head in the ministry of internal affairs, confessed his participation in the murder and reportedly named its initiator, but this name has not been made public at this writing. In late July 2009, the remains of Gongadze's skull were found.

By far the most widely publicized event of the new century, the event that actually made Ukraine better known in the entire world than ever before, was the Orange Revolution of November 2004–January 2005. Whether or not it was indeed a revolution is open to debate, particularly given that no change of power elites took place. In November 2004, Viktor Yanukovych, then Prime Minister, was declared the winner of the presidential elections, which had been largely rigged, as the Supreme Court of Ukraine later ruled. The official results and the electoral fraud caused a public outcry in support of opposition candidate, Viktor Yushchenko (the leader of the Our Ukraine faction of the Ukrainian parliament and also former Prime Minister), who challenged the results and led the peaceful protests. Although born in the easternmost part of Ukraine, Yushchenko had his electoral base for the most

part in western Ukraine and in Kyiv and was seen as the proponent of Ukraine's ultimate detachment from Russia and joining Western structures, including EU and NATO. Yanukovych's electoral base was almost completely in eastern and southern Ukraine, and he was seen as a proponent of close ties with Russia. Protests began on the eve of the second round of voting, as the official count differed markedly from exit poll results which gave Yushchenko up to an 11 percent lead, while official results

FREEDOM

Thousands took to the streets of Ukraine to carry through the "Orange Revolution," writing a new chapter in the history of the Ukrainian people's struggle for greater freedom and independence.

gave the election win to Yanukovych by 3 percent. Beginning on November 22, 2004, massive protests started in cities across Ukraine: the largest, in Kyiv's Independence Square, attracted an estimated 500,000 participants, who on November 23, 2004, peacefully marched in front of the headquarters of the Ukrainian parliament, many wearing orange or carrying orange flags, the color of Yushchenko's campaign coalition. On November 24, 2004, Yanukovych was certified as victor by the Central Election Commission, and the peaceful protests rose to new heights. By many estimates, on some days they drew nearly one million people to the streets despite freezing weather. Ukraine's Supreme Court finally broke the deadlock and on December 3, 2004, invalidated the official results that would have given Yanukovych the presidency. The court ordered a new run-off be held on December 26, 2004. The result of the revote was 52 percent in favor of Yushchenko and 44 percent in favor of Yanukovych. Thus, doubly pro-western (i.e., representing western Ukraine and leaning to the West geopolitically) Yushchenko became president of Ukraine.

Much has been said and written about the involvement of outside forces in the Orange Revolution. On Yanukovich's side was Russia, with President Putin of Russia congratulating Yanukovych (twice) on his electoral victory while the results of the election were still being contested. On Yushchenko's side, the U.S. State Department, the National Democratic Institute for International Affairs, the International Republican Institute, Freedom House, George Soros's Open Society Institute, the Polish government, and the rogue Russian oligarch Berezovsky (in asylum in the U.K.) were all widely reported to have been involved.

Side by side with Viktor Yushchenko, another political leader, Yulia Tymoshenko, a beautiful and energetic woman, emerged as a rising star in Ukrainian politics. Tymoshenko's plaited hairstyle became iconic at the time of the Orange Revolution. [According to some sources, Lesya Ukrainka's (a Ukrainian writer, 1871–1913) hairstyle inspired the over-the-head braid.] Prior to her political career, Yulia Tymoshenko was a successful but controversial businesswoman in the gas industry, which made her wealthy. In 2001, she was arrested on charges of forging customs documents and smuggling of gas between 1995 and 1997 (while president of United Energy Systems of Ukraine) but was released several weeks later. The following year Tymoshenko was involved in a mysterious car accident that she survived with minor injuries. During this time, she founded the Yulia Tymoshenko Bloc that received 7.2 percent of the vote in the 2002 parliamentary election. She is the head of the Batkivshchina (Fatherland) political party. Tymoshenko first became Prime Minister in 2005 and served between January 24 and September 8. Several months into her government, numerous inner conflicts inside the post-Revolution coalition began to damage Tymoshenko's administration. On September 8, 2005, after the resignation of several senior officials Yulia Tymoshenko's government was dismissed by President Viktor Yushchenko during a live TV address to the nation. She was appointed prime minister yet again on December 18, 2007. Unlike Yushchenko, who is a free marketer and avowedly pro-Western, it is hard to pinpoint the political views of Yulia Tymoshenko. For the most part she is referred to as a populist ready to tell people what they want to hear. Tymoshenko's 2007 article in *Foreign Affairs* titled "Containing Russia" is for the most part regarded as her attempt to endear herself to the West and to win applause among Yushchenko's electorate at a time when the president's popularity was declining. Indeed by 2007, Yushenko's own rating was in the single digits, and his faction, Our Ukraine, won less than 14 percent of the seats in the Ukrainian parliament versus more than 22 percent won by the bloc of Yulia Tymoshenko. The election, however, was won by Viktor Yanukovych's Party of the Regions with over 32 percent of the vote. That meant that only together could Tymoshenko's and Yushchenko's factions outweigh that of Yanukovych. Hence the resumption of their mutual alliance despite worsening personal relations between them.

Political geographers marvelled at the fact that the electoral base of each of the three leading Ukrainian politicians—Yushchenko, Tymoshenko, and Yanukovych—forms a contiguous area composed of adjacent and/or nearby regions. Thus, the 2006 electoral map (See, for example, The Economist July 7, 2007, p. 50) shows that as a national-level politician, Tymoshenko dominates central Ukraine, with Yanukovych dominating eastern and southern Ukraine, and Yushchenko having strong positions in just three regions of western Ukraine. The political geography revealed by the 2007 parliamentary elections was about the same, but the vote split was different: the biggest gains were on the side of Tymoshenko's bloc, which received almost 31 percent of the vote; Yanukovich received 34 percent and Yushchenko 14 percent. The 2007 parliamentary elections resulted in significant voter consolidation around big players, while small parties lost much of their electoral support.

DEMOCRACY AND GEOPOLITICS

Two ideological undercurrents are discernible when international political commentators interpret developments in Ukraine. One of them frames analyses in terms of democracy versus autocracy; and the other as leaning toward the West versus leaning toward Russia. Most Western commentators observe that following the Orange Revolution, Ukraine embraced democracy more consistently than before, and freedom of press and lack of heavy-handed government interference in the media bear testimony to that. U.S. President Bush was an ardent supporter of Ukraine's joining NATO, and Mr. Yushchenko remained adamant that joining "Euro-Atlantic" structures (i.e., NATO and the European Union) is the major goal of the Ukrainian foreign policy. However, more than 60 percent of Ukrainians, according to various polls, are against joining NATO; and even though joining the EU is seen as a more acceptable option, with 44.7 percent in favor of joining and 35.2 percent against it (according to a December 2008 poll), the opposition is quite significant even in regard to the EU. During the December 2008 NATO foreign ministers' meeting, Ukraine was not offered a Membership Action Plan, despite some optimistic pronouncements in that regard made during the April 2008 NATO summit in Bucharest. In addition to the United States under Bush, the most vigorous proponents of Ukraine's NATO membership were Poland and the Baltic States; the staunchest opponents were such Europe's heavyweights as Germany, France, and Italy. Ukraine's prospect of joining the EU is similarly dim, particularly in light of the recent initiative of offering an Eastern Partnership expressly for six post-Soviet countries, including Ukraine, as an alternative to full membership.

Because Russia's steadfast opposition to Ukraine's incorporation into the Western structures has been, by some accounts, one of the most significant reasons why Ukraine's bid for NATO membership was rebuked, it would be tempting to assume that both above-mentioned undercurrents—a quest for democracy and a quest for NATO membership—are related. In other words, because Western countries are so much more democratic than Russia, Ukraine's integration into Euro-Atlantic structures would spell the triumph of Ukrainian democracy. But this line of reasoning is debatable. First, to the dismay of West Europeans, Bulgaria and Romania did not show significant improvement after joining the EU and NATO in precisely the areas where Ukraine appears to be most vulnerable, namely, corruption and organized crime. Furthermore, Ukraine is faulted for its "venal elites" (Wilson 2002) and its decadent political culture. The war of words (at times profane and possibly not limited to words alone) between President Yushchenko and Prime Minister Tymoshenko and their respective entourages is so demeaning to Ukraine that doubts regarding their statesmanship and political maturity arise spontaneously. Thus, when in February 2009 Tymoshenko signed an agreement with Russia that put an end to an acute Ukraine-Russia dispute about natural gas, a dispute that affected much of Europe, Yushchenko likened that agreement to the 1939 Molotov-Ribbentrop non-aggression pact. During the February 10, 2009 meeting of Ukraine's National Security Council, Yushchenko and Tymoshenko publically exchanged accusations whose substance and tone would perhaps be more fitting for a peasant market squabble or a circuit court trial of a pickpocket.

Doubts as to whether or not the fight for democracy is inherently related to the geopolitical orientation of Ukraine are reinforced by how some Western pundits describe Ukraine. Case in point: the January 2009 lecture of William Green Miller, former U.S. Ambassador to Ukraine, at the Washington-based Kennan Institute. On the one hand, Miller argued that "for all its difficulties, Ukraine remains the most democratic state of the former Soviet Union." Yet on the other hand, "thus far the leaders of the Orange Revolution have failed to carry out their pledges," and today "the major threats to Ukraine's democratic possibilities are from within" as corruption and its attendant ills, conflicts of interest, abuse of office for private gain and unfair advantage are the main enemies of

Kyiv: Ukraine's Ministry of Foreign Affairs. Displayed on the left is the yellow-blue national flag of Ukraine. Displayed on the right is the flag of the European Union (of which Ukraine is not a member).

a healthy Ukraine. "In the current political terrain," remarked Miller "money buys elections, votes, judicial outcomes, places in the universities, beds and treatments in hospitals, favors, benefits, political power, and control." Following this diatribe, Miller nonetheless recommended that the Obama administration voice its support for Ukraine's eventual entry into NATO. However, if major threats to Ukrainian democracy are indeed from within, then touting NATO membership as a vehicle of democratization hardly makes sense.

In a 2009 article about Belarus, Balasz Jarabik wrote that even president "Lukashenka [of Belarus] has proved to have greater national responsibility and integrity than the entire Orange elite in Ukraine." Jarabik is a Slovak author who is head of "Pact Ukraine" in Kyiv. With such financial donors as USAID and the EU, he is well attuned to even the slightest

fluctuations in his donor agencies' official standpoints. Considering Lukashenka's reputation as "Europe's last dictator," Jarabik's pronouncement displays as much Western disappointment with Ukraine as one can have. Ukraine's post-Soviet society appears to suffer from some of the same deficiencies as its Russian counterpart: limited civility and a low level of mutual trust. On top of that, statecraft seems to be lacking in Ukraine. In addition, Ukrainians are split on many crucial issues confronting their country, and the disagreements are not so much between individuals as between large, territorially compact groups of Ukrainians. The differences are manifest in language, historical memory, and visions for Ukraine's future. Even history textbooks for secondary schools in the west differ from those in the east. Western Ukrainian textbooks portray some World War II actions as a conflict

between pro-national forces and Bolsheviks, whereas eastern Ukrainian textbooks emphasize the positive consequences of the Soviet period.

RECENT DEVELOPMENTS

Alongside the still unsuccessful attempt to integrate into Western structures, Ukraine has frequently made appearances in international news reports in conjunction with (a) its repeated disputes with Russia about natural gas, (b) the impact of the global economic crisis on its economy, and (c) its upcoming (January 2010) presidential elections.

Since 2005, there have been three major Ukraine-Russia disputes with significant implications for the rest of Europe. Ukraine is the main transit corridor for Russia's natural gas export. In 2004–2005, about 80 percent of Russian gas exports to the EU were made through Ukraine. Two-thirds of Russia's Gazprom revenue comes from the sale of gas that crosses Ukraine. Ukraine's own annual gas consumption in 2004–2005 was around 80 billion cubic meters (bcm), of which around 20 bcm was its own production, 36 bcm was bought from Turkmenistan, and 17 bcm was received from Russia as a payment for Russian gas transit. The remaining 6–8 bcm was purchased from Russia, which until recently sold natural gas to Ukraine at prices significantly lower than to West European customers.

The first gas dispute broke out when Russia raised the price from $50 to about $160 per 1,000 cubic meters in 2006. The government of Ukraine had agreed to a gradual increase of prices in return for increased gas transit fees and a change in payments for transit from barter to cash. In May 2005, it was revealed that the 7.8 bcm of gas which Gazprom had deposited in Ukrainian storage reservoirs during the previous winter had not been made available to the company. It remained unclear if the gas was missing, had disappeared due to technical problems, or had been stolen. This issue was resolved in July 2005 with an agreement among Russia's Gazprom, Ukraine's Naftohaz, and RosUkrEnergo, a shady intermediary registered in Switzerland. However, negotiations between Gazprom and Naftohaz over gas prices and a new gas supply agreement failed; on January 1, 2006, Gazprom started to reduce the pressure in the pipelines from Russia to Ukraine. Although Russia terminated only supplies to Ukraine, a number of European countries saw a drop in their supplies as well. The European Commissioner for Energy, Austria—which held EU Presidency at the time; and several affected countries warned against blocking gas deliveries. The supply was restored in early January 2006, after a preliminary agreement between Ukraine and Gazprom was signed. A five year contract was signed, although prices were set for only six months. According to the contract, the gas was sold not directly to Naftohaz, but to the RosUkrEnergo. The price of natural gas sold by Gazprom to RosUkrEnergo rose to $230 per 1,000 cubic meters; after mixing cheaper supplies from Central Asia (which constituted two-thirds of the total) it was resold to Ukraine at a price of $95 per 1,000 cubic meters. On January 11, 2006, presidents Vladimir Putin and Viktor Yushchenko announced that the conflict had been resolved.

Dispute broke out again in October 2007 when Gazprom threatened to cut off gas supplies to Ukraine because of an unpaid debt of $1.3 billion. This problem appeared to have been settled within days, but in January 2008 Gazprom warned Ukraine it would again reduce its gas supplies if a $1.5 billion gas debt was not paid.

A third gas crisis was sparked by the failure to reach agreement on gas prices and supplies for 2009. By late 2008, Ukraine owed $2.4 billion to Gazprom, and Gazprom demanded that this amount be paid before the commencement of a new supply contract. Partial payments followed, but Ukraine and Gazprom were unable to agree on a price for 2009. As a result, deliveries of natural gas to Ukraine were cut off completely on January 1, 2009. Transit deliveries to the EU, however, continued. President Yushchenko sent a letter to the president of the European Commission proposing the European Union's involvement in settlement of the dispute. Possibly as a result of Ukraine's using gas destined for East Central Europe, Hungary, Romania, and Poland reported that pressure in their pipelines had dropped; and Bulgaria reported a diminished supply as well as diminished transit to Turkey, Greece, and Macedonia. On January 5, 2009 Russian Prime Minister Putin recommended that Gazprom reduce supplies via Ukraine to Europe by the amount of gas Ukraine had allegedly taken since deliveries ended on January 1. On January 7, all Russian gas flow through Ukraine was terminated. A major drop in supplies of Russian gas, starting on January 7, 2009, was reported by several countries, with Bulgaria, Moldova, and Slovakia being affected the most. Talks between Naftohaz and Gazprom resumed on January 8, 2009, but supplies to Europe were not restored for another twelve days.

On January 17, 2009, Russia held an international gas conference in Moscow with EU representatives and Yulia Tymoshenko of Ukraine in attendance. The summit did not achieve any solution to the crisis, and negotiations continued bilaterally between Prime Ministers Putin and Tymoshenko. Early in the morning of January 18, after five hours of talk, Putin and Tymoshenko reached a deal on restoring gas supplies to Europe and Ukraine. They agreed that Ukraine would start paying European prices for its natural gas, less a 20 percent discount for 2009, and would pay the full European market price starting in 2010. In return for the discounts Ukraine agreed to keep its transit fee for Russian gas unchanged in 2009. The two sides also agreed not to use intermediaries. On January 19, the head of Gazprom, Alexei Miller, and the head of Naftohaz, Oleh Dubyna, signed a 10-year agreement on natural gas supplies to Ukraine for the period of 2009–2019. Gas supplies restarted on January 20 and were fully restored the next day. During the hiatus in gas supply residents of Ukrainian cities bought up all the electric heaters and micro electric stoves available in hardware stores. More importantly, according to the EU Commission and Presidency, the gas crisis caused irreparable and irreversible damage to customers' confidence in both Russia and Ukraine.

Asked why Estonia and Poland can settle their natural gas disputes with Russia without scandal, but Ukraine and even Belarus cannot, a Russian political commentator, Vadim Dubnov, used a Darwinist explanation: "For Russia, Poland and Estonia are aliens, and so arguments with them are a different genre. As for Ukraine and Belarus, those are the same kind as us, and intra-species conflicts always assume the most acute form."

Tensions between Russia and Ukraine continued as it became clear that Ukraine could not pay for natural gas supplies in 2009–2010. Ukraine asked Russia for about $5 billion in credit, but Russia seems to prefer payments in kind, by which it means the assets of some major Ukrainian enterprises. Some analysts interpret the latest natural gas dispute as a form of further deindustrialization of the post-Soviet space, this time notably of eastern Ukraine, whose economy has been dealt a double blow, first when global demand for steel declined with the global recession and second with price hikes on Russia's natural gas. Ukraine appears to be the country hit by the global recession the hardest. From January to April 2009, its industrial output declined by a whopping 31 percent compared with the identical period of 2008. This compares grievously with declines in Russia (16%), and even more so, Belarus (3.4%). The Ukrainian national currency experienced 50 percent devaluation in relation to the U.S. dollar. In December 2008, the exchange rate had been 5 hryvnia for $1, in September 2009, it was 8 hryvnia for $1.

The first round of new Presidential elections took place on January 17, 2010. Eighteen contenders participated, including the incumbent Victor Yushchenko who placed fifth with only 5.45% of the vote. No one won more than 50% of the vote. This led to the February 7 showdown between the two leaders in the race, Victor Yanukovych, who ended up with 35.32% of the popular vote, and Yulia Tymoshenko, with 25.05%. The election's only sensation was the third place position achieved by Serhiy Tihipko with 13.05% of the vote. Tihipko, who had had a successful career in the Young Communist League (Komsomol) under the Soviets, had later been a private banker, Governor of Ukraine's Central Bank, the economics minister, and the manager of Yanukovych's 2004 presidential campaign.

Shortly after the first round of elections, on January 22, 2010, outgoing President Yushchenko declared a World War II-era nationalist leader, Stepan Bandera, a Hero of Ukraine. Victor Yanukovych promptly denounced the move, stating that granting Bandera "hero status" was a step that would divide Ukraine. Indeed, like another member of the Ukrainian Insurgent Army, Roman Shukhevych, who had been declared a Hero of Ukraine earlier by Yushchenko, Bandera is glorified in western Ukraine but in the rest of the country he is considered a traitor for his collaboration with the Nazis. Furthermore, Bandera stands accused of massacring Poles and Jews. Almost immediately, in February 2010, lawsuits were filed in the District Administrative Courts in Donetsk and Luhansk, two regional centers in eastern Ukraine, seeking to overturn Yushchenko's order to award the Hero of Ukraine title to Stepan Bandera. In Luhansk, the leader of the regional administration himself filed a lawsuit. And in the U.S., the Simon Wiesenthal Center issued a protest. In a letter to Oleh Shamshur, Ukraine's Ambassador to the United States, Mark Weitzman, the Wiesenthal Center's Director of Government Affairs, expressed "deepest revulsion at the recent honor awarded to Stepan Bandera, who collaborated with the Nazis in the early stages of World War II, and whose followers were linked to the murders of thousands of Jews and others. It is surely a travesty when such an honor is granted right at the period when the world pauses to remember the victims of the Holocaust on January 27." President Kaczynski of

Poland also blasted his Ukrainian counterpart for the glorification of Bandera. "An estimate of activities of the Organization of Ukrainian Nationalists and the Ukrainian Insurgent Army is categorically negative in Poland," Kaczynski said in his statement. He added that the OUN and the UIA carried out large-scale massacres of the Polish people in the eastern territories of the former Rzech Pospolita. "These killings raise unambiguous protest in Polish society."

Proclaiming Bandera a Hero of Ukraine may have contributed to even more polarization of the Ukrainian electorate. In the subsequent showdown, the western part of Ukraine consolidated around Yulia Tymoshenko as one of the former leaders of the Orange Revolution, and the east and south around Victor Yanukovich, the 2004 electoral winner whom that revolution deposed. On February 7, 2010, in the second round of presidential elections, Yanukovych won with 48.95% of the vote against Tymoshenko's 45.47%. The 2010 electoral map of Ukraine reflects a deeply divided country. Yanukovych won in just ten regions (oblasts) of Ukraine, which, however, happen to be the most densely settled ones. These regions form a compact zone in the southeast. Tymoshenko won in the remaining fifteen regions and in Kyiv. The following numbers describe the degree of geographic polarization of the popular vote. In the Donetsk region, 90.5% of the electorate went for Yanukovych, and in the neighboring Luhansk region almost 89% did. These are the two easternmost regions of Ukraine. In contrast, in Lviv, Ternopil, and Ivanovo-Frankivsk oblasts, 86%, 88%, and 89%, respectively, voted for Tymoshenko. These three oblasts are in the western part

of Ukraine. Unfortunately for Tymoshenko the three together have only a few more residents than does Donetsk Oblast by itself.

Yulia Tymoshenko did not concede the election and filed lawsuits against alleged falsifications of the vote in favor of Yanukovych, but there is little probability of overturning the results this time. International observers deemed the elections honest; and many Western dignitaries, including President Obama and Herman Van Rompuy, the first president of the European Council, congratulated Yanukovych on his victory. Ironically, the 2010 electoral results are an exact replica of the 2004 results, when Yanukovych won the same 49% of the vote versus Yushchenko's 46% but then faced a third round not stipulated by the Constitution. Yushchenko was able to ride the wave of the Orange Revolution and the enthusiasm that it had sparked in part of Ukraine and in the West. But by 2010 this enthusiasm had expired and so had the American and West European zeal for the forceful promotion of democracy. These two factors seem to be the main difference between the outcomes of the 2004 and 2010 presidential elections in Ukraine.

HEALTH/WELFARE

The Ukraine national Ministry of Health has the responsibility for managing health policy. However, most health care is delievered in facilities owned and managed at the regional and district level, and funded by the respective tiers of government from allocations provided by the Ministry of Finance or raised locally.

Belarus

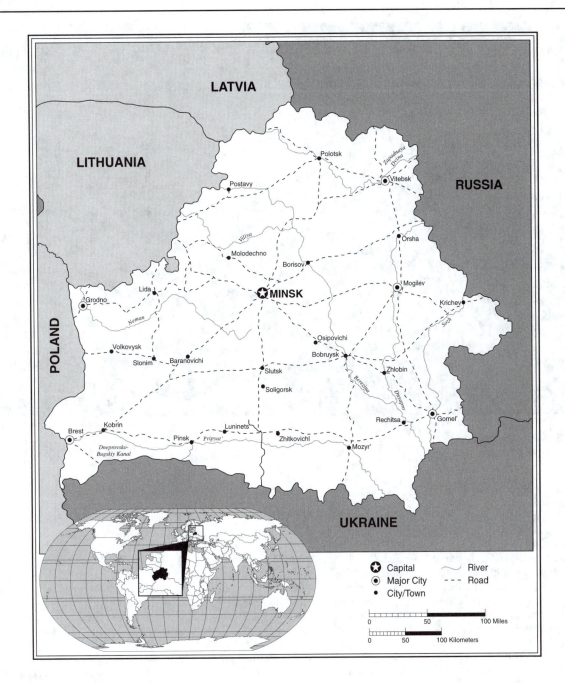

Belarus Statistics

Area in Square Miles (Kilometers):
 80,154 (207,600)
Capital (Population): Minsk
 (1,830,000)

Population

Total (millions): 9.7
Percent Urban: 74
Per Square Km: 47

Rate of Natural Increase (%): −0.3
Ethnic Makeup: Belarusian 81.2%;
 Russian 11.4%; Polish 3.9%;
 Ukrainian 2.4%; other 1.1%
 (1999 census)

Health

Life Expectancy at Birth: 65 (male);
 76 (female)

Infant Mortality Rate: 5/1,000
Total Fertility Rate: 1.4

Economy/Well-Being

*Currency ($ US equivalent on 14 October
 2009):* 2738 rubles = $1
2008 GNI PPP Per Capita: $12,150
*2005 Percent Living on Less than US$2/
 Day:* <2

Belarus Country Report

Independence Square in Minsk, Belarus

Belarus is a medium-sized (207,600 sq km) European country with a population of 9,671,900 million people (2009). It claims to be in possession of the geographical center of Europe, as conventionally defined (from the Atlantic coast to the Urals), although identical claims are also made by Lithuania and Ukraine. Belarus consists of six regions, called oblasts, whose capitals with their corresponding 2009 population numbers in parentheses are Brest (318,000), Grodno (338,200), Gomel (488,100), Minsk (1,829,100), Mogilev (372,000), and Vitebsk (347,500). A regional center, Minsk is also the national capital of Belarus. In 1921–1939, when Belarus was split into eastern (Soviet) and western (Polish) parts, Grodno and Brest regions in their entirety were under Polish administration as were the western part of Minsk Oblast and the northwestern part of Vitebsk Oblast.

Belarus is the only post-Soviet country which committed itself to some sort of reintegration with Russia. The so-called Union Treaty between Russia and Belarus was signed in 1999. Although such critical provisions of that treaty as common currency, a constitution, and supremacy of union structures over national governments have never been implemented, Belarus and Russia have tight military cooperation, their mutual border is transparent, and Russian citizens on their personal and business trips to Belarus do not normally get a sense of being abroad. Instead, they feel like they are traveling to their Soviet past, oftentimes to the better landscapes of that past as Belarus is better groomed than Russia, particularly if one compares the provincial (outside capital cities) areas of both countries.

According to Belarus's census of 1999, ethnic Belarusians accounted for 81.2 percent of Belarus's population, Russians for 11.4 percent, Poles for 3.9 percent, and Ukrainians for 2.4 percent. Between 1989, when the last Soviet census was conducted, and 1999 the share of people identifying themselves as Belarusians had grown (from 77.8 percent to 81.2 percent) while the share of other major ethnicities declined. This preponderance of ethnic Belarusians stands in stark contrast to the insignificant use of the Belarusian language.

Belarusian is a Slavic language distinct from Russian and Polish, which are its close relatives; indeed, each is closer to Belarusian than they are to each other. The same applies to Ukrainian, which is also midway between Polish and Russian. Based on a statistical analysis of coincidental lexemes in the translations of identical texts, such as the Bible and the UN Declaration of Human Rights, Ukrainian qualifies as the closest relative of Belarusian. Belarusian has been one of the official languages of Belarus since the Belarusian Soviet Socialist Republic (BSSR) was proclaimed (1919). In 1924–1939, Russian, Yiddish, and Polish were also given official status in the republic, but since 1939 Belarusian has shared its official status only with Russian. In 1992, Belarusian was proclaimed the sole official language. However, based on the national referendum of 1995, Russian was re-introduced as one of two official languages of Belarus. This reintroduction gained support from 83.3 percent of the voters.

As for the actual use of Belarusian, available surveys are more in line with numerous field observations made in Belarus than are official census statistics. Thus, a 1999 representative survey of 1,081 Belarusian adults selected randomly throughout the country and polled by a Belarusian research firm contracted by the U.S. Department of State showed that as few as 12 percent of respondents spoke Belarusian at home and 7 percent at work. "In Minsk, where four-fifths of the inhabitants are of Belarusian nationality, almost everyone usually speaks Russian at home (86%) and at work (90%)."[1] Voters in the March 19, 2006, presidential election, assessed by the March 26–April 6, 2006 national survey conducted by the major polling firm of the Belarusian opposition, also revealed the modest role of Belarusian. Thus, among those who voted for the major opposition candidate, Milinkevich, Belarusian speakers accounted for 3.2 percent, and those speaking Russian and Belarusian intermittently, for 11.6 percent. Among voters for Alexander Lukashenka, the incumbent president, the corresponding groups accounted for 3.7 percent and 14.7 percent. All the remaining voters identified themselves as Russian speakers.

Just as in Ukraine, where up to 18 percent use *surzhyk*—a mixture of Ukrainian and Russian, in Belarus up to one-quarter of the population uses *trasianka* (literally, a mixture of hay and straw), which is a blend of Russian and Belarusian. But Belarus is different from Ukraine in that Ukrainian is embraced by a much larger share of Ukraine's population than Belarusian by the population of Belarus. Also, whereas in Ukraine there is a vivid spatial divide, with a preponderance of Ukrainian speakers in western Ukraine and a preponderance of Russian speakers in urban areas of eastern and southern Ukraine; there is no equivalent divide in Belarus where the dominance of Russian is across the board.

Unlike Russia and Ukraine, Belarus is not blessed with natural resources. The only local resource available in commercial quantities and that is exported is potassium. Neither is Belarus blessed by fertile soils. And yet, according to the United Nations' Development Program, Belarus is ranked 67th as a country with "high human development." Thus Belarus is ahead of Russia (73) and Ukraine (82). Among the former Soviet republics, only the three Baltic States are ahead of Belarus. On GDP per capita, Belarus was ahead of Russia up until the most recent spike in oil prices (2005–2008). During that spike Belarus fell behind Russia but was still ahead of Ukraine and all the other post-Soviet republics save the Baltic States.

PRE-SOVIET HISTORY AND BELARUSIAN IDENTITY

According to Nina Mečkovskaya, "the principal problem of Belarusian history has been the problem of cultural and political survival . . . 'in the shadow' of Russia and Poland. It is an unfavorable geopolitical fate to be the object of Russian and Polish assimilation and of two powerful and mutually antagonistic expansions."[2] Perhaps the most phenomenal feature of Belarusians has been the long-lasting absence of a common name that would be perceived as the token of their collective identity. The term Weisse Rusen (White Russia or Belaya Rus/ Belarus) was first used by German authors in the fourteenth century. During the reign of Ivan the Third (1462–1505), this term was appropriated by the official language of the Russian court. In 1654, "White Russia" was included in the official title of the Russian monarch (as the Tsar of Great, Small, and White Russia). The meaning of the term is not quite clear. What is clear, though, is that for many centuries the term Belarus was used as a toponym (a place name), not a marker of ethnic identity. Only by the end of the nineteenth century did some pioneers use it in that capacity.[3] Prior to the essentially top-down imposition of the toponym *Belarus* and ethnonym *Belarusian* during the Soviet period of 1919–1991, most regional residents of Slavic background introduced themselves as "*tuteishiya,*" which means "locals." This lasting anonymity is truly exceptional

because a common name is the most basic indicator of belonging to a group.

Currently, 67 percent of Belarus's population lives in cities, but urbanization was grossly delayed, and as recently as 1926 the share of urbanites among ethnic Belarusians (8%) yielded to that of other ethnicities within the republic, notably, Jews, Poles, and Russians. A middle-class intelligentsia that "would invite the masses into history" was late in coming. As a result, in the beginning of the twentieth century, residents of Belarus had a barely discernible sense of separate ethnic identity, and Belarusian nationalists did not seem to have much following among the predominantly peasant Belarusian masses. Most importantly, no sense of *shared* identity among the social classes had been forged in Belarus before the Communist Revolution in the Russian Empire. The upper and even middle (merchant and craftsman) strata pledged allegiance almost exclusively to Russian, Polish, and Jewish causes. [Being part of the Jewish pale of settlement established by the Russian Empire, Belarus was characterized by a heightened share of Jews in its population prior to World War II (8.2% of the total population of Soviet Belarus in 1926 versus 1.8% in the Soviet Union at large; and 8.8% of the total in Western Belarus, according to the 1931 Polish census). This share was particularly high in the urban populations. For example, in the city of Minsk in 1926, Jews accounted for 41 percent of the total; in Gomel 44 percent; in Vitebsk 37.5 percent; and in

(Courtesy of Grigory Ioffe)

A Christian Orthodox procession in Minsk, Belarus

115

Mogilev 34 percent. The 1897 census statistics in regard to Jews in Belarus had been even higher.

The idea of a Proto-Belarusian identity grew out of the folkloristic research of some Vilna (today's Vilnius, the capital of Lithuania) University professors and students, notably Jan Barszczewski (1790–1851) and Jan Czeczot (1796–1847), whose language of everyday communication was Polish. On the basis of linking their folklore research with literary and official documents of the Great Duchy of Lithuania (GDL; 1253–1569), they came to the conclusion that they had inherited a cultural-historical legacy that had all the trappings of a tradition distinctive from that of the Poles. As they uncovered the historical past of local Slavs and became aware of the cultural rebirth of other Slavic communities such as Czechs, Serbs, Croats, Bulgars, and Slovenes, these Catholic intellectuals became convinced that the formula "*gente Rutheni, natione Poloni*" (a Pole of Russian descent) did not quite fit their ethnic domain. This idea was subsequently refined by Frantsyšek Boguševič (1840–1900), a poet, who in 1891 appealed to his fellow countrymen that they were Belarusians and that their land's name was Belarus.

The emergence of a Belarusian *national* idea (a step forward compared with the awareness of ethnic distinction) matured in the *Nasha Niva* literary circle, which

(Courtesy of Grigory Ioffe)

Novogrudok, Belarus: A Catholic Church's interior shows banners bearing inscriptions in Polish.

in 1909–1915 published the eponymous newspaper in Vilna. The city of Vilna, where *Nasha Niva* was edited and published, played a significant role as the meeting place of nationally-conscious Belarusians, and so from the perspective of Belarusian nationalism, Vilna was its most significant center. The first ever Belarusian-language elementary school was opened in that city in 1915. The subsequent loss of Vilna, first to Poland and then to the newly-emerging Lithuanian state as a result of the 1921 Riga Treaty and the 1939 border rearrangements, respectively, may have been harmful for the Belarusian national cause. However, at the turn of the century and up until World War II, Vilna was dominated by Poles and Jews. Polishness was promoted by the Catholic Church and the character of the local university, one of the principal centers of Polish nationalism. While Jews could not possibly raise any national claim to the city, it was in Vilna that Ben Yehuda (born as Eliezer Perelman in a shtetle of Luzhki, currently in Vitebsk Oblast of Belarus) set out to revive Hebrew. Later, his son, Ben-Zion, who became known by his penname, Itamar Ben-Avi, became the first person in modern times to speak Hebrew as his native language because it was the only language spoken in Ben Yehuda's family. Vilna thus meant many things to many people.

Situated between Poland and Russia both geographically and linguistically, the early promoters of the Belarusian national idea identified themselves in opposition to either one or the other of Belarus's expansionist neighbors. Today, for obvious reasons, the bogeyman to disassociate oneself from is Russia, but historically Russia and Poland were used in that capacity intermittently, as springboards of sorts. In fact, Poland was to play this role *first* because the Belarusian national idea developed amidst Polish-speaking intellectuals who began to define themselves in opposition to that country. In their opposition, however, they stood a chance of falling into the embraces of Russia. Similarly, rebounding from Russia at a different point in time, Belarusian nationalist thinkers had to be on the lookout lest they become too Polish: a peculiar pendulum effect.

The pendulum analogy is irresistible; it is ensconced in the spontaneous imagery of the language used to describe the early stages of Belarusian nationalism. Thus, according to Zakhar Shybeka, the outcome of the 1839 conversion of the Uniates into Orthodox means that Belarusians were "pulled away as it were from the Catholic Poles but drawn dangerously close to

the Orthodox Russians."[4] Mečkovskaya writes that in Belarus, in the late 1800s-to-early-1900s, "anything that was elevated above the illiterate peasant existence, be that church, school, or officialdom, automatically became either 'Russian' (and Orthodox) or 'Polish' (and Catholic)."[5] Correspondingly, the upwardly mobile local people with Slavic backgrounds had two options for self-identification, Russification or Polonization. For the most part, the Orthodox (the majority of Belarusian speakers) chose the former option and the Catholics the latter. The ideological blueprint of Russification was the so-called West-Rusism, a theory that emphasized Belarusian peculiarity only within the confines of the Russian cultural universe. The most prominent author and promoter of this theory was Mikhail Koyalovich (1828–1891). Born in the Belarusian-speaking area's extreme west, wherein the Orthodox were a minority, Koyalovich was imbued with a lofty mission of Russian Orthodoxy. Born ten years later in an area dominated by the Orthodox peasants, Konstanty (Kastus) Kalinowski (1838–1864) was imbued with a lofty mission of Polish Catholicism for the enlightenment and liberation of that peasantry. In a primordialist sense–that is, assuming that nation comes first and nationalism later, both Koyalovich and Kalinowski were Belarusians. But they were committed to dragging the "Belarusian pendulum" in opposite directions: for Koyalovich, the Belarusians' natural home was Russia; for Kalinoswki it was Poland.

Before the Communist revolution, Belarus was one of the poorest regions of European Russia. Belarus had little manufacturing and was beset with rural overpopulation. In agriculture, yields were meager, in part due to acidic and poorly-drained soils. The first industrialization wave (1880s) that affected many regions of European Russia largely skirted Belarus. Only a few large enterprises were born, such as the iron foundry *Yacobson, Lifshits, and Co* and the machine-building firm *Tekhnolog* in Minsk, a flax mill and an optical factory in Vitebsk, three wood processing factories (producing plywood and matches), a foundry, and a cigarette wrapper factory in Pinsk, a brewery in Mogilev, and a cigarette factory in Grodno.[6] In the first decade of the twentieth century, in Belarus there were 800 industrial establishments that employed 25,000 workers. Most of these production units were small and primitive wood and food processors.[7] Still, in 1912, they contributed 24.4 percent of Russia's output of oak sleepers for railway lines, 23.5 percent of plywood, 44.8 percent of matchwood, and 14.5 percent of wallpaper. Despite this evidently high level of specialization in wood processing, the output of alcoholic beverages was the number one contributor of local revenues, with wood processing being number two. And the per capita industrial output in Belarus was one half of that in Russia as a whole.[8]

A POST-WAR SUCCESS STORY

Although several sizable manufacturing plants emerged in Belarus during the implementation of the first two Soviet five-year plans, the scope of Belarus's pre-war industrialization was dwarfed by that in Ukraine and the Russian Federation, and correspondingly the pull of urban centers was small. In 1940, only 21 percent of Belarus's population lived in cities and towns, compared with 34 percent in Russia and in Ukraine.[9] By all accounts, Belarus's location along the western frontier of the Soviet Union was deemed strategically vulnerable.

During 1941–45, Belarus experienced arguably more devastation than any other country affected by World War II. At least one-quarter of the entire population perished, including 750,000 Jews. On the eve of the war, the population of Belarus was 9.2 million people; in 1945, it was only 6.3 million. Out of 270 towns and rayon (rayon is a name of a minor civil division in Belarus, Russia, and Ukraine) centers 209 were demolished, including Minsk, Gomel, and Vitebsk, in which 80–90 percent of all pre-war buildings were destroyed.[10] According to Soviet monetary assessments of the war-inflicted damage, Ukraine sustained the largest destruction.[11] However, Ukraine is much larger than Belarus and pre-war industrial investment in Ukraine had been far more significant, which must have heightened the value of what was then exposed to destruction. In per capita terms, the war-inflicted loss of property appears to have been higher in investment-poor Belarus, which underscores the extraordinary scale of its devastation.

Belarus's industrial spurt began with post-war reconstruction. The newly obtained *cordon sanitaire* of satellite states along the western border of the Soviet Union changed Moscow's perception of Belarus's location. It was no longer vulnerable. Belarus was now the locus of the major transit routes linking Russia with East-Central Europe. Later on, the significance of these routes increased even more, as the Soviet Union began to sell its oil and gas to the West, and received consumer goods and food in return. From the late 1950s on, Belarus emerged as one of the major Soviet manufacturing regions, emphasizing tractors, heavy trucks, oil

processing, metal-cutting lathes, synthetic fibers, TV sets, semi-conductors, and microchips. Much of Belarus's high-tech industry was military-oriented.

Between 1970 and 1986 fixed assets (an indicator of the scope of completed investment projects) grew by 360 percent in Belarus. Over the same time, Ukraine featured a 270 percent growth and the entire Soviet Union 311 percent. As a result, Belarus led all 15 republics in growth rate of national income per capita (240 percent in 1970–86) and was second only to Armenia in the growth of national income during the same period. Belarus also led all the republics in total labor productivity growth and in industrial labor productivity growth in particular. On the eve of the 1990s Belarus had one of the better managed regional economies, with an unusually high share of export-oriented enterprises: more than 80 percent of industrial output was exported to other republics or foreign countries. The worn-out phrase that Belarus was an assembly workshop of the U.S.S.R. is not exactly accurate. Belarus's specialization was on R & D *and* assembling high-tech products. The important feature was that almost all the personnel for R & D were trained within Belarus, which among other things explains why there are so few migrants in Belarus and the share of ethnic Belarusians is so high. In 1986 Belarus, which 70 years before had had no institutions of higher learning at all, was second only to Russia in the number of college students per 1,000 residents. The quality of Belarusian institutions of higher learning was among the highest in the U.S.S.R.

In the 1980s, more than half of the industrial personnel of Belarus worked for enterprises with over 500 employees. Most of the large-scale processing and assembly operations were located in Minsk and the eastern part of the republic. The industrial core of eastern Belarus took shape due to its transit location. Three transportation axes crossed in this region: between Moscow and the most economically advanced of East European satellites of the U.S.S.R., between Leningrad and Ukraine, and between the Baltic ports and Ukraine. Ten manufacturing giants and dozens of their smaller subsidiaries form the industrial core of eastern Belarus. These enterprises fall into four branches: mechanical engineering, petrochemical, radio-electronic, and ferrous metallurgy. Mechanical engineering is represented by six giants: MTZ (Minsk; tractors), MAZ (Minsk; trucks), MoAz (Mogilev; self-propelled scrapers and earth movers and trailers for underground works), Gomsel'mash (Gomel; harvesting combines, mowers, and sowing

machines), MZKT (Minsk; an offspring of MAZ; heavy duty tractor trailers), and BELAZ (Zhodino, Minsk region; heavy trucks for mining operations). The technological cycles of all these factories are entwined. The petrochemical industry is based on two refineries: NAFTAN, based in Novopolotsk, Vitebsk region, and Mozyr NPZ in Mozyr, Gomel region. NAFTAN is the largest refinery in Europe, with a processing capacity of 20 million tons of crude oil a year, while Mozyr NPZ can process up to 12 million tons a year. The refineries are located on two different pipelines from Russia. The products of NAFTAN are transported through pipelines from Novopolotsk to Ventspils suited for gasoline and diesel fuel. So, the Latvian port of Ventspils appears to be the major transshipment site for NAFTAN. The Mozyr NPZ receives crude oil from the pipeline *Drouzhba;* gasoline and other products are then delivered to Central Europe by tank-trucks and rail. The combined capacity of the two Belarusian refineries exceeds domestic demand by a factor of three, so export has been the major function of Belarus's refineries from the outset. Several chemical plants connected to the major refineries by local pipelines operate in Novopolotsk, Polotsk, Mozyr, Mogilev, and Grodno.

The leading enterprise in radio-electronics in Belarus is Minsk-based *Integral,* offering a broad array of automotive and power electronic products (e.g., monochip voltage regulators and temperature sensors), timers, sensors, microcontrollers, LCD-drivers, plasma-panel drivers, ICs (integrated circuits) for electronic contact cards, ICs of transporters, ICs for consumer electronics, including watches, thermometers, and TV sets. *Integral* is technologically linked with other enterprises like Minsk-based *Gorizont,* the CIS's largest producer of TV and radio sets and also a producer of systems for satellite and cable TV. Yet another industrial giant is the free-standing Belarusian Metallurgic Plant in Zhlobin, Gomel region, whose principal products are steel cord and steel wire. Other components of Belarusian industry, mostly based in eastern Belarus, are textiles and the production of potassium fertilizers. The latter represents the only production cycle located in and controlled by Belarus in its entirety, as Soligorsk district in the Minsk region is where potassium is mined. All industries other than the production of potassium are deeply integrated with Russia and to some extent with Ukraine. They either process raw materials from Russia (petrochemical industries) or depend on parts and semi-finished products from

Russia and Ukraine (as do all plants in the mechanical engineering sector, which receive up to 80 percent of their parts from outside Belarus). Some Belarusian industries are attached to major consumers in Russia, as are factories producing electronic and optic devices for the Russian army and enterprises producing household appliances such as refrigerators and TV sets.

For a long time, the main obstacle to agricultural development in the southern part of the republic (both west and east) was the prevalence of marshes in the poorly drained Polessye lowland, which accounts for 40 percent of the entire land area of Belarus. Straddling the border between Belarus and Ukraine, this is Europe's largest concentration of freshwater marshes. Here the patches of dry land on the few elevated areas between the tributaries of the Pripet River (the axis of the entire lowland) used to contract significantly during high water periods. A costly state program of land improvement was launched in the 1960s. A massive artificial drainage system proved to be particularly successful in the western part, in the region of Brest, because land reclamation projects were implemented in areas that still had vibrant rural communities.

In the Brest region, wherein land reclamation schemes proved most successful, and in the Grodno region, with a much smaller scale of land improvement (but with the highest *natural* fertility of the soil in Belarus), 16 giant cattle-fattening and pig farms were built in the late 1970s and '80s. A large segment of the local population became dependant on supporting this highly centralized animal husbandry, which resembled rhythmic industrial operations more than traditionally seasonal peasant farming. In terms of quality of life in the countryside, intensity of land use, and output per unit of land, western Belarus in the 1980s was equal to or higher than the Baltic Republics (the all-Soviet leaders in agriculture).

A country of dismal workshops and unproductive wetlands at the beginning of the twentieth century, Belarus 70 years later was dominated by large-scale industry and vastly modernized agriculture. Despite the ingrained flaws of the Soviet model of economic development, Belarus was an undeniable Soviet success story. All the impulses and/or driving forces of Belarus's achievements, and their side effects as well, have been of Soviet vintage. Minsk was a true symbol of this Soviet-style success. It demonstrated an astounding pace of population growth: 509,000 people in 1959, 917,000 in 1970, 1,331,000 in 1980 and 1,543,000 in 1987. No other large city in the entire Soviet Union grew so fast.

SOVIET HISTORY AND BELARUSIAN IDENTITY

The Belarusian language was codified in 1918 when Branislau Tarashkevich published the first textbook of Belarusian grammar. He based his version of standard Belarusian on dialects spoken in the southwestern part of the Belarusian ethnographic area. Just as in Ukraine, where a Ukrainization campaign was launched in the early 1920s, so in Belarus a Belarusification campaign started at the same time. Belarusification met with grassroots' resistance, especially in the easternmost part of the republic and in the south. Not only did extremely few people in Gomel and Vitebsk speak Belarusian, but the rural villagers residing around those cities spoke dialects that were close to Russian and occasionally (in Gomel province) Ukrainian. The 1933 Belarusian language reform engendered a natural backlash. On the one hand, it was a typical Soviet campaign in that it resorted to heavy-handed ideological ammunition and had a drumbeat irrelevant to the actual purpose of reform. In the decree of the Belarusian Council of the People's Commissars of August 26, 1933, the alleged Belarusian National Democrats (which had become a code phrase for traitors as early as 1930) were accused of "intending to tear away the Belarusian literary language from the language of the Belarusian working masses and of thus creating an artificial barrier between the Belarusian and Russian languages." On the other hand, the 1933 reform discarded grammar options whose genesis had been southwestern dialects, which at the time were spoken beyond the borders of the BSSR in Poland, and reoriented the dialectal basis of the Belarusian literary norm to the southeastern dialectal zone, which was genetically close to southern Russian dialects. To some extent, therefore, the reform pursued the goal of making standard Belarusian closer to the actual speakers.

In the meantime, in western Belarus, the Belarusian national cause did not have good prospects for success under the exceedingly unitary and assimilatory policies of Polish administrations. Proficiency in the native language was higher in western Belarus than in eastern Belarus. But, while in the east, Russification was more from below, more spontaneous and did not cause inner protest; in the west, the policy of Polish authorities caused conscious resistance, which could express itself in a desire to preserve the language. With time, the official stance of Polish authorities in regard to Belarusian worsened even more because communist propaganda from across the border incited insurgence.

Comparing Polish and Soviet attitudes toward Belarusian national aspirations in the 1930s, Nicholas Vakar observed that whereas repressive Polish authorities mainly targeted the symbols of Belarusian cultural separateness, repressive Soviets targeted the people who promoted those symbols.[12] Indeed, beginning in 1929, the Minsk-based Belarusian cultural elite were dealt a severe blow when many found themselves behind bars and many lost their lives for alleged bourgeois nationalism and espionage for Poland. Yet the Belarusification campaign did not cease. "Even after a purge began of the Belarusian nationalist elite in 1929, the Soviet authorities replaced the old leadership with younger Belarusians. Such trappings of cultural nationalism as Belarusian language in the schools, administration, and literature continued to receive official support."[13] This resulted in an apparent paradox: In 1928, there were 30 Belarusian-language newspapers, and in 1938—that is, after much of the national elite was purged—there were 149.[14] Russian-language newspapers numbered but 48. "[T]he output of books in Belarusian (including works in geology, physics, and other non-humanities topics) in the BSSR was seven times higher than in 1928."[15]

After the devastating Second World War,[16] ethnic Belarusians became the ethnic majority in all the cities of Belarus, initially because of the drastic reduction in the number of Jews and later also because of greatly accelerated rural migration. Nevertheless, mass adoption of the Russian language by Belarusian urbanites (and later by rural folks as well) continued unabated: Rural migrants in cities preferred to send their children to Russian language schools and adopted Russian themselves as their language of everyday communication. By the late 1960s, one-third of all secondary school students in Belarus, almost exclusively urbanites, did not take Belarusian even as a separate subject at the Russian-language schools they attended. In the mid-to-late 1960s, 1970s, and much of the 1980s, it was practically impossible to encounter a conversation conducted in standard Belarusian in the streets of the Belarusian capital.

In the late 1980s, fresh winds from Moscow stirred nationally-conscious people in Belarus to action. As at other crucial junctions in the region's history, the stimulus for change came from outside. Moreover, Belarus's Communist Party leadership was one of the most resistant to democratic change in the entire Soviet

Union. Ales Adamovich, a prominent Belarusian writer who in the 1980s relocated to Moscow, labeled his homeland the Vendée* of Perestroika. As Alexander Motyl wrote in his 1987 book, *Will the non-Russians rebel?* "The Belorussian contribution to dissent has been virtually nonexistent." However, change emanating from Moscow was irresistible, and several local initiatives developed as a direct result of it, including the creation of the Belarusian Popular Front (BPF) modeled after similar movements in the Baltic States and in Ukraine. The BPF committed itself to mass and swift linguistic Belarusification and to shaping Belarusian identity so the country could mentally detach itself from Russia. This task proved to be more difficult in Belarus than in any other post-Soviet country.

It is tempting to portray the evolving views on Belarusian identity as a perpetual fight for meaning. Just as in Russia proper, where the rivalry between the Westernizers and the Slavophiles did not vanish in the Communist revolution but was reduced to an undercurrent, so in Belarus calling things by their proper names has been and still is an exception. And yet the ongoing struggle for the meaning of Belarusianness appears to unfold according to the old script first written in the very end of the nineteenth century, when Belarusian distinctiveness began to assert itself in opposition to the twin threats, Russification and Polonization. Because throughout much of the twentieth century Russification was decidedly more of a threat than Polonization (particularly after all of Belarus became unified as the BSSR), Belarusian nationalists grouped in the BPF positioned themselves as avid Westernizers, the heirs of the GDL, a consistently European entity in contrast to inherently Asiatic and despotic Russia.

Of course in the early 1990s this idea was not new. It is then all the more essential to realize that throughout the twentieth century Belarusian "Westernizers" seldom rose to a position of power. In fact, this happened only three times, and each period was brief and marked by external supervision and controversy. The first time the Westernizers made a splash was in 1918–1919, when the Belarusian People's Republic (BPR) was proclaimed under the

German military occupation. The BPR introduced the white-red-white flag and a coat-of-arms bearing the form of "pursuit" (*Pahonia* in Belarusian) by a knight mounted on a racing horse. As the emblem of the GDL, the latter represented a link to what was referred to as early Belarusian history. After the BPR was rejected by the Bolsheviks, Poland reemerged on the political map of Europe (1919), and the war broke out between it and Soviet Russia. In the wake of that war the areas populated by Belarusians were divided in 1921 between Poland and Russia. The BSSR was proclaimed in the eastern section of "ethnographic Belarus." Within the BSSR, Belarusian Westernizers (a.k.a. nationalists) and the backers of West-Rusism represented two mutually hostile groups. Both had local roots, and tried to curry favor with the communist regime. Initially the regime favored the nationalists. First, Great Russian chauvinism had been given a bad name by Lenin himself, and in the early 1920s fighting it was still on the communist agenda. Second, in order to undo territorial losses, Moscow decided to cast the BSSR as the "true" Belarusian home (as opposed to Western Belarus where Belarusians suffered from discrimination by the Poles). Last but not least, religion was now considered the "opium of the people," and West-Rusism had been a brainchild of the Russian Orthodox Church. As a result, the Westernizers (a.k.a. nationalists) received official support and obtained leverage out of proportion to the size and influence of their group. By the late 1920s, however, the backers of West-Rusism had regrouped. They could no longer appeal to the authority of the Russian Orthodox Church. Instead they appealed to the Soviet state, this bastion of "proletarian internationalism," and labeled Belarusian nationalists as Polish spies. Although nationalists themselves had some success in casting their more numerous opponents as Russian chauvinists and closet Orthodox Church supporters, the Soviet reincarnation of West-Rusism gained the upper hand. Many proponents of the Belarusian national idea were condemned as *natsdems* (national democrats, a code word for an ideological corruption of true Leninism) and were then exiled to the deep interior of Russia. Those arrested or exiled in 1930–31 were, for the most part, subsequently released, but many were imprisoned yet again in 1937–38. This time, Stalinist repressions were more relentless, and most prisoners labeled *natsdems* lost their lives.

The Westernizing platform briefly resurfaced under the supervision of German occupiers, this time the Nazis. When Germany invaded the Soviet Union in 1941,

they treated Belarus as "nothing more than a vague geographical term."[17] Only after the assassination of the general commissar of occupied Belarus, Wilhelm Kube, in September of 1943, and following the overall success of the Soviet-led guerilla activity, did the Germans decide to play the card of Belarusian nationalism. The white-red-white flag and the coat of arms of Pahonia were in use again by the local authorities, who were appointed by, and worked under, the close supervision of the occupiers. In 1944, most local Nazi collaborators left Belarus with the retreating German army.

To this day, in the minds of many ordinary Belarusians, the collaboration with occupiers is a stigma. Hence, attempts by some Westernizers to "rationalize" collaboration. For example, Jerzy (Yury) Turonek, a Belarusian historian based in Poland, claims that "the attitude of the German war-time administration toward the Belarusian question was better than the attitude of the Soviet state. This is reflected in the fact that under occupation in the capital of Belarus, only Belarusian-language schools functioned, whereas after its liberation only Russian-language schools did. This qualification cannot be shattered by reference to genocide by the Germans; in Belarus, the number of victims of [Stalinist purges] was no less than the number of crimes committed by the SS."[18]

A contrasting opinion is rendered by Yury Shevtsov, who is a Belarusian historian much like Turonek, only from the other side of the ideological barricade.

> In Belarus, the war was very brutal and the cultural-political polarization of Belarusians turned out to be tough and across-the-board. During the crushing defeat of the Nazis, the proponents of the non-Soviet version of Belarusian identity were killed or left the country, forming the core of Belarusian emigrant communities in the West. The hatred of the victorious version of Belarusian culture toward the collaborationists is usually automatically transferred to historic Belarusian symbols that those collaborationists used and on everything that is linked with the non-Soviet version of Belarusian identity and ideology including literature and (sometimes) Belarusian language. There is also reciprocal rejection by Belarusian emigrants in the West of all aspects of Belarus's life after the war. This split of the nation has not been overcome.[19]

---

*The Vendée is an area in west central France remembered as the place where the peasants revolted against the French Revolutionary government in 1793. In the writings of Karl Marx regarding revolutionary struggles in various countries, he uses the term "a Vendée" as meaning "a focus of persistent counter-revolutionary activities." Thus, the Vendée of Perestroika is the place most inimical to Gorbachev's reforms.

In 1988, Zianon Pazniak, an avid Westernizer and the BPF leader, discovered a mass grave in Kuropaty, near Minsk, a grave that was traced to the Stalinist terror,[20] not to German occupiers, as the authorities of Soviet Belarus had strived but failed to prove. In the atmosphere of Gorbachev's Perestroika, this discovery ushered in the third and most recent period of the Westernizers' drive to power.

THE EFFECT OF CHERNOBYL

The Chernobyl nuclear power plant is located 7 km south of the Belarus-Ukraine border. Seventy percent of all radionuclides (radioactive isotopes) discharged during the meltdown of one of its four reactors on April 26, 1986, were deposited in Belarus. Twenty-three percent of Belarus's land acquired a level of contamination in excess of 1 Ci/km^2. In Belarus, the pockets of extremely high radioactivity, exceeding 15 Ci/km^2 or 185 kBq/m^2, cluster in the southeast. One of the two most visible clusters (Fig.) is in close proximity to the power plant, mostly in the south of Gomel Oblast. The second cluster, more remote from ground zero but actually larger than the first, is astride the border between Gomel and Mogilev Oblasts and the national border of Belarus and Russia (Fig.). Most probably the contaminated cluster straddling the border between Belarus and Russia resulted from natural rainfall shortly after the explosion.

A veil of secrecy was kept over the catastrophe up until the spring of 1989, but particularly throughout the first weeks after April 26, 1986. Not only did this secrecy inhibit and preclude coordination of efforts of numerous agencies but it also sowed seeds of distrust in any Chernobyl-related information that would at some later date emanate from public sources.

Consequently, wild rumors and exaggerations abounded. Despite the ensuing proclivity to attribute any death whatsoever or in fact any illness occurring in Chernobyl-affected areas to the 1986 catastrophe, the only definitive proof of that cause-and-effect relationship pertains to thyroid cancer. In southeastern Belarus, the actual rate of this disease was more than double the natural rate. The highest incidence of thyroid cancer was recorded among children and attributed to the radioactive isotope iodine-131, whose half-life is only 8 days. Thus, children exposed to radiation shortly after the blast absorbed the largest doses of that isotope. Many of them drank locally-produced milk with high iodine-131 content. With the passage of a couple of weeks from the disaster, soil rather than air became the primary source of danger. However, the absorption of radioactive isotopes with a half-life of 29–30 years, such as those of caesium-137 and strontium-90, takes more time and can be prevented or significantly reduced by precautionary measures.

The breakup of the Soviet Union and the ensuing withdrawal of all-Union funds initially earmarked to combat the effects of Chernobyl have arguably exacerbated the situation, and so has the fact that in 1992–1995 the Republic of Belarus experienced a steep economic decline. When quality of life and health care deteriorate for reasons not related to Chernobyl, it becomes difficult to properly assign causes and effects, which leads to disorientation and frustration. From 1991 to 2003, the Republic of Belarus spent $13 billion on fighting the effects of Chernobyl. In 1991, a whopping 22.3 percent of the entire state budget was directed to that purpose.[21] Thereafter, the share of budget allocations remained high but declined gradually to 6.1 percent in 2002. For any country, this financial burden would be difficult to sustain, but it was particularly so during the precipitous 1992–1995 economic decline.

International aid was solicited by Belarus and delivered by multiple countries and agencies. In order to channel Western aid, particularly that from private sources, scores of charitable organizations emerged in Belarus. The foundation *Children of Chernobyl* became the most well known of them. This agency alone dispatched over 60,000 Belarusian children to Germany in the summer of 1994.[22] There, Belarusian children stayed temporarily with families and in summer camps.

Alongside the economic decline of 1992–1995, one particular set of local problems made combating the consequences of Chernobyl more difficult. This is what the UN document calls "culture-specific ways of expressing distress."[23] In plain language, this is alcoholism. Widespread in Russia, Ukraine, and Belarus, this disease claims relatively more victims in rural areas. The Chernobyl-affected areas are overwhelmingly rural, with low personal incomes and few employment opportunities outside agriculture. The pervasive myth that alcohol somehow helps rid one's organism of radiation exacerbated the problem. David Marples describes cases of alcoholism he encountered during his 1995 tour of Chernobyl-affected areas, including alcohol abuse by the cleanup workers. More recent observations are not reassuring. Medical doctors interviewed by Vasily Semashko during his 2006 trips to the southeastern part of Gomel Oblast woefully stated that alcohol abuse in those areas is by far a bigger problem than radiation.

It must be added that the younger and more educated part of the population left the affected areas. As a result, the share of the elderly and resigned shot up. If only in part, therefore, the overwhelming feeling of hopelessness has roots in this change of the population structure. No less disturbing are the problems caused by a low level of civility and public trust. Chernobyl did not cause those problems; it just made them more vivid. For example, Marples refers to cases of hostile attitude toward the evacuees, corruption, money-laundering through artificially boosting the costs of construction of new housing for the victims of Chernobyl, false claims of being a victim of Chernobyl in order to go to Spain and the Netherlands for recuperation, and "negative competition"[24] between Belarus-based charitable foundations whereby each of them tried its best to undermine the efforts of the other. According to Vasily Semashko, *all* houses abandoned by the evacuees in the exclusion and primary evacuation zones have been plundered. Those plunderers are, for the most part, residents of nearby villages. Early on, much of the theft of abandoned property was perpetrated by police dispatched to guard the perimeter of the exclusion zone. Initially, the police teams worked in two week-long shifts, following which they were sent back to non-affected areas for recuperation. Throughout those shifts, many of those policemen drank heavily and robbed abandoned homes. It thus appears that while many people suffered from Chernobyl, many others took advantage of them. In that climate, longing for a strong hand that would establish order became popular in Belarus.

POST-SOVIET HISTORY AND BELARUSIAN IDENTITY

In 1992, the white-red-white flag and the state emblem of the GDL (man armed with a sword, mounted on a horse, ready to defend his fatherland) became the official insignia of independent Belarus; and the Belarusian language was proclaimed the only official language of Belarus. But as early as 1995, a modified Soviet Belarusian flag and a slightly modified version of the national emblem made a comeback. Along with the ruble as the unit of national currency and Russian as one of the official languages of Belarus, these symbols unmistakably reflect Belarus's closeness to Russia, the bulwark of the former Soviet Union. Thus, Zianon Pazniak's idea that all the deep-seated underpinnings of Russophile leanings could be undone forcefully and swiftly through a state-sponsored assault on popular ways of thinking did not materialize. In 1996, Pazniak emigrated to the United States. Having left the country, he lost his high moral ground even among intellectuals, the only group with which he had once had some appeal.

That all three episodes of the Westernizers' triumph were so brief may, of course, be attributed to the ploys of Russian colonialists. Such reasoning, however, appears to be shallow. It shifts attention away from the inherent weaknesses of the Belarusian national movement itself, as well as from the fact that not one but several different and opposing forces have been scrambling to flesh out Belarusian identity from the outset. Today, Belarusians are faced with two sets of national symbols. They are profoundly different and backed by different mythologies. The first set of symbols is inherited from the GDL, and the second is of Soviet vintage. Belarusian officialdom, some prominent historians and linguists, and apparently a large part of rank-and-file Belarusians cling to the second set. The anti-Nazi guerilla war is at the heart of this Russophile variety of Belarusian symbols. It was indeed a major epic story. At least until the late 1970s, all of Soviet Belarus's high ranking officials had been recruited from the Soviet-led network of the 1941–44 anti-Nazi underground, a glorified group.

As if two national projections for Belarus, pro-Russian and pro-Western, were not one too many, a third national image has emerged, largely associated with Alexander Lukashenka, Belarus's president from 1994 to this day. Some Belarusian Westernizers called this third national projection Creole. They borrowed this label from Mykola Ryabchuk, a Ukrainian philosopher who spent about ten years calling into question dichotomies like Russians/Ukrainians, Ukrainian speakers/Russian speakers, and nationally-conscious "patriots"/mankurts.[25] For Ryabchuk, Creoles are those Ukrainians who enthusiastically support Ukrainian statehood yet speak Russian as their primary language and distance themselves from other sociocultural aspects of Ukraine-ness.[26]

THE PHENOMENON OF ALEXANDER LUKASHENKA

Who is Mr. Lukashenka? It depends on who you ask. Until recently, for Western politicians, he was "the last dictator of Europe." The opinions of Western Belarus-watchers from among social scientists are more nuanced than those of politicians.

For example, David Marples professes "little doubt that the incumbent president [of Belarus] would have led by a considerable margin in a free election with an accurate vote count."[27] This is because "the degree of popular support for Lukashenka is quite impressive."[28] And Ronald Hill recognizes that "Lukashenka reflects popular values and aspirations."[29] More often than not, however, Lukashenka's popularity is seen as a nuisance: his supporters are just unable to see their own good.

In Russia, Lukashenka is a "divider, not a uniter." National-patriots idolize him while liberals despise him and tirelessly ridicule his Belarusian accent. Andrei Okara, a Ukrainian author based in Russia, attributes Lukashenka's popularity to the fact that "his rhetoric, behavior, and politics match the Belarusian-peasant archetype."[30] Other Russian authors analyzing Lukashenka's personality also attribute Lukashenka's mannerisms, particularly his habit of thrashing members of his government in front of TV cameras, to a communal peasant ethos. From 2003 to 2006, however, reviling Lukashenka was not condoned in major Russian media outlets; only the most liberal of them, like the radio channel Echo Moskvy did not toe the line. However, by the end of 2006, the mainstream Russian media resumed the innuendos. For example, in January 2007, when the Russia-Belarus trade war was in full swing, an article in *Izvestia* called Lukashenka a "conceited parasite."[31]

The nature of the caricatures reveals that neither Russia nor the West is prone to take Lukashenka for what he most likely is: a national leader whose policies reflect his nation's circumstances and mindset. Lukashenka's rule should instead be viewed in the context where it truly belongs, that is, in the context of Belarusian history and identity and not as some abstract political form (dictatorship) deceitfully imposed on an allegedly benighted people whom better positioned and better informed outsiders seek to enlighten and liberate.

Despite the fact that one of the three co-signers of the Belavezh Accords,* which was the death sentence for the Soviet Union, was a Belarusian (Stanislav Shushkevich), most of his fellow countrymen were ill-prepared for independence.

---

*The Belavezh Accords is the agreement which declared the Soviet Union effectively dissolved and established the Commonwealth of Independent States in its place. It was signed at the government retreat near Viskuli in Belovezhskaya Pushcha on December 8, 1991, by the leaders of Belarus, Russia, and Ukraine.

While a few Minsk-based intellectuals were able to convert the newly emerging freedom into some sort of social capital that materialized in contacts with the West and in its financial support, most Belarusians saw their lifelong savings evaporate and their quality of life plummet. By 1994, most industrial enterprises in Belarus were about to discontinue their operations because of the disruption of Russia-controlled supply lines. In 1995, Belarus's GDP was 33.9 percent lower than in 1991. Predictably, it had decreased by about the same degree as Russia's, whose GDP was 34.6 percent lower.[32] In both Russia and Belarus, most people intensely disapproved of privatization of industrial and other enterprises, but in Russia it proceeded anyway. As early as 1992, 40 percent of Russia's GDP was produced in the private sector, and by 2006, that figure was at 65 percent. In contrast to that, in Belarus, the share of the private sector in the GDP was 10 percent in 1992, and since that time it has increased to just 25 percent.[33]

Back in 1994, when the transition paths of Russia and Belarus began to diverge, stoppages of factories, empty grocery stores, and a steep rise in the cost of living resulted in huge rallies in usually calm Minsk. Because industry had stopped as a consequence of severed supply lines from Russia, Belarusians demanded that these ties be restored. Unlike in Russia, though, the elites and "ordinary people" are in agreement on many issues. In the words of Wilson and Rontoyani, "to the overwhelming majority of the Belarusian political elite, the costs of an economic strategy aimed at reducing interaction with Russia were immediate, certain, and of such a magnitude as to make such a choice unthinkable."[34] The premonition of impending chaos and doom—preponderant in the Belarusian society—was largely responsible for bringing to power a strongman, Alexander Lukashenka, formerly director of a state farm in Mogilev Oblast and a self-proclaimed crusader against corruption. Lukashenka stopped voucher privatization (which had barely begun) and secured subsidized transport, utilities, and free health care and education. By that time (1995), most industrial workers worked barely 2–3 days a week (as plants ran out of supplies and could not dispose of their output), but they could perhaps eke out a living for a year or so at the most with the aid of their own kitchen gardens. In about a year (by 1996) most industrial giants resumed their full capacity work schedule, mostly due to the restored ties with Russia. That was the pledge on which Lukashenka had been elected in 1994, and that was the promise he made good on. Since 1996, Belarus has been experiencing impressive economic growth: 7 percent in 2003, 11.4 percent in 2004, 9.4 percent

DEVELOPMENT

Since 1996, Belarus has been experiencing impressive economic growth: 7 percent in 2003, 11.4 percent in 2004, 9.4 percent in 2005, 9.9 percent in 2006, 8.6 percent in 2007, and 10.0 percent in 2008.

in 2005, 9.9 percent in 2006, 8.6 percent in 2007, and 10.0 percent in 2008.

Nevertheless, until 2005, Western media and to a large degree the Russian media assessed Belarus's economic situation as abysmal. The phraseology used by academics included "the myth of economic revival,"[35] and "'shampoo paradise' of Brezhnev's 1980s,"[36] (to which Belarus of the late 1990s was likened). One attempt to implant a modicum of objectivity into the analysis of economic growth in Belarus was described as painting "a picture of an economic utopia."[37] Because no alternative analysis was offered, one could discern only one reason to maintain requisite skepticism: Alexander Lukashenka is a bad guy, and so everything under him must be bad. In 2005, the tone had to change after the World Bank and the International Monetary Fund published their Belarus reports. While critical of Belarus's economic policies, the reports disposed of any suspicion that economic growth in Belarus may have been contrived. According to the World Bank, "in contrast to some other CIS countries, the patterns of growth in Belarus have been much more beneficial for labor" since "benefits from recent growth were broadly shared by the population. Growth in labor earnings amounted to about 53 percent of the total GDP growth over 1996–2003. Poverty rates declined substantially, while inequality remained rather stable and moderate. The poverty headcount ratio . . . was more than halved—from 28.6 percent of the population in 1996 to 17.8 percent in 2004, which meant that about 2 million people moved out of poverty."[38]

To be sure, the emergence of the current global financial crisis has placed most national economies (and their analysts alike) in uncharted waters. Based on sheer dynamics of output, Belarus has thus far (summer 2009) sustained less industrial decline than Russia and particularly Ukraine, although some worrying signs engage economists' attention as

Belarus seems to be failing to introduce the painful but necessary changes in the structure of its economy and continues to release industrial products (e.g., tractors and trucks) in quantities far exceeding the vastly reduced demand. As a result, those products are filling up warehouses, stadiums, and open spaces. In May 2009, Russia postponed the transfer of the last tranche of its $2 billion loan to Belarus, allegedly because Belarus's government crisis action plan was inadequate. And although this move met with starkly-worded dismissal by Lukashenka, some independent and not heavily politicized Belarusian economists are not so resolute. "It is difficult to say what is more in Russia's stance—crude blackmail or refined calculation," writes Elena Rakova. "However, [by] conducting current economic policy—that is, refraining from budget cuts and from terminating some enterprises, endlessly borrowing money and postponing painful reforms, the government undermines the economic independence of Belarus. There won't be any business as usual anymore; the time for strategic decisions and shocks has arrived."

It is indeed unclear at the moment how the country will weather the current recession, but Lukashenka's previous and lasting success at sustaining the image of a stellar chief executive officer of Belarus (among one half of Belarusians at the very least) is still worth pondering. One cannot understand the reasons behind that success without responding to the broader question, namely why in much of the non-Western world people desire order more than democracy. This issue requires separate and more broad-based analysis (e.g., of the kind contained in some books of Fareed Zakaria)*, but it is important to understand that Lukashenka's image in Belarus is, first and foremost, that of the guardian of order. Additionally, he presided over twelve years of economic growth and he is seen by about two-thirds of Belarusians as the true father of Belarus's statehood. Whether or not Lukashenka is truly a dictator is debatable, but it is beyond doubt that he is an authoritarian leader. What is equally beyond doubt, however, is that being an authoritarian leader fits the East Slavic version of non-Western political tradition. So far, no leader deviating from that tradition (for example, Mikhail Gorbachev of the Soviet Union or Viktor Yushchenko of Ukraine) has

been able to gain lasting popularity in the region itself. That in some foreign countries those self-proclaimed democrats have been more popular than at home actually reinforces the case.

Lukashenka's continuing grip on power is effectively abetted by a motley group of lackluster and exceedingly weak political opponents. Observing the relationships within and between those opponents to the extent they are reflected by the opposition media and the Belarusian Service of Radio Liberty, one cannot but get the impression that the struggle *within* the Belarusian opposition for visibility and favors from the West is more important to the leaders of opposition parties than the grand idea of unseating Alexander Lukashenka. This impression is particularly hard to discard now as the party bosses are questioning the overall leadership of Alexander Milinkevich, one of two 2006 opposition presidential candidates and the darling of some Western dignitaries.[39] It seems unlikely that most, if any, Belarusian party leaders would survive the test of unfettered political life. In a peculiar way they are both suppressed and legitimized by one and the same agency, the ruling regime of President Lukashenka.

For much of the 1990s, Belarusian Westernizers repeatedly accused the Belarusian masses of being nationally indifferent. It now appears, however, that many Belarusians who speak "trasianka," which is phonetically Belarusian and lexically Russian, are quite patriotic and nationalistic. As described by Uladzimer Abushenka, these people are midway in their sociocultural evolution. For them, things Russian no longer belong in "we," but they cannot be assigned to "they" yet. Similar ambiguity typifies their attitude to things Belarusian. Creole consciousness is a kind of extrapolation of "tuteishasts," i.e., a Belarusian variety of localism.[40] Creole is essentially a pre-national consciousness. Igar Babkou believes that it has long existed in areas that straddle a kind of cultural divide where the peripheries of adjacent cultural zones come together. Belarus is just such a place.[41]

The framers of the Creole projection are associated with Lukashenka's Presidential Administration. They supervise the development of what is called the "state ideology of the republic of Belarus." The major components of this ideology are the historic attachment of Belarus to Russia, the role of the Great Patriotic War of 1941–1945 that cemented this bond, a communal and anti-entrepreneurial ethos, and (however ironic it sounds in conjunction with national ideology) an anti-nationalist sentiment directed squarely against the Westernizers.

---

*In particular , Fareed Zakaria, *The Future of Freedom,* New York: W.W. Norton and Co, 2003.

Only in this context can one appreciate the peculiar reference to Lukashenka as "the main anti-Belarusian nationalist of Belarus."[42] Because in Russian "nationalism" is given a bad name, "anti-nationalist" might be translated as "averse to xenophobia," if in fact the term did not also include aversion to Belarusian nationalists of the Westernizing strand, whose spiritual mentors such as Radaslau Astrousky, Fabian Akinchyts, and Yaukhim Kipiel collaborated with the Nazis.[43] It is little wonder that so much has been made of this collaboration by Lukashenka staff propagandists. After all, more than one-quarter of Belarus's population perished in World War II.

The "state ideology of the Republic of Belarus" is still under construction. If anything, still developing interpretations pertain more to the past than to the future, which is envisaged as a patrimonial welfare state with elements of a market economy. But there is no codified narrative of the pre-Soviet past. The message of Creole nationalism is contained, for example, in Lukashenka's September 23, 2004, speech before a student audience at Brest State University. In the beginning of that speech, Lukashenka described the confusion and despair that possessed his fellow countrymen in the early 1990s when disruption of supply lines from Russia and a shrinking Russian market brought most Belarus factories to the verge of closure. He then turned to external influences on Belarus exerted at that time.

Two outside forces wanted to sway us. On the one hand, Russia was trying to shape Belarus's choice, but it did so in an unpersuasive and unsystematic way because it was itself going through confusion and vacillations. On the other hand, the West that won the cold war was aggressive and businesslike. The West's message to us was: Quickly conduct privatization, unhook yourself from Russia, and jettison the Russian military; then we will accept you and assist you financially. Those were the conditions that the West presented me with. . . . Not only was this influence from without. Also, inside the country a pro-Western party, the Belarusian Popular Front—and it was the only political party at the moment—pushed us in the same direction. Why then didn't Belarus go to Europe? . . . Well, in the first place because, in contrast to Poland and the Baltic States, Belarus never—I

dare say, never ever—has been part of Western culture and the Western way of life. Yes, we were subjected to the influence of Western culture within Rzeczpospolita and the GDL. That influence, however, was short-lived. They did not succeed in implanting the Western ways then, and they probably cannot succeed today. . . . Yes, we were, are, and will be an inalienable part of pan-European civilization, which is a mosaic of different cultures. But to the Catholic-and-Protestant . . . civilization, Belarus and Belarusians, who are predominantly Orthodox and for centuries coexisted in the same political setting with Russia and Russians, are alien.[44]

In more recent programmatic texts of Creole nationalism, cultural attachment to Russia coexists uneasily with aversion to Russian style capitalism. Thus, according to Anatoly Rubinov, now deputy speaker of the upper chamber of the Belarusian parliament, "today's Russia is no longer a country with which we lived together. In Russia, the order of things, the priorities, have changed, and they are subject to different influences. . . . Powerful corporate structures exert influence there; they have narrow and self-serving financial concerns, and their interest in Belarus is quite pragmatic. The major interests of some of Russia's political and financial groups are in the West."

Distancing from Russia and yet clinging to it is very much at the heart of Alexander Lukashenka's geopolitical brinkmanship. For example, he has made it a habit to assert that Russians and Belarusians are one and the same people, and yet he is vehemently opposed to the unification of Belarus and Russia.

RECENT DEVELOPMENTS

In 2004 Lukashenka conducted a referendum to enable him to change the article of Belarus's constitution that set a limit of two terms in office for the president. Elected twice, in 1994 and 2001, Lukashenka was wanting to rule for a third time, and elections were set for 2006. More than 50 percent of the vote was required to adopt the constitutional amendment. Official statistics indicated that 79.4 percent of registered voters approved the measure as opposed to the 48.4 percent approval rating indicated by exit polls administered by the Lithuanian arm of the Gallup Institution.[45] At the time of the referendum not a single opposition figure enjoyed the support of

more than 1.5 percent of the electorate.[46] On October 2, 2005, Alexander Milinkevich was nominated the major opposition candidate by the congress of the opposition forces attended by members of five parties, including the BPF, the United Civic Party, and a splinter Communist Party. Milinkevich, who has a doctorate in physics, served as vice-mayor of Grodno responsible for culture and education from 1990 to 1996. The other opposition candidate who participated in the 2006 presidential election was Alexander Kozulin, the former rector (president) of Belarusian State University and minister of education who had had a falling out with Lukashenka.

To anyone versed in developments in Belarus (not just in what Western media had said about Belarus), the prospects of a Lukashenka landslide in the 2006 presidential elections appeared certain. In hindsight, the most impressive forecast was issued by the late Alexander Potupa, chairman of Belarus's Union of Entrepreneurs. In an interview with the Latvian Russian-language newspaper *Telegraf* on September 14, 2005 (a fairly early date!), Potupa indicated that "in 2006, Lukashenka's victory is as assured as it was in 2001. He will win no less than 80 percent of the vote. But one has to understand that he would win 2:1 (or about two-thirds of the vote) even without vote rigging."[47] Potupa's forecast proved impeccable not only in terms of the official count but also, most probably, in terms of the degree of distortion of the actual vote cast. Indeed, according to repeated polling by the opposition sociologists, Lukashenka won 63 percent—65 percent of the popular vote, not 83.0 percent as per official results, whereas Milinkevich won 18 percent—21 percent of the vote, not 6.1 percent as per official results. Kazulin's official vote count was 2.2 percent; according to independent sociologists, it was within the 5 percent—7.3 percent margin.

On March 7, 2008, the U.S. Treasury Department announced sanctions against Belneftekhim, a Belarusian petrochemical conglomerate, citing lack of democratization as demanded by America's Belarus Democracy Act. First adopted by the U.S. Congress in 2004 and reauthorized in 2006, the Act had become the major document of Western foreign policy in regard to Belarus. Minsk responded by recalling its ambassador from Washington for consultation and by issuing a recommendation that the United States recall its own ambassador and reduce its embassy staff in Minsk to six to match the Belarusian embassy staff in Washington. When no reaction from the U.S. embassy in Minsk came, Karen Stuart, the U.S. ambassador, was summoned to the Ministry of Foreign Affairs and put on notice that should she not leave within two days, she would be declared persona non grata. Stuart promptly left on March 12; and the same sequence of events unfolded for the remaining thirty-three U.S. diplomats, twenty-seven of whom departed by April 30, leaving only six staff to run the embassy.

Throughout the entire tussle, an impression of poorly coordinated diplomacy on the American side was fostered. U.S. assumptions were wrong on at least three counts. First, Belarus was apparently expected to be aggrieved over demotion or even loss of diplomatic ties with Washington. This expectation, however, was well off the mark, and had any expert versed in Belarus been consulted, such a miscalculation would certainly have been averted. The second consequence of the diplomatic conflict initiated by the United States was a significant boost in the domestic popularity of President Lukashenka. Third, the American side had initiated the fracas almost simultaneously with the European Commission's opening of its Minsk office. Although Europeans had long—but half-heartedly—been toeing the American line on Belarus, as the end of the Bush administration neared, they quickly lost any incentive to do so.

The change of heart on the part of European institutions gave rise to yet another set of developments. In October 2008, the European Union suspended travel sanctions on Lukashenka (in place since 1998) and most other Belarusian government officials, signaling a significant rapprochement between Belarus and Europe. This development has been greeted with dismay by most members of the constantly feuding Belarusian opposition. The concessions extracted from Belarus in exchange for this cardinal change in European policy were so laughably minor that they seemed more a way for Europe to save face than a genuine compromise. To be sure, Europe's new stand vis-à-vis Belarus has been conditioned by several factors, the growing irritation with the Bush policy being only one. Other impetus came from the increasing importance of transit of Russian fuel via Belarus and Belarus's own economic success, confirmed by the World Bank and the IMF. Yet another issue was the Russian-Georgian war. Increasingly leery of Russia's aggressive behavior, Europeans came to the conclusion that their American-inspired policy of cogent democracy promotion in Belarus was not only ineffective but counterproductive, since it tended to push Belarus further into the embrace of the Russian bear. In his interview with *Agence France Presse* on November 24,

2008, Lukashenka explicitly recognized Europe's misgiving as a driving force in Europe's new Belarus policies: "For some reason, the European Union is concerned that we may lose our independence. I am not sure, though, why are they so concerned about that." Incidentally, since October 2008, Lukashenka has given more interviews to Western, mostly European, media than during much of the prior fourteen years of his presidential career.

Even more importantly, in 2009, Lukashenka singlehandedly achieved a total reversal in how the West treats him. Although the inertia of previous approaches is still noticeable in the media, contacts at the highest level and generosity of international financial institutions speak for themselves. In February 2009, Xavier Solana, the EU's High Representative for Common Foreign and Security Policy, visited Minsk and conducted face-to-face negotiations with Lukashenka. Belarus was invited to the EU summit in Prague in May 2009 (the Czech Republic held the chairmanship of the EU until June 2009) in what is called the 27+6 format, that is, the full members of the EU, plus Georgia, Armenia, Azerbaijan, Ukraine, Moldova, and Belarus. Because of some anti-Lukashenka rhetoric by the hosts, Lukashenka sent deputy premier Semashko to participate in that important meeting. On April 27, Lukashenka held a historic meeting in Rome with the Pope and then with the Italian Prime Minister Berlusconi. In June 2009, the IMF decided to increase its credit line to Belarus by $1 billion just at the time Russia was having second thoughts whether to send the second tranche of its own loan to Belarus (as Belarus had not, contrary to Russian expectations, recognized statehood of Abkhazia and South Ossetia, two secessionist Georgian enclaves). In early September 2009, Hans Dietrich Genscher, the patriarch of German foreign policy, went to Minsk and had a lengthy talk with Lukashenka. Finally, in late September, Lukashenka's official visit to Lithuania took place. Because Lithuania is a NATO and EU member, this visit's significance goes far beyond just neighborly ties. Thus the Western policy of ostracizing Lukashenka, practiced since 1996, failed without any serious concessions from Lukashenka.

Two conclusions seem to be in order. First, the U.S. policy of forceful democracy promotion is waning. This policy never recognized the cultural specificity of targeted countries and rested on the arrogance of policy makers and their self-righteous moralizing and messianic zeal more than on any practical foundation. For the most part, democracy promotion was driven

by poorly disguised geostrategic and geopolitical interests. In fact, Western policy toward Belarus was directed by geopolitical interest during the estrangement stage (1996–2008) and continues to be so during the current rapprochement. During the former period, policy makers considered Lukashenka a *kolkhoznik** whose public support was gained exclusively by authoritarian methods and could easily be overturned by "democrats" if they are generously aided by the West. Today, geopolitical interests revolve around preventing Russia's annexation of Belarus.

HEALTH/WELFARE

In October 2009, the Health Ministries of Belarus and Turkey signed an agreement of cooperation. It will facilitate the exchange of experts and information on medical science and practices between the two countries.

Second, if the West is indeed interested in having Belarus retain its statehood, it ought to recognize the country as a unique entity and support, and be informed by, well-founded Belarusian studies. This is worth doing because Belarus is one of the most economically successful of the post-Soviet states. It is a country that has defied numerous clichés and acquired statehood not after but before nation-building progressed to a mature stage. (Nation-building is usually at the core of the process of state formation.) Successful Belarusian nation-building is truly in the West's interests, and the current political regime of Belarus is at the forefront of this process whether some politicians and pundits like it or not.

NOTES

1. R. B. Dobson, *Belarusians Gravitate toward Russia: A Pull of Russian Language, Media Remains Strong* (Washington, DC: Office of Research, Department of State, October 11, 2000), 2–3.

2. Nina B. Mečkovskaya, *Belorusskii Yazyk: Sotsiolingvisticheskie Ocherki,* Muenchen: Verlag Otto Sagner 2003: 61.

3. Viacheslav Nosevich, op. cit.:11; Curt Woolhiser, "Constructing national identities in the Polish-Belarusian borderlands," *Ab Imperio* (Kazan', Russia), No. 3, 2003: 293–346.

4. Zakhar Shybeko, op. cit.: 2.

*Literally, *kolkhoznik* means a member of a collective farm. At the same time this is a derogatory term (meaning an uncouth or rustic person) when applied to an urbanite, especially a head of state.

5. Nina Mečkovskaya, op. cit.: 28.

6. Golubovich, V.I. (Ed.), *Ekonomicheskaya Istoriya Belarusi*, Third Edition, Minsk: Up Ecoperspectiva 2004: 156.

7. "Belorusskaya Sovetskaya Sotsialisticheskaya respublika," *Bol'shaya Sovetskaya Entsyklopedia*, Moscow 1969–1978, accessed via www.rubicon.ru/bse.

8. V.I. Golubovich, (Ed.), *Ekonomicheskaya Istoriya Belarusi*, Third Edition, Minsk: Up Ecoperspectiva 2004: 169.

9. *Naseleniye SSSR 1987*, Moscow, Finansy I Statistika 1988: 16.

10. *Posledstviya Velikoi Otechestvennoi Voiny dlya Belarusi*, www.president.gov.by/gosarchives/vov/spravpos.htm.

11. *Narodnoye Khoziaistvo SSSR za 70 let*, Moscow: Finansy I Statistika 1987 contains the following monetary estimates of damage: Ukraine, 285 billion rubles and Belarus, 75 billion rubles in 1941 prices (p. 45). According to Naseleniye SSSR (Moscow: Finansy I Statistika 1987), the 1940 population of Ukraine was 41,340,000 people, and Belarus was home to 9,046,000 people (p. 9).

12. Nickolas Vakar, *Belorussia: The Making of A Nation*, Cambridge: Harvard University Press 1956.

13. Nickolas Vakar, op. cit.: 59.

14. *Ibid.*

15. Curt Woolhiser, "Metalinguistic Discourse, Ideology, and 'Language Construction' in the BSSR, 1920–1939," the English original of an article published in French as "Discours sur la langue, idéologie et 'édification linguistique' dans la RSS de Biélorussie, 1920–1939." In *Le discours sur la langue en URSS à l'époque Stalinienne*, edited by Patrick Sériot, 299–337. Cahiers de l'ILSL no. 14 (Lausanne: University of Lausanne, 2003).

16. Belarus did not regain its 1939 population (8.9 million people in current borders) until 1969. "Owing to the effects of the war, Belorussia lost a greater percentage of its population than any other region of the USSR; between 1939 and 1951, the population of Belorussia declined by 12.7%" Ralph S. Clem, "Belorussians," in *The Nationalities Question in the Soviet Union*, ed. Graham Smith (London: Longman, 1990), p. 113.

17. Nickolas Vakar, op. cit.: 263.

18. Jerzy Turonek, *Bialorus pod okupacja niemiecka* (Warsaw: Ksiazka i Wiedza,1993), p. 241.

19. Yury Shevtsov, *Obyedinionnaya Natsiya: Fenomen Belarusi*, Moscow, Yevropa 2005: 75.

20. A detailed account can be found in David Marples, *Belarus: A Denationalized Nation*, Amsterdam: Harwood 1999: 54–60.

21. *Chernobyl's Legacy: Health, Environmental and Socio-Economic Impacts and Recommendations to the Governments of Belarus, the Russian Federation and Ukraine.* New York, NY: United Nations, 2005.

22. Marples, David R. *Belarus from Soviet Rule to Nuclear Catastrophe*. New York: St. Martin's Press, 1996.

23. *Chernobyl's Legacy: Health, Environmental and Socio-Economic Impacts and Recommendations to the Governments of Belarus, the Russian Federation and Ukraine.* New York, NY: United Nations, 2005.

24. Marples, 1996: 72.

25. Creatures whose historical memories were surgically removed from their brains; they are pictured by Chingiz Aitmatov in his 1980 hallmark novel *The Day Lasts More Than a Hundred Years.*

26. Mykola Ryabchuk, *Vid Malorossii do Ukrayiny*, Kyiv: Kritika 2000.

27. David Marples, "Color revolutions: the Belarus case," *Communist and Post-Communist Studies*, 2006, Vol. 39: 363.

28. *Ibid.*: 361.

29. Ronald Hill, "Post-Soviet Belarus: In search of direction," Stephen White, Elena Korosteleva, and John Loewenhardt (Eds.), *Postcommunist Belarus*, Lanham: Rowman & Littlefield Publishers 2005: 1–16.

30. Andrei Okara, "Belarus v Otsutstvii Tret'yei Al'ternativy," *Russkii Zhurnal* 14 November 2001, www.russ.ru/politics/20011114-oka-pr.html.

31. Vitaly Ivanov, "Ne nuzhno zabluzhdadtsia naschiot Belorussii," *Izvestia*, 18 January 2007; www.izvestia.ru/comment/article3100239/index.html.

32. *CIS in 2004*, Moscow: Interstate Statistical Committee of the CIS 2005: 25.

33. EBRD Transition Report; Structural indicators: www.ebrd.com/country/sector/econo/stats/sci.xls; last accessed 3 January 2007.

34. Andrew Wilson and Clelia Rontoyani, *Security or Prosperity: Belarusian and Ukrainian choices*, Robert Legvold and Celleste Wallander (Eds.), *Swords and Sustenance: The Economics of Security in Belarus and Ukraine*, Cambridge, MA: MIT Press 2004: 43.

35. David R. Marples, *Belarus: A Denationalized Nation*, Amsterdam, Harwood 1999: 40.

36. Margarita M. Balmaceda, "Myth and Reality in the Belarusian-Russian Relationship," *Problems of Post-Communism*, Vol. 46, No. 3, 1999: 12.

37. David Marples, "Europe's last dictatorship: the roots and perspectives of authoritarianism in 'White Russia,'" *Europe–Asia Studies*, Vol. 57, No. 6, September 2005: 896.

38. *Belarus: Window of Opportunity to Enhance Competitiveness and Sustain Economic Growth*, Washington, D.C.: The World Bank 2005; www.web.worldbank.org/WBSITE/EXTERNAL/COUNTRIES/ECAEXT/BELARUSEXTN.

39. See the account of elections and the aftermath in Chapter 7 of Grigory Ioffe, Understanding Belarus and How Western Foreign Policy Misses the Mark, Lanham: Rowman and Littlefield 2008.

40. Uladzimer Abushenka, "Mickiewicz kak 'Kreol': ot 'Tuteishikh Geneologii' k Geneologii Tuteshastsi," accessed www.lib.by/frahmenty/sem-abuszenka.htm.

41. Igar Babkou, "Genealyogiya Belaruskai Idei," *Arche*, 3 (2005): 136–165.

42. Alexander Feduta, "Nedoliot," *Nashe Mneniye*, 3 August 2005, www.nmnby.org.

43. See John Loftus, *The Belarus Secret*, New York: Alfred Knopf 1982. A few alleged collaborators who found themselves in the West were tried and convicted. Anton (Andrei) Savoniuk, who in 1999 was given two life sentences for his role in a 1942 massacre of Jews, died in a British prison in November 2005.

44. Alexander Lukashenka, *Stenogramma Vystupleniya pered Studencheskoi Molodiozh'yu Brestchiny*, 23 September 2004; accessed at www.president.gov.by.

45. Yanina Bolonskaya, "Nezakazannye tsifry," *Belorusskaya Delovaya Gazeta*, No 1483, 26 November 2004.

46. Bykovsky, Pavliuk, "Reiting Lukashenko i yego konkurentov," www.ru.belaruselections.info/current/2006/sociology/0022756.

47. Svetlana Martovskaya, "Pochemu Belarus' golosuyet za Lukashenko," *Telegraf*, 14 September 2005; accessed at www.inosmi.ru/translation/222228.html.

Moldova

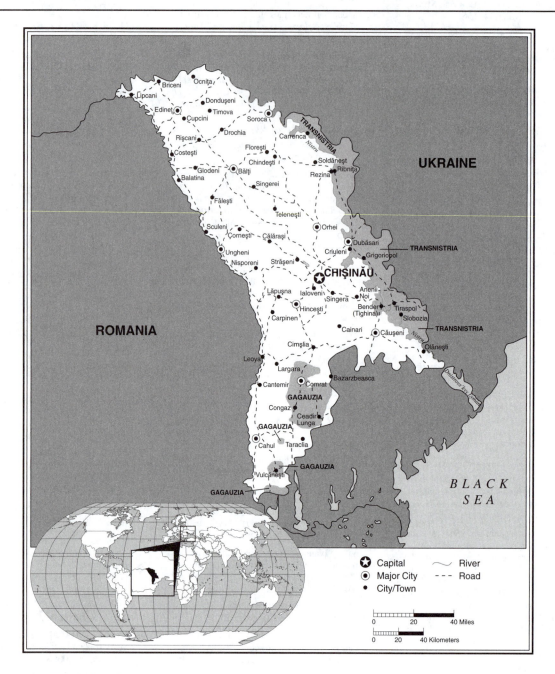

Moldova Statistics

Area in Square Miles (Kilometers):
13,067 (33,843)
Capital (Population): Chisinau (610,150)

Population

Total (millions): 4.1
Per Square Km: 122
Percent Urban: 41
Rate of Natural Increase (%): −0.1

Ethnic Makeup: Moldovan/Romanian
78.2%; Ukrainian 8.4%; Russian 5.8%;
Gagauz 4.4%; Bulgarian 1.9%; other
1.3% (2004 census)

Health

Life Expectancy at Birth: 65 (male); 73
(female)

Infant Mortality Rate: 12
Total Fertility Rate: 1.3

Economy/Well-Being

*Currency ($US equivalent on 14 October
2009):* 11.1 Leu = $1
2008 GNI PPP Per Capita: $3,210
*2005 Percent Living on Less than US$2/
Day:* 29

Moldova Country Report

A country squeezed between Ukraine and Romania, Moldova has a land area of just 33,800 sq km and a population of 4.1 million people (2009). The capital city is Chisinau, where one-fifth of the entire population lives. By hydrographic description, Moldova is largely situated between the two north-south flowing rivers, the Prut and the Dniester (Nistru). While the Prut is a tributary of the Danube and constitutes the border between Moldova and Romania, Dniester's middle and lower reaches are all *within* Moldova. The border between Moldova and Ukraine runs a little bit east of Dniester and parallel to it. In the extreme south, Moldova has access to the Danube along a tiny, 600-m long strip but not to the Black Sea, so the country is landlocked.

Physiographically, Moldova is an undulating lowland with the highest elevation just 430 m above sea level. It possesses the naturally fertile soils typical for a steppe region. Because Moldova is located within the southwestern margin of the Eurasian steppe belt, dominant westerly and southwesterly winds bring about moderate precipitation, so the local steppe is not arid. A combination of fertile soils, gentle slopes, rare droughts, and short and mild winters have long made Moldova an important agricultural region excelling in production of cereals, grapes, and other fruits. Since industrialization and urbanization of Moldova commenced relatively late, it is one of just four countries of Europe (besides Albania, Slovenia, and Bosnia) in which rural villagers still outnumber urbanites.

Ethnic Moldovans account for 71.5 percent of the population, including among them self-identified Romanians (1.8%). The distinction between Moldovans and Romanians is a matter of debate with strong political overtones. By all accounts, they have identical ethno-linguistic roots and there is no grammatical and little lexical difference between Romanian and Moldovan literary standards. Even dialectal distinctions are reportedly small and such as pertain to virtually all languages. Both countries claim the same authors as national writers, notably Mihai Eminescu (1850–1889), the most well-known Moldovan-Romanian writer whose poetry is translated into many languages. The linguistic commonality of Moldova and Romania was reinforced in 1989 when Moldova switched to Latin script from Cyrillic, which had been used earlier under the Soviet rule. (Romanians themselves

underwent this same change in 1860.) Moldovan-Romanian is a Romance language rooted in what is called Vulgar Latin; it was introduced by Roman warriors in the first century A.D. to the area populated by the Dacians, who spoke a now extinct Indo-European language. Somehow Roman warriors managed to assimilate linguistically the much more numerous Dacians during just two centuries of Roman domination over the area which roughly coincides with modern Moldova and Romania.

Even the official documents of the Republic of Moldova are conflicted about the issue of Moldovan-Romanian distinction. Thus, the Constitution of Moldova states that the Moldovan language is the official language of Moldova, while the Declaration of Independence of the Republic of Moldova names Romanian the official language. The 1989 State Language Law speaks of a Moldo-Romanian linguistic identity. In 2003, the government of Moldova adopted a thesis which states that one of the priorities of the national politics of the Republic of Moldova is the insurance of the existence of the Moldovan language. The split between Romanian and Moldovan may be entirely political in nature. It was first instituted when the Moldavian[1] Soviet Socialist Republic was formed, and it is now sustained by the reluctance of the current Moldovan political elite to lose its status in case of a putative merger with much larger Romania. As one proceeds east from the current Romanian ethno-cultural core, one encounters two degrees of detachment from it. East of the Prut River, the Moldovan identity, as distinct from the Romanian one, exists despite linguistic commonality. Then, east of the Dniester (Nistru in Moldovan and Romanian) River, in a narrow strip squeezed between the river itself and the border with Ukraine, a Transnistrian identity occurs; it gave rise to the split of Transnistria from Moldova proper in the early 1990s (see the following). Despite a separate identity, Moldovans form the largest ethnic group in the Transnistrian region of Moldova; they account for 32 percent of the population in that region, whereas Ukrainians account for 29 percent and Russians for 31 percent. However, both of the latter two East Slavic groups primarily use Russian, as do many Moldovans in Transnistria, so Russian speakers summarily outnumber Moldovan speakers in Transnistria. Therefore, the aforementioned spatial (west-east) trend may

be described in different terms: as one proceeds east from the current Romanian core, one encounters two successive gradations of Russification with a clear peak in Transnistria. Transnistrian identity, it may be argued, is one of the few remaining vestiges of Soviet identity.

Because history, as complex as it is, may furnish justification for nearly every ethno-political ambition, it can be argued that, although currently Moldova is in every respect a much weaker state than Romania, the Moldovan identity is actually older than the Romanian one. Moldova was a medieval principality (subsequently divided among Russia, Austria, and Turkey), yet no common name for Moldovans or the people of Wallachia, Transylvania, etc.—that is, other regions that comprise modern Romania—existed before the mid-1800s. Thus, reference to age of identity may, however awkwardly, justify separation between Moldova and Romania. An opposing view could state with equal truth that the Republic of Moldova is just the northeastern part of historical Moldova, the southwestern (and larger) part of which actually lies in Romania, where it forms the province of Moldavia. It may thus be aserted that no reason for separation exists now that the Soviet Union is gone. Moldovan-Romanian relationships are a conundrum; abundant evidence justifies almost any political division or lack thereof within the Moldovan-Romanian ethnic field.

Besides Ukrainians and Russians, the ethnic minorities of Moldova include Gagauz and Bulgarians. The use of the Russian language in Moldova is well out of proportion to the number of ethnic Russians; in fact, the Russian language is widely used in all levels of the society and state. Gagauz make up a compactly settled Turkic-speaking community and are frequently referred to as "the only Christian Turkic" people. This, however, is a mistake, since Chuvash and Yakut (Sakha) ethnicities, living in their respective homelands in the Russian Federation, are also Turkic-speaking and Christian. In the past, a Jewish community was numerically strong in Moldova. According to the 1897 Russian census, Jews accounted for almost 12 percent of the entire population of Bessarabia (which is what Moldova was then called) and 37 percent of all urbanites. Even before the Holocaust, in which most Moldovan Jews perished, Jews had migrated from the area en masse due to poverty and pervasive anti-Semitism. In fact, the 1903 Kishinev pogrom became one of the most publicized massacres of Jews in the Russian Empire (49 dead, 586 wounded and maimed; and 1,500 houses destroyed). Just as other pogroms, this one was triggered by a rumor of a ritual killing of a Christian boy by the Jews. As it turned out later, the boy was killed by his own cousin, but this was revealed long after the massacre. Currently, Jews account for just 0.12 percent of Moldova's population.

The above mentioned toponym "Bessarabia" was used in the Russian Empire in reference to present-day Moldova and the strip of the Black Sea coast immediately south of Moldova, now belonging to Ukraine. The term comes from Bessarab-the-Great, the ruler of Wallachia and Moldova from 1310–1352. After the Communist Revolution of 1917, the day-to-day usage of the term was discontinued.

Most ethnic Moldovans are Orthodox Christians. The existing parishes belong to two different Orthodox jurisdictions, one centered in Moscow, Russia, and one in Bucharest, Romania. The majority of parishes belong to the former.

The following issues deserve to be elaborated upon in view of their supreme significance for Moldova: political history; economy; secessionism; Moldovan-Romanian relationships; and the April 2009 failed attempt at a colored revolution.

POLITICAL HISTORY

With the exception of Transnistria, the area of the modern Republic of Moldova was part of the Principality or Princedom of Moldova (Tara Moldovei) which emerged in the fourteenth century. The Principality also included Romania's current province of Moldavia as well as Bukovyna and part of the present-day Odessa region of Ukraine. The capital of the principality was the city of Iasi (now the capital of the Romanian province of Moldavia and located close to the Romanian-Moldovan border). Beginning in the 1600s, the Principality of Moldova progressively fell under the rule of the Ottoman Empire. Later it became a bone of contention between the Ottoman Empire and another expansionist state, Russia. In 1812, the northeastern part of the principality, known as Bessarabia, was annexed by Russia, but the rest of the principality remained a vassal of the Ottoman Empire until the formation of an independent Romania in 1859. Since then, the western part of the former Principality of Moldova has been integral to Romania.

The territory of the present-day Republic of Moldova was under Russian control from 1812 to 1918. Following the 1917 Communist Revolution in Russia and amidst the ensuing civil war, the Moldovan Democratic Republic formed and joined Romania. When Moldova was part of Romania, the eastern border of that country was the Dniester (Nistru) River. However, in 1924, on the *eastern* bank of the Dniester River, a so-called Autonomous Moldavian Republic was declared (within the Ukrainian SSR). As a buffer entity its existence helped the Soviet Union sustain its claims on Moldova and insist on the illegality of the 1918 "annexation" by Romania. In 1940, the Soviet Union incorporated Moldova, in accord with the U.S.S.R.'s August 1939 non-aggression (Molotov-Ribbentrop) pact with Germany, a secret protocol which assigned former Bessarabia—as well as the Baltic States and the eastern part of mid-war Poland—to the sphere of influence of the Soviet Union. Thus, in 1940, the Moldavian SSR was established with its capital in Chisinau (Kishinev). When in 1941 Germany launched war against the Soviet Union, Romania (an ally of Nazi Germany) reoccupied Moldova. In 1944, the Red Army reentered Moldova and the Moldavian SSR was reestablished. The independent Republic of Moldova is entirely the result of the dissolution of the Soviet Union in December 1991. The Soviet Moldovan ruling elite was well-integrated into the Soviet elite, a fact reflected in the rotation of power within the communist cadre. Thus, Piotr Luchinsky (Petru Lucinschi), president of Moldova from 1997–2001 and an ethnic Moldovan, had in 1986–89 worked as the Second Secretary of the Communist Party of Tajikistan, a role almost invariably assigned to the most loyal—and usually ethnically Russian—officials. Some ethnic Moldovan writers began to write in Russian, such as Ion Druţă (Ion Drutse), who acquired considerable fame with some of his Russian-language novels. In cities, particularly in Chisinau, Russian became the language of everyday communication, whereas Moldovan was mostly spoken in the countryside. No secessionist movement developed in Moldova, although under Gorbachev's Perestroika, the Popular Front demanded that the Moldovan language switch to Latin script and be recognized as the only official language. The Front, however, was popular only among intellectuals and students in the capital city of Chisinau; the rest of Moldova was barely involved. Consequently, the 1991 independence resulted entirely from external circumstances.

However, once the Chisinau political elite that took shape during the Soviet period received national status for Moldova, they became committed to its retention. The first president of independent Moldova was Mircea Ion Snegur. In the Soviet era, he was known as Mircha Ivanovich Snegur. Prior to independence, he was Chairman of the Supreme Soviet of the Moldavian SSR. And prior to that, he was a secretary of the Central Committee of the Communist Party of the Moldavian SSR. In December 1991, running unopposed, he won the presidential election. However, Snegur's opposition to immediate reunification with Romania led to a split with the Moldovan Popular Front, which demanded reunification. He later regained the support of the Popular Front when he confronted secessionists in Transnistria. In the presidential election of 1996, Parliamentary speaker Petru Lucinschi surprised the nation with an upset victory over Snegur in a second round of balloting. Known in the Soviet era as Piotr Kirillovich Luchinsky, Lucinschi, had earned an even higher status under the Soviets than Snegur. Thus, in 1989, he became the First Secretary of the Communist Party of Moldavia. In that capacity, in 1990–91 he was also a member of the Politburo, a body at the very top of the Communist Party of the Soviet Union (CPSU); he was then Secretary of the CPSU's Central Committee. Luchinschi's tenure as president of Moldova coincided with a particularly steep decline in living standard that followed the 1998 financial crisis in Russia. Although it is unlikely that any of Lucinschi's actions contributed to that, in 2001 he was unseated by left-leaning parliamentarians clinging to the Party of Communists of the Republic of Moldova. This party was reinstituted in 1993 after having been outlawed in 1991, and it is now the largest political party in Moldova. In 2000, the Parliament adopted an amendment to the Constitution that transformed Moldova from a presidential to a parliamentary republic in which the president is elected by 3/5 of the votes in the parliament and no longer directly by the people. Only 3 of the 31 political parties passed the 6 percent threshold needed for a seat in Parliament during the February 25, 2001 elections. Winning 49.9 percent of the vote, the Communists gained 71 of the 101 MPs, and on April 4, 2001, elected Vladimir Voronin as the country's third president.

Voronin is of mixed origin; his father was an ethnic Russian from the Irkutsk region (in eastern Siberia), and his mother was ethnic Moldovan. Voronin's highest position during the Soviet era was Minister of Internal Affairs of the Moldavian SSR (which means he was head of the national police force) with the rank of general. Unlike his predecessors, Voronin has presided over a period of steady economic growth (2001–2008). In 2005, he was reelected for a second four-year term, which was supposed to be his last term according to the Constitution. However, in late July 2009, Voronin continued to hold office as acting president of Moldova due to the stalemate in the Moldovan parliament which prevented the election of a new president in April 2009 by the required 3/5 majority. New parliamentary elections were scheduled for July 29, following which a new attempt to elect a new president was to take place. Voronin is a skilled political manipulator (perhaps resembling Belarus's Lukashenka) who manages to both ingratiate and distance himself from Romania and Russia, two forces in a constant tug of war for Moldova. While Russians effectively control the secessionist region of Transnistria and want to buy up the most lucrative assets of Moldova, Romanians are threatening Moldova's independence by denying that there is any difference between Moldovans and Romanians and by offering Romanian citizenship to all residents of Moldova—a status many Moldovans have opted for, particularly since Romania's accession to the European Union. The fact that Voronin is the first communist in Europe who has ever been democratically elected as president of a country should not be overplayed. Himself a wealthy man controlling—in part through his son—much of Moldova's media, Voronin is no ideological purist. It should be remembered, though, that he presided over a poor country with a low level of urbanization, a country whose many people fondly remember the Soviet era as a kind of a golden age.

ECONOMY

Because within the Soviet Union Moldova received more financial transfers from the federal budget than it contributed, living standards in Moldova declined significantly after the dissolution of the Soviet Union. Moldova is now Europe's poorest country with a per capita GNP of just $1,260 in 2007 (Moldova, *Country Profile 2009*, The World Bank, 2009). Agriculture and agro-processing activities account for 34 percent of the country's GDP. Moldova must import most of its energy, including all fuel, from Russia. Moldova's major specialization is the wine industry; the country has 147,000 hectares of vineyards, most of which are used for commercial production. Before the dissolution of the Soviet Union, 40 percent of Moldova's industrial output and 90 percent of electricity were produced in the secessionist region of Transnistria. Between 1991 and 1999, Moldova's GDP declined by 60 percent; growth resumed in 2000 and has been robust (6%–8% a

year from 2004–2008), but the GDP has not yet reached its 1990 level. More than 25 percent of the economically active population has left the country in search of better opportunities abroad; Moldovans work in various countries from Russia to Portugal. Regrettably, Moldova is a source country for women and girls trafficked for the purpose of commercial sexual exploitation and men trafficked for forced labor. According to an International Labor Organization (ILO) report, Moldova's national Bureau of Statistics estimated that there were likely over 25,000 Moldovan victims of trafficking for forced labor in 2008. Moldovan women are taken primarily to Turkey, Russia, Cyprus, the UAE, and other Middle Eastern and Western European countries. Men are forced to work in the construction, agriculture, and service sectors of Russia and other countries. There have also been some cases of children trafficked to neighboring countries for begging. Internal trafficking of girls and young women is also prevalent from rural areas to Chisinau (*U.S. State Dept Trafficking in Persons Report,* June, 2009).

SECESSIONISM/SEPARATISM

Despite a fairly small size and a solid ethnic majority, Moldova faces separatism in two regions. The smaller and seemingly averted secessionist threat is centered in two administrative districts of southern Moldova (Comrat and Ceadir-Lunga) where a small, 127,000–strong Gagauzian minority is concentrated. Most Turkic-speaking and predominantly rural Gagauz were frightened by the possibility of Moldova joining Romania, a possibility that seemed quite real in 1990–1991. It should be pointed out that the second, third, and so forth largest ethnic groups of Soviet republics (e.g., Armenians, Abkhaz, or Ossetians in Georgia; Uzbeks in Kyrgyzstan; Tadzhiks in Uzbekistan; not to

mention ethnic Russians in every republic) tend to view ethnic majority nationalists in their own republics as a threat and the all-union political establishment as their protector. Such a view was not entirely without grounds, and it explains why Gagauz saw unification with Romania (which would have augmented the strength of what they perceived as a single Moldo-Romanian community) as an undesirable event. As a result, many Gagauz supported the Moscow coup attempt in August 1991, which might have led to the restoration of the Soviet Union; and Gagauzia declared itself independent on August 19, 1991, thus straining relations with Chisinau. However, when on August 27, 1991, the Moldovan parliament voted on whether Moldova should become independent, six of the twelve Gagauz deputies in the Moldovan parliament voted in favor, while the other six abstained. As a consequence, the Moldovan government toned down its pro-Romanian stance and paid more attention to minority rights. In February 1994, President Mircea Snegur promised the Gagauz autonomy. In 1994, the Parliament of Moldova awarded to "the people of Gagauzia" (through the adoption of a new Constitution of Moldova) the right of "external self-determination" should the status of the country change. In other words, if Moldova decided to join another country (by all accounts, that would be Romania), then Gagauzians would be entitled to decide by means of a self-determination referendum whether to remain part of the new state or not. On December 23, 1994, the Parliament of the Republic of Moldova accepted the "Law on the Special Legal Status of Gagauzia," resolving the dispute peacefully. This date is now a Gagauzian holiday. Gagauzia stands as a "national-territorial autonomous unit" with three official languages, Moldovan, Gagauz, and Russian. Three cities and twenty-three communes were included in the Autonomous Gagauz Territory. This was all localities in which Gagauz constitute over 50 percent of the population as well as those localities with populations composed of 40–50 percent Gagauz in which a majority of voters expressed a desire to be included during the referendums to determine Gagauzia's borders. Turkey financed the creation of a Turkish cultural center in Comrat and a Turkish library.

The second secessionist threat is by far more serious and far-reaching for the Republic of Moldova. It has to do with Transnistria, a narrow strip between the Dniester (Nistru) River and the border with Ukraine. Transnistria, whose land area is just slightly over 4,000 sq km and the population is slightly more than half-a-million

Eternitate memorial complex in Chisinau. The monument commemorates victims of the fight for independence and territorial integrity of Moldova.

people, is part of a region which used to be—from 1924 to 1940—the Moldavian Autonomous Republic, a Moldovan land separate from Romania. (The remainder of that region subsequently became part of Ukraine.) The capital city was Tiraspol, now capital of the unrecognized Pridnestrovian (Transnistrian) Moldavian Republic, an area which does not want to be part of the Republic of Moldova with its capital in Chisinau. Historically, the most important difference between Transnistria and the rest of Moldova is that Transnistria was never integrated with Romania with the exception of four years during World War II when Romanian forces were stationed there. On the contrary, a Soviet mentality has deep roots in Transnistria, and Russian is spoken on an everyday basis by a far greater share of the population than in Moldova proper. The possibility of Moldova's joining Romania in 1991 was perceived as a big threat by most people in the region. Despite its small land

area, in 1990 Transnistria accounted for 40 percent of the industrial output of the entire Moldavian SSR and 90 percent of its electricity output. Transnistria is home to a large hydroelectric power station, an iron-and-steel plant (responsible for 60% of budget revenues of Transnistria), a large textile factory, and a brandy-producing enterprise.

Tensions between the Transnistrian communities and Chisinau first arose in 1989 in the wake of the new language law adopted in the Moldavian SSR, according to which Moldovan was proclaimed the only official language in the republic, although it was still part of the U.S.S.R. After their objections to the new law were denied by the Moldovan Supreme Soviet, its Transnistrian members and official municipal leaders of Tiraspol, Bendery, and Dubossary joined forces and in September 1990 announced the formation of a separate Pridnestrovian Moldavian SSR within the Soviet Union. Moscow did not

uphold the separatist effort, but it criticized Chisinau authorities for discrimination against non-Moldovan speakers. The first bloodletting took place in November 1990 when Transnistrians blocked the bridge across the Dniester River and Chisinau's police force was dispatched to remove the blockade. Three people were killed and 16 wounded in the ensuing skirmish. When on March 17, the all-union referendum took place on whether to retain the Soviet Union, the Moldavian authorities in Chisinau effectively boycotted the referendum, whereas in Transnistrian communities the referendum occurred in orderly fashion and about 98 percent of the voters said yes to the Soviet Union. Tension between Transnistria and the rest of Moldova came to a head during the August 19–21, 1991, coup in Moscow. Transnistrian authorities supported the coup which aimed at undoing Gorbachev's reform and restoration of a strong central authority.

Bendery, Transnistria, Moldova: The monument commemorating the victims of the fight for Transnistria's independence.

On August 23, the Communist Party of Moldavia was disbanded, and the Chisinau police arrested the MPs from Transnistria. Igor Smirnov, chairman of the self-proclaimed Supreme Soviet of Transnistria, was apprehended by Moldovan security agents in Kyiv, Ukraine. In response, a "congress of people's deputies" of Transnistria decided to set up its own militia. This decision meant that the available police forces would no longer report to Chisinau. When Moldovan authorities attempted to restore the line of command, the Transnistrian militia resisted. Because this resistance enjoyed popular support, the Moldovan police had to release Igor Smirnov and other Transnistrian representatives held in the Transnistrian city of Dubossary.

During a December 13, 1991, fight between the Moldovan police and Transnistrian militia, four policemen and three militiamen were killed. The Transnistrian authorities subsequently made a PR stunt out of the fact that the three victims of the skirmish belonged to three different ethnicities, Moldovan, Ukrainian, and Russian. It is important to understand that the confrontation between Soviet-style internationalism and Moldovan nationalism is the major refrain of the entire secessionist effort. By the very end of 1991, when scores of volunteers from Russia and Ukraine arrived in Transnistria to help avert attacks by the Moldovan police, the overall situation was not yet out of control. In mid-December 1991, two meetings between President Snegur of Moldova and the separatist leader Smirnov were held in Chisinau; they resulted in the creation of a joint commission, in removing roadblocks, and mutual withdrawal of armed forces.

Tensions still were not diffused. Beginning in early March 1992, when a team of Transnistrian militiamen was ambushed while riding to the place of a reported accident (which, as it later turned out, had not occurred), sporadic clashes developed into an all-out-war. In April 1992, detachments of Moldovan police shelled two Transnistrian towns. Fourteen thousand industrial workers in Transnistria were mobilized to defend their separatist enclave; all of them received weapons. Two bridges across the Dniester River were blown up on orders of Transnistrian authorities to prevent the advance of Moldovan military units. At the same time, the bridges at the city of Rybnitsa and the Dubossary hydroelectric power stations were vigilantly protected. Sometime during the spring of 1992, members of Russia's Fourteenth Russian Army Division, long deployed in Transnistria, dropped their erstwhile policy of neutrality in the conflict and began to aid the Transnistrian militiamen and volunteers. At the same time, the Moldovan army was joined by 18,000 reservists. The bloodiest battle of the war was for the city of Bendery, waged June 19–21, 1992. The Moldovan attack on the city was defeated only with the help of the Russian army.

In Chisinau, leftist parties campaigned for the ouster of the prime minister and the minister of defense, both of whom resigned. In July 1992, Russians brokered some sort of a truce. On July 21, President Snegur of Moldova signed an agreement with President Yeltsin of Russia "On principles of resolving an armed conflict in the Transnistrian region of the Republic of Moldova." While the leader of the separatists, Smirnov, was present during the signing of this agreement, the very title of it suggests that, at least officially, Russia did not support the idea of Transnistria's separation from Moldova.

Overall, during the height of the conflict—its military stage, which seems to have ended in 1992 after the siege of Bendery—both sides lost about 1,000 lives, including 400 civilians. Some 4,500 people were wounded. Human losses are more or less equal on both sides, but because the war was waged on the Transnistrian land, the separatist region alone suffered material damage, including partial or complete destruction of 1,290 homes, three schools, and 15 health care institutions.

Over the course of the 1990s, Russia facilitated a number of meetings between the leaders of Moldova proper and Transnistria during which the principles of rapprochement were discussed and at times even agreed upon, but no reconciliation was ultimately achieved. All these discussions were aimed at the restoration of Moldovan territorial integrity through federalization of the country, with Transnistria obtaining some special status within Moldova. In the meantime, people from both banks of the Dniester River restored mutual contacts and were able to travel back and forth with no problem. At the beginning of the new century, tensions between Transnistria and Moldova proper rose several times, once when Moldovan authorities allegedly prevented the flow of industrial exports from Transnistria and then by mutual jamming of mobile phone network operations. Since 2003, the European Union and the United States have been trying to bring the two sides of the conflict to mutual understanding. The so-called Kozak Memorandum (named after Dmitry Kozak, a member of Putin's administration), also in 2003, was about to achieve some federal arrangements within a united Moldovan state, but at the last minute Chisinau decided not to accept it.

Between 2000 and 2008, no meetings of Tiraspol and Chisinau leaders took place. In 2008, President Voronin of Moldova and President Smirnov of the unrecognized Transnistrian Moldavian Republic did meet in Moscow, but no breakthrough was made. Instead, the rift between Transnistria and Moldova proper assumed an air of permanency and their mutual pretences have become clichés. Thus, Moldova claims that Transnistria is a criminal enclave engaged in illegal trade in weaponry, and Transnistria claims that Moldova is a weak agrarian country which will sooner or later become part of Romania. "Most probably, the future of Moldova is within Romania. But without us," said Vladimir Yasenchak, foreign minister of Transnistria in his April 2009 interview with the Russian daily *Izvestia.* It may well be that at some opportune moment, Transnistria will join Russia, becoming an exclave much like Kaliningrad. A September 2006 popular referendum in Transnistria can be used as the basis for such a solution. During the referendum, people were asked to respond to two questions: (1) Do you support the tack to independence of the Transnistrian Moldavian Republic and its subsequent joining of the Russian Federation? and (2) Do you agree with giving up on independence of the Transnistrian Moldavian Republic and with subsequent joining the Republic of Moldova? The results: (1) 97.1 percent yeas and 2.3 percent nays; 2) 3.4 percent yeas and 94.6 percent nays. There is a possibility of replaying this referendum any time with quite predictable (and by no means rigged) results. What is interesting is that the issue of joining Ukraine, a vast country that separates Transnistria from Russia, is not even contemplated. There are no grassroots demands for that sort of development despite a significant Ukrainian community in Transnistria. The region is indeed a bastion of Russified Soviet culture and sees Russia as its grand protector. Apparently, Russia has not been willing to go along lest it infuriate the West; so the issue of Transnistria joining the Russian Federation has not yet been put on the agenda in the Russian Duma. One cannot rule out, however, that this is just a matter of time. Yet another legal ground behind putative unification of Transnistria with Russia is that the Molotov-Ribbentrop Pact, which had wrested Moldova from Romania, was declared null and void by the Romanian parliament as early as 1991; and it was also condemned by part of the political spectrum in Chisinau. But if the Pact that laid the foundation for former Bessarabia's becoming part of the Soviet Union is nullified, then why should Transnistrians be committed to a link with the political space between the Prut and Dniester rivers? This is a rhetorical question that Transnistrian separatists repeatedly pose. Indeed, Transnistria was not part of that space (i.e., current Moldova) prior to 1940.

MOLDOVAN-ROMANIAN RELATIONSHIP

Pan-Romanianism, a notion that the border along the Prut River is artificial and Moldovans and Romanians are one people speaking the same language, has been a consistent part of Romanian politics. Between 1989 and 1992, the notion was shared in Moldova, which adopted the same Romanian tricolor as its national flag, named its currency unit a lei (identical to the Romanian currency unit), and switched the Moldovan language to Latin script, thus eliminating the only difference between it and Romanian. Indeed, the Moldovan Popular Front has been and remains committed to unification between Romania and Moldova. Romania was the first country to recognize Moldova as an independent state—indeed, only a few hours after the declaration of independence was issued by the Moldovan parliament. But even before the dissolution of the Soviet Union on May 6, 1990, Romania and the Moldavian SSR had lifted restrictions on travel between the two entities, and hundreds of thousands of people crossed the Prut River, which marked their common border. At the time, it was generally expected in both countries (especially in Romania, though to some extent in Moldova, too) that they would soon be united.

The unification, however, did not occur. Moreover, since 1994 when the Romanian parliament protested against Moldova joining the Commonwealth of Independent States, a Russia-centered alliance of the former Soviet republics, there have been tensions between the two countries. While the media have mostly focused on the different historical experiences of Moldovans and Romanians and—on the Moldovan side—on the condescending attitude of Romanians to their Moldovan brethren as too Russified and in need of special aid to undo the outcomes of Russification, the real problem seems to be the reluctance of the Chisinau political elite to lose their national status and become provincial leaders, at best, in a much larger Romanian polity. While pan-Romanianism struck roots among university students and intellectuals, larger groups of primarily rural Moldovans did not, and still do not, share the pan-Romanian identity. For older generations, it is still important that, under the Soviet rule, living standards in Moldova were higher than in Romania; and those able to visit their relatives across the Prut River during Soviet times repeatedly testified to that.

The situation changed after the dissolution of the Soviet Union and even more so with Romania's 2007 accession to the

European Union. From 1991 to 2005, 95,000 citizens of Moldova applied for Romanian citizenship (without giving up their own), by the beginning of 2009, close to 800,000 Moldovans had applied for Romanian citizenship, which is more than one-fifth of Moldovans living between the Prut and Dniester Rivers. The fact that the Romanian side is actually encouraging Moldovans to apply for Romanian citizenship has led to even greater tensions between Romania and Moldova. Those tensions came to a head during the April 2009 events in Chisinau.

A MOLDOVAN COLOR REVOLUTION?

Civil unrest began on April 7, 2009, in major cities of Moldova (including the capital Chisinau and Bălţi) after the results of the 2009 Moldovan parliamentary election were announced. The demonstrators claimed that the elections, which saw the governing Party of Communists of the Republic of Moldova (PCRM) win 60 seats out of 101, were fraudulent and demanded a recount, a new election, or resignation of the government. Similar demonstrations took place in other major Moldovan cities, including the country's second largest, Bălţi, where over 7,000 people protested. While hardly any election in post-Soviet country is entirely free of violations, the observers from the Organization for Security and Cooperation in Europe did not uncover any significant or systematic violations. The protesters organized themselves using the online social network service, Twitter. In Chisinau, where the number of protesters rose above 15,000, the demonstration escalated into a riot on April 7. Rioters attacked the parliament building and presidential office, breaking windows, setting furniture on fire and stealing property. For a brief period of time, the Romanian flag flew atop the presidential palace. To be sure, the difference between the Romanian and Moldavian flags is slight; each of these is a vertical tricolor of blue, yellow, and red, but the Moldovan flag carries a coat of arms (an eagle holding a shield charged with an aurochs) on the center bar on the obverse side of the flag only and the Romanian flag does not.

President Voronin of Moldova displayed political wisdom when he decided that the national army and/or police refrain from shooting into the rioters. Fairly soon, most citizens of Moldova distanced themselves from violent protesters, more than 200 of whom were arrested and charged with attempting to overthrow legitimate power. A vote recount did not result in any change in the distribution of the popular vote: 49.48 percent (60 seats) went to communists. Three major opposition parties received votes as follows: Liberal Party, 13.13 percent; Liberal Democratic Party, 12.43 percent; and the Alliance Our Moldova, 9.77 percent. Voronin, however, accused Romania of instigating and organizing the protesters. He declared the Romanian ambassador persona non grata and introduced visas for Romanians visiting Moldova. However, it was unlikely that this new border crossing regime would stand because of the objections from the European Union, which referred to Moldova's earlier decision to waive visas for *all visitors from the EU.* While the political opposition and the protestors were indeed pro-Romanian, no evidence of Romania's organizing role in the violent protests in Chisinau has been discovered. Early comparisons of the events in Chisinau with the Orange Revolution in Kyiv, Ukraine, were subsequently disqualified, as Chisinau protestors did not show the signs of well-prepared and organized resistance so typical of the Orange Revolution.

In June 2009, the newly elected Moldovan parliament failed to elect a new president of Moldova. Sixty-one votes in favor of one candidate were required, but the communists were able to deliver only sixty for Zinaida Grechanaia (a communist and also the prime minister). New parliamentary elections were scheduled for July 29 in order to overcome the impasse, with Voronin serving as acting president until the new parliament is able to elect his successor. Voronin appealed to the European Union to convince Romania to refrain from exerting undue influence on the upcoming elections. Committed to retaining Moldovan independence, the Chisinau political elite are eager to be accepted by the European Union, but "only through Brussels, not Bucharest."

The July 29 elections did in fact change the distribution of seats in the parliament. Communists won only 49 parliamentary seats (out of 101), down from 60 seats in April. While their faction is still the largest in the parliament, it no longer commands a majority. Collectively, four opposition parties gained 53 seats and managed to elect the non-communist Speaker, Mihai Gimpu, who is the leader of the Liberal Party. Upon becoming Speaker, Gimpu said that while the reunification with Romania remains his personal goal he is not going to enforce it given the lack of unity on the issue. On September 11, 2009, Vladimir Voronin resigned, and Gimpu took over as the acting head of state, as four non-communist factions have not yet figured out how to make one of their own people president for which purpose the minimum of 61 parliamentary votes is needed. A visa requirement for Romanian visitors was rescinded, and Gimpu issued an appeal to Russia to withdraw its military from Transnistria. It remains to be seen what, if anything, will ordinary Moldovans gain from this new political configuration. One thing, though, is clear. Moldova is a place where elections have a meaning not quite different from that in the West. While no longer unique, this is still fairly uncommon for the post-Soviet space, and Moldovans may take pride in that.

HEALTH/WELFARE

The reform of the health system of the Republic of Moldova started in 1995, when the World Bank and the Moldovan government started negotiating the enhancement of the health system and the subsequent adoption of the Law on Health Protection in March 1995. Since then, new principles regarding health financing and organization of primary and secondary health care have been established and a private health sector has begun to take shape.

NOTE

1. In the Soviet Union, the country was called Moldavia or Moldavian SSR. Moldavia is also the name of a province in Romania.

Armenia

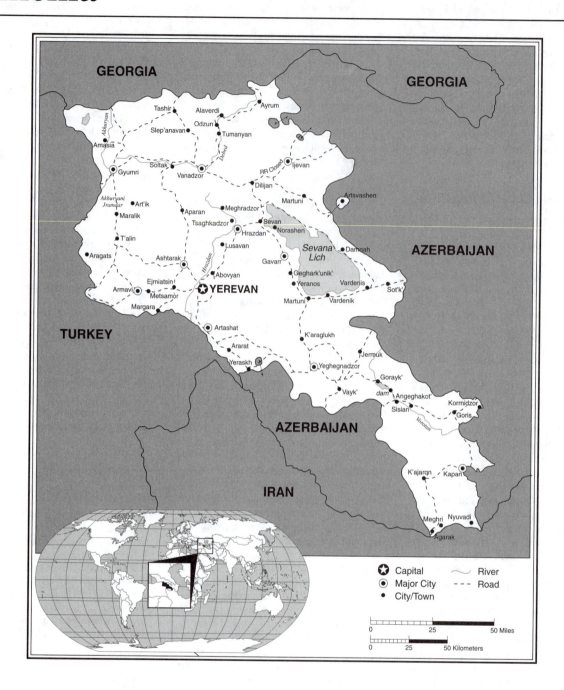

Armenia Statistics

Area in Square Miles (Kilometers):
  11,500 (29,800)
Capital (Population): Yerevan
  (1,356,000)

Population

Total (millions): 3.1
Per Square Km: 104
Percent Urban: 64

Rate of Natural Increase (%): 0.6
Ethnic Makeup: Armenian 97.9%; Yezidi
  (Kurd) 1.3%; Russian 0.5%; other
  0.3% (2001 census)

Health

Life Expectancy at Birth: 68 (male); 75
  (female)

Infant Mortality Rate: 25
Total Fertility Rate: 1.7

Economy/Well-Being

*Currency ($US equivalent on 14 October
  2009):* 385.4 drams = $1
2008 GNI PPP Per Capita: $6,310
*2005 Percent Living on Less than US$2/
  Day:* 43

Armenia Country Report

(© Andrew Sevag Behesnilian)

Armenia: The twelfth century Khor Virap monastery in the shadow of Mount Ararat, upon which Noah's Ark had supposedly once come to rest.

Armenia is a small, landlocked, mountainous country, with an extremely old, though interrupted, tradition of independent statehood. The first Kingdom of Armenia was established around 600 B.C. Armenia lies in the highlands surrounding the biblical mountain of Ararat, upon which, according to the Book of Genesis, Noah's Ark came to rest after the flood. Armenia was the first state that made Christianity its official religion; this happened in A.D. 301, 36 years before the Roman Empire did the same. Over subsequent centuries, the ethnic territory of Armenians was conquered by Romans, Byzantines, Arabs, Mongols, Persians, Turks, and Russians. The ethnic territory in question is within the so-called Armenian Highland, which extends far into the modern state of Turkey. Mount Ararat itself, whose image is sacred to every Armenian and integral to Armenia's coat of arms, is actually in Turkey. It is observable from the Ararat Valley of Armenia, wherein lies the capital city of Yerevan, one

of the world's oldest sites with continuous settlement (it was founded in 782 B.C.). The Republic of Armenia today is just the northeastern part of historical Armenia.

This republic has a land area of only 29,800 sq km, and was the smallest union republic of the Soviet Union. At least one-third of its population of 3.2 million people reside in the city of Yerevan. Although much of Armenia is at high elevations (one-half of it is above 1800m and only one-tenth is within the 600–1000m range), it was the second most densely populated Soviet republic (after Moldova). Within the Soviet Union, Armenia stood out in terms of its ethnic homogeneity, a pattern particularly typical of Old World countries whose modern jurisdiction falls well short of the historical area where the respective ethnic majority once lived. (In this regard, a natural parallel would be Hungary.) Today, ethnic homogeneity (more than 97 percent of the population are ethnic Armenians) is several percentage points higher than it was during

the late Soviet period, largely because of the wholesale outmigration of the Azerbaijani (Azeri) minority and of most ethnic Russians as well. The Armenian language is Indo-European. Linguists typically view Armenian as an independent subunit of the Indo-European language family, though some scholars assign it to the same group as Greek and the now extinct Indo-Iranian languages. The Armenian language uses its own unique script. This is how the name of the country is spelled Հայաստան (pronounced Hayastan).

In what follows, several issues of utmost importance for the Republic of Armenia will be outlined: Armenian-Turkish relations; the formation and development of the Armenian Soviet Socialist Republic (1936–1991); Armenian-Azerbaijani relations and the conflict over Nagorno Karabakh; population and economy of independent Armenia (1991–2009); Armenia and the Armenian diaspora; and government and political conditions.

ARMENIAN-TURKISH RELATIONS

Originating in Central Asia, Turkic tribes had begun to migrate en masse to the Armenian Highland and further onto the Anatolian plateau in the tenth century. Turkic migration came in waves and encroached upon indigenous local populations, including Armenians. In 1071, the Turks defeated the Byzantine army at Manzikert, an Armenian city in Anatolia. After the Turks overran Anatolia and implanted their Islamic culture and civilization, Armenian communities continued to flourish under relatively tolerant Ottoman rule for centuries, either as minority populations in urban areas or as exclusively Armenian settlements in rural areas.

However, during the half century leading up to World War I, the Armenian populations of Anatolia endured increasingly more brutal persecution under Sultan Abdul Hamid II. As the Ottoman Empire declined, its political leadership either authorized or tolerated violent and reckless attacks on the Armenian population, which attracted harsh criticism from various Western nations whose missionary communities in Anatolia witnessed several massacres. Violence against Armenians was condemned by several heads of state, including American President Grover Cleveland. While it remains unclear to what extent this violence was organized by the Turkish government, Cleveland noted that "strong evidence exists of actual complicity of Turkish soldiers in the work of destruction and robbery." Later, in 1909, from 15,000 to 30,000 Armenians perished in the Adana Massacre, a wholesale slaughter of the Armenian community residing in the city of Adana in southeastern Turkey. The awakening of Turkish nationalism and the perception of the Armenians as a separatist, European-controlled entity contributed to the violence, as did the fact that the Armenian minority was more prosperous than the Turkish majority.

But even the Adana Massacre paled beside what was yet to transpire under the republican regime of the Young Turks. This new political regime was initially perceived by Armenians as the long expected improvement over the monarchical governments in Turkey. However, from 1915 to 1917, at least 600,000 Armenians were massacred (the highest estimate is at 1,500,000) and hundreds of thousands of survivors were forcibly evicted from their land in northeastern Turkey. These events came down in history as the first genocide of the twentieth century, preceding the Jewish Holocaust by 2.5 decades. Mass forcible migrations of Armenians from Turkey during those years and thereafter laid the foundation for the Armenian diaspora (see the following). The fact of Armenian genocide

at the hands of Turks is recognized by most European and American scholars and by quite a few governments, including those of Russia, France, Italy, Switzerland, and Canada. In 1951, the United States government officially recognized the Armenian Genocide in a document submitted to the International Court of Justice (ICJ), also known as the World Court. This document, filed by the United States with the ICJ, is included in the May 28, 1951, ICJ Report titled, "Reservations to the Convention on the Prevention and Punishment of the Crime of Genocide." The specific reference to the Armenian Genocide appears on page 25 of the ICJ Report: "the inhuman and barbarous practices which prevailed in certain countries prior to and during World War II, when entire religious, racial and national minority groups were threatened with and subjected to deliberate extermination. The practice of genocide has occurred throughout human history. The Roman persecution of the Christians, the Turkish massacres of Armenians, the extermination of millions of Jews and Poles by the Nazis are outstanding examples of the crime of genocide."

In the view of many Armenians, however, this convoluted and hard-to-come-by statement falls short of explicit recognition of the Armenian Genocide by the U.S. Congress, which does not want to complicate its relations with Turkey, an important American ally. As a result, the State Department has an ongoing policy of not using the word genocide and opposing Congressional bills which mention the term genocide in reference to Armenians. However, numerous U.S. states (42 states so far) and cities have independently recognized the Armenian Genocide.

The Armenian Genocide Memorials Database (*www.armenian-genocide.org/memorials.html*) provides information about memorials and monuments dedicated to the memory of the victims of the Armenian Genocide. The Armenian National Institute (ANI), to date, has identified 135 memorials in 25 countries. Many are documented with multiple images. The database is organized according to country and the monuments listed alphabetically according to their civic location. The database is a work in progress and ANI welcomes any additional information or digital images that viewers can provide to augment and improve the documentation on the memorials. The largest of the monuments is, appropriately, in the Armenian capital.

Dedicated in 1968, Yerevan's Armenian Genocide Monument is visited by hundreds of thousands of Armenians each April 24, the anniversary of the start of the bloody slaughter. On that day, hundreds of thousands of local residents gather there in sorrowful silence. The monument is

located in Tsitsernakaberd (Swallow's Fortress) Park and has two sections: massive basalt slabs bending in grief over an eternal flame, and a pointed mast rising high, symbolizing the rebirth of the Armenian people. Integral to the memorial is also an underground museum with library and archive sections. A basalt wall bears the inscriptions of the names of the regions, towns, and villages of historic Western Armenia (the regions where the tragedy occurred and which are now within Turkey). A central memorial structure consists of a circular area that shelters the eternal flame memorializing all the victims of the tragedy; it stands as the Memorial Sanctuary. The eternal flame is housed under 12 tall, inward-leaning basalt slabs forming a circle. The shape of these walls simulates traditional Armenian *khatchkars* (cross-bearing carved memorial steles covered with rosettes and botanical motifs).

The official position of the Turkish government and a dominant opinion of Turkish society are that losses sustained by ethnic Armenians in Turkey did not amount to genocide. Rather, they were integral to World War I, in which Armenians and Turks found themselves on different sides of the frontline. Consequently not only Armenians sustained losses but Turks as well. The Turkish educational system continues to offer this alternative view of the events in its public schools and through many of its governmental websites. In recent years, the Armenian question has been increasingly discussed in Turkey at conferences and universities. The law does not prevent debates on the topic; freedom of speech and freedom of thought are guaranteed by Turkish law. However, due to the nature of Article 301 of the Turkish Penal Code, people claiming an Armenian Genocide can still be accused of "insulting Turkishness." Over 80 authors have faced prosecution for "insulting Turkishness." Orhan Pamuk, a famous Turkish novelist and Nobel Prize winner, was one of those against whom criminal charges have been pressed for his outspoken comments on the Armenian Genocide.

Two practical implications of tense Armenian-Turkish relations are that no diplomatic relations exist between Armenia and Turkey and that the Armenian-Turkish border has so far remained closed—the latter also relates to Armenian-Azerbaijani relations (see the following), since Turks see Azeris as close relatives (indeed, the languages of Turks, Azerbaijani, and Turkmen are close and mutually intelligible). As a result, Armenia can be reached only through its borders with Georgia and Iran.

Some promising signs appeared in September 2008, when a Turkish president visited Armenia for the first time in history.

The visit of the head of Turkish state (Abdullah Gul) was formally associated with a qualifying game for the 2010 world soccer cup between the national teams of Armenia and Turkey. Security for Gul's trip was tight. Attack helicopters escorted his jet on arrival and hundreds of demonstrators lined the streets of Yerevan. The two presidents, Gul of Turkey and Sargsyan of Armenia, who attended the game together, expressed hope at their post-game meeting. After receiving Gul, Armenian President Serzh Sargsyan told a news conference, "We hope we will be able to demonstrate goodwill to solve the problems between our countries and not transfer them to future generations." Gul, alongside Sargsyan, said he was "leaving optimistic." He told reporters after the game, which Turkey won 2–0, "If we create a good atmosphere and climate for this process, this will be a great achievement, and will also benefit stability and cooperation in the Caucasus." Sargsyan said he would attend the return match in October 2009 and that the invitation to do so suggested Gul "also has some expectations that there will be some movement between these two meetings." (Reuters: www.alertnet.org/thenews/newsdesk/L6345007.htm).[1] Indeed, on October 11, 2009, in Zurich, Switzerland, Armenia and Turkey signed landmark protocols aimed at ending their long-time hostilities and normalizing bilateral relations. The protocols were signed by Armenian Foreign Minister Edouard Nalbandian and his Turkish counterpart Ahmet Davutoglu. According to the protocols, the two countries will establish diplomatic ties and open long-sealed common borders. They will also try to solve their historical dispute over the massacres of Armenians under Ottoman rule. But the two countries' parliaments have to ratify the protocols before they can take effect, and this will not be an easy task on either side. Those present at the signing ceremony included such heavyweights as U.S. Secretary of State Hillary Clinton, Russian Foreign Minister Sergei Lavrov, French Foreign Minister Bernard Kouchner and the European Union's foreign policy chief Javier Solana. Switzerland has acted as mediator in the process to normalize bilateral relations between Armenia and Turkey for over a year.

ARMENIA AS A SOVIET REPUBLIC

Although the Russian army gained most of Ottoman (Western) Armenia during World War I, their gains were lost with the Communist Revolution of 1917, the civil war of 1918–1921, and intervention of the Entente forces (Britain and France) on the anti-Communist side. In April 1918, Russian-controlled Eastern Armenia, Georgia, and Azerbaijan attempted to merge into the Transcaucasian Democratic Federative Republic. This federation, however, only lasted one month, till May 1918 when all three parties decided to dissolve it. As a result, Eastern Armenia became independent as the Democratic Republic of Armenia (DRA) in May 1918. The DRA's short-lived independence was fraught with war and a mass influx of refugees from Western or Ottoman Armenia. Still, Britain and France, shocked by the actions of the Ottoman government, sought to help the newly-found Armenian state through relief funds. At the end of the war, Britain and France attempted to divide up the Ottoman Empire. Signed between the Allied and Associated Powers and Ottoman Empire at Sèvres, France, in August 1920, the Treaty of Sèvres included a commitment to sustain the DRA and to attach the former territories of Ottoman Armenia to it. On the basis of the Treaty of Sèvres, the Turkish national movement (The Young Turks) under Mustafa Kemal Atatürk declared itself the legitimate government of Turkey and replaced the monarchy centered in Istanbul with a republic centered in Ankara.

However, weeks before the Treaty of Sèvres was signed, war broke out between the Turkish nationalist forces and the DRA. According to one interpretation, Turkish nationalist forces started the war by invading the DRA in July 1920. The alternative explanation of the beginning of hostilities is that the DRA initially moved its forces into Oltu, a district of Erzurum province of Turkey. Anyway, Turkish forces seized Armenian territories (with the cities of Kars and Ardahan) that Russia annexed in the aftermath of the 1877–1878 Russo-Turkish War and occupied the city of Alexandropol (present-day Gyumri). The war ended with the Treaty of Alexandropol (December 2, 1920), which forced the DRA to cede more than half of its pre-war territory, including the lands of western or Ottoman Armenia granted to the DRA by the Treaty of Sèvres just three months earlier. To make matters worse, the Soviet Red Army invaded Armenia in late November 1920 and within days entered Yerevan. The short-lived Armenian republic collapsed.

Thus Turkey and Russia proceeded to control two parts of historical Armenia. Out of these two mutually hostile powers, Russia was seen as preferable, as no hostile activity even remotely reminiscent of Turkey's attitude had ever been practiced by Russians in regard to Armenians as an ethnic group. Moreover, in the eyes of many Armenians Russians were fellow Christians and therefore saviors. Also, it is within a Russia-dominated polity that the Armenian national home was to be recreated; apart but still truly Armenian land.

Many Armenians subsequently took active parts in Soviet economic and political campaigns and many became world famous as Soviet citizens. Those of note are Anastas Mikoyan (1895–1978), a Soviet statesman and Politburo member; Artem Mikoyan (1905–1970), his brother and one of two chief designers of the Mikoyan-Gurevich, better known as MIG, aircraft; Aram Khachaturyan (1903–1978), a world-famous composer, and many others.

From 1922 to 1936, Armenia was part of the Transcaucasian Soviet Federative Socialist Republic or TSFSR), but Armenia retained its separate jurisdiction within it, as did Georgia and Azerbaijan. Areas assigned to each of these entities were in most, but not all cases, based on ethnic majorities of their populations. There were some notable exceptions to that rule, and all of them had to do with Armenians. Thus, Nakhchivan region with a significant Armenian population (up to 45% of the total at the beginning of the twentieth century) became politically part of Azerbaijan even though in regard to Azerbaijan, Nakhchivan is an exclave, not contiguous with Azerbaijan proper. And yet no symmetrical solution was applied to Nagorno-Karabakh*: with its overwhelming Armenian majority, this region was not assigned to Armenia, even though it could have become an Armenia-subordinated exclave just like Nakhchivan is to Azerbaijan. Likewise, the Javakh or Javakheti region became part of Georgia despite an Armenian majority in the region. When the TSFSR was disbanded and Georgia, Armenia, and Azerbaijan became full-fledged union republics, these territorial arrangements were not altered. Although they probably had more to do with existing economic ties than with covert discrimination against the Armenians, they planted the seeds of future conflicts.[2] It is worth remembering, though, that at the turn of the century Tbilisi and particularly Baku were significant economic centers, while Yerevan could not possibly rival them, and so the area gravitating to Yerevan was set small.

While this may sound unorthodox, Armenia became one of the Soviet Union's success stories in terms not only

---

*Because Nagorno-Karabakh is an internationally recognized part of Azerbaijan, its location is shown on the corresponding map on p.143 as part of the country report devoted to Azerbaijan. Note that all place names within Nagorno-Karabakh have their Azeri and Armenian versions, and only the former version is reflected on that map. For example, the Armenian name of Nagorno-Karabakh's capital is Stepanakert, whereas it is shown as Xankandi on the map on p.143. Because Nagorno-Karabakh is *de facto* under the Armenian control and also because Stepanakert was the city's name throughout the Soviet period, it is in fact a more well-known name than Xankandi.

of economic development. Despite all the ingrained limitations of central planning and lack of political freedoms, the worst trauma in Armenian history was left behind. For the first time in modern history Armenians could build their national home in relative calm. Significant investment was assigned to a public education system, in large measure using the Armenian language, and to health care. Over the course of the Soviet period, Armenia enjoyed economic growth twice as rapid as the Soviet Union at large. New energy producing installations included hydroelectric power stations on the Razdan and Vorotan rivers; the Yerevan thermal energy plant; and ultimately the Metzamor nuclear power plant. While today the nuclear station is considered inherently unsafe and will probably be closed in the near future, when the station was commissioned in the 1970s it was considered to be a major achievement. Industry of the Armenian SSR specialized on electric machinery, mechanical engineering, and electronics. All the industrial plants depended on suppliers in other parts of the Soviet Union, notably in Russia. Only copper mining and production could be fully sustained on a local basis due to the abundance of copper in Armenia as well as construction materials, notably, volcanic tuff and marble. Because there are few flatlands in Armenia, its demand for cereals and livestock was to a significant extent met by imports from other republics. In exchange, the Armenian SSR produced more than its "fair share" of brandy and wine. Many slopes were terraced to support orchards and vineyards. At picturesque Lake Sevan, several major resorts developed, as did a high quality ski resort at Tsakhkadzor. Practically all fuels were delivered to Armenia from the neighboring republic of Azerbaijan.

The city of Yerevan became a true symbol of the Armenian SSR's success. A small, provincial town with 29,000 people at the beginning of the twentieth century, it became a modern metropolis of 1.3 million despite complex topography, including a 500-meter range of elevations within the city and the deep Razdan gorge that cuts across the city. In the 1950s and 60s, thousands of ethnic Armenians were repatriated to the Armenian SSR. In 1946, the Supreme Soviet of the U.S.S.R. adopted a decree according to which Soviet citizenship was to be bestowed on the Armenian repatriants. While not all actually returned to Armenia permanently, it is important to realize that the recreation of the Armenian national home was an issue dealt with positively at the highest level of the Soviet regime. Indeed, the national monument commemorating the victims of genocide

was built during the heyday of the Soviet Union.

The end of the Soviet period was darkened by two tragic events of different natures: the disastrous December 1988 earthquake and the worsening of Armenian-Azerbaijani relations and beginning of the conflict over Nagorno-Karabakh.

The earthquake, with a magnitude of 7.2 on Richter scale, had its epicenter near the city of Spitak in northwestern Armenia. The quake, augmented by lack of proper seismic-resistant designs and by poor quality construction materials, resulted in the deaths of 25,000 people; more than half-a-million people lost their dwellings. The disaster assumed such a scale that the Soviet Union asked the United States for humanitarian assistance for the first time since World War II.

ARMENIAN-AZERBAIJANI RELATIONS AND THE WAR OVER NAGORNO-KARABAKH

Armenian-Azerbaijani relations are even more difficult to discuss than Armenian-Turkish relations. At least in the latter case there is some degree of international consensus as to who was the victim and who was the perpetrator, although the Turkish side always claims this consensus is the result of a coordinated plot of the wealthy Armenian diaspora in the West. In the former case even this degree of consensus is missing, although few sources would deny the fact of the Armenian pogrom in the Azeri city of Sumgait in 1988 or in the Azeri capital of Baku in 1990. And yet other events, like the treatment of the Azeri population in Armenia and, most importantly, the thorny and unresolved issue of Nagorno-Karabakh are subject to diametrically opposed interpretations.

Because this book also contains an essay on Azerbaijan, the Azeri perspective is related in that essay (see pp. 146–149). As for the traditional Armenian standpoint, there is hardly a distinction between Turks and Azeri. Over the course of several centuries Azeri Turks, who descended from Central Asian nomads, squeezed the autochthonous Armenians out of their land. Yet they were jealous of the Armenians' labor ethics, ethnic solidarity, and upward mobility. Whenever Armenians lived as an economically successful minority in Azeri-dominated communities this jealousy fueled anti-Armenian sentiment.

During much of the Soviet period, however, ethnic grievances were swept under the rug. The negative consequence of this was that differences could not be openly discussed. The positive result was that mutual hostility was kept in check.

A large part of the Armenian diaspora lived in Azerbaijan, where, according to the last Soviet census in 1989, there were almost 400,000 Armenians (compared to 475,000 back in 1979), including almost 200,000 living in the city of Baku. With the exception of Nagorno-Karabakh, a semi-autonomous region inside Azerbaijan, most Armenians in Azerbaijan were urbanites. In contrast, most Azeri in Armenia (161,000 in 1979 and 85,000 in 1989) were rural villagers. Today, there are just 30 Azeri in Armenia (!), according to the data from the UN High Commissar for Refugees. As for Azerbaijan, in 1999 there were just 645 Armenians, with 120,000 more in Nagorno-Karabakh, which is internationally recognized as an integral part of Azerbaijan but is de facto controlled by Armenia. Almost no Armenians remain in Nakhchivan, an Azeri exclave which used to have a large Armenian population. This is a quite remarkable case of mutual disengagement preceded by centuries of living in mixed communities. So, what was the course of events that led to this disengagement?

As mentioned above, Nagorno-Karabakh existed as an enclave with an Armenian majority (140,000 people) and Azeri minority (48,000 people) and very few members of other ethnicities. The enclave was part of Azerbaijan, although one stretch of its western border is just four kilometers from Armenia. Armenians have long believed that the Baku-based government conducted discriminatory policies towards Armenians in Nagorno-Karabakh (and beyond); and this, among other things, is fraught with the potential of altering the Azeri-Armenian ratio in the enclave. Indeed, during the post-World War II decades, the share of Azeri in Nagorno-Karabakh's population steadily increased.

Beginning in 1985, Soviet President Gorbachev's policy of openness unleashed long suppressed hostility between Armenia and Azerbaijan. In late 1987, the Armenians of Nagorno-Karabakh collected over 80,000 signatures to petition the Soviet government about the secession of the region from Soviet Azerbaijan and its reunification (Miatsum, in Armenian) to Armenia. This petition was categorically disapproved by the government and public opinion in Azerbaijan. In January and February 1988, reports of anti-Azeri violence in Armenia reached Baku as did a report about an Azeri-Armenian skirmish in the Azeri town of Agdam, in which two Azeris were killed. Following these reports, in February, a rally gathered at the central square of the Azeri city of Sumgait. There are conflicting reports about the role of the city's official leaders in the massacre of the Armenian minority residing in that city. This is because often, the rioters knew where Armenians lived and forced their way into the apartments. There are hundreds of media and eye-witness reports of atrocities committed by the frenzied mobs, including gang rape and beatings with icepicks, pipes, and pieces of sharpened metal. The official death toll released by the Procurator General (tallies were compiled based on lists of named victims) was 32 people (26 Armenians and 6 Azeris). However, eyewitnesses reported a much larger number. Many insist that at least 200 people were killed.

In February 1988, Nagorno-Karabakh called on the Armenian and Azerbaijani Supreme Soviets (parliaments) to approve the transfer of the region to Armenia. With no official approval received, Karabakh seceded from Azerbaijan on July 12, 1988. Azerbaijan declared the act illegal according to the Soviet Union's Constitution, which stated that the borders of a republic could not be changed without its consent. Moscow imposed martial law in some areas in September and deployed Interior Ministry troops in November and army troops in May 1989. The Azerbaijan Popular Front (PF) began a rail blockade of Armenia and Karabakh, restricting food and fuel deliveries. In November–December 1989, massacres of Armenians occurred in several cities of Azerbaijan. Following these events, most Armenian residents of Baku began to leave the city, relocating with their relatives in different parts of the Soviet Union and/or becoming de facto refugees. By the beginning of 1990, only 30,000 to 40,000 Armenians, mostly women and retirees, remained in Baku (out of the population of about 200,000 recorded in the 1989 census).

On January 13, 1990, the Armenian pogrom in Baku began. According to Armenian sources, about 300 people were killed; other sources report lower numbers. Sources differ as to what immediately sparked the massacre, but at that time tensions between Armenians and Azerbaijani had reached such a level that virtually any piece of news or public statement could have sparked violence. The massacres lasted until January 19, at which point the Soviet army entered the city. Most victims died from beating and knife wounds. Many Armenians, including the world's thirteenth chess champion, Garry Kasparov (who was born into a mixed Armenian-Jewish family in Baku), claimed that pogroms in Baku were well organized. Unlike a small town where people live in single family residences and where it is common knowledge where families of minorities live, Baku is a large city with about 1.8 million residents. In a Soviet city of that size, virtually everyone lived in multistory apartment buildings. That rioters nevertheless knew exactly which apartments were occupied by Armenians, claimed Kasparov, could only have meant that some municipal authorities supplied residency information to the rioters [www .yerkramas.org/news/2008-12-06-2489].

Meanwhile, the Nagorno-Karabakh conflict was developing. The August 1991 Moscow coup attempt ended the ambiguous Soviet role in the conflict, but also ended hope of an imposed settlement. In September, Moscow declared that it would no longer support Azerbaijani military action in Karabakh. Azerbaijan nullified Nagorno-Karabakh's autonomous status and declared direct rule on November 26, 1991. On December 10, 1991, a Karabakh referendum chose independence.

The breakdown of the Soviet army led to the nationalization, sale, and/or theft of arms from its installations and soldiers, enabling Armenians to obtain weapons. The December 1991 demise of the Soviet Union and the withdrawal of Soviet troops from Karabakh (completed in March 1992) further endowed Nagorno-Karabakh forces with arms and prompted an early 1992 Armenian offensive. The two-year long hostilities, whose detailed account may be found in a Congressional Research Service paper (www.au.af.mil/au/awc/awcgate/crs/ib92109.pdf), resulted in 30,000 deaths and the displacement of roughly one million people. The Armenian forces won the war, and so most (85%) of Nagorno-Karabakh fell under Armenian control, nominally under the control of the self-proclaimed state of Artsakh (the Armenian name for Nagorno-Karabakh). In addition, five adjacent districts of Azerbaijan proper and parts of two other districts are under Armenian control. The overall area under Armenian control accounts for about 20 percent of the internationally recognized territory of Azerbaijan. Most Azerbaijani residents of this area abandoned their homes and relocated to other parts of Azerbaijan. A cease-fire brokered by Russia went into effect in May 1994.

Despite the ceasefire, fatalities due to sporadic armed conflicts between Armenian and Azerbaijani soldiers continued. As of August, 2008, the United States, France, and Russia (the co-chairs of the OSCE Minsk Group) were mediating efforts to negotiate a full settlement of the conflict, proposing "a referendum or a plebiscite, at a time to be determined later," to determine the final status of the area, the return of some territories under Karabakh's control, and security guarantees. Ilham Aliyev, President of Azerbaijan, and Serzh Sargsyan, President of Armenia, traveled to Moscow for talks with Russian President Dmitry Medvedev on November 2, 2008. The talks ended with the three Presidents signing a declaration confirming their commitment to continue talks. The Armenian and Azerbaijani presidents have met twice since then, most recently (July 28, 2009) in St. Petersburg, but no solution of the Karabakh problem has yet been found.

The Armenian-Azerbaijani relations remain permeated by a mutual sense of grievance and animosity so passionate and so difficult to reconcile that very few cases in recent (post-World War II) history can offer a parallel. Perhaps the Israeli-Palestinian conflict and especially war in Bosnia and the relationships between Serbs, Croats, and Muslims in that republic furnish some similarities.

INDEPENDENT ARMENIA: POPULATION AND ECONOMY

Armenia's population grew rapidly albeit at a declining rate through the last four decades of the Soviet Union's existence. While natural increase in Armenia was lower than in Muslim republics, it was higher than in other non-Muslim parts of the Soviet Union. For example, in 1983, it was as high as 1.78 percent. Net migration was also positive mostly due to ethnic Armenians' migration from other republics of the Soviet Union. Some researchers believe that the 1989 census underestimated the population of Armenia by some 300,000 because it was conducted soon after the disastrous earthquake and quite a few people had temporarily left the republic (www.demoscope .ru/weekly/2003/0131/analit06.php). The official count of this last Soviet census for Armenia was 3.3 million. Because it was about the same or even smaller 19 years later [on April 1, 2008, the official population count in Armenia was 3.2 million (www.am.spinform.ru/people .html)], it is obvious that there has been significant outmigration from Armenia in the meantime. As to how significant this outmigration has been, there is no consensus, but the highest estimate (1.5 million) circulating in the media is most probably a significant overstatement. However, even half that much outmigration would be quite significant for a country the size of Armenia. The largest group of post-Soviet migrants from Armenia lives and works in Russia, but others have settled elsewhere, including Cyprus, Greece, France, etc.

People have been leaving due to a significant drop in living standards and narrowing employment opportunities. Lerman and Mirzakhanian (2001) reported that by the late 1990s, agriculture had jumped to

30 percent of the total GDP and 40 percent of total employment, whereas in 1991, the corresponding figures were 20 percent and 10 percent respectively. Coupled with the fact that the GDP contracted from 1991 to 1995 and fell nearly 60 percent from 1989 to 1993, this means that Armenia lost many industrial jobs. This was chiefly a result of the breakdown of former Soviet trading networks and the time it takes to develop new trading ties. Armenia's economy also suffers from the economic blockade by Azerbaijan and Turkey. In this regard, the 2007 inauguration of the first section of the Iran-Armenia pipeline was a major breakthrough for Armenia, which had previously received all its fuel from one source, Russia, still Armenia's foremost trading partner.

In spite of the Nagorno-Karabakh conflict, the government of Armenia has been able to carry out wide-ranging economic reforms that have paid off in dramatically lower inflation and steady growth. Armenia has registered strong economic growth since 1995, with double-digit GDP growth rates every year since 2002.

DEVELOPMENT

The structure of Armenia's economy has changed substantially since 1991, with sectors such as construction and services replacing industry and now also agriculture as the main contributors to economic growth.

The structure of Armenia's economy has changed substantially since 1991, with sectors such as construction and services replacing industry and now also agriculture as the main contributors to economic growth. The diamond processing industry, which was one of the leading export sectors in 2000–2004 and also a major recipient of foreign investment, has faced a dramatic decrease in output since 2005 due to raw material supply problems with Russia and an overall decline in the international diamond market. Other sectors driving industrial growth include energy, metallurgy, and food processing.

According to the U.S. State Department's background notes, Armenia has been more successful than Azerbaijan in settling displaced people. Roughly 4,700 of 360,000 ethnic Armenians who fled Azerbaijan since 1988 remain refugees, while approximately 572,000 of the estimated 800,000 ethnic Azeris who fled during the Karabakhi offensives still live as internally displaced persons in Azerbaijan

(according to the Internal Displacement Monitoring Centre, quoting Azeri government statistics, June 2008) (www.state.gov/r/pa/ei/bgn/5275.htm).

ARMENIAN DIASPORA

A diaspora is "a people with common national origin who reside outside a claimed or an independent home territory. They regard themselves or are regarded by others as members or potential members of their country of origin (claimed or already existing), a status held regardless of their geographical location and citizen status outside their home country" (Shain 1994). Until recently, only Jewish, Greek, or Armenian groups were referred to as diasporas.

The existence of an Armenian diaspora goes back as far as the end of the fourteenth century and, according to some scholars, even earlier. Thus, according to Tölölyan, "the first Armenian diaspora communities emerged in the eleventh century in the Crimean peninsula (now in Ukraine) and reached the peak of their prosperity in the fourteenth and seventeenth century in what are now Poland, Ukraine and Moldova; over time others developed in the adjacent territories of what are now Romania, Hungary and Bulgaria. . . ." Tölölyan also mentions several waves of Armenian migration outside the homeland, mostly because of power struggles between dominant powers on Armenian territory.

However the Armenian diaspora only grew to noteworthy size after World War I as a result of Ottoman deportations of the Armenian population. Initially, most Armenians displaced by Turks found themselves either in Russia (or Russian-controlled areas) or in the Middle East. Most Armenian migrants who found themselves in Russia stayed there, because during much of the Soviet period emigration was disallowed; but quite a few members of Armenian communities in the Middle East (in Lebanon, Syria, Iran, and Iraq) subsequently migrated to other countries, most notably to France and the United States. Today, estimates of the overall size of the Armenian diaspora are between seven and nine million people. That means that many more Armenians live outside than inside Armenia. The largest Armenian diaspora is in Russia (about one million people) and the second largest in the United States (about 800,000 people and up to one million, according to some sources). In the views of some observers, it is the successful lobbying activity of the Armenian National Committee of America that explains why the Republic of Armenia is the largest per capita recipient of American

foreign aid (www.anca.org/press_releases/press_releases.php?prid=1584). In Western Europe the greatest number of Armenians live in France, where there are more than 300,000. A large (some 250,000 people) Armenian community exists in Georgia, although some Armenian scholars believe that naming the Armenian community in the Georgian region of Javakh (adjacent to Armenia) a diaspora is a mistake. Then, of course, there is the large Armenian community of Artsakh or Nagorno-Karabakh. In the Middle East, Iran and Lebanon each has close to 200,000 Armenians. Between 40,000 and 70,000 Armenians remain in Turkey, tellingly not in historical Armenia, but in Istanbul. In addition, the existence of several hundred thousand crypto-Armenians (that is, people with Armenian roots who hide or suppress their identity) is suspected in northeastern Turkey.

In a 2009 article by Bahar Baser and Ashok Swain in the online journal *Caucasian Review of International Affairs* the point is made that at times the Armenian diaspora adopts and promotes policies that do not benefit the Republic of Armenia. The authors are referring to the rejection by the diaspora of any gesture that might lead to reconciliation with Turkey. Leaders of the Armenian community in Turkey, such as the late Hrant Dink and Etyen Mahcupyan, were reportedly critical of that kind of radicalism. So too is Mesrob II, the eighty-fourth patriarch of Turkey's Armenian Orthodox community, who has said that the Armenian Genocide Resolution pending in the U.S. Congress would have a negative effect, as it would disrupt both the relations between Turks and Armenians in Turkey and between Turkey and Armenia. "Disrupting relations," however, may come across to some as a weak argument considering the antagonism of some Turks against Armenians in Turkey. This was vividly displayed in the tragic January 2007 death of Hrant Dink at the hands of a young Turkish radical. The message that this assassination sends is that even a devout Armenian advocate of Turkish-Armenian reconciliation who is ready to make important concessions is not acceptable to extremist Turkish nationalists. To their credit, however, about 100,000 mourners, mostly ethnic Turks, marched in Istanbul in protest of the assassination, chanting "We are all Armenians" and "We are all Hrant Dink."

GOVERNMENT AND POLITICS

Considering the precarious geopolitical situation of Armenia whereby the borders with two of its three neighbors are blocked, Armenia has precious little freedom of maneuver in choosing its allies. Under these circumstances ties with Iran appear vital as are traditional ties with Russia. In 1997, Armenia and Russia signed a far-reaching friendship treaty, which calls for mutual assistance in the event of a military threat to either party and allows Russian border guards to patrol Armenia's frontiers with Turkey and Iran. Russia has a military base near the Armenian city of Gyumri. After the removal of Russian bases in Georgia and the 2008 Russian-Georgian war, the significance of that base has increased in Russia's eyes. A bilateral treaty provides for the Russian military occupation of the base for 25 years; but Armenian authorities have said that, if needed, this time-frame can be reviewed and prolonged. Russia does not pay the Armenian government for use of the Gyumri military base, and Armenia maintains all public utilities, such as water, electricity, etc.

[The following paragraphs are from the US State Department's Background Notes (www.state.gov/r/pa/ei/bgn/5275.htm)]

Armenians voted overwhelmingly for independence in a September 1991 referendum, followed by a presidential election in October 1991 that gave 83 percent of the vote to Levon Ter-Petrossian. Ter-Petrossian had been elected head of government in 1990, when the Armenian National Movement defeated the Communist Party. Ter-Petrossian was re-elected in 1996 in a disputed election. Following public demonstrations against Ter-Petrossian's reportedly weak policies on Nagorno-Karabakh, the President resigned under pressure in January 1998 and was replaced by Prime Minister Robert Kocharian (who was from Nagorno-Karabakh). Kocharian was subsequently elected President in March 1998. Following the October 27, 1999 assassination in Parliament of Prime Minister Vazgen Sargsyan, Parliament Speaker Karen Demirchian, and six other officials, a period of political instability ensued during which an opposition headed by elements of the former Armenian National Movement government attempted unsuccessfully to force Kocharian to resign. Riding out the unrest,

Kocharian was later reelected in March 2003 in a contentious election that the Organization for Security and Cooperation in Europe (OSCE) and the U.S. government deemed to have fallen short of international standards.

As a result of the May 2007 parliamentary elections, and with the February 2008 decision by the Country of Law Party to join the governing coalition, 113 seats out of the 131 in the National Assembly are held by pro-government parties. The sole opposition faction in parliament, the Heritage Party, holds seven seats. The remaining members of parliament are independent, although most of these are aligned de facto with the pro-government parties. The unicameral National Assembly has 90 seats which are elected by proportional representation (party list), and 41 are single mandate districts.

The government of Armenia's stated aim is to build a Western-style parliamentary democracy as the basis of its form of government. However, international observers have been critical of the conduct of national elections in 1995, 1999, and 2003, as well as the constitutional referendum of 2005. The new constitution in 2005 increased the power of the legislative branch and allows for more independence of the judiciary; in practice, however, both branches remain subject to political pressure from the executive branch, which retains considerably greater power than its counterparts in most European countries.

Armenia held presidential elections on February 19, 2008. The elections, while originally deemed by the OSCE's Office for Democratic Institutions and Human Rights (ODIHR) to be "mostly in line" with OSCE standards, were later seen to be marred by credible claims of ballot stuffing, intimidation (and even beatings) of poll workers and proxies, vote buying, and other irregularities. Recounts were requested, but ODIHR observers noted "shortcomings in the recount process, including discrepancies and mistakes, some of which raise questions over the

impartiality of the [electoral commissions] concerned."

Mass protests followed the disputed vote. For 10 days, large crowds of pro-opposition demonstrators gathered in Yerevan's downtown Freedom Square. Police and security forces entered Freedom Square early in the morning of March 1, 2008, ostensibly to investigate reports of hidden weapons caches. This operation turned into a forced dispersal of demonstrators from Freedom Square by massed riot police. Following the clearing of Freedom Square, clashes erupted in the afternoon between massed demonstrators and security personnel, and continued throughout the day and evening, leading to ten deaths and hundreds of injuries. President Kocharian decreed a 20-day state of emergency in Yerevan late on March 1, which sharply curtailed freedom of media and assembly. Dozens of opposition supporters were jailed in the wake of the violence, in proceedings that many international watchdog groups have criticized as politically motivated. Armenia's media freedom climate and freedom of assembly remained poor overall, though somewhat improved after the state of emergency was lifted. Serzh Sargsyan took office as President in April 2008.

HEALTH/WELFARE

The government of Armenia has been making concerted efforts to increase public expenditure on health. In recent years there have been steady improvements in this area, however public expenditure channeled from the state budget to the health sector as a percentage of GDP still falls short. In 2006 the Government of Armenia spent 1.64 percent of its GDP on the health sector (Source: Armenia Mid-Term Expenditure Framework 2007–2009).

The National Assembly launched a parliamentary ad hoc commission tasked with an inquiry into the events of March 1–2. The ad hoc commission showed early promise, despite concerns about its pro-government composition. The commission members summoned senior government officials to testify in public hearings, and subjected them to probing questions. This effort was expanded by a presidential directive on October 23, 2008 that formed an independent fact-finding group tasked to support and report to the ad hoc commission. It was composed of members appointed in equal numbers by ruling and opposition parties. These initiatives to uncover the truth about the March 1 events have been welcome, albeit imperfect, steps to provide public accountability.

FREEDOM

Armenia's media freedom climate and freedom of assembly has somewhat improved since the March 1, 2008 demonstrations and their subsequent dispersal took place in Freedom Square.

NOTES

1. However, such is the climate of Armenian-Turkish relations that shortly after Gul's visit he was accused by a member of the Turkish parliament of having Armenian roots on his mother's side. In his response, Gul asserted that he "respects the ethnic background, different beliefs and family ties of all [his] citizens." He also furnished assurances of the indisputable Turkishness of his maternal roots. That it is perceived as defamatory to call someone an Armenian in Turkey testifies to the fact that the normalization of Armenian-Turkish relations is not going to be easy (www.forum.hyeclub.com/showthread.php?t=12228). In all fairness, though, being called a Turk in Armenia is equally bad.

2. It is important to remember, though, that numerical majority cannot *always* underlie demarcation of borders. Thus, at the turn of the century ethnic Armenians outnumbered Georgians in the Georgian capital of Tbilisi and almost equaled Azerbaijani in Baku (now the capital of Azerbaijan) wherein both Azerbaijani and Armenians were dwarfed by Russians. But Armenians who lived in those cities were part of the Armenian diaspora; for the most part they or their ancestors migrated to those cities from western Armenia. This was much like the predominance of Jews in many Ukrainian and Belarusian towns. In all those cases ethnic frontiers were perhaps rightly drawn on the basis of the rural population's ethnic affinities. As a matter of fact, in the beginning of the twentieth century, Azerbaijanis outnumbered Armenians even in Yerevan.

Azerbaijan

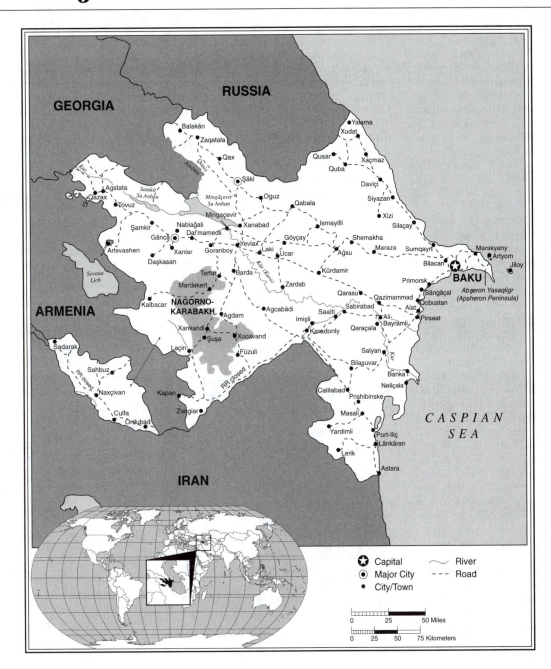

Azerbaijan Statistics

Area in Square Miles (Kilometers):
 33,436 (86,600)
Capital (Population): Baku (1,917,000)

Population

Total (millions): 8.8
Per Square Km: 101
Percent Urban: 52
Rate of Natural Increase (%): 1.2

Ethnic Makeup: Azeri 90.6%; Dagestani
 2.2%; Russian 1.8%; Armenian 1.5%;
 other 3.9% (1999 census); Note:
 almost all Armenians live in the
 separatist Nagorno-Karabakh region

Health

Life Expectancy at Birth: 70 (male);
 75 (female)

Infant Mortality Rate: 11
Total Fertility Rate: 2.3

Economy/Well-Being

*Currency ($US equivalent on 14 October
 2009):* 0.8 manat = $1
2008 GNI PPP Per Capita: $7,770
*2005 Percent Living on Less than US$2/
 Day:* <2

Azerbaijan Country Report

Along with Armenia and Georgia, the Republic of Azerbaijan is part of the Trans-Caucasus, a region squeezed between the Caucasus mountain range and the southern border of the former Soviet Union, of which the Trans-Caucasus used to be part. Unlike Armenia and Georgia, both overwhelmingly Christian and proud of having introduced Christianity as the state religion earlier than most states, Azerbaijan has a Muslim majority, although it is now a secular republic with separation between religion and state. With a land area of 86,600 sq km Azerbaijan is the largest republic of the Trans-Caucasus. Its population of 8.8 million people is also the largest in the region, although there are reasons to believe that this population number may be significantly overstated (see the following). Part of Azerbaijan is the Nakhchivan exclave (5,400 sq km and almost 0.4 million people) separated from the Azerbaijani "mainland" by Armenia. The city of Baku (population about 2 million) is the capital of Azerbaijan.

As an ethnic community, the Azerbaijani or Azeri people developed as the result of a merger between the autochthonous Caucasian Albanians—a Christianized group, and Turkic Muslim migrants from Central Asia, who since the eleventh century have been assimilating local groups. However, while linguistically the Azeri people are the closest relatives of the ethnic majority in Turkey—so much so that the late Azerbaijani leader Heydar Aliyev spoke of "one people and two states"—culturally the Azeri people are equally close to Iran. For most of its known history, Azerbaijan was ruled by Iran. Consequently, the Azeris or Azeri Turks are Shiite Muslims in contrast to Turks in Turkey, who are Sunnis. In fact, the Republic of Azerbaijan occupies only the northern part of the Azerbaijani ethnic field. Many more Azeri people live in northwestern

(© Farhad Kalantari)
Azneft square in Baku at night.

Iran and account for perhaps 25 percent of the entire population of Iran. Even the current Supreme Leader of Iran, Ayatollah Ali Khamenei, is half Azeri. However, during several decades of living in such different polities as the Soviet Union and Iran, the Azeri peoples on the two sides of the border drifted apart; and relatively few contacts between both parts of the Azeri ethnicity were maintained, this despite the existence of an Iranian consulate in the Azerbaijani capital of Baku throughout the entire post-World War II period. Thus, in Iran, Perso-Arabic script is used to write the Azeri language, whereas in the Republic of Azerbaijan the Latin script is now in use; and formerly, from 1939 to 1991, when Azerbaijan was part of the Soviet Union, Cyrillic script was used. Another important divide between northern and southern Azerbaijan is in the share of people practicing Islam. Obviously, in the Islamic Republic it is significantly higher than in the former Soviet republic.

FROM THE INCORPORATION INTO THE RUSSIAN EMPIRE TO THE 1917 COMMUNIST REVOLUTION

In the eighteenth and the early nineteenth centuries, the territory of today's Republic of Azerbaijan consisted of feudal states, so-called khanates. With the passage of time these khanates were gaining more and more independence from the Persian monarchy. The process of incorporation of the khanates into the Russian Empire lasted from 1803 until 1828 and followed two Russo-Persian wars. Most Azeris had been peasants working for their landlords. Unlike in Russia proper, where the government since 1861 lent peasants money to buy land from large estate owners, in Azerbaijan, feudal relationships with limited release of local labor for the needs of industry lasted until the 1917 Communist revolution and beyond. This explains, in part, why the local oil industry depended to a significant degree on a migrant labor force that included Armenians and Russians.

The development of the oil industry around the city of Baku in the last quarter of the nineteenth century was the single most momentous event for Azerbaijan. Although the real oil boom started in 1871, its extraction for local needs had been practiced for two centuries prior to that, and knowledge of oil seepage on the Absheron peninsula (where the city of Baku is located) existed for much of

recorded history. After the Baku khanate became part of the Russian Empire (1803), commercial oil extraction was based on the so-called *otrupnaya sistema,* that is, a system of leasing out an oil well for four years with fixed payments to the state treasury. This system restrained the development of oil industry, yet it continued with several modifications until the early 1870s. After it was abolished, the Absheron oil lands were parceled out to local and Russian-born investors.

[The following six paragraphs are from "The History of Oil in Azerbaijan" by Natig Aliyev (1994) at www.azer.com/aiweb/categories/magazine/22_folder/22_articles/22_historyofoil.html]

In most oil-extracting countries, the oil-industry dates back to the moment mechanical boring equipment begins operations. The first oil well in the world was drilled in Baku in 1847 at Bibi-Eybat oil field under the direction and initiation of Russian Engineer Semenov. In 1850, oil extraction in the world had reached about 300 tons. By 1881, it had grown to 4.4 million tons. By 1891, it was 22.5 million tons, of which 9.5 million tons came from the United States and 11.4 million tons from Russia, of which 95 percent was extracted from Azerbaijan.

By the end of the nineteenth century, Baku's fame as the "Black Gold Capital" spread throughout of the world. Skilled workers and specialists flocked to Baku. By 1900 the surroundings of Baku had more than 3,000 oil wells of which 2,000 were producing oil at industrial levels.

Foreign companies were particularly active in the oil business. One of these companies was created by Ludwig and Robert Nobel. In 1873, they established the Branobel or Nobel Brothers Oil Extracting Partnership. The company brought the first tanker to the Caspian Sea, in order to reduce transport expenses. The ship, which they named Zoroaster, was designed and built in Sweden at the Motall Shipbuilding Factory in 1877. Because of the success of that first tanker, the Nobel Brothers built an entire fleet of tankers, giving names to the ships such as Moses, Spinoza, and Darwin. The Nobel Brothers also were first to introduce railway tanks for oil transportation. In 1878 they built a pipeline which reduced the expenses of transportation by five times and paid for itself within a single year. The Rothschild Company and Shell lead by Samuel Markus were also involved in oil production in Baku. More than 50 percent of the oil extraction and 75 percent of the oil production commerce were held by these three foreign companies.

Oil turned Baku into a center of world oil commerce and enabled it to exert an incredible influence on the economic development in the entire region of the Caucasus. In 1897–1907, the longest pipeline in the world (at the time) was built from Baku to Batumi on the Black Sea Coast, a distance of 883 kilometers. The diameter of the pipeline was 200 mm and it was equipped with 16 pumping stations.

Prior to that, in 1883 a railway line from Baku to Tbilisi was commissioned, enabling the oil to be transported by trains. The oil business generated a flurry of bank activity and various financial societies and organizations were created. In 1884, the oil barons in Baku established their own organization, the Council of Oil Producers, for the discussion of oil business. They had their own magazine, *Oil Business,* a library, school, hospital, and pharmacy. For six years, the Council of Oil Producers was directed by Ludwig Nobel (1884–1890).

The oil industry greatly influenced the architectural appearance of Baku as a modern city. Administrative, social, and municipal institutions were established which, in turn, made decisions about the city's illumination, roads, streets, buildings, telephone stations, and horse-drawn trolleys. Gardens and parks were laid out and hotels, casinos, and beautiful stores were built. As the largest industrial center of the Trans-Caucasus, Baku saw the rise of a radical anti-tsarist political

left, which presided over large-scale strikes by the Baku working class in 1905–07 and thereafter.

Although the oil industry developed on Azerbaijani land, it was in many ways an extraterritorial phenomenon, as ownership of oil wells, proceeds from oil extraction, and even to some extent employment in that industry were tied to the outside world. That doesn't necessarily mean "foreign" in a strict sense, since all the workers were Russian citizens, but many had roots outside Azerbaijan itself. This was reflected in the ethnic makeup of the population of Baku. According to *Granat* (a Russian pre-revolutionary encyclopedia), in 1903, of the 155,000 residents of Baku, Azeris accounted for just 36.4 percent, Russians 33.9 percent, and Armenians 17 percent. The socioeconomic development contrasts between Baku, a large city with capitalism in industry, and the rest of Azerbaijan, with its medieval agriculture, was stunning, a situation that lingered well into the Soviet period and even beyond.

AZERBAIJAN AS A SOVIET REPUBLIC

After the collapse of the Russian Empire during World War I, Azerbaijan, together with Armenia and Georgia, became part of the short-lived Trans-Caucasian Democratic Federative Republic. When the republic dissolved in May 1918, Azerbaijan declared independence as the Azerbaijan Democratic Republic (ADR). This declaration was made in Tbilisi (Georgia) because the city of Baku was at the time governed by a revolutionary Soviet, in which a major role was played by non-Azeris (mostly ethnic Armenians, Russians, and Jews). The leader of the Baku Soviet was Stepan Shaumyan, an ethnic Armenian. The ADR solicited the support of Turkey. Under threat of Turkish military intervention, the Baku Soviet distanced itself from local ethnic Azeri organizations, particularly from Musavat, a political party dedicated to autonomy for Azerbaijan. In March 1918, clashes between local Azeris and armed forces loyal to the city Soviet took place in which from 3,000 to 12,000 Azeris were killed. Later (in July 1918) when a Turkish military contingent approached the city, the members of the Baku Soviet fled Baku on a ship bound to Astrakhan. Twenty six of those Bolshevik leaders were captured by a joint force of the Russian (anti-Bolshevik) Provisional Government and the British military mission and shot by a firing squad on the eastern coast of the Caspian Sea (modern day Turkmenistan). When on September 15, 1918, the combined forces of the ADR and Turkey

captured Baku, an Armenian massacre followed which was perceived as revenge for the March massacre of the Azeris. After the Soviet regime was reinstalled in Azerbaijan, the 26 Baku commissars were honored as martyrs by the Soviet government, which reburied them in a central Baku park in 1920. During the entire Soviet period, a cult of the 26 Baku Commissars, assassinated by counter-revolutionaries, existed. They were celebrated in every history textbook, and a major memorial to them was built in Baku. Incidentally, only two of them (Mir-Hassan Vezirov and Meshadi Azizbekov) were ethnic Azeris.

The Bolshevik Red Army recaptured the city of Baku in April 1920 and established the Azerbaijan Soviet Socialist Republic. In 1922, Azerbaijan (along with Georgia and Armenia) became part of the Trans-Caucasian SFSR (TSFSR), a constituent member of the newly-established Soviet Union. In 1936, TSFSR was dissolved and Azerbaijan SSR became one of the constituent member states of the Soviet Union. During World War II, Azerbaijan supplied much of the Soviet Union's oil on the Eastern Front.

Although the significance of Azerbaijan as the source of oil for the entire Soviet Union declined in the wake of oil discoveries elsewhere (first between the Volga River and the Urals and then in Western Siberia), oil extraction continued to be the core of Azerbaijan's industry throughout the Soviet period. In Sumgait, an industrial satellite of Baku, iron and steel production based on iron ore from Dashkesan (in the western part of Azerbaijan) developed as did production of pipes for oil pipelines. An aluminum refinery using raw alumina found not far from the city of Gyandzha was also located in Sumgait. All of that made Sumgait an important industrial center with a heavily polluted environment. In agriculture, Azerbaijan became the only republic of the Trans-Caucasus with significant cotton cultivation. Vegetables, fruits, and wine were also produced in Azerbaijan and exported to other Soviet republics, notably Russia. As a source of power, oil was supplemented by hydroelectricity from three power stations on the Kura River, while the artificial lakes created by the three dams provided necessary water for irrigation, as much of Azerbaijan is semi-arid.

Throughout the Soviet period many Azeris made significant contributions to different areas of Soviet life, particularly culture. The operatic and popular singer Muslim Magomayev and two writers and film directors, Magsud and Rustam Ibragimbekovs, have become best known. Azerbaijan's most respected statesman, Heydar Aliyev, was not only born and raised during the Soviet period but was molded by such inherently Soviet establishments as the KGB, with which he was affiliated from the age of 18 until the age of 46 (1941–1969), and the Communist Party. Aliyev became First Secretary (Leader) of the Communist Party of Azerbaijan in 1969 and worked in that position until 1982, when he was moved to Moscow, where he became a full member of the Politburo. In 1987, he was forced to resign by Gorbachev, who apparently wanted to cleanse the Politburo of Brezhnev loyalists. Aliyev was definitely such a person. However, the 1987 ouster turned out to be but a temporary withdrawal of Heydar Aliyev from politics. He later reemerged as the leader of independent Azerbaijan.

Because in Ronald Reagan's terms the Soviet Union was "the evil empire," it has become habitual in America to see the Soviet impact on constituent non-Russian entities in a negative light. But that 70-year long stretch of history abounded in events which cannot be painted black. When some of the ethnic borderlands were incorporated into the Russian Empire, they experienced the various stages of socioeconomic development much earlier than the empire's heartland. Azerbaijan is a case in point. Unlike its neighbors Armenia and Georgia, which had been independent states in the past, Azerbaijan had almost no prior experience with independent statehood. By the beginning of the twentieth century, the national identity of the Azeri people had not yet fully developed, and, therefore the national movement was weak and enjoyed but meager grassroots support. If Russia was itself economically backward compared to West Europe, Azerbaijan was even more so. Torn between Iran and the Russian Empire, it had only those two polities as potential niches for its national home. It is, of course, debatable which of the two parts of Azerbaijan has ultimately fared better, the Iranian or the Soviet. However, in terms of literacy, secular education, and the social status of women, as well as in terms of living standards, the Soviet part of Azerbaijan had a noticeable edge. Throughout the Soviet period, the republic experienced significant progress in economic development as did other parts of the Trans-Caucasus; and, considering its very low starting point, progress was even greater than in the other places.

The end of the Soviet period brought about a still unhealed national trauma resulting from the Nagorno-Karabakh conflict and the eviction of Azeris from Armenia, Nagorno-Karabakh, and, subsequently, from seven rayons (districts)

adjacent to Nagorno-Karabakh. Today two countries and two ethnic groups, Azeris and Armenians, perceive each other as arch-foes despite having lived peacefully side by side for centuries.

AZERBAIJANI-ARMENIAN RELATIONS

According to a 1989 census, 390,505 Armenians lived in Azerbaijan, down from 475,486 in 1979 and from 483,500 in 1970. The largest groups resided in Baku (close to 200,000) and in the Nagorno-Karabakh autonomous region (in which Armenians in 1989 accounted for 76 percent of its total population of 192,000 people). In addition, Armenians were the ethnic majority in the Shaumyan rural district adjacent to Nagorno-Karabakh and were an important minority in each and every significant Azeri city. In Armenia, in 1989, there were 84,860 ethnic Azeri residents, down from 160,841 in 1979. Mixed marriages of Armenians in Azerbaijan and Azeris in Armenia were not rare but were not particularly numerous either. Therefore, the downward trend in population numbers did not result from mutual assimilation; rather, it reveals that the outflow of Armenians from Azerbaijan and Azeris from Armenia began well before the Armenian-Azerbaijani conflict came to a head.

The active phase of the conflict likely originated on February 20, 1988, when the elected ethnic Armenian deputies of the Nagorno-Karabakh Soviet solicited Moscow to change the political map of the Trans-Caucasus by incorporating Nagorno-Karabakh and the nearby Shaumyan district into Armenia (Azerbaijan would thus lose those areas). Moscow and Baku rejected the appeal, but the Armenian movement for Nagorno-Karabakh's secession from Azerbaijan had already gained momentum. Skirmishes between Armenian and Azeri youths took place in various locations, and some ethnic Azeris were forced to leave both Nagorno-Karabakh and Armenia and to settle as refugees in Sumgait and Baku. Massacres of Armenians on February 26–28, 1988, in Sumgait and on January 13–20, 1990, in Baku followed. These events essentially eliminated the Azeri minority in Armenia and the Armenian minority in Azerbaijan, because all the survivors left their places of residency. In January 1990, the Soviet military entered Baku and installed martial law. As a result of clashes with the residents of Baku, 168 people were killed, most of them Azeris. Even before the military intervention, one particular initiative of the Kremlin incensed many Azeris. Namely,

in January 1989, a State Committee, with Gorbachev loyalist Arkady Volsky as its head, was established to administer Nagorno-Karabakh directly from Moscow. Although the committee was disbanded as early as November of the same year, Azeris viewed that short-lived institution's function as preparation for the extraction of the Azerbaijani region from Azerbaijani jurisdiction forever.

During the military phase of the conflict over Nagorno-Karabakh (1991–1994), Azerbaijan lost close to 20 percent of its territory. Nagorno-Karabakh itself, five adjacent rural districts in their entirety, and parts of two other districts fell under Armenian military control. In 1992, the Armenian Karabakh military killed several hundred Azeri dwellers in the town of Khojaly located within Nagorno-Karabakh. All the Azeris from that region and adjacent rural districts of Azerbaijan proper who survived the war became refugees. Altogether close to 800,000 refugees now reside in various areas of Azerbaijan.

Formally speaking, the areas lost by Azerbaijan are under the control of Nagorno-Karabakh forces, but the Azerbaijani government does not recognize the Armenian state of Artsakh (which is what Armenians call Nagorno-Karabakh) and believes that Armenia and Azerbaijan are the only sides in the conflict. No country, not just Azerbaijan, not even Armenia, has recognized Artsakh to date. On May 5, 1994, due to Russia's mediation, an agreement between the warring parties was signed in Bishkek, Kyrgyzstan, to establish a truce along the frontline between the Armenian and Azeri forces so that peaceful means of solving the conflict could be pursued. With minor violations, the truce has been maintained since May 12, 1994.

Since that time, Azerbaijan has set the restoration of its territorial integrity as its major national goal. Several meetings between the leaders of Armenia and Azerbaijan and their foreign ministers have not resolved the problem. At this writing, the latest summit between the national leaders, Ilham Aliyev and Serzh Sargsyan, took place in St. Petersburg, Russia, in July 2009.

By some accounts Azeri identity has been strengthened by the conflict with Armenians and by the popular perception of the loss of 20 percent of Azerbaijani land as a national trauma. Needless to say, the Azeri perspective on the conflict and on the whole background and history of the Azerbaijani-Armenian tensions is worlds apart from the Armenian perspective. Of these two contrasting perspectives, the Azeri one is less known in the outside world. A 2006 Russian-language book, *The Karabakh Conflict: the Azeri View,* released by the Moscow publisher Yevropa, contains a systematic account of what prominent Azeris (those close to the government and those belonging to political opposition) think about the issue. The book also includes the full text of two official documents, the March 26, 1998 decree of President Heydar Aliyev of Azerbaijan about the genocide of the Azeri people and the subsequent appeal of the Milli-Mejlis, the Azeri parliament, to international institutions.

Aliyev's 1998 decree turns the tables on Armenian grievances. The authors of the decree borrowed language from those grievances, but the victim, in their view, is the Azeri ethnicity and the perpetrators are ethnic Armenians. More than nine-tenths of the decree is a preamble relating the official Azeri view of the history of Azeri-Armenian relations. The very end of the decree establishes March 31 as the day of Azeris' genocide at the hands of Armenians—in commemoration of the 1918 Baku massacre—and contains an appeal to the Milli-Mejlis to conduct a special session on Azeri genocide. It is, therefore, the preamble that presents most interest, as it describes a lasting Armenian and joint Armenian-Russian conspiracy to propagate a negative image of Azeris the world over and, above all, to cleanse the Azeri land of its rightful and legitimate owners in order to create a vital space for Armenians.

According to the decree, the course of tragic events goes back to the Russian-Persian treaties of 1813 and 1828, which split Azerbaijan between Iran and Russia. Soon thereafter, mass resettlement of Armenians from northeastern Turkey commenced to the Azeri-populated lands, including not only today's Republic of Azerbaijan but also much of today's Republic of Armenia (where, in the beginning of the twentieth century, Azeris also outnumbered Armenians); concomitantly, a false history of Armenians was concocted. The first clashes between the Armenian newcomers and Azeris, clashes in which thousands of Azeris were killed, occurred from 1905 to 1907. The epicenter of the clashes was the city of Baku. Acts of genocide were committed in March–April 1918 under the auspices of the Baku Commune, the Bolshevik-led local government, in which many ethnic Armenians were represented. According to the decree, thousands of Azeris were murdered by Armenians all across Azerbaijan; and this time the task of those who masterminded the clashes was to cleanse the entire Baku gubernia (a territorial unit in the Russian Empire) of Azeris. The short-lived (1918–1920) Azerbaijani Democratic Republic that existed during the

1918–1920 hiatus between the two phases of Bolshevik domination established March 31 as a day of national mourning and created a commission to investigate crimes against Azeris and inform the outside world of them. This work was not finished when the ADR fell under Bolshevik control once again. By authoring or signing off on such a text Aliyev, whose biography is inseparable from the Soviet communist regime, thus distances himself from the Soviet period. The Aliyev decree goes on to describe the 1920 assignment of Zangezur to Armenia with the aim of creating a wedge between much of northern Azerbaijan and its Nakhchivan part. Subsequently Azeris were squeezed out of Zangezur. After World War II, Armenians reportedly promulgated a 1947 Soviet law about the resettlement of the Azeri population from the Armenian SSR to Azerbaijan; and so from 1948 to 1953, several hundred thousand Azeris were uprooted. The emergence of the Nagorno-Karabakh conflict in the late 1980s is described as an anti-constitutional act of separatism; and the eviction of Azeris from their land, including the 1992 massacre in Khojaly, is also referred to as Azeri genocide. "All the tragedies of Azerbaijan throughout the nineteenth and the twentieth centuries," states the decree, "were accompanied by land grabs and were elements of a genocidal policy carefully and persistently conducted by the Armenians in regard to the Azeris." The decree criticizes the leadership of the Azerbaijani SSR, which in the late 1980s did not act decisively and failed to impress the Azeri perspective on Azeri-Armenian tensions on the Soviet government.

In a subsequent (March 31, 1998) resolution of the Milli-Mejlis, the same wholesale qualification of Armenians as perpetrators of genocide on Azeris is contained, more statistics of killings are presented, and the events of 1905–1907, 1918–1920, 1948–1953, and 1988–1993 are declared as manifestations of one and the same persistent and ominous policy conceived and conducted by Armenian chauvinists who were aided and abetted by the "reactionary forces of the great powers." The Milli-Mejlis appealed to the United Nations, OSCE, the Council of Europe, and the CIS to recognize Azeri genocide at the hands of Armenians.

A strong conspiracy streak is similarly contained in the responses of Azerbaijani politicians to questions posed by the authors of *The Karabakh Conflict* and highlighted in publications of the Azeri media. For example, an article by R. Suleymanoglu published in the January 20, 2006, issue of the Baku daily *Zerkalo* claims that the Armenian pogrom

in Sumgait (February 1988) was masterminded by the Armenians themselves. "In order to stir up disorder, the organizers managed to create a nourishing environment for the bestial acts. Liquor, imported cigarettes with narcotic ingredients, and various psychotropic substances were distributed among youths, particularly those from vocational schools. Provocateurs fluent in the Azeri language—mostly Armenians—were implanted amid the Azeri population. They tried to convince people that in Nagorno-Karabakh and Armenia, Azeris were being killed, Azeri girls were being raped, and corpses of Azeris have been already delivered to the railway station in Baku. . . ." Perhaps a less directly worded but essentially similar conspiracy allusion in regard to Sumgait massacres is contained in the interview of Ayaz Mutalibov, the last Azeri Communist Party boss and the first president of independent Azerbaijan, who now lives in Moscow. Mutalibov claims that Armenians in Sumgait had no reason to worry but "somebody wanted blood, which is a powerful catalyst of all revolutions." Yet the same Mutalibov resorts to a more rational explanation of the Azeri pogrom—at the hands of Armenians—in Khojaly (February 1992), which he attributes to the Armenians' thirst for revenge for Sumgait and Baku.

Some important points made by prominent Azeri interviewees are as follows.

- During the Soviet period, Armenians had their cultural centers abroad, which developed and implanted in the Armenian masses a certain ideology that differed from the Soviet one and aimed at undermining the Soviet Union.
- Some Armenian political scientists developed a model in which Armenia, like Israel, should be used by the United States as a weapon against the Muslim world.
- The United States is not impartial in Armenian-Azeri conflict: in October 1992 a special 907th amendment to the Freedom Support Act adopted by the U.S. Congress banned direct U.S. aid to Azerbaijan. Although the amendment has been suspended on an annual basis since 2001, it has not been repealed.
- During the initial period of the Nagorno-Karabakh conflict, Armenia won and Azerbaijan lost the information war.
- Armenians and Azeris are not as ethnically incompatible as the president of Armenia wants everybody to believe. In Russia and in Georgia both groups maintain mutually beneficial business and personal relationships.

- Roots of Armenian-Azeri tensions are 100 years old; they first arose when the Armenian population of the Ottoman Empire raised the Armenian question.
- During the several decades of the Soviet Union's existence, tensions were kept in check in large measure owing to Heydar Aliyev, the great son of Azerbaijan. Armenians took advantage of his ouster from the Politburo. Gorbachev in particular was surrounded by Armenian advisors such as Georgy Shakhnazarov (for some reason, the book calls him David Shakhnazaryan) and Abel Aganbegyan, who advised Gorbachev on matters such as Nagorno-Karabakh.
- Nagorno-Karabakh was a Kremlin tool with which to enforce compliance of the republics. Thus, Azerbaijan could be threatened with the possibility that Nagorno-Karabakh would be taken away from it if it misbehaved, and Armenia that Nagorno-Karabakh would not be allowed to join it.
- In case Armenia agrees with Nagorno-Karabakh autonomy status within Azerbaijan, a mutually beneficial corridor, Agdam–Stepanokert–Lachin–Goris–Nakhchivan, can be established under international supervision (Agdam is a town in Azerbaijan, next door to Nagorno-Karabakh; this town gave its name to a cheap sweet wine which used to be extremely popular all over the Soviet Union; Stepanokert is the capital of Nagorno-Karabakh; Lachin is a town in Azerbaijan located in the narrow zone between Nagorno-Karabakh and Armenia; Goris is a town in Armenia; and Nakhchivan is the capital of the eponymous Azeri exclave separated from the main body of Azerbaijan by Armenia.)

Having familiarized oneself with the Azeri and Armenian (see the essay about Armenia) perspectives on the Azeri-Armenian conflict one cannot escape the conclusion that achieving any peaceful resolution to this conflict will be no easier than solving the conflict between Jews and Arabs in the nearby Middle East.

POST-SOVIET AZERBAIJAN: POLITICS, ECONOMY, AND POPULATION

The end of Soviet rule was preceded by turbulence at the helm of power in Baku—somewhat unusual for that tier of Communist power—whereby five leaders succeeded each other within the scope of nine years. After Heydar Aliyev's 1982

move to Moscow, Kyamran Bagirov assumed leadership of the republic. He was ousted in 1988 over his inability to pacify Nagorno-Karabakh. From May 1988 to January 1990, Abdurakhman Vezirov ruled Azerbaijan. By some accounts he fled to Moscow because of his inability to control the situation in Baku. Interestingly, Vezirov subsequently reemerged as the Russian ambassador to Pakistan. In Baku, Vezirov was replaced by Ayaz Mutalibov as the First Secretary of the Communist Party. Mutalibov then became the first president of independent Azerbaijan. In September 1991 he ran unopposed and obviously "won" the election. But as early as March 1992, Mutalibov was forced to resign his post because of his inability to prevent the massacre on the Azeris in Khojaly.

At that time, the most influential political grouping in Azerbaijan was the Popular Front headed by Abulfaz Elcibey. The Popular Front was a political movement similar to the eponymous movements in the Baltic States, Moldova, Ukraine, and Belarus. Such movements rode the wave of ethnic nationalism and were opposed to Soviet-style socialism and to membership in the Soviet Union, a multinational state. Although Mutalibov dissolved the Communist Party in September 1991, ex-communists were still a dominant force in the Azeri parliament. In May 1992, they attempted to restore Mutalibov to the presidency. But the armed forces loyal to the Popular Front prevented this from happening. As a result, Mutalibov left the country and has been residing in Moscow ever since. Subsequently, Abulfaz Elcibey was elected President of Azerbaijan in June 1992. Elcibey has so far been the only leader of Azerbaijan without roots in communist *nomenklatura*. Moreover, Elcibey, an Arabist by training, was a dissident who once (in 1975) had been imprisoned for disseminating his political views. In essence, Elcibey was a passionate pan-Turkist, a proponent of the unification of northern (formerly Soviet) and southern (Iranian) Azerbaijan, a devout secularist in the tradition of Kemal Ataturk, and vehemently anti-Iranian and anti-Russian. If anything, Elcibey antagonized Iran for whom the pan-Turkic and therefore separatist incitement of its Azeri citizens (by the president of Azerbaijan) came across as a worst-case nightmare. According to Svante Cornell, a Swedish expert on Azerbaijan, this has been "the curious legacy of the Elcibey era: an Islamic state, Iran, ended up supporting Christian Armenia against Muslim Azerbaijan." Although Iran's support for Armenia fell short of any military involvement, it has become Armenia's most important

trading partner and its gateway to the outside world.

Based in the Popular Front, Elcibey's government proved incapable of either credibly prosecuting the Nagorno-Karabakh conflict or managing the economy; and many government officials came to be perceived as corrupt and incompetent. Growing discontent culminated in June 1993 in an armed insurrection in Ganja, Azerbaijan's second-largest city. As rebels headed by colonel Gusseinov advanced virtually unopposed on Baku, President Elcibey fled to his native province of Nakhchivan. After several years in virtual exile, he was able to return to Baku and head the political opposition. He had credible odds for success during the 2000 elections, but, Elcibey came down with a grave illness and died in a Turkish military hospital in 2000.

As president, Elcibey was succeeded by Heydar Aliyev, a veteran of Azeri and Soviet politics. Following his ouster from Politburo in 1987, Aliyev spent three more years in Moscow and then returned to Baku. In the early 1990s, it had seemed that his political star had set in view of his deep association with the Soviet regime. In pursuit of a new power base Aliyev moved to Nachchivan, where he was born, and there he became the Chair of the Supreme Soviet (parliament) of the Nachchivan Autonomous Republic and established a new political party *Yeni (New) Azerbaijan*. So, by the time Elcibey was about to flee to Nakhchivan, Aliyev was ready to move from Nakhchivan to Baku. In the meantime, longing for a strong national leader, Baku was readying to give another try to the political veteran. In June 1993, Heydar Aliyev was elected Chair of the Supreme Soviet of Azerbaijan (soon to be renamed the Milli-Mejlis) and acting president. The day of Aliyev's enthronement was subsequently declared a day of national salvation; since 1998, it has been celebrated as a national holiday of Azerbaijan. In October 1993 Aliyev was popularly elected president, and then reelected (with 76% of the vote) in 1998. However, international monitors of Azeri presidential elections, both parliamentary and presidential, usually ascertain that they do not meet a number of standards for democratic elections.

Heydar Aliyev developed heart problems, and in 1999 in an American hospital in Cleveland, OH, he underwent coronary bypass surgery. In August 2003, after his second heart attack, he again left the country, first for a clinic in Turkey and then again for Cleveland, OH, where he died on October 25, 2003. The plane which transported the ailing Aliyev from Ankara, Turkey, to Cleveland belonged, incidentally, to the Russian Ministry for Emergency Situations (www.newsru.com/arch/world/12dec2003/aliev.html). Earlier Heydar Aliyev had agreed to participate in the new presidential elections of 2003 but then withdrew because of his poor health. Instead, he and his political clan decided to transfer power to his son, Ilham. At the time, the Azeri constitution required that in case of poor health or other reason which made it impossible for the president to perform his duties, power should be transferred to the speaker of the parliament. Because Ilham Aliyev did not hold that position, the constitution was quickly amended to have the prime minister become acting president in cases of emergency. While Heydar Aliyev's son Ilham was not a prime minister either, becoming one—unlike becoming a speaker—did not require an election. Quite simply, Arthur Rasizade, Azerbaijan's prime minister resigned his post in early August 2003, and Ilham Aliyev was appointed in his stead. Rasizade became first deputy prime minister, but he was promptly returned to his earlier post once Ilham Aliyev was elected president on October 15, 2003. Just as his father had earlier, Ilham Aliyev won 76 percent of the popular vote.

Ilham Aliyev was born in 1961. He graduated from the Moscow State Institute for International Relations, where he subsequently (in 1985) defended his doctoral thesis in history and worked in teaching and research positions at his Moscow alma mater until 1992. In 1991, he registered a private business (in Moscow) which dealt with trade exchange with Turkey. In 1994, after his father became president of Azerbaijan, Ilham Aliyev moved to Baku, where he became vice president of the State Oil Company of Azerbaijan. From 2001–2003, that is, immediately before becoming president, Ilham Aliyev headed the delegation from the Milli-Mejlis to the Parliamentary Assembly of Europe. In October 2008, Ilham Aliyev was reelected with 88.7 percent of the popular vote cast in his favor.

The Azeri economy grew from 1996 to 2008 at a staggering rate of 13.6 percent a year. This growth was fueled by new discoveries of oil, production sharing agreements with major international oil companies, and the 2006 opening of the Baku-Tbilisi-Ceyhan pipeline, which now transports crude oil 1,760 km (1,094 miles) from the Azeri-Chirag-Guneshli oil field in the Caspian Sea to the Mediterranean Sea. The oil is pumped from the Sangachal Terminal close to Baku via Tbilisi, the capital of Georgia, to Ceyhan, a port on the southeastern Mediterranean coast of Turkey. It is the second longest oil pipeline in the world (the longest being the Druzhba pipeline from Russia to central Europe). This important pipeline allowed Azerbaijan to transport its oil to European customers while circumventing Russia's network of pipelines. Azerbaijan also transports oil via the Old Baku-Supsa (Georgia) and Baku-Novorossisk (Russia) pipelines. In the natural gas industry, the most important event was the 1999 discovery of the Shah Deniz field, the largest gas field discovered by British Petroleum in many years. The Shah Deniz gas plant at Sangachal Terminal started up in 2007 and made Azerbaijan into a major gas producer. Stage 1 of the Shah Deniz project is now complete and supplies Europe with 8 billion cubic meters of natural gas via the South Caucasus pipeline, which runs parallel to the Baku-Tbilisi-Ceyhan oil pipeline.

Azerbaijan's publicized attempts to diversify its economy beyond oil and gas extraction and utilization have not yet succeeded. As a result of limited employment opportunities, Azerbaijan effectively exports its labor force. The number of Azeris working in Russia alone is estimated at 2 million people. Just in the city of Moscow and the Moscow region more than 1.2 million Azeris are registered. Large Azeri diasporas appeared in Ukraine (about 500,000) and in Turkey; Turkey is a place for living and work for a very significant part of the Nachchivan exclave's population. Because many Azeris have naturalized in Russia and other countries, it is believed that the official census estimate of Azerbaijan's population is an enormous overstatement. The Russian demographer Arsenyev claimed in a 2002 publication (in the authoritative *Demoscope Weekly*) that the actual population of Azerbaijan was only 3.2–3.3 million people, not the 7.9 million reported in the 1999 Azeri census.

Important processes occur in areas of national identity and historical memory. Thus, the issue of changing the name of the ethnic community from Azerbaijani to

Azeri Turks or simply Turks is seriously debated. Farid Alekperli, a reputable historian, claims in his 2009 publication that the ethnonym "Azerbaijani" is a Stalinist fabrication dating only to the 1930s, whereas for centuries the people identified themselves as Turks. He maintains, though, that the term Azerbaijani should be retained as a token of citizenship, not ethnicity (www.zerkalo.az/2009-08-08/politics/158-).

In January 2009, the spacious memorial to 26 Baku commissars was demolished. "Having a monument to the 26 commissars, who were mainly Armenians, in the very centre of Baku is the same as if there was a monument to the SS in the middle of Tel Aviv," said Khikmet Gadzhizade, a former ambassador to Russia and a senior member of the Musavat party, which is in opposition to the government but which supported the removal. "The people who were buried there were participants in terror against the population of the country and guilty of the death of thousands of Azerbaijanis," he said. In fact, only eight of the commissars were Armenians; the rest were Georgians, Jews, Latvians, and Greeks, as well as two Azeris. But their leader Stepan Shaumyan, a communist legend and ally of Bolshevik leader Vladimir Lenin, was Armenian, thus casting an ethnic light on the group as a whole. The conflict with Armenia is definetely more than just a conflict over land.

Georgia

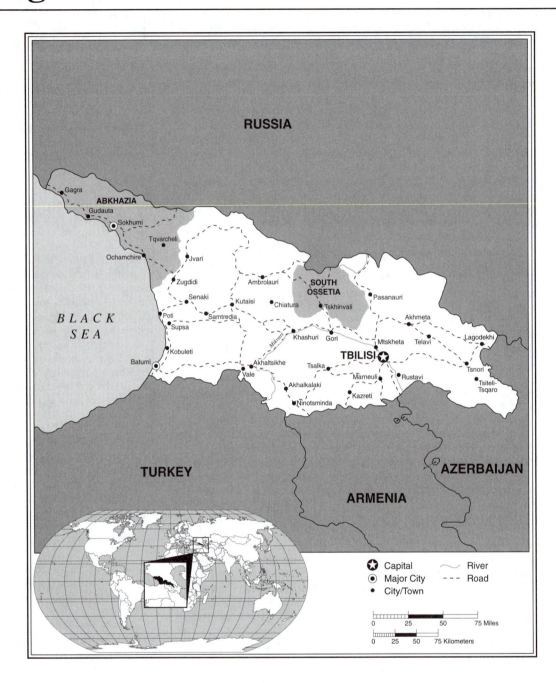

Georgia Statistics

Area in Square Miles (Kilometers):
26,911 (69,700)
Capital (Population): Tbilisi
(1,480,000)

Population

Total (millions): 4.6
Per Square Km: 66
Percent Urban: 53

Rate of Natural Increase (%): 0.2
Ethnic Makeup: Georgian 83.8%; Azeri
6.5%; Armenian 5.7%; Russian 1.5%;
other 2.5% (2002 census)

Health

Life Expectancy at Birth: 71 (male);
79 (female)

Infant Mortality Rate: 13/1,000
Total Fertility Rate: 1.4

Economy/Well-Being

*Currency ($US equivalent on October 14,
2009):* 1.67 lari = $1
2008 GNI PPP Per Capita: $4,850
*2005 Percent Living on Less than US$2/
Day:* 30

Georgia Country Report

Located in the western part of the Trans-Caucasus, the Republic of Georgia has a land area of 69,700 sq km and a population of 4.6 million people (2009 estimate). The capital of Georgia is Tbilisi, whose population is close to 1.5 million people. Ethnic Georgians account for 83.8 percent of Georgia's current population. Since the early 1990s, two separatist entities, Abkhazia and South Ossetia, have defied the Georgian government despite having been located within the borders of the Georgian SSR, which after the dissolution of the Soviet Union became the internationally recognized jurisdiction of the Republic of Georgia. After Georgia's August 2008 war with Russia, Abkhazia, and South Ossetia were recognized by Russia as independent states. The only other countries which have so far extended their recognition to these entities are faraway Nicaragua and Venezuela.

The South Caucasian or Kartvelian language group to which modern Georgian belongs is not known to be related to the other major language families of the world (e.g., Indo-European, Altaic, Semitic, etc.). The Georgian language uses its own unique script; for example, this is how the word "Georgian" or Kartuli is spelled ქართული.

Georgia is located between the (Greater) Caucasus and the Lesser Caucasus mountain ranges and controls the southern and northern slopes of those latitudinal ranges, respectively. The Likhi Range, which lies entirely within Georgia, is of meridional (north-south) extent and connects the Greater and Lesser Caucasus, thereby splitting Georgia into two parts. The western part is the Kolkheti Lowland, an area with fertile soils and a humid subtropical climate, which is also typical for the entire Black Sea coast of Georgia. The eastern part has a dry subtropical climate. With significant breaks in elevation, which ranges from the sea level to 5,202 meters on Georgia's highest mountain peak, a considerable amount of surviving forest, terraced and lowland farming, and old Christian Orthodox churches, Georgia is often described as the most beautiful part of the Trans-Caucasus.

PRE-SOVIET POLITICAL HISTORY

Historically, modern Georgia traces its roots to two ancient states, Colchis (or Kolkheti) in the western part of today's Georgia and Iberia (or Kartli) in the eastern part, in which lies Tbilisi. Both states formed several centuries before the birth of Christ. From 200 B.C. to 200 A.D., both proto-Georgian states waged wars with Romans, Armenians, and Pontic Greeks and were partially conquered. Christianity became the state religion in Eastern Georgia as early as 331 A.D. and in Western Georgia in the sixth century. Subsequently, the Georgian states were invaded and conquered by Byzantines, Persians, Arabs, and Turks. Georgia's Golden Age began under King David IV, the Rebuilder, who consolidated Georgian lands under his rule (1089–1125), having wrested them from the Turks, whose major forces were at the time diverted to struggle with the Crusaders. Since the time of David IV, Tbilisi has been recognized by Georgians as their capital. Georgia's golden age lasted for more than a century after David the Rebuilder's death; and during the reign of Tamar, David's great granddaughter (1184–1213), Georgia reached its zenith in terms of its territorial expanse and influence. It is at that time that the Georgian national epic poem, *The Knight in the Panther's Skin,* was written by Shota Rustaveli, a poet whose name is dear to every Georgian's heart. Georgia's golden age was interrupted by the Mongolian invasions that began in the 1220s, although in the fourteenth century, under George V, Georgia was briefly recreated within its pre-1220s borders. From 1386 to 1403 it survived eight invasions by Tamerlane (Timur), the Central Asian Turkic warrior. Throughout the fifteenth century, Georgia was a weak Christian country surrounded by the Muslim world on all sides; and in the 1460s, it split into four kingdoms. Those kingdoms fell within Turkish and Iranian spheres of influence.

By the time Russia created its first outposts in the northern foothills of the Caucasus in the seventeenth century, it was perceived by Georgian kings as a possible Christian savior. Since that time, they appealed to Russia for military help and proposed to join forces against Turkey and Iran. In the late 1600s, a Georgian colony emerged in Moscow which played a significant role in the subsequent establishment of friendly Georgian-Russian relations. In 1783, the East Georgian Kingdom (Kartli and Kakheti) signed a treaty in Georgievsk, a Russian stronghold in the North Caucasus, according to which eastern Georgia became a Russian protectorate. However, when Russian-Turkish war broke out in 1787, Russian forces abandoned Georgia. Russia also failed to protect Georgia against the 1795 Persian attack during which the city of

Tbilisi was burned to ashes in retaliation for Georgia's closeness to Russia. Only in 1799, did Russia establish a military presence in Georgia, and in 1801, East Georgia was annexed by Russia. "On the whole Georgian nobility was content with its equalization in rights with the Russian aristocracy" (Gachechiladze 1995: 27), which was only possible due to the religious bond between Russia and Georgia; both were not only Christian but belonged to the same Christian denomination. Many Georgians voluntarily "entered Russian military service, many of them reaching the highest ranks." Despite some uprisings against the new colonial administration, "Georgia appeared to be the most reliable strategic foothold of the Russian Empire. . . . Georgians for the most part were the Tsar's loyal subjects" (Ibid). The incorporation of the western part of today's Georgia into the Russian Empire occurred in stages from 1810 until 1878 in the wake of the Russian-Turkish wars. In the words of Revaz Gachechiladze, "The Russian Empire accomplished the 'gathering of the Georgian lands' that had been the dream of the Georgian people, who always will be grateful to Russia for this," even though "the unification of Georgia was never a conscious aim of the rulers of the Empire" (Ibid: 26). For Russia, obtaining this strategic foothold was instrumental in the consolidation of its imperial possessions in the south. In that regard, the religious affiliation of the majority of the Georgians played a most important role; belonging to the Orthodox community was always considered a token of loyalty in tsarist Russia. Not only was the hostile Muslim stronghold of Turkey located immediately to the south of Georgia, but Muslims also dominated Russia's North Caucasus. With the incorporation of Georgia, most of those Caucasian Muslims became sandwiched *between* Orthodox Christians—Russians to the north and Georgians to the south.

Several developments with major implications for the future of Georgia occurred between the establishment of Russian administration over Georgia and the Communist Revolution of 1917. Among these developments, one must mention population exchange in some peripheral regions of Georgia, the emergence of a peculiar inter-ethnic division of labor, development of industry and transportation, progress in the national consolidation of Georgians, and the revolutionary movement.

Population exchange affected areas with Muslim populations. Among Georgian speakers, those in the southwest of the country (Ajaria) had converted to Islam during several centuries of Turkish domination. There was another significant Muslim community in Abkhazia. Although in Georgia proper (that is, in Georgia without Abkhazia and South Ossetia), the 1897 census enumerated as many as eleven different groups, all of them closely related linguistically. But Abkhaz or Abkhazian is a separate ethnicity, and the Abkhazian language is not even a distant relative of Georgian. All the linguistic relatives of Abkhazians reside along the northern slopes of the western Caucasus in Russia. They consist of Abazins (the closest relatives) in the current Karachai-Circassian Republic of Russia, as well as Circassians and Adyghes. The map of the overall distribution of those related languages in the middle of the nineteenth century (which can be found at www.circassianworld.com/northwest.html) shows that current area of Abkhazia had a large degree of linguistic commonality precisely within those areas north of the Greater Caucasus. When Russia fought Turkey for domination in the Caucasus, it mistreated Caucasian Muslims, considering them Turkey's fifth column. Hence, many were offered refuge in and some were forced to leave for Turkey. As a result, the North Caucasus was reduced to a system of "Bantustans" wherein speakers of the northwest Caucasian languages were either surrounded by Russian speakers (like Adyghea) or deliberately enclosed in single civil divisions with speakers of unrelated languages (as are Kabardians and Balkars; Circassians and Karachai). Between 1866 and 1878, thousands of Abkhazians and their linguistic relatives left for Turkey. As to how many Abkhazians left, there is no consensus. Georgian scholars believe that forced migrants were in the low tens of thousands at best (Gachechiladze 1995: 81), but the Abkhazians (Inal-Ipa 1990: 35) claim that some 100,000 left their homeland. All agree, however, that those who left accounted for at least half of the entire Abkhazian ethnicity; and the abandoned areas were mostly, though not entirely, coastal—that is, those areas most coveted by Russians. Because large areas along the coast were emptied, the migration of Greeks and Armenians (who were being squeezed out by the Turks) to these areas was encouraged, as was the migration of some Russian religious sects. As for Abkhazians, those who had been Christian or had converted to Christianity mostly remained in their homeland but generally well inland from the coast.

In Georgia proper, a peculiar division of labor developed between the ethnic communities. Georgians were either landed aristocracy or peasants. What united these otherwise different social groups were language and common antipathy to trade. Trade, finances, and to a significant degree craftsmanship were traditional niches for Armenians, whose communities sprang up in every Georgian city, sometimes even outnumbering Georgians. Thus in 1864, in Tbilisi, Armenians accounted for 47.4 percent of the population, Georgians for 24.76 percent, and Russians for 20.47 percent. Twelve years later, the corresponding data were 40.7 percent, 23.6 percent, and 21.24 percent. (S. Chktetia, *Tbilisi in the 19th century,* Tbilisi: AN GSSR 1942: 161, 165). In the words of Gachechiladze, "the Georgian nobility, rather carefree and non-industrious, and still thinking in feudal terms that their major business was war (and large feasts between wars!), very often became dependent on Armenian creditors and usurers and tended to blame all their misfortunes on them" (Gachechiladze 1995: 29). At the same time, bureaucrats in Georgia were for the most part Russians or Russified Germans and Poles.

In 1865, construction of the Trans-Caucasian railway line commenced. The first commissioned stretch of that line was from Poti on the Black Sea to Zestafoni in west-central Georgia. Subsequently, Tbilisi was connected with Poti and Batumi, as well as with Baku, so Black Sea ports became connected with the major oilfields on the Caspian Sea. In 1900, the entire TransCaucasian railway network was connected to that of European Russia. In the last quarter of the nineteenth century, large-scale industry such as textile, metal working, production of leather goods, tobacco, brandy, and the extraction of coal (in Tkibuli in NW Georgia and in Abkhazia) and manganese (in Chiatura in north-central Georgia) began to develop in Georgia. In the 1890s, Georgia accounted for 50 percent of the world output of manganese. While no single industrial center in Georgia rivaled not-so-distant, oil-rich Baku, Tbilisi—or Tiflis as the city was known in tsarist Russia (and, in fact, in the Soviet Union until 1936), grew as a significant center of Georgian, Armenian, and cosmopolitan culture.

The development of transportation and large scale industry had two conflicting effects, that of fomenting the national consolidation of Georgians and that of fomenting working class consolidation. The mutually conflicting nature of these effects lies in the fact that bonds of organized labor transcended ethnicity simply because large numbers of industrial workers in Georgia were Armenians and Russians. While class consolidation was highest on the agenda of Russian Marxists of all ethnic backgrounds, the idea of national consolidation was high on the agenda of Georgian intellectuals. They wanted to overcome a situation whereby most ordinary people still identified themselves as, first and foremost, Megrelians, Svans, Kakhetians, Mesketians, Dzhavakhis, etc.—in reference to different parts of the country and their respective dialects. Such splintered identities did not allow people to realize their common goals and overcome the colonial status of their country. The most respected intellectual leader of Georgia's national movement was the poet Ilya Chavchavadze (1837–1907). Not impervious to national agitation, many Georgians were nevertheless leaning away from it and toward the anti-tsarist, socialist revolutionary movement which was on the rise in the last quarter of the nineteenth century in Russia as a whole.

The Russian Social Democratic Labor Party (RSDLP), a Marxist group which in 1903 split into the more radical Bolsheviks and the less radical (in the tradition of European social democrats) Mensheviks, enlisted quite a few Georgians. Most of them joined the Mensheviks, who in 1906 won all Georgia-representing seats in the Russian Duma (parliament). But there were exceptions, the most notable being Ioseb Dzhugashvili, who subsequently became known to much of the world by his Russian pseudonym, Stalin. Stalin was born in 1878, the son of a cobbler in the town of Gori in north-central Georgia. After attending an Orthodox seminary he discovered the writings of Vladimir Lenin and decided to join the Marxist revolutionaries. After RSDLP's split he joined the Bolsheviks, for whom he raised a lot of money through extortion, bank robberies, and armed hold-ups. Whereas Georgian Mensheviks did not support and embrace the 1917 Bolshevik revolt in Petrograd, Stalin was one of its major organizers.

During World War I Georgia found itself close to the frontline after Turkey joined the war on Germany's side. As a result of a power vacuum, Georgian Mensheviks took the lead and, having joined their ideological brethren in Armenia and Azerbaijan, formed (in April 1918) the Transcaucasian Democratic Federal Republic, separate from Russia. This republic, however, survived only one month because of inter-ethnic tensions. As a result, in May 1918, Georgia, with its social democratic (Menshevik) government, declared its independence as the Georgian Democratic Republic (GDR). This republic invited contingents first of German and subsequently of British troops to help it to sustain its fragile statehood in face of the Bolshevik Red Army's onslaught from the northwest and the Turkish invasion from the southwest. In December 1918, the number of British troops in Georgia amounted to 25,000. However, the British vacillated between rendering support to Georgia and to the counter-revolutionary Russian army (the Whites) under the command of Denikin, which sought to restore the territorial integrity of the former Russian Empire. But by January 1920, Denikin's army had been crushed by the Bolsheviks. In April 1920, the Bolsheviks took power in neighboring Azerbaijan, and the GDR faced imminent invasion by the Red Army. The GDR managed to survive for almost a year, as the Red Army was strained by war with Poland and did not risk another military campaign at the same time. For a time Soviet Russia and GDR even exchanged embassies. But in February 1921, the Red Army entered Tbilisi. By March, all power was in the hands of the Bolsheviks, and the GDR ceased to exist. For Georgia, one advantage of this development was that the Red Army recaptured Turkish-occupied Ajaria, populated by Georgian Muslims, thus once again consolidating Georgian lands under one administration. From the Bolshevik side, such ethnic Georgians as Stalin (Dzhugashvili), Sergo Ordzhonikidze, Philip Makharadze, and Abel Yenukidze played decisive roles in the establishment of Soviet power in their homeland.

GEORGIA AS A SOVIET REPUBLIC

Georgia was incorporated into the Transcaucasian SFSR, which united Georgia, Armenia, and Azerbaijan. Tbilisi was set as the TSFSR's capital. The TSFSR was split into its component parts in 1936, and Georgia became the Georgian SSR.

In order to begin to understand Georgia's fortunes in the Soviet Union, one has to keep the most basic political clichés at bay. To be sure, Georgia was not spared the evils of Soviet socialism. Industry was nationalized, and agriculture collectivized. In economy, the same profit-redistribution style model wherein efficient production units bailed out inefficient ones was installed as everywhere else in the U.S.S.R. And politically Soviet power was as repressive in Georgia as elsewhere. An opinion once popular in Russia that Georgia was spared purges because Stalin was an ethnic Georgian does not hold water either. Repressions in Georgia were as brutal as elsewhere in the Soviet Union, and many prominent intellectuals and, in fact, Communist Party leaders fell victim to them.

Yet there are several important qualifications pertaining to the Soviet period in Georgian history. First, it was within the Soviet Union that Georgia ultimately

(© Marko Petrovic www.trekearth.com)
Old Tbilisi.

experienced ethno-national consolidation and ethnic Georgians became the majority in their vastly expanded and beautiful capital Tbilisi for the first time in a couple of centuries. Second, although Stalin did not shield Georgians from repressions, Georgians' loyalty to Soviet power was enhanced by the fact that their fellow countryman was at the very helm of power in the Soviet Union. Georgy Nizharadze, who analyzed some persistent traits of the Georgian psyche, wrote that in Georgia Stalin was loved "because he exemplified willfulness free of any constraints whatsoever," and this was a commendable feature in the eyes of many Georgians (Nizharadze 1999). According to Nizharadze, a typical "Georgian is usually loyal to power (e.g., that of a sultan, a shah, or an emperor) exercised from some fairly distant place but resists local power" (Ibid). Stalin was in charge of a distant power center (Moscow is more than 2,000 km away from Tbilisi) and he was a fellow Georgian. In late April–early May of 1945, several groups of Soviet troops hoisted the red banner on the roof of the Reichstag in the vanquished German capital Berlin. Only two of the troops, a Russian Mikhail Yegorov and a Georgian Militon Kantaria, were singled out by the Soviet ideological apparatus as those who did the job—despite the fact that by some accounts they were actually part of the second crew that mounted a red flag on the Reichstag and they were accompanied by other troops. The symbolism of a victorious Russian-Georgian duo was certainly elected to please Stalin, who during the most difficult days of the war appealed to Russian nationalism in the first place, but gestures like this were certainly taken to heart in his native Georgia. Obviously, the thought that Russia and Georgia would someday emerge as arch-foes, as they did prior to and particularly in the wake of their 2008 military conflict (see the following), would have

never occurred to Stalin even in his wildest nightmares.

Surely, Stalin was revered in the Soviet Union, not solely in Georgia. When many Russians, Ukrainians, and people of other ethnicities learned that Stalin was condemned as a thug by someone no less than the Communist Party leader Nikita Khrushchev, this came as a great shock. (It is a different matter how and why this genuine devotion to Stalin as the Father of the Peoples could coexist with widespread if informal knowledge of repressions—quite a bit of research has been done on this subject.) But only in Georgia did people rally in protest against Stalin's denunciation for four straight days and nights until, on March 9, 1956, the rally was drenched in blood; by some accounts up to 100 people perished after being shot by security forces. "That spontaneous outburst of protest was motivated solely by a desire to defend the symbol of Georgian identity" (Nizharadze 1999). Since at least the early 1920s, the Soviet Union had not seen such a show of disobedience. It commenced on March 5, on the third anniversary of Stalin's death. Khrushchev, who had on certain occasions played the role of Stalin's jester, was now the Communist Party leader; he had delivered his initially secret speech to the delegates of the twentieth Congress of the Party just two weeks prior. In it he condemned Stalin's "personality cult." Residents of Tbilisi who gathered in the city park on the Kura River near a giant statue of Stalin could not fathom how such a venerable name could be so debased. By some accounts the security forces started shooting into the crowd on the fourth day of the rally, when a column of demonstrators was dispatched from the park to the main post office to send a telegram to the Kremlin and personally to "Stalin's friend Molotov" (Bezhanishvili 2008).

But Georgians en masse overcame the national trauma related to Stalin's denunciation, particularly since, unlike the rest of the Soviet Union, they were allowed to keep many (but not all) Stalin monuments intact. Moreover, Georgians remarkably adjusted to the latter phase of Soviet socialism. Nizharadze even claims that Georgians became "the darlings (balovni) of Soviet power," and that the period from the 1960s to the 1980s was "one of the most carefree periods in our history . . . when peace and basic needs were secure, sources of getting money numerous, cultural life resembled a horn of plenty, and the streets were full of smiling and well-meaning people." Revaz Gachechiladze also admits a high level of well-being in Soviet Georgia, "which was one of the most flourishing Union republics under Soviet power." Moreover, "it was apparent [to him] that the country was living beyond its means" (Gachechiladze 1995: 117). There is plenty of anecdotal evidence on that account, and most visitors from Russia, Ukraine, or Belarus were usually stunned by the number of brick mansions in Georgia's countryside, quite exquisite by Soviet standards. The informal, descending order of quality of life in the Soviet Union of the 1970s and 80s that most people who made business and recreational trips shared was (1) Moscow, Leningrad, the Baltic republics, and Georgia; (2) Ukraine, Moldova, Belarus, and Armenia; (3) Russia outside its two capitals and Kazakhstan; and (4) Azerbaijan and the Central Asian republics, except Kazakhstan. According to Gachechiladze, the average 1989 current personal bank deposit was 1,734 rubles for the Soviet Union at large, whereas in Georgia it was 2,459 rubles (Gachechiladze 1995: 122). Note that the highest official salary at the time was 500 rubles a month and the vast majority earned a third of that or less. With 74 personal vehicles per 1000 population in 1989, Georgia yielded only to the three Baltic republics (Gachechiladze 1995: 128), where integral indicators of economic efficiency and return on investment were much higher than in Georgia. In 1992 Andrei Illarionov—long before he became Putin's economic advisor and then one of his fiercest critics—published his calculations of 1990 per capita GDP, in 1985 U.S. dollars, and per capita consumption in each former union republic. Illarionov's data showed the significant leveling effect of the GDP's redistribution. Thus, according to those data, Russia's per capita GDP was 1.6 times that of Georgia, whereas Russia's per capita consumption was only 1.3 times higher. Even if one factors in the share of personal consumption related to heating needs—much higher in Russia than in subtropical Georgia—the per capita consumption ratio would not match anecdotal evidence of Georgia's relative wealth.

While it may well be that anecdotal evidence neglected to notice pockets of poverty in Georgia and so the widespread perception of Georgian wealth was exaggerated, there were two meaningful channels of income redistribution within the U.S.S.R.—channels whose validity is beyond doubt but whose reflection in the official Soviet statistics could only be partial at best. Both channels favored Georgia due to its unique (within the Soviet Union) physical geography—a republic in the humid subtropical periphery of the coldest country on earth. One component of this advantage had to do with Georgia's Black Sea coast, which used to host vacationers from all over the Soviet Union annually from May to October. While many of these vacationers spent time in corporate, governmental, or trade union sanatoria and vacation homes, many more rented houses, apartments, rooms, or just single beds. That means that millions of Soviet rubles ended up in the pockets of fortunate coastal real estate owners. For the most part, this additional income was not taxed and could easily double, triple, or quadruple the owner's working income. To be sure, the Georgian coastline was not altogether unique in the Soviet Union as a magnet for summer vacationers. To the north of the Georgia-controlled stretch of the Black Sea coast lay the Russian stretch, which from Sochi almost to Novorossiisk offered conditions and facilities not unlike Georgia's. And there was also the southern coast of the Crimea, which after 1954 belonged to Ukraine. However, neither in Ukraine nor (especially) in Russia did a lucrative recreational economy account for such a sizable share of the total economy and benefit such a large segment of the overall population. A potential rival of the Black Sea resorts, the Caspian coast (in Russia, Azerbaijan, and Central Asia), was decidedly less picturesque, largely undeveloped, and did not attract many visitors.

The second component of Georgia's physiographic advantage was agriculture specialized on such subtropical products as tea, peaches, tangerines, and wine, as well as on early vegetables. Although most farmland belonged to collective and state farms, most families in the Georgian countryside maintained their own orchards; and the relationships between collective and household farms were symbiotic. The gist of the matter, however, was in the nature of product-specific retail systems leading to extremely variable profit margins. From this perspective it makes sense to compare potatoes produced on a Russian farm and peaches grown in Georgia. Today one can go to any *rynok* (bazaar) or supermarket in the city of Moscow and realize that peaches may be only twice as expensive as potatoes. But this was not so when the Soviet Union existed. Potatoes were a necessity, a staple of diet for people of average means, so they were sold by state-run grocery stores for a state-controlled price of 0.1 ruble a kilo; potatoes producers had a hard time making ends meet as a result. But peaches—delivered from Georgia to Russia or Ukraine—were a luxury, and they were sold either by cooperative stores or on the so-called *kolkhoznye rynki* (urban based collective farm bazaars) for 3 rubles a kilo. This 30-times difference in price exceeded the production cost disparity by a factor of 12 at the very least. Again, Georgia was not altogether unique as a place where particularly lucrative agricultural products were grown, but land suitable for such products accounted for a much larger portion of the total land area of Georgia than in, say, more arid Azerbaijan, let alone rocky Armenia; and it benefitted a larger segment of the population. As for Central Asian farms, they were farther from the major urban centers of the European section of the Soviet Union and not as well connected with them. Georgian farmers and subsequently Georgian middlemen first appeared in Moscow's bazaars in the 1960s, and for two decades afterwards all southern salespeople were dubbed *Gruziny* (which in Russian means Georgians), even when the actual share of Gruziny declined as many other ethnicities joined the game. All in all, Georgians benefited from the division of agricultural labor, perhaps not even realizing that it owed its existence to the larger entity, the Soviet Union, to which Georgia belonged. Semi-closed, largely cold, and with a long suppressed entrepreneurial spirit, that entity took away freedom, but in its stead it offered benefits that suited many people well. Many Georgians were clearly among the beneficiaries. Besides the division of labor as such, Georgians were "remarkably well adjusted to the latter-day totalitarian Soviet reality" (Nizharadze 1999), more specifically to a situation whereby shadow economy, bribes, and connections were commonly used as ways of meeting consumer demand.

With its displays of personal wealth, striking natural beauty, and legendary hospitality, Georgia was loved by Soviet intellectuals of all ethnic backgrounds and specializations. It was deemed a great success to be able to organize or to participate in some all-Union, better still international conference, symposium, or festival on Georgian soil. Georgian hosts knew how to create a friendly atmosphere with good local wine, grilled lamb, herbs, tasty vegetables, and of course lengthy but exciting and often philosophical toasts. Georgian poetry, translated by Russia's own best poets, Georgian movies, theater, and Georgia's fascinating capital elicited almost universal adoration. And, while Georgia had its dissidents, no pro-independence sentiment (let alone movement) developed in Soviet Georgia on the scale of those in the Baltic States or Western Ukraine, and that enhanced the ambience.

MINORITY PROBLEMS AND THE END OF THE SOVIET POWER

Although in the late 1980s nothing pointed to the fast approaching end of this idyll, one phenomenon did cloud the horizon. In personal communications, Georgia's ethnic minorities complained about discrimination and lack of prospects in what came across to them as an exceedingly ethnocratic Georgian society. Unlike other republics of the Soviet Union, there were no complaints about anti-Semitism in Georgia. Instead, best known were the grievances of Georgian Armenians, who in the not so distant past had outnumbered Georgians in Tbilisi but by the end of the Soviet Union's existence accounted for just 12 percent of its population. Practically all upwardly mobile Armenian youths saw no prospects for social advancement in Georgia; they enrolled in colleges outside Georgia (for the most part in Russia), and did not return after graduation. Some Armenians switched to Georgian names and saw a favorable change in attitudes toward them. Somewhat less known outside Georgia were Abkhazian grievances, perhaps because Abkhazians were not as numerous a minority as Armenians; but almost anybody who vacationed on the northern (and most picturesque) stretch of Georgia's Black Sea coast, in the Autonomous Republic of Abkhazia, heard something of such grievances. As for Ossetians, their complaints were even less known, yet as far as Ossetian sources are concerned, they appear no less significant. Unlike Armenians, though, Abkhazians and Ossetians once enjoyed formal territorial autonomy in Soviet Georgia; and so within each respective autonomous region, they were titular nationalities. By and large, in the Soviet Union belonging to the titular nationality of a certain civil division used to be an important, if informal factor of career advancement. From the Abkhazian and Ossetian points of view, these advantages were denied them in Soviet Georgia. The recent historical memories of both minorities harbor even more inflammatory accusations.

From the Abkhazian side, while there is a hairsplitting debate about the actual status of the Abkhazian Republic from 1921 to 1931 and what role, if any, Stalin played in that republic's 1931 incorporation into Georgia, the major grievances are related to what happened afterwards. According to the Abkhazian sources tapped by Alexander Krylov in his Russian-language book *Religion and Traditions of the Abkhazians* (Moscow: Institut Vostokovedeniya 2001), organized settlement of ethnic Georgians in Abkhazia commenced in the mid-1930s and was accompanied by replacement of Abkhazian and Russian by Georgian in public education and legal documents, replacement of Abkhazian officials by ethnic Georgians, and falsification of Abkhazian history. As was mentioned earlier, Abkhazia had been depopulated due to forced migrations to Turkey during and in the wake of the Russian-Turkish wars of

the nineteenth century. However, according to Abkhazian sources, mass settlement of Georgians in Abkhazia occurred during the Soviet period. The same sources claim that, although beginning in the 1950s the most odious forms of discrimination against ethnic Abkhazians were abandoned, the resettlement campaign went on. Indeed, in 1989, Georgians accounted for 45 percent of Abkhazia's population, whereas Abkhazians accounted for only 17.8 percent; other numerous groups were Armenians (14.6%) and Russians (14.3%) (Gachechiladze 1995: 84). Flare-ups of anti-Georgian unrest in Abkhazia took place in 1957, 1964, 1967, and 1978—a fairly unusual phenomenon for the pre-perestroika Soviet Union. But if this was the case even under repressive Soviet leaders, then what would one expect during the short-lived Gorbachev era, when due to his policy of openness and the elimination of censorship pent-up ethnic grievances made it to the surface of public life in all multiethnic regions of the U.S.S.R.! Indeed, Georgian-Abkhazian clashes occurred in 1989; and following the 1990 election of Zviad Gamsakhurdia as Chairman of Georgia's Supreme Soviet, the Supreme Soviet of Abkhazia adopted a resolution calling for secession from the Georgian SSR and the recreation of the Abkhazian SSR, which had existed from 1918 to 1921. Ethnic Georgian deputies of the Abkhazian Supreme Soviet obviously did not vote in favor of that resolution.

Grievances from the Ossetian side have been along similar lines. It appears that the first violence flared up between ethnic Georgians and Ossetians as early as 1920, and then there was a long hiatus during which mutual grievances accumulated, only to rekindle violence in 1989–90, following which the Georgian Supreme Soviet annulled Ossetian autonomy. Since that time, "South Ossetia" has been referred to as the "Tskhinvali region" by the Georgian media, Tskhinvali being the name of that region's capital city.

Needless to say, the Georgian view of tensions and conflicts between ethnic Georgians and Georgia's minorities is entirely different. This standpoint is well reflected in the already quoted 1995 book by Gachechiladze, although he concedes "the mistakes of Georgian ideologists sometimes displaying unpardonable deafness to the pain, hopes, and interests of the 'other' peoples of Georgia" (Gachechiladze 1995: 172). However, in regard to Armenians, the author points out that many of them "became integrated into Georgian society. But as the Armenian Republic lay next door, the best conditions for promotion always lay abroad" (Gachechiladze 1995: 90).

As for Abkhazians, Gachechiladze's point is that they had always been small in numbers and yet were assigned government posts in Abkhazia well out of proportion to their numbers. Moreover, it is the Russian policy of divide and rule which is at fault, as the Russification policy had been conducted in Abkhazia since tsarist times. Envisaging the fast disappearance of the Abkhazian language, Russian administrators preferred to see it replaced by Russian, not Georgian. On Ossetian claims, Gachechiladze's view is even more dismissive. Ossetians are referred to as late comers who had been migrating from their actual homeland immediately north of the Great Caucasus and arriving in the Georgian (southern) foothills of the Caucasus "only" since the seventeenth and eighteenth centuries. "As landless newcomers they were welcomed by the Georgian feudal lords, who were in permanent need of cheap labor" (Gachechiladze 1995: 86). By 1989, only 39.2 percent of all Ossetians in Georgia lived within their autonomous region, whereas the majority were spread out in the rest of Georgia and well integrated. In 1886, not a single Ossetian resided in the township of Tskhinvali, but by 1989, it had become 74.5 percent Ossetian. There were 90 schools teaching in the Ossetian language in Ossetia, contrasting with no such schools at all in the Russian Federation's Autonomous Republic of North Ossetia prior to 1988. Thus the ethnic needs of Ossetians in Georgia were met, and everything would have been calm were it not for external incitement for their separatism.

This incitement is attributed to Russia. It is telling that, in Georgia, the pro-independence movement did not just ride the wave of a general liberalization of the Soviet regime under Gorbachev; to a significant degree the movement was a response to the ethnic separatism of Abkhazians and Ossetians. After March 1989, when several thousand Abkhazians demanded secession from Georgia and the restoration of union republic status to Abkhazia, a series of unsanctioned meetings took place across Georgia claiming that the Soviet government was using Abkhazian separatism in order to prevent Georgia from becoming an independent country. On April 4, 1989, tens of thousands of Georgians gathered in downtown Tbilisi. The protesters, led by several prominent dissidents, including Zviad Zamsakhurdia, organized a peaceful demonstration and hunger strikes, demanding the punishment of Abkhazian separatists and restoration of Georgian independence. After several days of demonstrations, the First Secretary of the Georgian Communist Party, Jumber Patiashvili, asked the Kremlin to send troops to restore order and impose a curfew.

Early in the morning of April 9, 1989, armored personnel carriers and troops under General Igor Rodionov surrounded the demonstration area. They received an order to disband and clear the avenue of demonstrators by any means necessary. The Soviet detachment, armed with military batons and spades, advanced on demonstrators moving along the Rustaveli Avenue, the main street of Tbilisi. The following stampede resulted in the death of 19 people, including 17 women. On April 10, the Soviet government issued a statement blaming the demonstrators for causing unrest and compromising the safety of the public. However, a parliamentary commission investigation of the events of April 9, 1989, in Tbilisi was launched by Anatoly Sobchak, a liberal member of the Congress of People's Deputies of the Soviet Union. After a full investigation and inquiries, the commission condemned the military, which had caused the deaths by trying to disperse demonstrators. On April 10, in protest against the crackdown, Tbilisi and the rest of Georgia went on strike and a 40-day period of mourning was declared. People brought masses of flowers to the place of the killings. A state of emergency was declared, but demonstrations continued.

In a referendum on March 31, 1991, Georgians voted overwhelmingly in favor of independence from the Soviet Union. With a 90.5 percent turnout, approximately 99 percent voted in favor of independence. On April 9, the second anniversary of the tragedy, the Supreme Soviet of the Republic of Georgia proclaimed Georgian sovereignty. Nevertheless, the actual secession of Georgia from the Union took place only after the U.S.S.R. was officially disbanded by the leaders of three Slavic republics—Russia, Ukraine, and Belarus—in December 1991.

POST-SOVIET DEVELOPMENTS AT A GLANCE

As a newly independent state, Georgia faced the double challenge of steep economic decline and separatism, by this time (end of 1991) armed with munitions from Soviet warehouses. Given the very specific role that Georgia played in the division of labor in the Soviet Union, whereby not only industrial goods but a fair amount of food used to be delivered from other republics, the state of the Georgian economy soon after independence was nothing short of catastrophic. Already in 1992, the GDP had contracted by 45 percent. For comparison, in Russia it contracted by

14.5 percent, in Belarus by 9.6 percent, and in Ukraine by 10 percent. In 1995, Georgia's GDP was 35.8 percent of its 1991 level, lower than anywhere else in the former Soviet Union; and industrial output, crippled by severed supply lines with Russia and other republics, was only 18 percent of its 1991 level (Medvedev 2005). The absolute volume of foodstuffs consumed decreased dramatically, especially of meat which went from 41 kg per capita in 1989 to 18 kg in 1992 and of milk and milk products, from 332 kg to 110 kg. In the succeeding years, the situation became even worse. According to a representative survey, at the end of 1993, the daily food of an inhabitant of Tbilisi consisted mostly of tea, bread, potatoes, and vegetables—mostly cabbage (Gachechiladze 1995: 117). In the words of Gachechiladze, cheap (i.e., state-subsidized) goods disappeared from retail outlets all over Georgia, as if a vacuum cleaner had gone through it. It is at that time that Georgia "rediscovered" its powerful southwestern neighbor, Turkey. Through a border crossing at Sarpi in Adjaria, "flows of smuggled non-ferrous metals from all over the former Soviet Union and timber from forest reserves of Georgia began to move towards Turkey. Conversely, a stream of food, garments, and hard currency started to flow from Turkey" (Gachechiladze 1995: 3). Apparently, this improved the situation in the consumer market only slightly, because hundreds of thousands of people who had lost their jobs began to seek employment and habitation abroad, for the most part in Russia, Turkey, and Greece. In Russia alone, the size of the Georgian diaspora is currently about 400,000 (www.saakachvili.narod.ru/diasp.html), which is roughly one-tenth of Georgia's population. Although economic growth resumed in Georgia in the mid-1990s after significant aid was delivered by international financial institutions and some Western countries, growth restarted from an exceedingly low level: in 1999, Georgia's GDP was still 46.7 percent that of 1991 (Medvedev 2005).

It has been observed on many occasions all across the world that tensions in multi-ethnic societies escalate when the economic situation worsens. So what might be expected of the separatist communities of Georgia, which showed unrest even during better times? In the case of Georgia, there was an additional irritant: both Abkhazia and South Ossetia—which had long felt antagonized by Tbilisi—border Russia, where the economy did not slump to the extent it did in Georgia and where the delivery of such public goods as retirement pensions and health care services was

maintained on a higher level. Ossetians have brethren in North Ossetia, a republic in the Russian Federation just north of the Greater Caucasus. And Abkhazians, as was shown earlier, have linguistic relatives in Russia. In addition to Ossetians and Abkhazians, the Armenian population of Javakhetia (Javakh in Armenian), bordering the Republic of Armenia, also became restless.

The double challenge of plummeting economy and rising ethnic separatism that Georgia faced from the very beginning of its independent existence required statesmanship of a kind the nascent country apparently could not deliver. Although technically the country was not young at all, as Georgia had exercised statehood in the past, that was a fairly distant past, and requisite skills of governance and economic management turned out to be extremely deficient attributes. As Lev Anninsky wrote in his review of the above-quoted publication of Georgy Nizharadze, "when everything ended and the time for sober thinking arrived, it became obvious that a minimum of security and well-being was the price to pay for giving up liberty; having now gained liberty, Georgians paid by giving up on [even] minimal well-being. There is no free lunch, that's what the problem is. And so when thousands of youngsters yell 'That's it! Enough! Down with X!' what is it exactly that they have in mind, down with dictatorship or down with liberty? This becomes clear as soon as a new 'sovereign' seizes the niche of an old one" (Anninsky 2004).

Indeed, Georgia's political leaders' major response to the immense challenges their young-old country faced was to reorient the country geopolitically from Russia to Europe and the United States. In particular they saw an opening in the natural desire of the West to obtain alternative (that is, not through Russia) access to Azeri and potentially Central Asian oil and natural gas. Georgia offered a transit route for both products' pipelines and became more of an object of Western interest than ever before in its history. Indeed, the construction of the Baku-Tbilisi-Ceyhan (BTC) oil pipeline and the Baku-Tbilisi-Erzurum (BTE) gas pipeline became major undertakings. The BTC pipeline project gained momentum following the Ankara Declaration, adopted on October 29, 1998, by President of Azerbaijan Heydar Aliyev, President of Georgia Eduard Shevardnadze, President of Kazakhstan Nursultan Nazarbayev, President of Turkey Süleyman Demirel, and President of Uzbekistan Islom Karimov. The declaration was witnessed by the United States Secretary of Energy Bill Richardson, who expressed strong support for the BTC pipeline. The intergovernmental agreement in support of the BTC pipeline was signed by Azerbaijan, Georgia, and Turkey on November 18, 1999, during a meeting of the Organization for Security and Cooperation in Europe (OSCE) in Istanbul, Turkey. The BTC pipeline was commissioned in 2005, and in May 2006, the BTE pipeline was commissioned.

Georgia's leaders envisaged not only significant investment from the West, both associated and not directly associated with these strategic pipelines, but also the incorporation of Georgia into such Western structures as the European Union and NATO. One important side effect of these developments would have been the undermining of ethnic separatism. As the economic situation in Georgia improved, it would become a truly Western country; and Abkhazians and Ossetians would see it as an advantage to stay within Georgia. Generous Western aid indeed arrived in Georgia. To appreciate the scale of this aid one may refer to the results of a "donors' conference" in Brussels on October 22, 2008. At that conference the United States confirmed it would be allocating $1 billion in aid to Georgia. The U.S. Congress subsequently appropriated about one-third of that amount for Fiscal Year 2009. In all, pledges of $4.55 billion in grants and loans over three years were made, with many of the funds coming from the International Monetary Fund and the World Bank, as well as the European Union (about $1 billion in grants and loans) and Japan ($200 million). Four-and-a-half billion dollars amounts to more than $1,000 per citizen of Georgia. This aid was allocated in the aftermath of the disastrous August 2008 war with Russia; but even prior to that, the scale of Western aid to Georgia was impressive. At one point (2004) it became known that the wages of all of Georgia's top government officials—ministers, deputies, the road police, and others—were singlehandedly paid by George Soros, an American billionaire financier. In the end, however, Georgia's leaders may have *over*estimated the potential implications of Western commitments, particularly the readiness to confront nuclear-armed Russia on Georgia's behalf. What they may have *under*estimated were the implications of Georgia's geographic location, the time required for profound socioeconomic change, and the extent and tenacity of ethnic grievances and separatist aspirations within their own country. In what follows, we will touch upon the political leadership dynamics of Georgia, two conflicts with separatists, and the August 2008 five-day war with Russia over one of the separatist regions, South Ossetia.

GEORGIA'S POLITICS

On May 26, 1991, Zviad Gamsakhurdia was elected the first president of independent Georgia. Gamsakhurdia was born into a distinguished family; his father, Konstantine Gamsakhurdia (1893–1975), was a famous Georgian writer. Zviad received training in philology and began a professional career as a translator and literary critic. Subsequently, he became engaged in civil rights activism, and in 1978 he was arrested and sentenced to three years in jail. He continued to be a noted dissident after his release from prison. In no other post-Soviet country did such a distinguished public intellectual come to power in the aftermath of the Soviet Union's breakup. Soviet intellectuals used to castigate their political leaders as uneducated parvenus. The assumption was that should an intellectual come to power, he or she would "civilize" the country. Unfortunately, successful statesmanship, though by no means incompatible with purely intellectual pursuits, requires a combination of organization, communication, and strategy skills as well as a profound sense of responsibility. From this perspective the refined intellectual Gamsakhurdia was not a successful leader. He stoked Georgian nationalism and vowed to assert Tbilisi's authority over the secessionist regions of Abkhazia and South Ossetia. Amidst passions fueled by lasting ethnic grievances, he made inflammatory statements, such as "subversive minorities" like the Ossetians "should be chopped up, they should be burned out with a red-hot iron from the Georgian nation . . ." (quoted in Robert English, *Georgia: The Ignored History,* NY Review of Books No. 17, November 6, 2008; available at www.nybooks.com/articles/22011). Needless to say, such statements only made it easier for separatists to make their case, and the offensive rhetoric continues to be quoted to this day.

Unable to either improve a disastrous economic situation or, much less, resolve inter-ethnic conflicts, Gamsakhurdia was overthrown in a bloody coup d'état that lasted from December 22, 1991 to January 6, 1992. The coup was instigated by part of the National Guards and by a paramilitary organization called "Mkhedrioni" or "horsemen." The country became embroiled in a bitter civil war which lasted three years. Gamsakhurdia escaped through opposition lines and made his way to Azerbaijan, where he was denied asylum. Armenia finally hosted Gamsakhurdia for a short period and rejected the Georgian demand to extradite him back to Georgia. In order not to complicate already tense relations with Georgia, Armenian authorities

allowed Gamsakhurdia to move to the breakaway Russian republic of Chechnya, where he was granted asylum by the rebellious government of General Dudayev. Gamsakhurdia returned to Georgia on September 24, 1993, establishing what amounted to a "government in exile" in the western Georgian city of Zugdidi. However, the "Zviadist" uprising quickly collapsed and Zugdidi fell to government forces, aided by Russian troops, on November 6. In December 1993, Zviad Gamsakhurdia died in circumstances that are still unclear. His remains were re-buried in the Chechen capital, Grozny. They were subsequently returned to Georgia for yet another reburial. Gamsakhurdia was interred alongside other prominent Georgians at the Mtatsminda Pantheon on April 1, 2007.

Gamsakhurdia was succeeded as president by the doyen of Georgian politics, Eduard Shevardnadze, who in 1992 returned to Georgia after a distinguished career (1985–1991) as Soviet foreign minister. In that capacity Shevardnadze had become popular in the West for his role in the détente which marked the end of the Cold War. He was credited with allowing the Soviet Union's European satellites to remove their communist governments rather than forcibly restraining any attempts to pursue a different course. Prior to 1985, Shevardnadze had served as the First Secretary of the Communist Party of his native Georgia for thirteen years. Incidentally, it was during that period that Gamsakhurdia was jailed as a dissident. In the late 1970s Shevardnadze managed to gain from the Kremlin an unprecedented concession in defense of the constitutional status of the Georgian language in Georgia.

In 1992, struggling with the immediate aftermath of the Soviet Union's breakup and with full-blown separatism, Georgia waited impatiently for Shevardnadze's return from Moscow. Upon his return, Shevardnadze joined two leaders of the anti-Gamsakhurdia coup to head a triumvirate called the State Council. Shevardnadze was appointed acting chairman of that entity in March 1992. When the Presidency was restored in November 1995, he was elected with 70 percent of the vote. He secured a second term in April 2000 in an election that was marred by widespread claims of vote-rigging.

During the war in the Russian republic of Chechnya on Georgia's northern border, Russia accused the Georgian leadership of harboring Chechen guerrillas on Georgia's soil. At the same time, Georgia accused Russia of rendering support to Georgian separatists. Which of these accusations matched reality, which implied action came first, and which was retaliatory are matters of debate. It is, however, indisputable that further friction with Russia was caused by Shevardnadze's close relationship with the United States. Under Shevardnadze's administration, Georgia became a major recipient of American foreign and military aid, signed a strategic partnership with NATO, and declared its ambition to join both NATO and the European Union. Russia did not like it either when Shevardnadze secured a $3 billion project to build the BTC pipeline through Georgia, arguably Shevardnadze's greatest achievement, because this pipeline was to be built expressly to circumvent Russia.

Under Shevardnadze's reign, Georgia suffered badly from the effects of crime and rampant corruption, often perpetrated by well-connected officials and politicians. This was not something entirely new, only the scale of public property theft and of kickbacks that Shevardnadze's closest advisers reportedly received for securing contracts with Western and Russian firms were unprecedented. It was estimated by some outside observers that Shevardnadze's inner circle controlled as much as 70 percent of the economy. Eventually, even his American supporters started to criticize him.

In November 2003, Shevardnadze was deposed by the bloodless Rose Revolution after Georgian opposition and international monitors asserted that the parliamentary elections were marred by fraud. The opposition protest reached its peak when Shevardnadze attempted to open the new session of parliament. This session was considered illegitimate by the major opposition parties. Supporters of two of those parties, led by Mikheil Saakashvili, burst into the session with roses in their hands (hence the name Rose Revolution), interrupting a speech by Shevardnadze and forcing him to escape with his bodyguards. Shevardnadze subsequently met with opposition leaders Saakashvili and Zurab Zhvania to discuss the situation, their meeting arranged by the Russian foreign minister. After the meeting, the president announced his resignation. That prompted euphoria in the streets of Tbilisi. More than 100,000 protesters celebrated the victory all night long, accompanied by fireworks and rock concerts. An important source of funding for the Rose Revolution was the network of foundations and NGOs associated with George Soros.

Mikheil Saakashvili, one of the three leaders of the Rose Revolution, was elected President of Georgia in 2004 at the age of 36. Born and raised in Tbilisi, Saakashvili graduated from the School of International Law of Kiev State University (Ukraine) in 1992. In 1993–1994

he attended Columbia University's Law School for one year, and in early 1995 he undertook an internship in the New York law firm of Patterson Belknap Webb & Tyler. He was then approached by Shevardnadze's emissary with an offer to enter public service in his native Georgia. In December 1995, Saakashvili won parliamentary elections, standing for the Union of Citizens of Georgia, Shevardnadze's party. In 2000, he became Minister of Justice. He initiated reforms in the Georgian criminal justice and prisons system. But in mid-2001 he accused two other ministers and the Tbilisi police chief of profiting from corrupt business deals. Saakashvili resigned in September, 2001, saying that he considered it immoral to remain a member of Shevardnadze's government. He declared that corruption had become the scourge of the Georgian government and that Shevardnadze lacked the will to deal with it. Having resigned his government post, Saakashvili founded a new center-right political party and hired some well-connected American advisors.

As a new president who won, stunningly, 96 percent of the popular vote, Saakashvili launched a series of reforms to strengthen the country's military and economic capabilities. Saakashvili's effort to reassert central Georgian authority in the southwestern autonomous republic of Ajaria led to a showdown with the local Abashidze ruling clan but in the end was successful. This success encouraged Saakashvili to intensify his efforts in breakaway South Ossetia and Abkhazia. Ajaria, however, bore scarce comparison with those separatist regions for the simple reason that Ajaria has consistently been an ethnic Georgian province albeit with a Muslim population. However, over the course of several decades, the Muslim community has been shrinking as younger generations either tend to be agnostic or convert to Orthodox Christianity. Unlike Abkhazians and Ossetians, for Ajarians awareness of their local roots is secondary to their Georgian identity.

Under Saakashvili, relations with Russia deteriorated even more than under his predecessor. Among other things, Saakashvili's government demanded that Russia's military bases (dating back to the Soviet era) in Batumi and Akhalkalaki be closed. In May 2005 Georgia and Russia reached a bilateral agreement by which Russian military bases were withdrawn. Russia fulfilled the terms, moving all personnel and equipment from these sites by December 2007, ahead of schedule. "Saakashvili grasped very early the importance of connections in Washington, and as president he has paid $800,000 to

Orion Strategies, a two-man lobbying firm run by Randy Scheunemann, a close advisor to John McCain" (Wendell Steavenson, *The New Yorker,* Dec 15, 2008: 68). The relationships between Saakashvili's government and the administration of George W. Bush became exceedingly cordial. During his May 2005 visit to Tbilisi, Bush called Georgia a beacon of liberty. "The American people will stand with you," Bush assured Saakashvili. While the tone of American press releases about Saakashvili's government was mostly favorable, voices of dissent were heard as well. Here's what Richard Carlson, former U.S. ambassador, former Director-General of the Voice of America, and a self-described Republican, wrote less than two weeks after Bush's visit to Tbilisi.

> I was in Georgia last month, and it is still colorful and still difficult, a poor country, poorer even than Haiti, with a new president but the same culture—one that cultivates a swaggering, prideful masculinity in its leaders who, since the fall of the Soviet Union, have been lionized by the U.S. foreign policy establishment and the Western press but who just as quickly seem to morph from lion to demon. A case in point is Eduard Shevardnadze, once the Soviet foreign minister, who was for more than a dozen years invariably described in the West as a stalwart friend of democracy and a liberal, honest fellow. Six months ago, he was ousted as the president of Georgia in a coup led by his young protégé, Mikheil Saakashvili, who is glorying in the same lavish treatment from the State Department and the media. They now paint him as honest, liberal, and democratic, while Shevardnadze is Bronx-cheered as corrupt and murderous, a brute who was forced from office by what Saakashvili (with an unerring eye for the sixties-sentimentality of the Western media) dubs "The Rose Revolution." (Carlson 2004)

Indeed, Saakashvili's democratic credentials were tested in October 2007 when a series of anti-government demonstrations were sparked by accusations of murder and corruption leveled against the president and his allies by one of Saakashvili's erstwhile associates and a former member of his government. The protests climaxed in early November 2007 and involved several opposition groups. The government's decision to use police force against the protesters evolved into clashes in the streets of Tbilisi on November 7. A declaration of a state of emergency by the president and restrictions imposed on some mass media sources led to harsh criticism of the Saakashvili government both in the country and abroad.

Saakashvili was nevertheless reelected as President with 53 percent of the popular vote in January 2008. It appears, however, that his major pledge to his constituency to defeat separatism and to restore the territorial integrity of Georgia has not been fulfilled. Moreover, today the probability of accomplishing these goals seems lower than ever before.

OSSETIAN SEPARATISM IN GEORGIA AND THE WAR WITH RUSSIA OVER SOUTH OSSETIA

Annulling South Ossetian autonomy in 1990 did not resolve the problem. On the contrary, Ossetian-Georgian tensions escalated into the 1991–1992 war that killed 3,000 people and generated two flows of refugees. Estimates of Ossetian refugee flow from South Ossetia and other Ossetian settlements in Georgia proper to North Ossetia (a republic in the Russian Federation) are wide ranging—from 50,000 to 100,000 people. More than 20,000 ethnic Georgians abandoned South Ossetia, now officially referred to (in Georgia) as the Tskhinvali region. In that mountainous region, no areas of contiguous Ossetian or Georgian settlement existed. Rather, there was an ethnic mosaic with clusters of ethnic Ossetian interspersed with clusters of Georgian villages. Without a continuous frontline, keeping peace was exceedingly difficult. Each side attributed atrocities to the other. The Ossetian side, for example, reported that in February 1991 Georgia terminated the supply of electricity to the city of Tskhinvali, and many old and sick froze to death. Writing in 1995, Gachechiladze stated that in 1993, relationships started to improve and that the Tskhinvali separatists' insistence on reunification of South and North Ossetia under Russian patronage was a bargaining chip to restore South Ossetian autonomy in Georgia. Indeed, in 1994, a Joint Control Commission was set up with Russian and Georgian peacekeepers deployed in the region; and in 1995 Georgia's government and Ossetian separatists signed the Russian and OSCE-mediated memorandum renouncing the use of force in the future, and some refugees returned to their homes. But tensions did not lessen. They flared up again in the summer of 2004 after the Georgian police shut down the Ergneti market, a

major trading point for tax-free goods from Russia. After several days of gun and mortar fire, a ceasefire agreement was signed, but it was soon violated. In January 2005, in his speech to the Parliamentary Assembly of the Council of Europe in Strasbourg, France, Saakashvili outlined his new propositions, including granting broad authorities to the South Ossetian parliament, and pledged Georgia's investment in improving the well-being of the region in exchange for the renunciation of separatism. Yet no rapprochement was achieved; and in November 2007 two rival elections were conducted in South Ossetia, one of which took place in Tskhinvali and villages under the control of separatists and the other one in settlements under Georgian control. They elected two leaders with diametrically opposite agendas, the separatist leader Eduard Kokoity pledging independence from Georgia and the pro-Georgia leader Dmitry Sanakoyev submitting a plan to Georgia's parliament for pacifying the region. Clashes and incidents escalated in the summer of 2008 amidst high-level diplomatic activities, yet no large-scale military contingents participated in those clashes until the night of August 7–8. From then on, the development of the conflict has become the object of intense international scrutiny. According to the Georgian side, 150 Russian tanks crossed the border into South Ossetia through the Roki Tunnel (a tunnel across the Greater Caucasus range), so Georgia launched a military offensive to restore constitutional order in the whole region. Specifically, a multiple rocket launcher system was used by the Georgian forces to shell the separatist capital of Tskhinvali. As a result, twelve Russian peacekeepers were killed, and there were multiple civilian casualties. The Russian and Georgian accounts of the course of events immediately leading to the full-scale conflagration of August 9–13, of course, differ.

Here is the timeline of events from a British publication of August 12, 2008.

August 7

Georgian troops launch a surprise attack on Tskhinvali, capital of South Ossetia, "to restore constitutional order". Russia is outraged, and later accuses Georgian troops of war crimes against civilians.

August 8

Russia pours troops into South Ossetia, vowing to defend Russian "compatriots." Fighting takes

place in Tskhinvali. Russian jets attack Georgian military bases. President Saakashvili of Georgia says that Russia and Georgia are now at war. Thirty Georgian and 21 Russian troops are reported killed. Thousands of South Ossetian civilians flee, others shelter in basements.

August 9

Russia captures Tskhinvali. Russian fighter jets drop bombs in Georgian city of Gori, 15 miles from border with South Ossetia. Sixty civilians are reported killed. Russia claims that 1,400 civilians killed in South Ossetia fighting and Georgia estimates 130. America lends military transporter aircraft to fly many of Georgia's 2,000-strong troop contingent out of Iraq to join the fighting at home. United States and EU peace envoys head for Georgia.

August 10

Georgia calls a ceasefire, Russia says the fighting is continuing. Russian fighters bomb outskirts of Tbilisi. Russian warships deployed along Georgia's Black Sea coast. Rebels in Abkhazia say they are mobilizing to attack Georgia. The United States criticizes Russia's "disproportionate" response. Most of Gori's 50,000 inhabitants flee in fear of a Russian invasion.

August 11

Moscow accuses Georgia of continuing to shell Tskhinvali despite its supposed ceasefire. Russian troops invade Georgia from Abkhazia and issue an ultimatum to Georgian troops in area to disarm. EU negotiators visit Gori and Tbilisi, where President Saakashvili signs a draft ceasefire. Georgian troops retreat to Gori, then abandon the city for Tbilisi. Saakashvili accuses Russia of trying to overthrow him and seize control of strategic oil pipelines. Russia accuses him of genocide; he accuses them of ethnic cleansing. UN warns of a growing refugee crisis.

August 12

Shelling kills five in Gori, including a Dutch television cameraman. Russia orders a halt to its invasion. Nicolas Sarkozy *[the French president]* holds talks with Russian leaders in Moscow to broker EU-backed cease-fire negotiation. First humanitarian aid flight lands in Georgia. (www.timesonline.co.uk/tol/news/world/europe/article4514939.ece).

On August 22, 2008, following a negotiated ceasefire between Georgia and Russia, Russia pulled its forces from Georgia proper back to Russia and South Ossetia, leaving military contingents disbursed through various areas as observation and security posts. These were withdrawn by October 8, 2008. The war left the mostly Ossetian city of Tskhinvali in ruins, ethnic Georgian villages burnt to the ground, and tens of thousands of people displaced. As a result of the war, the checkerboard of Georgian-inhabited and Ossetian-inhabited towns and villages ceased to exist because the separatists consolidated the area under their control and declared an independent state of South Ossetia. It was recognized by Russia on August 26, 2008.

As of this writing, exactly one year after the war over South Ossetia ended, three points are worth making.

First, in the above-mentioned timeline Russia justifies its actions by the need to defend "compatriots." Who are these people aside from the peacekeepers who were killed during Georgia's August 8 attack? Ninety-five percent of South Ossetians appear to hold joint Russian-Ossetian citizenship. Gareth Evans, formerly Australia's foreign minister, avers that Russia's responsibility to protect its own citizens cannot be invoked, particularly if there is reason to suspect that citizenship may have been granted with an eye toward making just such a claim (Gareth Evans, "Putin Twists UN Policy," *The Australian,* Sept 2, 2008). The Russian argument refers to humanitarian reasons. As thousands of refugees from South Ossetia inundated the Russian republic of North Ossetia, granting them citizenship was a way to provide them with access to social services, including health care and pensions. Faced with a choice to seek Russia's patronage or to accept Georgia's authority, the overwhelming majority opted for the former.

Second, what was the casualty count of the war? Former German Chancellor Otto von Bismarck is credited with an observation that people never lie so much as after a hunt, during a war, or before an election. On August 9, 2008, the official representative of the South Ossetian separatist government talked about 1,600 civilian deaths on the Ossetian side. The same representative adjusted that number on August 20, claiming there were 1,492 deaths. One year after the war, the Investigative Committee of the Russian Federation confirmed 162 civilian deaths. As far as casualties among the Russian military are concerned, Georgia's Media News agency in August 2008 reported 1,789 deaths. The official data of the Russian Defense Ministry: 74 killed, 19 missing in action, and 171 wounded. The most recent data of the Investigative Committee: 67 military deaths. As for Georgia's casualties, data changed as well, although the order of their magnitude stayed the same. Altogether, there were 397 deaths, including 168 military. On the eve of the war, the total population of South Ossetia was 72,000 people (down from 98,000 in 1989), of whom 64 percent were Ossetians and 25 percent Georgians. One year after the war the population of South Ossetia is about 60,000, which means that the number of Georgians who fled South Ossetia is about 12,000.

Third, if one follows the European and American media commentaries that have appeared since the end of the war, one can observe both continuity and change. On the continuity side, Georgia is seen as the underdog in the conflict and this warrants humanitarian and other assistance to that country. Russia is seen as a bully who applied a disproportional force. Not a single country, except Russia itself, Nicaragua, and Venezuela, recognized South Ossetia (as well as Abkhazia) as an independent state; all or most of the international community believes that territorial integrity of Georgia must be upheld. And yet the style of commentaries changed with regard to Mikheil Saakashvili. He is now usually seen as an opportunist unrealistically staking his hopes on Western willingness to confront Russia on Georgia's behalf. As early as November–December 2008, the almost universal consensus in the Western media had become that Saakashvili actually started the war. "Privately, every Western diplomat I spoke with," wrote Wendell Steavenson in *The New Yorker,* "said that the Georgian attack on Tskhinvali was a mistake." The cold war paradigm of Russia's military threat, in terms of which the West saw the Russia-Georgia conflict, is now reflected upon and occasionally reassessed. "After years of provocations by Georgia and Russia alike, Georgia launched an attack in August against the separatist enclave of South Ossetia. The attack was stymied by a large-scale invasion and the defeat of the Georgian army on its home soil," wrote *The New York Times* in December 2008. Moreover, George Bush's pledge to incorporate Georgia into NATO has now been put on the back burner. "The US and the EU have concluded Mr Saakashvili contributed to his own misfortune by letting himself be provoked," wrote Stefan Wagstyl and Elizabeth Gorst in the *Financial Times* of London in August 2009. "Discontent is rife, as are complaints about Mr Saakashvili's heavy-handed rule and the folly of going to war." Perhaps the highest point (to date) in reassessment of the 2008 Russian-Georgian war over South Ossetia has been the September 2009 report commissioned by the Council of the European Union. More than 30 European military, history, and legal specialists—headed by Swiss diplomat Heidi Tagliavini—compiled the document. This much-anticipated European Union inquiry into the August 2008 war in Georgia concludes that Georgia ignited the conflict by attacking separatists in South Ossetia, but that Russia had provoked violence in the enclave for years and exploited its consequences. The report found no evidence that there was a Russian invasion under way on August 7, when Georgia ordered the shelling of the South Ossetian capital, Tskhinvali. It says Georgia broke international law by using force against Russian peacekeepers stationed in the city, and that Russia's army had legal grounds to defend the peacekeepers. To be sure, the report also says that Russia "went far beyond the reasonable limits of defense" in undertaking a drive outside South Ossetia that violated international law and was "not even remotely commensurate with the threat to Russian peacekeepers." But this sort of evenhandedness constitutes the biggest setback for Mikheil Saakashvili who has long counted on the unconditional support of the West and on similarly unconditional repudiation of Russia. Saakashvili's image both at home and abroad is now seriously tarnished, and the opposition, which includes some of Saakashvili's Rose Revolution partners, may be gaining influence.

ABKHAZIAN SEPARATISM

Prior to the war in South Ossetia the Georgian army had been trained by Western military instructors and was reasonably well equipped, but the brunt of the separatist war in Abkhazia fell in 1992–1993, when Georgian forces were not well prepared. At that time, they faced not a regular army but an Abkhazian militia, and yet they did not succeed. It is highly unusual that a minority accounting for less than 18 percent of the population of a separatist region (and that was the share of Abkhazians in Abkhazia in 1989) was able to organize and mount an offensive that drew out the much more numerous Georgians, who should have been protected by their central government. Abkhazians, however, were supported by many volunteers from the so-called Confederation of Peoples of the Caucasus (CPC), an umbrella group uniting a number of indigenous groups of the North Caucasus, which included but was not limited to the linguistic relatives of Abkhazians. Among those volunteers was Shamil Basayev. Little-known at the time, he later became one of the leaders of Chechen separatists in Russia. The CPC, however, was a newborn group at the time, and following the war in Abkhazia it quickly lost ground. Be that as it may, the united forces of the Abkhazian militia and the CPC managed to defeat Georgian forces in Abkhazia. Thus Sukhumi, the capital of Abkhazia, fell on September 27, 1993; and Shevardnadze, who had arrived in Sukhumi to lead the Georgian army, narrowly escaped death. The Abkhazian and CPC forces committed numerous atrocities against ethnic Georgians in Sukhumi and elsewhere. Most Georgians who survived the war—more than 200,000 people—were forced into exile. One of those refugees appears to have been Militon Kantaria, who in 1945 had hoisted a red Soviet banner on the roof of the German Reichstag. Kantaria fled to Tbilisi and then moved to Moscow where he died in 1993. Later, from 40,000 to 60,000 Georgians returned to the southernmost district of Abkhazia, which prior to the war had been populated almost entirely by ethnic Georgians. The fact of the ethnic cleansing of Abkhazia of its Georgian residents has been recognized by numerous international institutions, including the UN General Assembly and Security Council. In April–May, 1998, the conflict flared up yet again, and an additional 20,000 Georgian refugees fled the southern part of Abkhazia.

The new government of President Mikheil Saakashvili promised not to use force and to resolve the problem by diplomacy and political talks. But as that government leaned ever closer to the West, Russia, which had hitherto maintained neutrality in the Georgian-Abkhazian conflict, began to render support to the Abkhazian side, for example, by commencing the mass issuing of Russian passports to Abkhazians upon request. As in the South Ossetian case this was motivated by humanitarian reasons. The following excerpt from Nicolai N. Petro's article justifying Russian intervention in Georgia highlights the passport situation as it applies to Abkhazia.

Sergei Bagapsh, then Abkhazia's prime minister, described meeting with Georgia's president Eduard Shevardnadze in 1998 to resolve the passport crisis. "Nobody cares about our need to import

medicines," he complained. According to Bagapsh, however, Shevardnadze angrily refused to issue any Georgian passports to Abkhaz, suggesting they make do with U.N. travel documents. Bagapsh told him that, in that case, "We will ask Russia to help—and in five years most of our citizens will have Russians [sic] passports." This is precisely what happened. Through the 1990s most of those who applied for dual citizenship did so mainly to receive basic state benefits, like pensions, and to be able to travel abroad. This changed in 2002 when Russian citizenship laws became more stringent. June of 2002 alone saw some 150,000 Abkhaz apply for Russian citizenship, just before the new law came into effect. Not many years later, Alexander Ankvab, Abkhazia's new prime minister, explained why so many Abkhaz prefer Russian citizenship to Georgian as follows:

"Russia helped us to survive. . . . When our passports, because of the position of the Georgian authorities, lost their legal status and our citizens could not leave the country . . . our elderly could not get a Georgian pension, even in the laughable amount of 10–60 rubles, but today from Russia 25,000 pensioners receive 40 million rubles in pension. More than a hundred seriously ill received free medical care in Moscow. This year Abkhazia received 200,000 free textbooks from Russia."

In late July of 2006 the Kodori crisis erupted, resulting in the establishment of the government of Abkhazia in Kodori (northeastern Abkhazia), a structure subordinated to Tbilisi and opposed to the separatist government in Sukhumi. On August 10, 2008, the war in South Ossetia spread to Abkhazia, where separatist rebels and Russian airplanes launched an attack to force Georgian troops out of the Kodori

Gorge, which they controlled. As a result of this attack, Georgian troops were driven out of Abkhazia entirely. On August 26, 2008, Russia officially recognized Abkhazia as an independent state. Unlike South Ossetia, which can hardly support itself, Abkhazia is in possession of a Black Sea coastline with dozens of former Soviet government resorts and with extensive tangerine plantations. Now the separatist authorities watch with anxiety as Russian tycoons buy up some of the best pieces of coastal land. Abkhazia is also next-door to the Russian city of Sochi, the site of the 2014 Winter Olympics. Significant investments in the construction of various facilities for that event are expected to spill over into Abkhazia.

As for Georgia, just as it did many times in its earlier history, it now finds itself in a precarious situation. Having survived a disastrous war, it depends on financial aid from powerful but distant and not entirely committed sources, no less than one-quarter of its population seeks its fortune abroad, and its quest for the restoration of its territorial integrity is suspended. After Russians consolidated Georgian lands in their empire-turned-Soviet-Union and—with the aid and at times under the stewardship of native Georgians—administered Georgia for two centuries, the country has eventually restored its independent statehood but has not yet found its bearings. Most importantly, it is yet to nourish and promote statesmanship on a par with David-the-Rebuilder and his great granddaughter Tamar, the legendary rulers. If Georgia succeeds in gaining mature and shrewd leaders, it will survive and resolve its problems. But if not, all bets are off.

Central Asia

Central Asia is composed of five countries, each of which has "stan" as the ending of its name. This ending means "place of" in Persian, so that Kazakhstan is the place of Kazakhs, Uzbekistan of Uzbeks, and so on. Besides the common ending of the country-level toponyms, Central Asian countries have other common features.

First, they all control some part of the world's largest closed drainage basin. Indeed, with the exception of the Irtysh, Tobol, and Ishim watersheds of northern Kazakhstan, all Central Asian rivers flow into either the Aral Sea or the Caspian Sea, salt lakes that have no outlets. On a topographic map, Central Asia looks like a giant (1.5 million sq miles) saucer with elevated edges, the highest elevations being along the southern edge, which is composed of some of the world's highest mountain systems.

Second, much of Central Asia is arid; desert and semi-desert are the two largest biomes in the region.

Third, Central Asia is the only macro-region of the former Soviet Union where all titular nationalities (Kazakh, Uzbeks, Turkmen, Kyrgyz, and Tajik) are people of Islamic background, and four out of five speak Turkic languages; Tajik, an Iranian language, is the exception.

Fourth, at the time of Central Asia's incorporation into the Russian Empire (the second half of the 1800s) this region experienced even earlier stages of socioeconomic development than Russia, which itself lagged behind Western Europe and North America. No homegrown capitalism had developed in Central Asia before it was conquered by Russia, and consequently capitalism was unknown prior to the dissolution of the Soviet Union.

Feudalism combined with pre-feudal nomadic communities dominated the region before state socialism was imposed on it.

Fifth, despite the fact that Central Asian economies are almost entirely creatures of the Soviet period and despite undeniable socioeconomic progress experienced by the region during the Soviet period, by the time of the Soviet Union's demise Central Asia was still the least developed and least urbanized part of it, and it had the most rapid population growth. In fact, all five union republics of Central Asia offered social services (like education, health care, and retirement pensions) on budgets that could not be sustained without financial transfers from the federal center. Only one union republic outside Central Asia, Azerbaijan, was in the same category. Although to some extent, such dependencies were conditioned by the Soviet pricing practices applied to components of economic output (whereby oil and gas were fairly cheap), there is no denying that Central Asia emerged from the Soviet period as largely a Third World region. This qualification may no longer apply to part of Kazakhstan, but by and large it remains valid for much of the region today, perhaps even more so than when the Soviet Union existed.

Many toponyms of Central Asia have multiple spellings in English. This is because some traditional spellings were transliterations from Russian and the Cyrillic alphabet, but today it makes more sense to transliterate geographic names from the languages of the Central Asian titular nationalities. This creates a kind of disorder of which the reader ought to be aware and which is only exacerbated by the frequent renaming of cities with changes in political control.

Uzbekistan

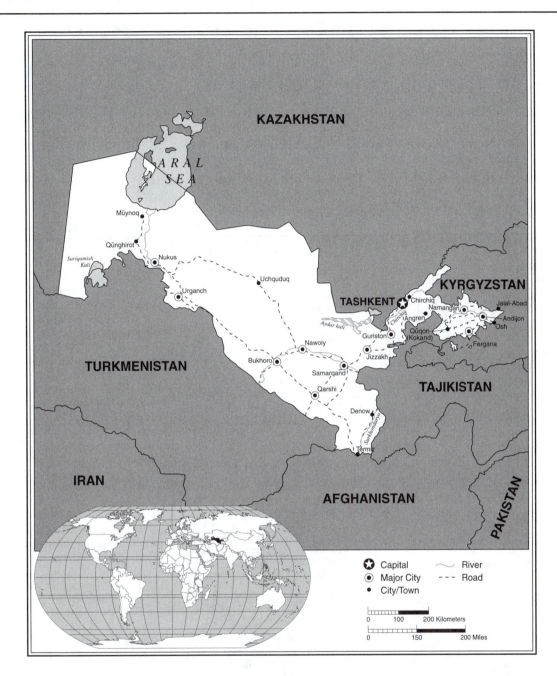

Uzbekistan Statistics

Area in Square Miles (Kilometers):
  172,741 (447,400)
Capital (Population): Tashkent
  (2,180,000)

Population

Total (millions): 27.6
Per Square Km: 10
Percent Urban: 36

Rate of Natural Increase (%): 1.8
Ethnic Makeup: Uzbek 80%; Russian 5.5%;
  Tajik 5%; Kazakh 3%; Karakalpak
  2.5%; Tatar 1.5%; other 2.5% (1996 est.)

Health

Life Expectancy at Birth: 65 (male);
  71 (female)

Infant Mortality Rate: 48/1,000
Total Fertility Rate: 2.6

Economy/Well-Being

*Currency ($US equivalent on 14 October
  2009):* 1498 som = $1
2008 GNI PPP Per Capita: $2,660
*2005 Percent Living on Less than
  US$2/Day:* 50

Uzbekistan Country Report

The Republic of Uzbekistan is sometimes called a doubly landlocked country. Indeed, it is one of only two countries in the entire world from which one needs to cross the territory of *two* other countries in order to reach an ocean or any of its flanking seas (The other such country is Liechtenstein). Uzbekistan has a land area of 447,400 sq km and a population of 27.6 million people (2009 estimate).

Uzbekistan is the most populous country of Central Asia; and its capital city, Tashkent, is Central Asia's largest city (with 2.2 million people). Much of Uzbekistan consists of the Kyzyl Kum, a mid-latitude desert which lies between two great rivers, the Syr Darya and the Amu Darya. This desert experiences wide temperature ranges in both its daily and annual patterns and a stark deficiency of water outside the few river valleys. The rest of the republic is made up of large intermontane valleys—Fergana, Chirchik-Angren (in the east) and Surkhan-Sherabad (in the south)—fringed with foothills and mountains such as the western Tian Shan and the Gissar range and the valley of the Zeravshan river, which used to be the major tributary of the Amu Darya, but no longer reaches it because of extensive water consumption in and around the ancient city of Samarkand, Uzbekistan's second largest

city. The Fergana Valley possesses particularly fertile soils due to two rivers, the Naryn and the Kara Darya, which unite in the valley near Namangan to form the Syr Darya. The human population is distributed extremely unevenly in Uzbekistan, and population density ranges from 7 people per sq km in the Navoiy Viloyati (*viloyati* is a civil subdivision) to 590 people per sq km (1,529 people per sq mi) in Andijan Viloyati. All three of the viloyatis in the most climatically hospitable and fertile Fergana Valley—Fergana, Namangan, and Andijan—are extremely densely settled. In fact, of the 164 civil divisions in the twelve CIS countries, in 2008 only six divisions had population densities in excess of 150 people per sq km. Five are Uzbekistani viloyatis; the sixth is the highly urbanized Moscow Oblast of Russia. In the viloyatis of the Fergana Valley, high population density is sustained in a predominantly rural setting without major metropolises—a fact that underscores agricultural pressures on the land. Outside the Fergana Valley, high population density is recorded around the city of Tashkent (in the Chirchik-Angren Valley) and in the only densely settled area of western Uzbekistan, in the Khorezm oasis (multiple spellings occur in English, including Chorezm, Khwarezm, or Xorazm, as it is now spelled in Uzbekistan

itself) centered around the city of Urgench on the Amu Darya.

The northwestern part of Uzbekistan, around the much diminished Aral Sea, is a semi-autonomous area called Karakalpakstan (or Qaraqalpaqstan). Within the Soviet Union it had been designated as the national home of Karakalpaks, an ethnic group distinct from Uzbeks and speaking a Turkic language which is closer to Kazakh than to Uzbek. Karakalpaks now account for roughly one-third of the local population; the rest are Uzbeks and Kazakhs.

UZBEKS: CONSTRUCTING AN IDENTITY

Ethnic Uzbeks now make up 80 percent of Uzbekistan's population. The ethnonym Uzbek derives from Uzbeg, the ninth Khan of the Golden Horde, who ruled from 1314 to 1341. Within the Golden Horde the initial Mongol substratum had been largely assimilated by Turkic-speaking people. A group of Turkic-speaking nomads using Uzbek as their name migrated to Mawarannahr, the land between the Syr Darya and the Amu Darya, from the steppes north of the Syr Darya and, in 1500, conquered the ancient cities of Samarkand and Bukhara from the descendants of Timur (Tamerlane). However, the ethnonym Uzbek was accepted by most Turkic-speaking people of present-day Uzbekistan only in the twentieth century largely as a result of the national delimitation policy implemented in 1924–25 by Soviet authorities and the subsequent top-down assignment of the toponym Uzbekistan to one of five Central Asian republics. The incursion of self-proclaimed Uzbeks was by no means the first Turkic incursion into the southern part of Central Asia (that is, Central Asia south of Kazakhstan). Turks first migrated here en masse with Mongols as part of Genghis Khan's armies in the early 1200s. Since that time, the Turkic-speaking population, initially nomadic, has been gradually supplanting the Persian/Iranian-speakers who had populated much of what is now Uzbekistan since the first millennium B.C. It was the Persians who around 700 B.C. founded Samarkand and Bukhara, two of the world's oldest cities, both of which benefited from trade along the Silk Road from China. Although Timur, or Tamerlane, who in 1369 made Samarkand the capital of his dominions, was a Turkic warrior born south of Samarkand, his successors on the throne became Persianized (i.e., adopted Farsi

Uzbekistan: History Museum (formerly Lenin Museum) in Tashkent

(© Vilen Schwartz)

(© Ronald Wixman)

Registan Square in Samarkand, Uzbekistan

as their language), so strong was the Iranian cultural supremacy in Mawarannahr. In fact, Iranian cultural influences continued to be strong in Bukhara and Samarkand for centuries thereafter, and even now Tajik, an Iranian language, is spoken by large groups in both cities. However, in the early 1920s, Moscow decided to assign Bukhara and Samarkand to Turkic-speaking Uzbekistan, not Iranian-speaking Tajikistan. Because belonging to the titular nationality of a republic was instrumental in one's career advancement and was a shield against discrimination, many Iranian-speaking residents of Bukhara and Samarkand preferred as early as the 1930s to be identified as Uzbeks, not Tajiks, in their internal passports.

By the beginning of the twentieth century, three Turkic-speaking groups found themselves within present-day Uzbekistan: (1) Sarts, sedentary people of the oases (often bilingual—Iranian and Turkic-speakers); (2) Turk-Karluks, who retained the semi-nomadic lifestyle of their ancestors, the Turkic tribes of Mawarannahr; and (3) Uzbeks, nomads who began to settle down only during the Soviet period. In 1924, the Soviet regime decreed that henceforth all settled Turks in Central Asia would be known as "Uzbeks," and that the term "Sart" be abolished as an insulting legacy of colonial rule. Indeed, according to an incorrect but widespread belief, the name Sart was derived from *Sart İt* ("Yellow Dog" in Turkic), an insulting term used for town dwellers by nomads;

and it was this supposed root which led the Soviets to abolish the term for being "derogatory." Ironically, the language that was designated as the Uzbek literary language was actually one of the languages spoken by the Sarts. The closest relative of that language within the Turkic group of the Altaic languages is Uighur, the language of the indigenous population of the Chinese province of Xinjiang. Uzbeks (as well as Uighur) are a mixed European-Asian racial type, with faces decidedly more "Caucasian" than their Kazakh and Kyrgyz neighbors.

As true of all peoples with a newly constructed identity, Uzbeks extended their history back in time and adopted glorious ancestors such as Timur, now a national Uzbek hero. Some major streets in Uzbekistan cities were named after him and huge monuments were erected in Tashkent and Samarkand when Uzbekistan became an independent country after the breakup of the Soviet Union. Another important member of the national pantheon is Alisher Navoiy (1441–1501), a brilliant poet who lived in Gerat, now in northwestern Afghanistan. Navoiy began to write in Persian and then switched to a Turkic language (referred to as Chagatai, which is believed to be the forerunner of Uzbek). In one of his treatises, "Argument between the two languages," Navoiy makes the point that Turkic-speakers have no reason to look down on their language as it is no less noble nor less prone to poetry than Persian.

The Uzbek historical narrative is in conflict with the Tajik one. In fact, they use different approaches to historical memory. For Uzbeks the focus is territory; thus anything that took place within present-day Uzbekistan is Uzbek by definition. In contrast, Tajiks, whose national territory falls well short of the former expanse of Persian culture in Central Asia, focus on a specific ethnicity and find their numerous precursors in Samarkand and Bukhara, thus encroaching on Uzbek historical memory.

FROM TIMURIDS TO NATIONAL DELIMITATION IN CENTRAL ASIA

Timur, also known as Tamerlane (1336–1405), was the fourteenth-century Turko-Mongol conqueror of much of Western and Central Asia and founder of the Timurid Empire and the Timurid dynasty in Central Asia. A military genius, Timur was born in Shahrisabz, 80 km south of Samarkand. Although the Timurid dynasty formally existed until 1857 as the Mughal dynasty of India, its initial Central Asian core severed its ties with the Timurid empire in the early 1500s and became divided into khanates, feudal entities ruled by khans. Within what is today Uzbekistan, the Khanate of Khiva and the Khanate of Bukhara were established. Subsequently, in the early 1700s, the Khanate of Kokand split off from the Khanate of Bukhara. Except for the Khanate of Khiva, these entities extended into present-day Tajikistan and Kyrgyzstan. In these feudal states, the language of high culture and government bureaucracy was Persian (referred to as Tajik) and so was the language of most of the urban population. However, outside urban areas different Turkic dialects were spoken, and many educated people were bilingual. In the 1700s, the khanates found themselves at the crossroads among expansionist centers of power—China expanding from the east, Russia from the north and west, and Persia from the south. Thus, the khans of Kokand were forced to pay tribute to the Qing dynasty of China between 1774 and 1798. And Khiva and Bukhara briefly (in the 1740s) belonged to Persia; however, the Persian Empire soon crumbled, not in the least due to an exhausting showdown with Russia in the Transcaucasus. With the Iranians/Persians ceasing their attempts to regain Central Asia and with the Chinese limiting their expansion to eastern Turkestan, where in 1761 they established their farthest outpost, the province of Xinjiang, the major rivals for Central Asia came to be Great Britain, the colonial master of India, and Russia. The strategic rivalry and conflict between the British Empire and the Russian Empire for supremacy in Central Asia came down in history as the Great Game, generally regarded as running more or less from the Russo-Persian Treaty of 1813 (a treaty humiliating for Persia) to the Anglo-Russian Convention of 1907.

Advancing from what is now Kazakhstan, Russian forces captured Tashkent in 1865 and two years later made it the center of their new Turkestan province (Turkestan being the name Russians used to denote Central Asia as a whole until the early 1920s). Samarkand became part of the Russian Empire three years later and the independence of Bukhara was virtually stripped away in a peace treaty (between Russia and Bukhara's Emir) the same year. Russian control now extended as far as the northern bank of the Amu Darya. The British regarded Russian expansion as threatening to their rule in India, and in order to thwart that expansion, Great Britain waged two disastrous wars in Afghanistan in the 1800s. Following Russia's 1884 attempt to establish an outpost south of the Amu Darya (in Afghanistan), the Anglo-Russian Boundary Commission was formed to delineate the northern frontier of Afghanistan. The commission led to Afghanistan's becoming effectively a buffer state between British India and the Russian Empire. The new boundary brought the southward expansion of Imperial Russia to a halt. Much of the area with a Turkic-speaking (Uzbek and Turkmen) population was left north of the border in Russia, while a substantial Turkic-speaking (Uzbek) minority found itself south of the border in Afghanistan.[1] From the acquisition of Central Asia in the 1860s to the Communist Revolution of 1917, Russia recognized the power of local rulers (emirs) as well as locally applied Islamic law. The local people were not considered full-fledged Russian citizens and thus were exempted from military service. Aside from relatively small but growing ethnic Russian colonies in Tashkent and Samarkand, there was little interaction between locals and Russians. The Central Asian revolt of 1916, in which many Uzbeks participated, was caused by the Russian government's decision to forcibly mobilize Central Asian Muslims to dig trenches along the frontlines of World War I. During that uprising, many Russian administrators were killed. The Soviet historiography subsequently had a hard time categorizing that uprising as anti-imperialist, because its leaders (Muslim clerics and landlords) did not fit the image of proletarians; instead, the 1916 uprising was described in Soviet textbooks as a popular revolt against colonial rule.

Perhaps the most significant pre-1917 development in the southern part of Central Asia was the construction of railway lines. The first of these, the Trans-Caspian line, was built in 1880–1896 and connected Krasnovodsk (a Caspian port in what is now Turkmenistan) and Samarkand. By 1899, it was extended to Tashkent and Andijan. In 1906, the Trans-Aral railway line (mostly in Kazakhstan) connected Tashkent and Orenburg.

Following the 1917 revolution, there was a flurry of short-lived political arrangements which ended with the 1918 establishment of the Turkestan Autonomous Soviet Republic within Russia with its capital in Tashkent. However, feudal monarchies in Khorezm and Bukhara (under a Russian protectorate since 1865) retained their power until 1920. The standoff between the Soviets and entrenched Sunni Islamic elite was particularly tough in Bukhara. Initially, the Soviets relied on a rebellious political group within Bukhara which called themselves Young Bukharans (Mladobukhartsy in Russian). The group consisted of some 200 activists, most of whom had experience of lasting contacts with Russians. For example, the leader of the movement, Faizulla Khodzhayev (1896–1938), first visited Russia at the age of eleven (with his father) and for a couple of years attended a secondary school in Moscow. Most Young Bukharans gravitated to secular Pan-Turkism; hence the name of the movement resembling Young Turks in the Ottoman Empire. Just as the Young Turks wanted to Europeanize Turkey by taking cues from Western Europe, so Young Bukharans saw allies in revolutionary Russia and tried to coordinate their activities with the Red Army. In March 1918, the Red Army marched to the gates of Bukhara and demanded that the emir surrender the city to the Young Bukharans. The emir responded by murdering the Bolshevik delegation and incited the population to a jihad against the Bolshevik "infidels." About 8,000 people were killed in the ensuing pogroms, including most Young Bukharans. Because the majority of Bukhara's population did not support the invasion, the Red Army force retreated to Soviet-controlled Samarkand outside the Emirate of Bukhara. But the emir had won only a temporary breathing space. In August 1920, the Red Army troops attacked the city of Bukhara, and the Emir, Alim Khan, fled to Dushanbe and later to Kabul, Afghanistan. On September 2, 1920, after four days of fighting, the emir's citadel was destroyed and Bukhara was seized by the Red Army. In 1920, the Khorezm People's Soviet Republic and the Bukhara People's Soviet Republic were created within the borders of the former khanates. These republics cut across the earlier established Turkestan Autonomous Republic. Along with the Kirgiz Autonomous Republic, the division of Central Asia was maintained until 1925. Since 1924, the Bolsheviks had been implementing a plan of national delimitation and creating five separate republics in Central Asia; Uzbekistan was one of them. This plan was adopted by the Politburo of the Communist Party on the grounds that it would be easier to govern and pursue Marxist-Leninist indoctrination within linguistically homogeneous groups led by their respective leaders, who had been co-opted by the Soviet regime. Not only were borders between the new territorial units to be drawn, but also *within* each such unit the homogenization policy was to be conducted through the public education system so that new socialist nations would take shape. The policy was justified as necessary by the fact that at the beginning of the twentieth century there was no single Uzbek identity, yet there were groups speaking closely-related Turkic dialects which were summarily labeled Uzbek. Based on this grouping, it was considered perplexing and inappropriate that out of the total number of people thus defined as Uzbeks, 66.5 percent lived in the Turkestan Republic, 22.2 percent in the Bukhara Republic, and 11.3 percent in the Khorezm Republic. The term "national delimitation," used in English-language accounts of the early Soviet period in Central Asian history, is a translation of the Russian term *natsionalnoye razmezhevaniye*. A more direct translation of that term might be "national disengagement;" that is, a separation of linguistic communities whereby no, or just a very few, "aliens" would find themselves on the "wrong" side of the newly drawn border. In other words, Uzbeks live only in Uzbekistan, Tajiks in Tajikistan, etc. Such a neat separation was hardly possible, particularly between Uzbeks (Turks) and Tajiks (Iranians), because in most of the oases of Central Asia, Turkic-Iranian coexistence had been the case for almost a millennium. Another reason why neat disengagement was impossible to achieve was because the evolving network of nodal regions (urban centers with their hinterlands) did not match ethno-linguistic geography. For these reasons the delimitation/disengagement plan had its critics. One of them, the famous Russian orientalist Vasily Barthold (1869–1930), in his *Note on Historic Relationship between Turkic and Iranian Ethnicities of Central Asia*, was mildly critical of the plan on the grounds that "the national principle [and] the way it was carried out during the 1924 national delimitation in Central Asia had been a product of the West European history of the nineteenth century and is totally alien to local traditions." But despite criticism, the delimitation idea was implemented in 1925. And while it did leave a sizable Tajik minority in Uzbekistan and Uzbek minorities in Tajikistan and Kyrgyzstan, the country of Uzbekistan and ultimately the Uzbek identity were conceived and pieced together through this early Soviet effort at ethno-political engineering.

Because of its central location in the macroregion of Central Asia—Uzbekistan is the only republic in that region that borders all four of the other republics—it required more effort at disengagement and border delineation than any other Central Asian republic. It is likely that Uzbekistan benefited from the Soviet authorities' favoritism in that Samarkand and Bukhara were assigned to it, although their assignment to Tajikistan would have been at least equally justifiable. Moreover, initially, Tajikistan was formed as an autonomous republic *within* Uzbekistan; only in 1929 did it become a separate union republic. The roots of pro-Uzbek favoritism may be sought in social stratification and class antagonism (unlike in the eastern or Pamir part of the former Bukhara Khanate, where Tajik was spoken by all strata, in Bukhara and Samarkand, it was the upper classes who were mostly Tajik-speakers), and in, however brief, a romantic relationship between Soviet communism and Pan-Turkism, the strangest of bedfellows. It was at the peak of this relationship, in November 1921, that Lenin dispatched Enver Pasha, a Turkish military officer and a leader of the Young Turk revolution, to Bukhara to help suppress an uprising against the local pro-Moscow Bolshevik regime. Even though Enver Pasha switched sides and became the leader of the anti-Soviet Basmachi and was killed in 1922, the idea of an alliance may have still lingered in some Kremlin minds. Bestowed with the most controvertible borders, Uzbekistan predictably experienced more territorial adjustments than other Central Asian republics; in 1929 Tajikistan was subtracted from it and in 1936, the Karakalpak Autonomous Republic was added. Adjustments of the border with Kazakhstan took place repeatedly until the early 1960s.

SOVIET UZBEKISTAN

In 1925, Samarkand was established as the capital of the Uzbek SSR. In 1930, Tashkent became the capital. Practically the entire economy of Uzbekistan is a creature of the Soviet period, although some industry was founded before the Communist Revolution—for example, the oil refinery in the city of Fergana was commissioned in 1908 to process local oil. From the very beginning, all planning and investment were subordinated to the grand task of cotton production, which required large-scale irrigation schemes, which included construction of long irrigation canals. One of the most famous and well-publicized of these irrigation works was the Great Fergana Canal, which drew water from the Naryn River, the largest tributary of the Syr Darya. The canal crosses several left tributaries of the Syr Darya and flows into the Syr Darya itself near Khujunt (former Leninabad) in Tajikistan. Of the total canal length of 345 km, 283 km lie within Uzbekistan, specifically in the southern part of the Fergana Valley. Beginning in late July 1939, 160,000 Uzbek kolkhozniks (collective farmers), assisted by several thousand Tajiks, were mobilized by the Communist Party to construct the canal. The main purpose of this endeavor was to draw water to irrigate the cotton fields of the Fergana Valley and thereby to achieve "cotton independence" for the Soviet Union. The last great construction project of the 1930s, the Great Fergana Canal amazingly took only forty-five days to build. Not surprisingly, there was a high fatality rate during the construction process.

The Great Fergana Canal set in motion forces that would result in the desiccation of the Aral Sea, one of the great ecological disasters of the second half of the twentieth century. This canal alone allowed the irrigation of 257,000 hectares of cropland mostly assigned to cotton. In 1940 and 1941, two more canals in the Fergana Valley were built. By the end of the Soviet period, 911,000 hectares of cropland in the Uzbek part of the Fergana Valley were irrigated. The water of the Amu Darya was diverted into the Amu-Bukhara and Amu-Karakul canals, which also supported irrigated cotton plantations. Large reservoirs were created on the Zaravshan and Kashka-Darya rivers, both tributaries of the Amu Darya.

Cotton became the alpha and omega of the Uzbek economy. During harvests, up to two million college and secondary school students—urban and rural dwellers alike—were mobilized for up to two months at a time to gather cotton. They worked seven to nine hours a day. Picking cotton bolls and putting them into sacks requires squatting for hours at a time under a Central Asian sun that is still hot in September. To avoid this annual chore would be unpatriotic and subject to public censure. In addition to students, many other people whose work was considered of lesser importance and thus worth interrupting for the sake of harvesting cotton were also mobilized.

Cotton harvested in Uzbekistan was subsequently processed at more than 100 cotton-cleaning plants, where fibers were separated from seeds. Nine-tenths of the fiber (1.6–1.8 million tons or two-thirds of the Soviet Union's total cotton fiber output) was subsequently shipped to textile factories outside Uzbekistan; but almost all seeds were delivered to Uzbekistan's oil mills, which once produced about 400,000 tons of cooking oil a year. Cotton accounted for roughly 70 percent of agricultural output; but Uzbek crop farming also produced melons, watermelons, and grapes for all-Union consumption, and animal husbandry specialized on pasture-based sheep breeding producing astrakhan fur and meat. Sheep grazed natural pastureland on the extreme periphery of oases and in the deserts beyond. Some heavy manufacturing plants, mostly in Tashkent, supplied cotton plantations with necessary equipment for cultivating and harvesting cotton. However, the cotton harvesters they produced were not of high quality, so most cotton was picked by hand.

Besides cotton production and its supporting industrial infrastructure, Soviet Uzbekistan boasted significant natural gas, coal, and non-ferrous metals production. The center of the natural gas industry was the town of Gazli, 106 km northwest of Bukhara in the Kyzyl Kum. The town was almost completely razed by a powerful earthquake on May 17, 1976. Owing to the time of the quake, 7:58 A.M., most adults were heading to work and children to schools (which used to start at 8:30 A.M.), preventing mass casualties. By the late 1980s, proven reserves of natural gas at Gazli were considered to be almost exhausted. A coalfield in Angren, about 75 km east of Tashkent, was vigorously exploited to supply several large power plants. Also, a series of hydroelectric stations was built on the Chirchik River in the vicinity of Tashkent (a mountainous river, the Chirchik is a right tributary of the Syr Darya). The largest enterprise of non-ferrous ore-mining and processing was located in Almalyk, 65 km south of Tashkent. A detailed construction plan for the enterprise, based on local deposits of copper, molybdenum, zinc, and lead, was first made in 1940; however, actual construction commenced only after World War II. By the mid-1960s, the Almalyk ore-mining and processing plant (OM&PP) was one of the largest in the Soviet Union.

Besides the above-mentioned metals, the plant also extracts gold from polymetallic ore, producing about 10 tons of gold per year. However, the OM&PP in the town of Nawoiy (in the Kyzyl Kum desert), based on the huge Muruntau strip mine nearby, produces seven times more gold than Almalyk. Operating since 1969, this enterprise has made Uzbekistan the world's ninth biggest producer of gold (and the world's first in per capita output of gold). The presence of gold is implanted in local toponymy: The name of the Amu Darya tributary on which the town of Nawoiy is located—Zaravshan—means gold-bearing in Tajik. One particular enterprise in Tashkent, a large aircraft factory which assembled Soviet *Ilyushin* passenger and cargo planes, did not evolve from local resources but depended totally on a supply of parts from Russia and employed a largely non-Uzbek labor force. The factory was the outgrowth of an aircraft production facility located in Khimki, a northern suburb of Moscow, which had been transferred to Tashkent in 1941, when the frontline was at Moscow's doorstep.

Despite significant efforts at economic development, Uzbekistan remained the second-poorest republic of the Soviet Union. In 1990, only in neighboring Tajikistan was the monetary estimate of personal consumption lower. Widespread, though hidden, unemployment was a scourge of the Uzbek countryside and small towns. Creation of jobs did not keep pace with rapid population growth. By 1990, 60 percent of the population still lived in the countryside, where the birth rate was 45 per 1,000—that is, almost the same as in sub-Saharan Africa; the death rate was about 8 per 1,000. Thus, in the 1980s, the population grew by almost half a million people every year. Although labor-deficient areas were abundant in not so faraway Siberia, Uzbeks were among the ethnic communities least apt to migrate. Aside from social mores such as intricate decision-making processes in extended families and a reluctance to live in a non-Uzbek environment, the insignificant spread of vocational schooling precluded the formation of a skilled labor force that could look for jobs outside the republic as, for example, many indigenous people from the Caucasus had done. Definitely suffering from overpopulation, the Uzbek countryside was particularly poverty-stricken in densely settled areas such as the Fergana Valley.

To anybody who has lived in the Soviet Union, its still fairly recent history expresses itself in taglines or clichés reflecting the most significant events preceding its collapse. In regard to Soviet Uzbekistan, these would be the Tashkent earthquake, the Cotton Affair (khlopkovoye delo), and the massacre of Meskhetian Turks.

THE TASHKENT EARTHQUAKE

At 5:23 A.M. on April 26, 1966, Tashkent was rocked by a strong earthquake. While the magnitude of the quake might not seem overly significant (5.2 on the Richter scale), because of the epicenter's significant depth (from 8 to 3 km) it caused powerful ground shaking (8-to-9 on 1-12 MSK-64 scale[2]). The zone of maximum destruction covered about ten sq km and included the city center. The relatively small number of casualties—eight deaths and several hundred injured—in a city of more than one million people can be attributed to the predominance of vertical (as opposed to horizontal) seismic waves, which precluded total collapse of even dilapidated adobe structures. However, after the main shock, the number of fatalities went up due to heart attacks during even insignificant aftershocks. As a result of the quake, over 300,000 were left homeless, and some 78,000 poorly engineered homes were destroyed beyond repair mainly in the densely packed areas of the old city where traditional adobe housing predominated. All Soviet republics and some foreign countries sent construction workers and urban planners to help rebuild a devastated Tashkent. They created a city looking less oriental than before; rather it was a "model Soviet city" of wide and straight streets, immense public spaces such as parks and plazas for military parades, fountains, monuments, and vast tracts with apartment blocks. About 100,000 new apartments were commissioned by 1970, many of which were filled with the families of the builders. At the time of the collapse of the Soviet Union in 1991, Tashkent was the fourth largest city in the country and a center of learning in the science and engineering fields. Despite the fact that outmigration of ethnic Russians from Uzbekistan commenced as early as the mid-1970s, about half of Tashkent's population was either ethnic Russian or Russian-speaking at the end of the Soviet Union's existence. The Tashkent urban agglomeration that formed included some thirty nearby towns, the largest being Yangiyol (28 km to the southwest of Tashkent) with over 70,000 residents. Accounting for just 8 percent of Uzbekistan's land area, the Tashkent agglomeration became home to one-third of Uzbekistan's total population and two-thirds of its urban population. In 1977, the first subway line was commissioned in Tashkent, which is still the only city in Central Asia with a subway.

RASHIDOV AND THE COTTON AFFAIR

As Uzbeks began to gain leading positions in society, they also revived unofficial networks based on regional and clan loyalties. These networks provided support and often profitable connections among Uzbeks, the state, and the party. While close and even distant relatives acquired certain government positions through birthright, outsiders could buy those positions. An unofficial price list existed for becoming, say, a local policeman, a clerk in a local retail trade department, or anything else up to district communist party leader. These and other positions were lucrative because of the custom of bribing their holders. To avoid a traffic ticket, to get a land use certificate or a building permit, to be treated by a reputable doctor—everything required bribes. Such practices were not by any means limited to Uzbekistan in the Soviet Union; but anecdotal evidence suggests that in that republic they became particularly pervasive under the leadership of Sharof Rashidov.

Rashidov was born in 1917 to a peasant family in Jizzakh. He graduated from Samarkand University as a philologist and worked as a teacher, journalist, and editor of a Samarkand newspaper. As a petty military officer he participated in World War II, was wounded, and returned home in 1942. Having published a book of poems, he became head of the Uzbekistan Writers Union in 1949, and subsequently published three novels (very Soviet or socialist-realist in style). He was elected to the post of Chairman of the Presidium of the Uzbek Supreme Soviet in 1950. In 1959, he became First Secretary of the Uzbek Communist Party, a post he held until 1983—an unusually long time.

To the common Soviet citizen Rashidov became synonymous with corruption, nepotism, and the Great Cotton Scandal of the late Brezhnev period. As production quotas rose, the Uzbek government responded by reporting miraculous growth in cotton output and record improvements in production and efficiency. Most of these records were eventually revealed to have been falsified. By some accounts, Rashidov committed suicide when the likelihood of his being brought to justice became imminent; by other accounts he died from a heart attack. An alternative theory of the Cotton Scandal claims that Uzbekistan actually did fulfill cotton production quotas and even exceeded them, but in the Soviet economy of the 1980s barter and shady deals made up for failures of the system to supply regions with necessary producer goods such as fertilizers, lubricants, etc. As a result, a lot of cotton was simply sold to illicit buyers.

"After the death of Rashidov, Moscow attempted to regain the central control over Uzbekistan that had weakened in the previous decade. In 1986 it was announced that almost the entire party and government leadership of the republic had conspired in falsifying cotton production figures. Eventually, Rashidov himself was also implicated together with Yuriy Churbanov, Brezhnev's son-in-law. A massive purge of the Uzbek leadership was carried out, and corruption trials were conducted by prosecutors brought in from Moscow. The Uzbeks themselves felt that the central government had singled them out unfairly; in the 1980s, this resentment led to a strengthening of Uzbek nationalism." (www.eastlinetour.com/uzbekistan/new_history.html)

THE MASSACRE OF MESKHETIAN TURKS

One of the most tragic manifestations of increasing nationalism was the massacre of Meskhetian Turks. From May 23 to June 8, 1989, 757 houses were burned and 103 people killed, including 52 Meskhetian Turks and 36 Uzbeks; 1,011 people were maimed and wounded. The authorities reacted passively. Their efforts became largely limited to an organized evacuation of the Meskhetian Turks. Thus, a decision was taken to move 17,000 Meskhetian Turks from the Fergana Valley—essentially the entire remaining Meskhetian Turk population of the valley—to several regions in Russia's heartland. But in February–March 1990, pogroms took place in the Tashkent region. The Meskhetian Turks then began to leave Uzbekistan en masse. Some refugees were able to leave their homes while being guarded by the authorities, but most fled on their own. By early 1991, more than 90,000 Meskhetian Turks had left Uzbekistan.

The Meskhetian Turk community hails from Meskheti, a region in southwestern Georgia. It is now part of Javakheti (or Javakh) which also has a sizable Armenian community. Meskhetian Turks originated from the intermarriage of migrants from nearby Turkey and locals who converted to Islam at the time when the area was under Turkish administration (1576–1829). Language and religion separated them from ethnic Georgians. In 1944, Meskhetian Turks became one of the groups sentenced to wholesale deportation from their homeland under the pretext that they had been collaborating with the Turkish intelligence. During World War II, Turkey maintained neutrality but was perceived by the Soviet Union as a hostile country. It is unclear whether any of these accusations held water in a situation when most able-bodied men had been enlisted in the Soviet army, and many had perished on the eastern front. In any case, collective punishment was readily meted out under Stalin, and so about 90,000 Meskhetian Turks were deported to Central Asia, where they had to settle in specially designated areas without permission to leave. This is how many of them came to find themselves in Uzbekistan. In 1956, they were acquitted of collaboration and constraints to their mobility were lifted. But unlike the much more numerous Chechens (also deported), who in the late 1950s were allowed to recreate their national home in the Caucasus, Meskhetian Turks were not welcome in Georgia. Their homes in Meskheti were by then occupied by Georgians, and the government of Soviet Georgia was not willing to accept them. (As a matter of fact, independent Georgia was equally reluctant; only in 2007 was a repatriation law adopted by the Georgian government to allow those Meskhetian Turks willing to apply for Georgian citizenship to do so until the deadline of January 1, 2009. By that time, however, too much water had flowed under the bridge, and it is unlikely there were many applicants.) Most stayed in Uzbekistan, and quite a few became successful and prosperous gardeners and marketers in suburban Tashkent and in the Fergana Valley. Although Turkic-speaking, Meskhetian Turks kept to themselves and did not intermarry with Uzbeks. As befits a migrant community, they developed bonds of mutual solidarity and survival skills which made them on average more enterprising and competitive compared with members of the ethnic majority. The same applied to Koreans deported to Uzbekistan and Kazakhstan from the Far East in the late 1930s and to other migrant communities, such as, for example, Greeks (from the so-called Democratic Army of Greece, a communist group and party to the 1944–49 civil war; many of them willingly emigrated to the Soviet Union and were settled in Uzbekistan) or Crimean Tatars (exiled to Uzbekistan from Crimea in 1945). As to why Meskhetian Turks and no other minority group fell victim to pogroms at the hands of Uzbeks, there are sundry conspiracy theories, but it is unlikely that any is valid.

Most probably, the root cause of the antagonism against the Meskhetian Turks should be sought in the anxious-and-aggressive mood in society that set in at the end of Gorbachev's Perestroika, a mood caused by truly tectonic political shifts combined with a dramatic worsening of the quality of life, which in Uzbekistan had been one of the lowest in the Soviet Union to begin with, particularly outside Tashkent. When fear of the state subsided, it stopped being the major target of grievances, and those grievances could now be easily rechanneled. Hatred of aliens became the most widespread outlet for pent-up aggression, not just in Uzbekistan. Under such conditions, any tension could trigger violence. Thus, the pogroms of Meskhetian Turks were reportedly caused by a fight over the high price of strawberries peddled by some Turkish gardeners in one of the open air markets in Fergana. Actually, a similarly insignificant episode led to Uzbek pogroms in nearby Kyrgyzstan at approximately the same time (see the essay on Kyrgyzstan).

POST-SOVIET UZBEKISTAN

In Uzbekistan, economic decline, common for the entire post-Soviet space in the immediate aftermath of the 1991 Soviet Union breakup, lasted only until 1995 and was the least steep in the Commonwealth of Independent States, of which Uzbekistan is a member. Thus, in 1995, Uzbekistan's GDP was 81.6 percent of that in 1991. For comparison, in Russia it was 65.4 percent, and in Kazakhstan 69 percent. However, one has to take into account that Uzbekistan was the second-poorest Soviet republic in terms of per capita consumption and had the second most rapid population growth. Thus in 1989, the rate of natural increase was 32 per 1,000 in the countryside where 60 percent of the population lived and 19 per 1,000 in urban areas. This means that at least 2.6 percent economic growth was required just to sustain the late-Soviet level of well-being. Indeed, from 1998 to 2003, economic growth amounted to 3 to 4 percent a year, thereafter rising to 7 to 8 percent. The total GDP volume of the last Soviet year (1991) was achieved in 2001, earlier than in most other post-Soviet countries, but by 2001, the population of Uzbekistan had become 25 percent larger than in 1989! In 2008, GDP in Uzbekistan grew 8 percent and kept on growing at the same rate even during the first half of 2009, despite global economic crisis (*The Economist,* July 21, 2009). The Uzbek economy may be insulated from global malaise because it is still largely state-controlled and has a strong emphasis on export substitution.

Macroeconomic data suggest that Uzbekistan is a typical developing country with 28 percent of labor engaged in agriculture and a per capita GDP (PPP) of $2,250 (2006)—well below neighboring Kazakhstan's. Under this level of per capita GDP, most rural dwellers likely live on less than $1.25 a day, the World Bank's current poverty threshold. Uzbekistan is a CIS country with the least per capita foreign direct investment because of a poor investment climate informed by tenacious corruption, frequent closures of borders, politically-motivated termination of joint ventures, and (until 2003) lack of full convertibility of the national currency.

Cotton continues to be the major Uzbek export, and students continue to be mobilized to harvest it much as during Soviet times. Since harvesters include pupils from grades as low as fifth, Uzbekistan is frequently cited as a country using compulsory child labor. During the post-Soviet period, the authorities have had to reduce the amount of cropland under cotton to boost food self-sufficiency, and land under wheat has expanded. In agriculture, the government assigns production quotas on cotton and wheat to household farms and requires them to sell these products to state buyers at prices well below market prices. The need for fresh water for irrigation has only increased, so not only does the Amu Darya still not reach the Aral Sea, but it also continues to shrink

(see a more detailed account in the essay on Kazakhstan); its Uzbek portion will be irretrievably lost in the foreseeable future.

Other than cotton, exported goods include natural gas, gold (an output of about 80 t per year), and low-priced automobiles, half of which are sold in Russia. The automobile industry is the only significant addition to Uzbekistan's economy since the breakup of the Soviet Union. Since 1992, a factory producing Korean Daewoo cars has operated in Asaka (near Andijan in Fergana Valley). This factory has since expanded and now also assembles Chevrolets, since Daewoo is now owned by GM. Recently, cooperation with Russia revived the *Ilyushin* aircraft plant in Tashkent. Of particular importance is the Il-76 cargo plane and its many modifications used by the military in many countries.

While the official unemployment statistics are low, by some accounts up to one-quarter of the total labor force of 15 million works abroad, with the largest contingents in Russia and Kazakhstan. The most widespread use of migrant workers from Uzbekistan is in construction. Apparently, it is the utter poverty and lack of opportunity at home that has eventually boosted Uzbek labor mobility, which had been extremely low in Soviet Uzbekistan. The government itself tries to find jobs for Uzbek youths abroad; with this in mind it recently concluded a labor agreement with the United Arab Emirates (*The Economist,* July 21, 2009).

Uzbekistan is a country with 40 percent of the population below the age of 16, even though the pace of natural increase has subsided by about 30 percent since the last Soviet census of 1989. By 2002, birth rates in the countryside, which now accounts for 63 percent of the population, had already declined from 32 to 20 births per 1,000 people and in cities from 19 to 12 per 1,000 population. Another change in population has been the outmigration of ethnic Russians and other groups of Russian-speaking people, including Koreans, Greeks, Tatars, etc. According to the 1989 census, 1,660,000 ethnic Russians lived in Uzbekistan, more than half of whom resided in Tashkent, which was largely a Russian-speaking city. Many Russians worked in industry, education, corporate research, and government. By 2006, about 1,000,000 ethnic Russians remained in Uzbekistan. Three out-migration waves, the largest of which occurred in the early 1990s, were fueled by fear of violence and discrimination. Anti-Russian violence did not materialize, but discrimination has been

quite noticeable, particularly in the areas of government and education. According to available surveys, most remaining Russians would like to leave but cannot for economic reasons. Selling a reasonably high quality two-bedroom apartment in Tashkent is possible for $5,000 or $6,000, a pittance compared to prices for an equivalent dwelling in any sizable city of Russia. And Russia's government support for resettlement has been weak or lacking altogether.

UZBEKISTAN POLITICS AS VIEWED THROUGH DIFFERENT LENSES

Politically, Uzbekistan is often described as the dictatorship of Islam Karimov, who was the last (appointed in 1989) leader of the Communist Party of Uzbekistan and has ruled independent Uzbekistan as its president since 1991. Karimov was born in 1938 in Samarkand to a mixed Uzbek-Tajik family. He has two college diplomas, as an engineer and as an economist. He worked for two large Tashkent-based enterprises, the Agricultural Machinery Plant and the Aircraft Plant. He subsequently climbed the ranks of the government apparatus, becoming the Finance Minister and then Head of the Central Planning Commission of Soviet Uzbekistan. The fact that Karimov became Uzbekistan's leader under Gorbachev reflects favorably on his qualifications. That is not to say that demagogues did not rise under Gorbachev. However, Karimov was never an ideologue; at the time of his promotion he was regarded as an able economist and administrator with extensive professional experience. Having become the leader of a specific political elite, however, he proceeded to play by its rules. Karimov was first elected president of Uzbekistan by the Supreme Soviet of that republic in December 1991. There were two candidates, and Karimov won 86 percent of the vote. In 1995, Karimov extended his term until 2000 through an internationally criticized referendum, and he was re-elected with 91.9 percent of the vote on January 9, 2000. Western observers criticized the election as neither free nor fair. The sole opposition candidate openly admitted that he entered the race only to make it seem democratic and publicly stated that he voted for Karimov. On January 27, 2002, Karimov won a new referendum extending the length of presidential terms from five to seven years; Karimov's current term, formerly due to end in 2005, was subsequently extended by parliament, which then scheduled elections for December 2007.

Western observers who criticize Uzbek leadership for dictatorial ways often point to the inertia of the remaining Soviet-style, strong, centralized leadership in which the substance of governance in Uzbekistan has not changed much from Soviet times even though the ruling party is now called People's Democratic—not Communist. While this observation is not incorrect, path-dependency is not a peculiarly Uzbek specialty. The origin of that path, however, is problematic, and there are legitimate doubts as to whether it is purely Soviet. After all, *each* Soviet republic organized its government in conformity with one and the same model, and yet no one would have confused Soviet Uzbekistan with, say, Soviet Estonia. Yes, there was one bureaucratic form, but it was filled with place-specific substance. The Cotton Scandal in Soviet Uzbekistan, for example, bore a distinctive Central Asian mark. Most upper and mid-level positions in Uzbekistan's government were closely related to clan politics, and they still are. Corruption was considered normal, and it still is. It has never been the same in Estonia. (Russia is somewhere in between these two worlds.) Now, eighteen years after the Soviet Union's demise, Estonia is one of the world's least corrupt countries (It ranks 28 among 179 countries.), and Uzbekistan is one of the most corrupt, near the very bottom of the ranking (number 175 out of 179). Although the Transparency International Index (TII) of perceived corruption is often influenced by politics (for example, when the U.S. Congress adopted the Belarus Democracy Act in 2004, Belarus's ranking abruptly slipped 25 positions lower on the TII than it had been prior to the bill's passage), being so far apart on a long scale is noteworthy. Despite the Communist Party's stranglehold on all matters of state importance, civility had a different meaning in Soviet Uzbekistan than in Soviet Estonia, and circles of trust in the two societies were structured in different ways. In Estonia, kinship and membership in a regional fraternity did not mean one thousandth of what it meant in Uzbekistan, and individual professional characteristics meant much more. Karimov's rise to the helm of power was actually a good omen for Uzbekistan because he was promoted on the basis of being a reputable professional.

Aside from civilizational contrasts per se, there were stark territorial contrasts in the quality of life in the Soviet Union. By now the differences have become gargantuan. The pockets of rural and small-town poverty exacerbated by staggering population growth that existed in Uzbekistan were not found in Estonia or even in Russia. Poverty breeds despair, despair breeds radicalism, and radicalism is shaped by local tradition. It is not true that in Soviet Central Asia all mosques were closed and clerics were not allowed to go to Mecca; but it is true that functioning mosques were limited in number and far apart, and those allowed to make a hajj were few and strictly controlled. After the fragmentation of the Soviet Union, Uzbekistan's border with neighboring Afghanistan was not as impenetrable as before; it was particularly porous in the early 1990s. All sorts of emissaries paid visits to Uzbekistan and funded radical preachers. That is why, after September 11, 2001, it was entirely natural for the secular (though ironically headed by a man whose first name was Islam) Uzbek regime to become a strategic ally in the United States' "War on Terrorism." By that time, the Islamic Movement of Uzbekistan (IMU) had already existed for three years. Funded by foreign sources and organized by two ethnic Uzbeks from the Fergana Valley—Jumaboi Khojayev (nicknamed Juma Namangani), a former Soviet paratrooper who had fought in Afghanistan as a member of the Soviet military contingent and Tohir Yuldashev, a local preacher—the forerunner of the Adolat movement, a Salafi Islamist group, had assumed civic authority in Namangan as early as 1991 and demanded that president Karimov impose sharia in Uzbekistan. Subsequently, the IMU was active during the civil war in Tajikistan and then in Afghanistan, where in 2001 Juma Namangani was actually appointed a deputy defense minister in the Taliban government. Earlier, in August 2000, the IMU had kidnapped four American mountain-climbers in the Kara-Su Valley of Kyrgyzstan, holding them hostage until they escaped. (Following this episode, the State Department included the IMU in its list of terrorist groups.) Still earlier, in February 1999, the members of the IMU detonated a series of explosions in Tashkent. Cars filled with explosives blew up in several places, including Independence Square, and also in front of the National Bank of Uzbekistan, in front of the State Committee for Security, in front of the headquarters of the Council of Ministers, and even near Karimov's residence. Sixteen people were killed and about 100 wounded. The target of one of these explosions was probably Karimov himself, as he had been expected to take part in a session of the Council of Ministers. In 1999 and 2000, a large group of the IMU militants tried to infiltrate the Fergana Valley but were stopped in Kyrgyzstan's mountains. Uzbek authorities responded by mass purges of Islamists. Several thousand found themselves behind bars.

Consequently, Karimov and his regime were quite sincere in their offer of post-9/11 support to the United States, as the threat of Islamic fundamentalism on Uzbekistan's own soil was real and significant. Americans needed a transit route to Afghanistan, until the summer of 2005, Uzbekistan hosted an 800-strong U.S. troop presence at the Karshi-Khanabad base, or "K2" which supported U.S.-led efforts in the 2001 invasion of Afghanistan. At that time, however, human rights activists were already criticizing the U.S. government for compromising its pro-democracy stand. This criticism was hard to ignore after the Uzbek government's brutal suppression of protests in Andijan in May 2005.

A trial of twenty-three local businessmen, accused of membership in banned fundamentalist organizations, took place in May 2005. During the night of May 12, armed protesters took control of the prison where the accused were kept and set free more than 1,500 inmates. Subsequently, the protesters took control of the regional administration headquarters. On the square in front of the headquarters, a rally was organized and speakers appealed to the government to stop prosecuting businessmen. On May 13, the Uzbek Interior Ministry and National Security Service troops fired into the crowd of protesters. Estimates of the number killed on May 13 are wide ranging—from 187, the official count of the government, to several thousand. Islam Karimov placed blame for the unrest on Islamic extremist groups. But Western critics said that this was merly a label that he used to describe political opponents as a pretext for maintaining a repressive state. The United States and the EU demanded an independent investigation and accused the Uzbek regime of applying disproportionate force. The regime qualified these demands and accusations as interference in Uzbekistan's domestic affairs. Both the EU and the United States imposed economic sanctions on Uzbekistan, including an embargo on defense equipment and official travel limitations. In November 2005, on the initiative of the United States, the UN General Assembly adopted a resolution condemning Uzbekistan for its rejection of international inquiry into events in Andijan. Seventy-four countries voted for this resolution, 39 voted against, and 58 abstained. Among those voting against the resolution were Afghanistan, Azerbaijan,

India, Kazakhstan, China, Russia, Saudi Arabia, Tajikistan, and Turkmenistan. The first country Islam Karimov visited after this show of international censure was China, which expressed wholehearted support for Karimov's policy. Incidentally, China battles an Islamic insurgency among Uighurs, close linguistic relatives of Uzbeks, in its province of Xinjiang. A significant rapprochement between Uzbekistan and Russia followed, even though prior to 2005, Uzbek-Russian relationships had been tense. On November 6, 2007, Karimov accepted the nomination of his party to run for a third term. Following the election on December 23, 2007 official results showed Karimov winning with 88.1 percent of the vote; the turnout rate was placed at 90.6 percent. Western observers criticized the elections as lacking a genuine choice.

Because the United States joined and even presided over the international chorus of critics, the Uzbek government ordered the United States out of its military base in the country. The base was officially closed in November 2005. The mutual estrangement was interrupted in early 2009 after neighboring Kyrgyzstan promised Russia that it would close its American base. Uzbekistan had not become more democratic by that time, but General David Petraeus promptly visited Tashkent and met with President Karimov to offer to renew cooperation. Although Kyrgyzstan had second thoughts and did not close its American base after all, U.S. military transit through Uzbekistan resumed.

While suppression of public protests in Andijan and mistreatment of the opposition are serious matters, it seems that in their unwillingness to take the sociocultural context of Uzbekistan into account, Western crusaders for democracy risk undermining the strategically important, authentic, and mutually beneficial alliances between the United States and Central Asian countries, alliances that may bring about democratic change in the long run. Whatever their defects, consistently secular regimes rule these countries, and their secularism is rooted in the Soviet period of their history. Although they are presiding over the least developed parts of the former Soviet Union and despite all the well-known flaws of Soviet-style governance and economic development, the Central Asian *stan*s do not lack for achievements. Their progress is particularly visible if these countries are compared not to those located well to the west of them (and in a different cultural realm) but to their closest neighbors, Afghanistan and Pakistan. Afghanistan is particularly relevant as a point of comparison because it shares not only the dominant religion (Sunni Islam) but also, to a significant extent, ethnicity with Uzbekistan and Tajikistan. Attitudes toward women and gender equality, overall literacy and education, the material basis of everyday life and the degree of interaction with the outside world (including hearths of technological innovations)—on these and other variables former Soviet Central Asia does not rank very high; but it ranks considerably higher than its southern neighbors. Yes, Uzbekistan is poverty-ridden and so is Tajikistan, but one should reserve judgment until one sees poverty in Afghanistan and Pakistan. Would anybody consider building a GM car factory in Afghanistan today? Uzbekistan already has one. According to a representative survey of 1,000 Uzbekistan citizens by the U.S. Institute of Peace, to the question, what political system would be best for Uzbekistan, 50 percent responded that "any system that would bring order to the country" and 12 percent said "Western-style democracy." That was in 1993. When identical questions were asked fourteen years later in a survey administered by Nancy Lubin and Arustan Joldasov, the ratio between the same answers was even higher at 56 percent to 12 percent (Lubin and Joldasov 2009). Islam Karimov may not be the worst case scenario for Uzbekistan after all; on the contrary, he may well be the best of available options.

FREEDOM

The EU and the United States have expressed concern over the recent imprisonment of the independent journalist Dilmurod Saiid. The trial was held behind closed doors and did not include the attorney for the accused. It bears striking similarities to other recent cases in Uzbekistan where independent journalists have been arrested and convicted on criminal charges not related to their journalistic activities.

DEVELOPMENT

Mining and agriculture are the backbone of the economy. Gold alongside cotton are the major export commodities. They represent roughly 20 percent of the total exports for the country.

HEALTH/WELFARE

The system of health care in Uzbekistan is comprehensive and services are provided mainly free of charge. Yet the overall efficacy of the Uzbek system was still relatively low as of 2000. The public often used hospitals for primary care.

NOTES

1. According to the CIA Factbook, Uzbeks now account for 9 percent of Afghanistan's population (which would mean about 2.9 million Uzbeks). The leader of the Afghan Uzbek community is Abdul Rashid Dostum, currently Chief of Staff to the Commander in Chief of the Afghan National Army.

2. The Medvedev-Sponheuer-Karnik scale, also known as the MSK or MSK-64, is a macroseismic intensity scale used to evaluate the severity of ground shaking on the basis of observed effects in an area of the earthquake occurrence.

Kazakhstan

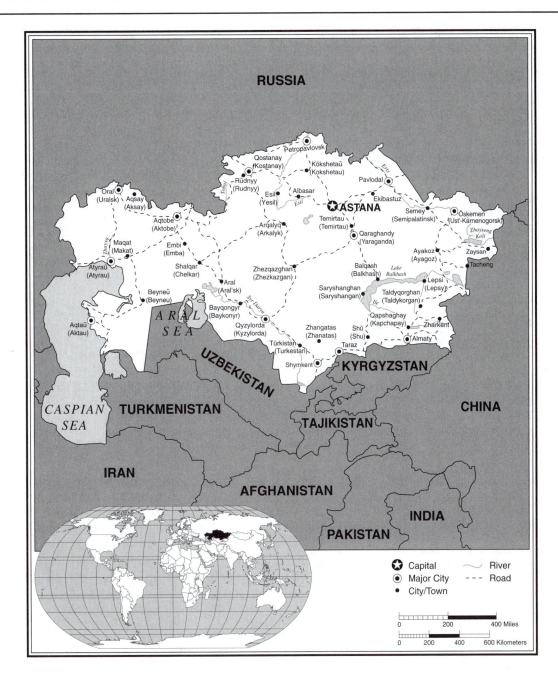

Kazakhstan Statistics

Area in Square Miles (Kilometers):
  1,049,155 (2,717,300)
Capital (Population): Astana (750,630)

Population

Total (millions): 15.9
Per Square Km: 6
Percent Urban: 53
Rate of Natural Increase (%): 1.3

Ethnic Makeup: Kazakh 53.4%; Russian 30%;
  Ukrainian 3.7%; Uzbek 2.5%; German
  2.4%; Tatar 1.7%; Uygur 1.4%; other 4.9%
  (1999 census)

Health

Life Expectancy at Birth: 62 (male);
  72 (female)

Infant Mortality Rate: 32/1,000
Total Fertility Rate: 2.7

Economy/Well-Being

*Currency ($US equivalent on 14 October
  2009):* 151 tenge = $1
2008 GNI PPP Per Capita: $9,690
*2005 Percent Living on Less than
  US$2/Day:* 17

Kazakhstan Country Report

The Republic of Kazakhstan is the world's largest landlocked country; with its land area of 2,717 thousand square km, it is the ninth-largest state in the world, larger than Western Europe as a whole. Because part of the conventional border between Asia and Europe is the Ural River, Kazakhstan has a European section (approx. 130,000 sq km). Kazakhstan's population of 15.9 million people is small compared with its land area, and population density is just 6 persons per square km, one eighth of the world's average. Since 1997, the capital city has been Astana (750,000 people); prior to that Almaty, Kazakhstan's largest city (with more than 1,000,000 population and situated in its southeastern corner), was the capital. Kazakhstan's longest (6,846 km) border is with Russia; there is hardly any physiographic, ethnic, or infrastructural discontinuity associated with that border. Indeed, a stretch of Russia's Trans-Siberian railway line cuts across northern Kazakhstan.

From the Caspian Sea (a salt lake) in the west to the Altai Mountains in the east and from the foothills of the Tian Shan to the southern reaches of the Siberian taiga, Kazakhstan's land is, for the most part, arid. Based on a depth of runoff indicator (the total runoff from drainage basin, divided by its area), Kazakhstan, with 42 mm, is the driest among all the former Soviet republics (www.ca-c.org/journal/13-1998/st_09_tursunov.shtml). Even Turkmenistan is less arid due to its 1300-km long and navigable Karakum canal. Kazakhstan has a variety of continental temperate climates, the yardstick of "continentality" being a large annual temperature range. In Kazakhstan, mean July temperatures vary between 19°C (66°F) and 28°C (82°F), and mean January temperatures are from −19°C (−2.2°F) and −2°C (28.4°F). The maximum temperature of 49°C (120°F) was recorded in south central Kazakhstan, and the minimum temperature of −57°C (−70.6°F) was recorded in its extreme north.

In northern Eurasia, biomes form largely latitudinal belts, and Kazakhstan hosts several such belts from mixed forest in the extreme north to steppe, semi-desert, and desert in the south. Kazakhstan hosts parts of such significant deserts as the Aral Karakum (to the northeast of the Aral Sea) and the Kzylkum (shared with Uzbekistan). The Kazakh steppe, a natural region mostly suited for agriculture, occupies one-third of the country. Kazakhstan has numerous lakes, most of which are saline,

such as the Caspian Sea, the world's largest salt lake, which is shared with Russia, Turkmenistan, and Azerbaijan; the Aral Sea, a salt lake shared with Uzbekistan; and Lake Balkhash, whose western part is fresh but eastern part is saline. Overall, the most hospitable areas in Kazakhstan lie along its perimeter, the southeastern corner of the country with the city of Almaty being arguably the most hospitable of all due to its lack of extremely cold winters and of strong winds. The areas with the harshest environment are in the central part of the country. Climatic factors have definitely affected population distribution, the most densely settled areas being located along the country's perimeter while the center is sparsely settled. Such a configuration of hospitable areas, along with the nomadic lifestyle of the indigenous population (until the 1930s), may explain the long-time absence of a strong national core in Kazakhstan. Much of the country is flat. However, in the westernmost part of Kazakhstan, on the Mangyshlak peninsula, the Karaghie depression drops to 132 m below sea level, and in the extreme east of Kazakhstan, in the Tian Shan mountains, Khan Tengri peak rises to 7,010 m above sea level.

Kazakhstan is exceptionally rich in fossil fuels (oil, gas, and coal) and metallic ores (copper, zinc, lead, uranium, gold, etc.) and has extensive area under cultivation.

KAZAKHS AND THEIR POLITICAL HISTORY

Kazakhs, most of whom until the 1930s were nomadic sheep herders, are descendants of three major groups: (1) Indo-Iranian Scythians and Sarmatians, the earliest population of the region; (2) Turkic tribes that migrated (around the fourth century) from their core area in the Altai Mountains; and (3) Mongolian tribes that first invaded the region in 1219. Modern Kazakhs owe their language to Turks, their racial type to Mongols, and their religion, Islam, to Arabs who briefly ruled Central Asia in the eighth century. Islam initially took hold in the southern part of Central Asia and thereafter gradually spread northward. Islam also took root due to the zealous missionary work of Samanid rulers, notably in areas surrounding Taraz (now in Jambyl Oblast of Kazakhstan), where a significant number of Kazakhs accepted Islam. Additionally, in the late 1300s, the Golden Horde propagated Islam amongst Kazakhs and other Central Asian

tribes. Beginning in the 1200s, masses of Mongols mixed with the local Turks and accepted their language. After the Mongolian Khanate of the Golden Horde was defeated by Tamerlane (1391), it split into several khanates (states) ruled by different offspring of Genghis Khan. One of these khanates was established on the Syr Darya. In 1460, several tribal leaders (sultans), disgruntled by the harsh rule of the khan, migrated with the people loyal to them to the Zhetysu region, located south of Lake Balkhash. Zhetysu or, in Russian, Semirechye, is literally the land of seven rivers, all flowing into Balkhash.

It is these migrants from the Syr Darya region to Zhetysu who first assumed the name Kazakh, which in Turkic languages means free. They created their state, the Kazakh khanate, and later extended it far to the west, incorporating nomadic groups in the western part of modern Kazakhstan, groups whose political organization also succeeded that of the defeated Golden Horde. The Kazakh khanate accomplished this expansion under Kasym Khan (1445–1521). The state was further consolidated under Tauke Khan, who ruled from 1680 to 1718 and under whose rule a sophisticated Code of Laws (Zhety Zhargy or Seven Laws) was developed and instituted. Though a state with a largely nomadic population, it had its capital in the town of Turkestan, in what is now South Kazakhstan Oblast. For ease of governance the state was subdivided into three territorial tribal unions, which Kazakhs called *Jus* (meaning a part or a side of; in some English language sources *jus* are referred to as hordes). These unions effectively succeeded the "civil divisions" of the Golden Horde. One *jus* was referred to as Great (south of Balkhash), the second one as Middle (north of Balkhash), and the third one as Little (western Kazakhstan). This division of Kazakhs into three tribal unions became etched in the Kazakh consciousness no less than their collectively belonging to the larger Kazakh whole. However, there were two select groups of Kazakhs, Tore and Kozha, who were not perceived as belonging to any particular *jus*. Rather, they were the upper stratum, standing above all tribal divisions. Tore were descendants of Genghis Khan, and Kozha were the descendents of the first Muslim preachers.

Beginning in the 1630s, the Kazakh khanate suffered repeated invasions from yet another nomadic state, Djungaria, which had its core in what is now the

Chinese province of Xinjiang. Djungaria also descended from the Golden Horde, but unlike the Kazakh Khanate was Mongolian (oirat) in language and had a strong, single central authority. The most important reason for the rivalry of these nomadic states was the exhaustion of existing pastures due to the growth of their flocks of sheep. In nomadic societies, land was held in common, but livestock was not, and so these societies were the first victims of what Garret Hardin christened "the tragedy of the commons."

In the 1600s and 1700s, Russia was vigorously expanding east. Peasants who fled their lords in Russia's heartland settled on lands to the southeast of it. Initially, nobody claimed those lands, and the various nomadic tribes that used them as pastures were either driven further east or assimilated by Russians. Russian peasants who evaded bondage called themselves Cossacks, the name having the same etymology and the same meaning (free) as Kazakhs. Only Cossacks did not evolve into an ethnic group that would be separate from Russians. Ultimately after several Cossack anti-government rebellions, they recognized the supremacy of the Russian tsar and became some of his most loyal subjects, instrumental in Russia's expansion. Thus Cossacks were set against Kazakhs, whom they proceeded to push out of their best lands. The first Cossack fortress in what is now Kazakhstan was founded in 1613 on the Jayik River, which the Cossacks renamed Ural. That fortress was called Jaitsky Gorodok. It subsequently became the city of Uralsk or Ural as it is called in independent Kazakhstan. Then Orsk (1735) and Orenburg (1743), now Russian cities close to Kazakhstan's border, were founded. From 1715–1720, three Russian fortresses along the Irtysh River were erected by Cossacks. These fortresses subsequently developed into the sizable cities of Omsk (1716; now a large Russian city), Semipalatinsk (1718; now called Semey in Kazakhstan), and Ust-Kamenogorsk (now called Oskemen in Kazakhstan). It was about that time that Tauke Khan asked Peter the First, Russia's tsar, to join the Kazakh Khanate to Russia as a kind of protectorate but without loss of autonomy. The request was not granted, as the Kazakh-Djungarian rivalry suited Russia's imperial goals. In 1718, the single Kazakh khanate ceased to exist under the onslaught of Djungarians and split into three jus. When the leader of Little Jus, closest to Russia's heartland, requested that the Russian tsarina, Anna, make it a Russian protectorate, she agreed. This move ushered in the incorporation of Kazakhstan into Russia. In 1740, the leader of Middle Jus, scheming to avoid surrender to the Djungarians, solicited and accepted citizenship from both Russia and China. In the 1750s, the Chinese army defeated the Djungarian stronghold; and it ceased to exist as a state. Many of the surviving Djungarians (oirats) migrated far west across the Volga River, where they formed the Kalmyk Khanate (now the Kalmyk Republic in the Russian Federation). In 1761, to replace Djungaria, the Chinese province of Xinjiang, meaning new border, was established.

As for Great Jus, for a time it found itself caught in a tug of war between Russia, China, and the still existing Kokand Khanate, which had its capital in the Fergana Valley, now in Uzbekistan. Russia established a foothold in the area by founding the Cossack fortress of Vernoye in 1855; it subsequently became the city of Alma-Ata, Soviet Kazakhstan's capital from 1927 to 1996. With Russia gaining the upper hand in the rivalry over Great Jus's land, the Russian-Chinese border became demarcated in two treaties in 1860 and 1864. This is now the border between Kazakhstan and China. When in 1866 the Russian army seized Chimkent (from the Kokand Khanate), all of present day Kazakhstan became part of the Russian Empire.

KAZAKHSTAN IN THE RUSSIAN EMPIRE

The incorporation of their land into the Russian Empire had multiple and far-reaching consequences for the Kazakhs. On the one hand, it opened them to European cultural tradition. Through learning Russian and being educated in Russian institutions of higher learning many Kazakhs gained knowledge about the rest of the world, and some became outstanding scholars themselves. Shokan Shinghisuly Walikhanuli (1835–1865), better known as Chokan Valikhanov, one of the first Kazakhs educated in the Russian language, is a case in point. During his short life, Valikhanov managed to graduate from a Russian military academy and serve as an officer in the Russian army; make lengthy folklore expeditions to Issyk-Kul and western Xinjiang and publish articles on the ethnography of Kazakhs and Kyrgyz; research the Kyrgyz epic poem Manas (see essay on Kyrgyzstan), write down and then translate some of its chapters into Russian; live and socialize in Saint Petersburg among the most enlightened people of his time; visit London and Paris; and befriend Fyodor Dostoyevsky, who in his letter of December 14, 1856, to Valikhanov wrote that "he had never ever been as drawn to anybody, including his own brother, as to you." Valikhanov was followed by other brilliant Kazakh scholars who subsequently became part of the Kazakh intellectual elite in tsarist Russia. Some researchers find it telling that almost all of these elite hailed from Middle Jus, which, due to its location (in central and northern Kazakhstan), had the largest zone of contact with Russia. Incorporation into Russia also consolidated all Kazakh lands within one polity and precipitated the development of industry. Some of Kazakhstan's mineral deposits were discovered during the tsarist period. Commercial coal-mining around Karaganda commenced as early as 1857 after a Russian merchant, Nikon Ushakov from Petropavlovsk, purchased a prospective coal-rich tract from two Kazakh families for 250 rubles. Even earlier, in 1786, the polymetallic (lead and zinc for the most part) mine was opened in Ridder in east Kazakhstan (named for Philip Ridder, an ethnic German geologist in the service of the Russian government). Also the Ekibastuz coalfield (in Pavlodar Oblast of northern Kazakhstan) has been under commercial exploitation since 1896—courtesy of the Russian merchant Artemy Derov. Both the Ekibastuz and the Karaganda coalfields are among the largest in the world. Finally, a copper ore dressing factory was commissioned in 1914 near Zhezqazghan in central Kazakhstan, based on a deposit known since the 1770s and first officially registered as such in 1847. Throughout the second half of the nineteenth century, Russian geologists discovered oil in several areas along the northern and eastern coasts of the Caspian Sea from the mouth of the Volga River to the Mangyshlak Peninsula. The commercial exploitation of the first oil well, near the mouth of the Emba River, was launched in 1899. Within the next 15 years several more oil wells in the Emba Valley were tapped.

On the negative side, Russian colonization undermined the traditional nomadic land use, as settlers willfully occupied the most fertile areas and pushed Kazakhs into more arid or otherwise less suitable pastureland. Whether or not the settlers used the patronage of the government (in many cases they did), the application of a first-come-first-served principle to the competition between sedentary and nomadic communities invariably favors the former and restrains the latter. Several times displacement of Kazakhs led to uprisings which were brutally suppressed by the Russian army. Mistreatment of the indigenous population by the Russians after the suppression of one of such uprising (in 1864) filled Chokan Valikhanov with such indignation that he resigned from the army. But the waves of migration from the west continued unabated. In 1906, the Stolypin agrarian reform was launched in Russia; it called for either increased parcellation of existing farmland (so communal peasants could become commercial farmers) or migration to newly colonized areas. Due to the overpopulation of the Russian and Ukrainian countrysides, migration was increasingly the preferred option; and so from 1906 to1912, 15.4 million hectares of land were assigned to 500,000 settlers, and Kazakhs were again pushed aside.

When Russia introduced parliamentarianism, the 1907 electoral law denied the indigenous peoples of Siberia and Central Asia electoral rights and also prevented them from being drafted. But during World War I those "aliens" (*inorodtsy*) aged 19 to 43 were recruited en masse for digging trenches along the frontline. The 1916 uprising under the leadership of Amangeldy Imanov was the Kazakh response. Imanov, a madrasah (a Muslim religious school)-educated Kazakh from Middle Jus, managed to recruit as many as 50,000 horsemen to fight the Russian military (Cossack) contingent and displayed the skills of a military tactician and planner. Imanov's makeshift army never actually surrendered, and after the war he switched his allegiance to the Bolsheviks. But while certain leaders of the Kazakh resistance were successful, overall the opposing forces were unequal and Russia prevailed. Close to 300,000 Kazakhs migrated to China.

BETWEEN THE 1917 REVOLUTION AND THE 1991 INDEPENDENCE

During the 1918–1921 civil war which followed the 1917 Communist Revolution, the Kazakh elite created a political party, Alash, which did not embrace Bolshevik (communist) ideas. Instead it sided with the Mensheviks (social democrats) and declared Kazakh-Kyrgyz autonomy (as Alash Orda) within Russia. In the 25 member Alash government, 10 seats were offered to Russians and members of other (non-Kazakh) groups. In early 1920, the Red Army recaptured all of Kazakhstan and crushed the autonomous unit; most of its leaders were subsequently executed. However, in August 1920, the Bolshevik government itself established the Kirgiz Autonomous Soviet Socialist Republic (ASSR) within the Russian Federation. At the time, Russians still called Kazakhs Kirgiz (whereas Kyrgyz were referred to as Kara-Kirgiz). The capital of that republic was set in Orenburg, a city on the Ural River and once a Cossack outpost. Its 1920 assignment as the Kazakh capital testified, however indirectly, to the

area's not yet being considered quite Russian. It had also been in Orenburg where two All-Kirgiz Congresses had convened in 1917 and established Allash Orda. After the 1925 "national delimitation" (see the essay on Uzbekistan), the Kirgiz ASSR was renamed Kazakh ASSR to reflect the name that the titular ethnicity ascribed to itself, and its capital was moved to the town of Ak-Mechet—which was renamed Kyzyl Orda or Red Horde—on the Syr Darya. Orenburg, with its surrounding province, was transferred to the Russian Federation. In 1927, the capital was moved again, this time to Alma-Ata; and in 1936 Kazakhstan became one of the union republics of the USSR. Importantly, it is the acquisition of that status that 55 years later allowed Kazakhstan—together with other union republics—to become a sovereign country in the wake of the Soviet Union's 1991 disintegration. (In the late 1950s, parts of Kazakhstan were transferred to Russia and Uzbekistan.)

Before Kazakhstan could even dream of independent statehood, it experienced vast and far-reaching changes that affected all aspects of life in the republic. Until the late 1920s, the Soviets tolerated local customs. Many Kazakhs could still learn their language in religious schools which used the Arabic script; elements of Islamic law were tolerated, and Friday was still the day off. Government policy then took a turn toward ideological purism whereby anything that deviated from communist teaching was considered an anachronistic evil subject to eradication. In Kazakhstan, this effectively meant the interruption of local cultural traditions. In Russia proper, when schooling became ideologically indoctrinated, this did not shatter the entire education system or alter the language of instruction. In Kazakhstan, however, the entire traditional education system was replaced by an increasingly Russified network of public schools. The Kazakh language was first transliterated to the Latin (1929) and then to the Cyrillic (1940) alphabet. This detached masses of young Kazakhs from the classic works of Kazakh and Pan-Turkic literature. The Russification of education in part had to do with ongoing changes in the ethnic makeup of Kazakhstan's population, whereby Kazakhs were set to become a minority in their own land. However, to some extent, Russification was rooted in an imperial-turned-Soviet white man condescension to what seemed to many Russians as an inherently backward and anachronistic Asian culture. Suffice it to say, by 1957 only one secondary school remained in the Kazakh capital of Alma-Ata with Kazakh as the language of instruction; and only

one institution of higher learning taught in that language—the Kazakh Women Pedagogical Institute which prepared elementary school teachers. However ironic, since Bolsheviks or communists were devout atheists, nothing comparable was possible in Christian Georgia, or Armenia, or in the Baltic republics (incorporated in 1940). Comparison with these republics, rather than with Belarus or Ukraine, makes sense because, unlike in Belarus or Ukraine, the languages of the titular nationalities of these republics are not close relatives of Russian, and neither is the Kazakh language.

In the late 1920s large-scale railway and industrial construction was launched in Kazakhstan. In 1927, rail connected Petropavlovsk and Kokchetav (Kokshetau), and by 1931 this line had been extended to Akmolinsk (nowadays, Astana). Also in 1931, the so-called Turkestan-Siberian railway line was commissioned, linking Alma-Ata (now called Almaty) with Semipalatinsk (Semey). In 1939–40, railway lines linking Akmolinsk and Karaganda, Iletsk and Urals, Rubtsovsk (in the Altai Krai of Russia) and Ridder, and Karaganda and Zhezqazghan were opened. For the most part, all these connections were used to transport resources extracted in Kazakhstan. So along with new transportation routes being built, the ore dressing facilities at the old sites (like Ridder with zinc and lead) and old coal mines (Karaganda) were restored and extended and new facilities were built, such as the Chimkent lead factory, the Balkhash and Zhezqazghan copper factories, the Ust-Kamenogorsk (Oskemen) lead-and-zinc factory, etc.

Such a scale of industrial construction boosted demand for labor. The result was organized migrations from Russia, Ukraine, and Belarus, whose countrysides and small towns had surplus labor. From 1931 to 1940, 559,000 people were sent to Kazakhstan from those republics. From the late 1920s on, the proportion of forced labor migrants increased, the first contingents consisting of the so-called kulaks, formerly wealthy peasants who had fallen victim to the collectivization of agriculture in the Slavic republics (that is, their farms were expropriated and they were exiled). As early as 1931, 70,000 people were assigned to labor camps established on twenty sites around Karaganda. They were poorly fed and worked in the coal mines. Altogether 189,000 kulaks with their families were exiled to Kazakhstan; by 1937, the number of exiles in forced labor contingents (so called *spetspereselentsy* or special re-settlers) in Kazakhstan had reached 360,000. By the early 1940s, a ramified network of gulags—high security correction facilities—had sprung up in Kazakhstan. Prisoners from all over the Soviet Union were kept in the gulags; most were victims of Stalinist purges, but some had received their sentences for felonies. Some of the Kazakhstan-based gulags were later described by Alexander Solzhenitsyn in his *Gulag Archipelago*. Such, for example, was a labor camp called ALZHIR, which is an abbreviation of the Russian for the "Akmolinsk Camp for Wives of Traitors to the Motherland."

In addition to prisoners in labor camps, in which people of all ethnicities and walks of life were kept, Kazakhstan came to host large groups of deportees representing single ethnic groups. Thus, after eastern Poland was occupied by the Soviets in 1939 (as a result of the Ribbentrop-Molotov non-aggression pact) and two parts of it were joined to Soviet Ukraine and Soviet Belarus, many ethnic Poles from these areas were exiled to Kazakhstan. In 1937, all ethnic Koreans from the Russian Far East were deported to Kazakhstan and Uzbekistan. When in 1941 Nazi Germany invaded the Soviet Union, the Autonomous Republic of Volga Germans was disbanded, and its residents of German descent were exiled wholesale to Kazakhstan, Siberia, and Central Asia. In 1943–44, several ethnic communities from the North Caucasus (Karachai, Balkar, Chechen, and Ingush) as well as Crimean Tatars were deported to Kazakhstan. Most of the deported groups numbered in the hundreds of thousands, so the overall population inflow due to deportations alone was close to one million people, with ethnic Germans (420,000 people) and Chechen and Ingush (about 400,000 people), comprising the largest groups of deportees. Note that in 1926, the overall population of Kazakhstan had been only slightly over six million.

A huge inflow of involuntary migrants was not the only factor of population change in Kazakhstan prior to and during World War II. Kazakhs, the titular nationality of the republic, sustained astounding losses as a result of starvation in 1932–33 and migrations to China, Mongolia, and other Central Asian republics. Hunger was caused by the collectivization of agriculture campaign, which in Kazakhstan arguably displayed an unusually high degree of ideological rigidity and cruelty. The campaign was conducted under the guidance of Philip Goloshchokin, who from 1925 to 1933 was in charge of the Communist Party of Kazakhstan. Born and raised in northwestern Russia, Goloshchokin joined the Bolshevik Party in 1903. In 1913, exiled by the tsarist government to Siberia, he befriended Joseph Stalin, who had been exiled to the same region. In 1918,

Goloshchokin was one of the immediate organizers of the 1918 execution of the Russian Tsar, Nicholas II, and his family in the city of Ekaterinburg. Appointed on Stalin's initiative to the helm of power in Kazakhstan, Goloshchokin set out to transform the nomadic sheep-breeding communities into sedentary collective farms based on nucleated villages that might resemble those in southern Russia. No attempt to work out some form of land use that would embrace Soviet collectivism and still adapt to local conditions was undertaken. Instead, the entire transformation was to occur by fiat. Livestock was rounded up from all those unwilling to surrender it and transported to designated prospective feedlots. Because it quickly proved impossible to sustain so many animals in dry steppe without roaming from place to place, most sheep and cattle were eventually butchered. As a result, from 1931 to 1933, only one-tenth of the original 40 million head of livestock remained in Kazakhstan. When Kazakh families tried to flee with their flocks and herds they were apprehended by an army dispatched to punish the so-called *basmachi,* the collective pejorative name (meaning bandits) that the Soviets applied to all enemies of Soviet power in Central Asia. As Kazakhs were deprived of their sheep, which had been their only source of subsistence, hunger spread. The estimates of 1931–33 hunger-related deaths in Kazakhstan are wide ranging. According to Alexander Alexeyenko, one of the most reputable demographers of Kazakhstan, the 1931–33 loss of ethnic Kazakh population of Kazakhstan amounted to 1,840,000 people or 47.3 percent of their 1930 total number! The heaviest losses were sustained in eastern Kazakhstan, where 64.5 percent of the Kazakh population of the region was lost. Approximately one-third (616,000 people) of that vanished population had managed to flee, including roughly 200,000 to China and Mongolia. Most of the remaining two-thirds starved to death. As a result of hunger and incoming and outgoing migration, the ethnic makeup of Kazakhstan's population changed. Whereas in 1926 ethnic Kazakhs accounted for 58.2 percent of the total population and Russians for 20.5 percent, in 1939 ethnic Kazakhs accounted for 36.4 percent of the total and Russians for 41.2 percent. It took Kazakhs until 1970 to restore their 1926 population numbers (3.9 million people). The Goloshchokin-led collectivization campaign was eventually considered a failure even by the Kremlin; in 1933 he was replaced by another outsider.

Following the June 22, 1941, invasion of the Soviet Union by Nazi Germany,

more than 140 industrial enterprises were transferred to Kazakhstan from Ukraine, Belarus, and western Russia; part of their personnel, more than 500,000 people, were also moved. This transfer contributed to the further development of local industry. During just the first one and a half years after war began, 25 new metallic ore mines, 11 ore dressing factories, and 19 coal mines were opened in Kazakhstan, as well as four new oil extraction sites and one refinery (in Guryev). Rail construction continued during the war as the significance of Kazakhstan as an industrial base far from the frontline rose immensely. After Moscow and Leningrad motion picture industries were moved to Alma-Ata, a local movie industry began to develop there.

During the entire post-war period, industrial development in Kazakhstan continued, particularly in extractive industries. Also, several hydroelectric stations were built on the Irtysh River. The network of institutions of higher learning first established in the 1930s expanded. Demand for a qualified cadre for these institutions was, to a significant extent, at least initially, met by those exiled to Kazakhstan and their offspring. Between the end of the war (1945) and the death of Stalin (1953), the system of gulags in Kazakhstan expanded even more to house thousands of Soviet military who had spent time in Nazi captivity and were therefore considered traitors. Just one "correctional" institution of that sort, the so-called StepLag (Steppe Camp) in northern Kazakhstan, had close to 200,000 prisoners in the early 1950s. When the labor camps were closed in the late 1950s, many of their former prisoners chose to stay in Kazakhstan.

In 1949, the Semipalatinsk Test Site (STS) opened as the primary testing venue for the Soviet Union's nuclear weapons. It was located in northeastern Kazakhstan some 150 km west of the town of Semipalatinsk (now Semey), with most of the nuclear tests taking place at various sites farther west and south. Between 1949 and the cessation of atomic testing in 1989, 456 explosions were conducted at the STS, including 340 underground (borehole and tunnel) shots and 116 atmospheric (either air-drop or tower shots). The site was officially closed in August 1991. As a result of the nuclear tests, about 2,000 sq km in the surrounding area sustained grave ecological damage.

The Baikonur Cosmodrome was founded in 1955. Originally built as a long-range-missile center, it was later expanded to include launch facilities for space flight. The Cosmodrome became the world's first and largest operational space launch facility. It is located in the desert steppes of Kazakhstan, about 200 kilometers (124 mi) east of the Aral Sea and north of the Syr Darya river. Many historic flights lifted off from Baikonur, including that of the first man-made satellite, Sputnik 1, on October 4, 1957; the first manned orbital flight by Yuri Gagarin on April 12, 1961; and the flight of the first woman in space, Valentina Tereshkova, in 1963. The Cosmodrome is leased by the Kazakhstan's government to Russia (currently until 2050). The rent is fixed at $115 million per year.

Though unquestionably grandiose, Baikonur was perceived inside and outside the Soviet Union as an all-Union endeavor that just happened to be placed on Kazakhstan's land. In contrast, the Virgin Lands Campaign was uniquely associated with Kazakhstan (and to some extent with adjacent West Siberia). Initiated in 1954 under Nikita Khrushchev, the Campaign was to open up vast tracts of virgin steppe to cultivation in northern Kazakhstan and in the Altay and Novosibirsk regions of Russia. Overall, from 1954 to 1960, 25.5 million hectares of pristine steppe were ploughed under. This is comparable to 1.3 times all the arable land of France or half of all the arable land in Canada. About 2 million people, mostly from Russia and Ukraine, were mobilized to settle in northern Kazakhstan and participate in the Virgin Lands Campaign. This time newcomers were arriving voluntarily, carried away by patriotic zeal and attracted by government incentives, such as a free rail ticket, free transport of significant amounts of personal property, a one-time handout of up to 1,000 rubles (a huge amount at the time), 10-year interest-free credit of up to 20,000 rubles to build a home and up to 2,000 rubles to buy cattle for one's own household farm, etc. Nearly all of the collective farms in the virgin lands grew the same one crop: wheat. Initially, the campaign seemed to be successful, as the 1956 output of grain in the Soviet Union beat all previous records. But by the 1960s, the soil had been exhausted of the nutrients most beneficial to wheat. Although the production of fertilizers in the U.S.S.R. had increased during this period, the loss of fertility was principally due to neglect of traditional crop farming techniques on drought-prone land. Before long, the lack of adequate measures to prevent erosion meant that a lot of topsoil blew away and barren land was left behind, something similar to what happened in the Dust Bowl of Texas and Oklahoma in the late 1930s. But while a significant amount of ploughed land was lost to cultivation, much was retained; and a gradual, if painful, adjustment of farming techniques followed, so the fact that today Kazakhstan

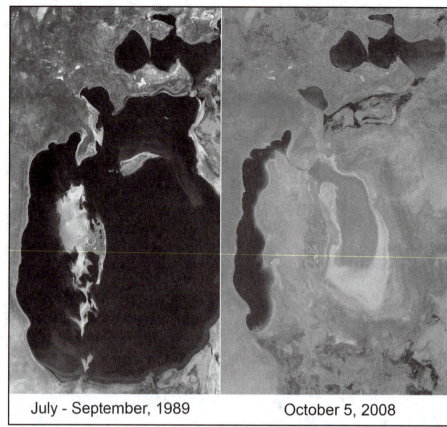

July - September, 1989   October 5, 2008

(NASA)

Satellite images of the Aral Sea (left: September 1989; right: October 2008).

is one of the world's major exporters of grain must be attributed to the Virgin Land Campaign. The city designated as the focal point of this program was Akmolinsk; in 1961 it was renamed Tselinograd—from "tselina," Russian for virgin land. (In 1992, it was renamed Akmola; and in 1997 it became Astana, the capital of independent Kazakhstan.)

The huge inflow of outsiders during and prior to the Virgin Lands Campaign meant that, despite high rates of natural increase, the share of ethnic Kazakhs in Kazakhstan was only 30 percent, according to the 1959 Soviet census. From that, its lowest level, this share started a gradual rise to 32 percent in 1970, 36 percent in 1979, and 39.6 percent in 1989 (the last Soviet census). From the mid-1970s on, this rise was due in part to the commencement of an outflow of ethnic Russians and other migrants and their descendents. As in other non-Russian republics, this outflow was conditioned by a heightened inter-ethnic competition for prestigious jobs; ethnic Russians for the first time were losing this competition to titular ethnicities, who were becoming more assertive and influential, particularly in all tiers of executive power, retail trade, health care, and education.

While the Virgin Land Campaign affected agriculture in northern Kazakhstan, prospects of agricultural growth in southern Kazakhstan were associated with irrigation. As in neighboring Uzbekistan and Turkmenistan, water was to be diverted from rivers flowing to the Aral Sea, a large salt lake and, until the 1950s, the world's fourth largest lake, occupying 68,000 sq km and having a maximum depth of 68 m. On the Kazakhstan side, the Syr Darya was the principal source of water for irrigation. Construction of irrigation canals for the sake of developing and expanding cotton plantations commenced in the 1930s and was given additional impetus in the 1960s. As a result of diverting water from the Amu Darya in Uzbekistan and the Syr Darya in Kazakhstan, these large rivers no longer reach the Aral Sea. The well-publicized ecological disaster that has seen the shrinking and poisoning of the Aral Sea was exacerbated by several other factors as well, such as dust storms spreading deposits of fertilizers and pesticides over a large area and the long-lasting (1942–1992) existence of a testing ground for bacteriological warfare on Vozrozhdeniya (Rebirth) Island—now a peninsula, the northern part of which is in Kazakhstan. Since 1961, the water level in the Aral Sea has been dropping with increasing speed from 20 to 90 cm per year. Prior to the 1970s, 34 fish species inhabited the Aral Sea, including 20 species of commercial importance. On Kazakhstan's coast were five fish processing factories and one factory producing canned fish. Now all of this is gone. In 1989, the Aral Sea split into two disconnected reservoirs, the Northern (Little) Aral and the Southern (Big) Aral. The overall area of the sea is currently 25 percent of what it was in the 1950s, the volume of water is one-tenth what it was prior to widespread irrigation, and the water level is 22 m lower than in the past. But there is still a commercial fishery in the Little Aral (all of which is in Kazakhstan) and, unlike fisheries in the Big Aral, it will probably survive.

The entire petroleum industry of Kazakhstan, now the backbone of Kazakhstan's relative prosperity and the major reason the country receives keen attention from the EU and the United States, is also entirely a creature of the imperial Russian and Soviet periods of Kazakhstan's history. To the Ural-Emba oilfield, first tapped before the Communist Revolution and developed in the 1920s and 1930s, other major discoveries were added, particularly in the 1960s, 70s, and 80s. Most discoveries were made in the western part of the republic—in Guryev (Atyrau), Aktyubinsk (Aqtobe), and Mangyshlak (Mangystau) oblasts. For example, the Tengiz oilfield in Mangystau Oblast, the sixth largest oil field in the world with an estimated volume of up to 25 billion barrels (4 km^3) of oil, was discovered in 1979.

The first ethnic Kazakh appointed to lead the Communist Party of Kazakhstan (CPK) for an extended period of time (that is, to lead the republic) was Zhumabai Shayakhmetov, who ruled from 1946 to 1954. Like members of Kazakhstan's intellectual elite eliminated during the Stalinist purges, Shayakhmetov was from Middle Jus. Until the Brezhnev era, most leading positions in Kazakhstan had been in the hands of ethnic Russians or Ukrainians, so the Jus factor did not play a significant role in political appointments. In 1960, Dinmukhamed Konayev was appointed First Secretary of the CPK. Konayev was from Great Jus by his father's line; his mother was a Tatar. Konayev graduated from the Moscow Institute for Non-Ferrous and Fine Metallurgy, and in the 1930s he worked as an industrial manager in eastern Kazakhstan. Konayev's first stint as CPK leader lasted only two years. However, in 1964 he resurfaced under Brezhnev (whom Konayev befriended during Brezhnev's own 1954–56 tenure in Kazakhstan) and stayed as Kazakhstan's leader until 1986, when he was accused of corruption and fired by Gorbachev. During Konayev's long second reign, many ethnic Kazakhs made it to high offices, and the proportions among the three Jus were carefully maintained. Thus, the office of the Prime Minister was assigned to a representative of Middle Jus, and the office of Chairman of the Supreme Soviet to someone from Little Jus. This was the descending order of status and real power in each Soviet republic.

When Konayev was forced into retirement, the intense power struggle between Nursultan Nazarbayev, Konayev's one time protégé also from Great Jus, Zakash Kamalidenov, Andropov's protégé from Little Jus, and another of Konayev's protégé, Saylaukhan Aukhadiyev from Great Jus, created a stalemate which the Kremlin broke by appointing a total outsider to the leading post. Thus Gennady Kolbin, an ethnic Russian and formerly the Communist Party boss in the Ulyanovsk region of Russia, was made the First Secretary of the CPK in December 1986. However, what used to be a matter of course before—that outsiders led Kazakhstan—did not come to pass in the new atmosphere of Gorbachev's Perestroika. Student riots under Kazakh nationalist slogans broke out in Alma-Ata on December 16 and 17, 1986. They were apparently quite unexpected by the Kremlin because, at the time ethnic Kazakhs accounted for less than 25 percent of the city's population. Yet the appointment of a non-Kazakh to the highest post in the country filled them with such a strong sense of injustice that unrest precipitated. The riots that ensued resulted in three deaths and more than 1,100 wounded, among whom were 365 demonstrators, the rest being police. Many students were expelled from colleges. While the riots were suppressed, they became the very first instance of mass unrest due to ethnic grievances in the Soviet Union, the forerunner of many more to surface soon thereafter.

The Kremlin corrected its mistake on June 22, 1989, appointing Nursultan Nazarbayev first secretary of the CPK. In April 1990, the Supreme Soviet of Kazakhstan elected him the first president of the republic. Following the August 1991 coup in Moscow, Nazarbayev left the Communist Party. On December 1, 1991, he was popularly elected president, winning (according to the official information) 98.7 percent of the vote. Nazarbayev has been Kazakhstan's leader ever since. Kazakhstan was the last Soviet republic to declare its independence on December 16, 1991.

POST-SOVIET KAZAKHSTAN

Kazakhstan's economy is now larger than the economy of the four other Central Asian states combined. But from 1992 to 1995, the country saw precipitous economic decline due to disrupted economic ties with Russia and other former Soviet republics. After all, the economy of Kazakhstan had been developed as a natural resource appendage of the Soviet Union's European section, so not only were markets for Kazakhstan's products largely outside Kazakhstan, but also sources of finished industrial goods for Kazakhstan were largely in Russia, Ukraine, and Belarus. All large enterprises in Kazakhstan had been managed by all-Union ministries; only agriculture, food processing, and the production of low-quality consumer goods were run locally. In 1992, Kazakhstan's GDP was 69 percent of that of 1991. The situation was so dire that food rationing was introduced in urban areas of Kazakhstan. A program of privatization of small businesses was introduced which turned out to be successful. As for large enterprises, Kazakhstan's leadership made a decision to sell them to Western firms and invited in Western economic management teams. In 1995, Kazakhstan received more than $2 billion in foreign direct investment plus $1 billion in low-interest loans. Ties with traditional markets were restored and new ties developed. Thus in 1996, Kazakhstan exported raw materials worth $4 billion, almost half of which went to Russia and another 38 percent to countries outside the former Soviet Union. By the end of 1997, 70 percent of economic assets were in private hands, for the most part in the hands of foreigners. Kazakhstan's largest foreign investor in those years was the United States, followed by the UK, Turkey, South Korea, France, and Japan. Russia joined this circle only in the new century. By September 1997, direct foreign investment in Kazakhstan's economy exceeded $7 billion, and investment commitments for the foreseeable future amounted to $60 billion. A country whose population was only 1/11 of Russia's attracted more foreign investment than Russia. Comparison with Russia makes sense because in

DEVELOPMENT

A program of privatization of small businesses was introduced which turned out to be successful. As for large enterprises, Kazakhstan's leadership made a decision to sell them to Western firms and invited in Western economic management teams. In 1995, Kazakhstan received more than $2 billion in foreign direct investment plus $1 billion in low-interest loans. Ties with traditional markets were restored and new ties developed.

Russia the industrial assets attractive for Western investors are also in the extractive industry. But whereas in Kazakhstan as early as 1997 the entire aluminum industry had been orderly transferred into the hands of Japanese, Korean, and Turkish firms and management teams, in Russia it was in the hands of domestic gangsters and criminals; and dozens were being killed in so-called aluminum wars. From 1994 to 1996, the 45 largest industrial enterprises in Kazakhstan were transferred into the hands of such corporations as Chevron, Glencore Energy, Exxon Mobile, Amoco, BP, etc. Such industrial giants as LG, Samsung, Bosch, Panasonic, and GM established production lines in Kazakhstan.

Kazakhstan produced approximately 1.45 million barrels of oil per day (bbl/d) in 2007 and consumed 250,000 bbl/d, resulting in net petroleum exports close to 1.2 million bbl/d. (www.eia.doe.gov/emeu/cabs/Kazakhstan/Oil.html). Kazakhstan's growing petroleum industry accounts for roughly 30 percent of the country's GDP and over half of its export revenues. As early as 1993, the Government of Kazakhstan created an international consortium to manage the development of the oil industry. Members of the consortium, aside from the government, are Royal Dutch Shell, STATOIL, Mobil, BP, TOTAL, and AGIP (www.kmg.kz/page.php?page_id=294&lang=1). One of the achievements of the consortium was the 2000 discovery of the Kashagan oilfield, located off the northern shore of the Caspian Sea near the city of Atyrau. The Kashagan field is the largest oil field outside the Middle East and the fifth largest in the world (in terms of reserves). The consortium operating the field has estimated the field's recoverable reserves at 13 billion barrels of oil equivalent, with total reserves-in-place around 38 billion barrels. Its exploitation began in 2007. Full-scale commercial production is not expected to commence until 2013 At the same time, no new significant discoveries of new oilfields have been made in Russia where Western investors were squeezed out.

Other important export-oriented sectors of national economy are natural gas and metallic ores. In agriculture, Kazakhstan is the seventh-largest producer of grain in the world, and wheat exports are an important source of hard currency for Kazakhstan.

Work is being done to restore, in part, the North (Little) Aral Sea. Irrigation works on the Syr Darya have been repaired and improved to increase its water flow, and in October 2003, the Kazakh government announced a plan to build Dike Kokaral, a concrete dam separating the two halves of the Aral Sea, the southern part [of] which is fast vanishing. . . . Work on this dam was completed in August 2005; since then the water level of the North Aral has risen, and its salinity has decreased. By 2006, some recovery of water level has been recorded—sooner than expected—stocks of fish have returned, and it now appears that rumors of the Aral Sea death were greatly exaggerated as far as its northern section is concerned.

In the new century, Kazakhstan has demonstrated robust economic growth, from double-digit in 2000 and 2001 to 9-10 percent per year from 2002 to 2008. Most international observers agree that post-Soviet Kazakhstan has been an economic success story.

Significant changes occurred in Kazakhstan's population. The steep economic decline of the 1990s, coupled with fear (sometimes justifiable) of ethnic discrimination, prompted many ethnic Russians and other non-indigenous groups to migrate from Kazakhstan. From 1990 to 1997, 1.2 million ethnic Russians left; 90 percent of them left for Russia. These migrants accounted for 14 percent of Kazakhstan's population. In more recent years, the intensity and pace of outflow has slowed, but still net outmigration of Russians from 1999 to 2004 amounted to 220,000. By some accounts, the migration potential of Russians in Kazakhstan is almost exhausted, and no more than 10–15 percent of those who remain in the country are likely to leave. Many Russians have made it into the growing middle class, and in many cases fear of mistreatment and discrimination turned out to be unfounded. Importantly, outmigration of ethnic Russians did not commence with independence of Kazakhstan but had started as early as the mid-1970s. It accelerated in the 1990s, and then slowed down. At the same time, out of almost one million ethnic Germans, nearly 800,000 have left for Germany.

Russians still account for almost one-quarter of Kazakhstan's population. Together with other groups which in Kazakhstan overwhelmingly use Russian (such as Ukrainians, Tatars, Germans, as well as many urban Kazakhs, etc.), they may represent no less than 40%. For the most part—Russians in Kazakhstan are urbanites. There is a striking spatial dimension to the distribution of Russians and Kazakhs. Kazakhs dominate the south and west, Russians are dominant in the north, east, and part of the center (Karaganda). Whereas in absolute terms outmigration of Russians has been most significant from Russian-dominated regions (whose overall populations had declined 17% from 1989 to 1999), the intensity of the outflow of ethnic Russians (as measured by the percentage share of those who left divided by the percentage share of Russians in the population in the beginning of the period) was higher from Kazakh-dominated regions. This is an indication that some level of inter-ethnic tension does in fact exist.

Overall, the trend of ethnic makeup of Kazakhstan's population is a fairly rapid growth of the share of ethnic Kazakhs. By 1999 they had reclaimed their majority status—53 percent (compared with less than 40% in 1989). And by 2009, their share had reached 63.3%, whereas the share of ethnic Russians had declined to 23.7%. Besides outmigration of Russians, the share of ethnic Kazakhs is boosted by two other factors. One of these factors is a higher, though declining, rate of natural increase of Kazakhs compared with Russians. Another is a relatively successful national program stimulating the return of ethnic Kazakhs from Mongolia and China. There is a yearly resettlement quota endorsed by the president of Kazakhstan. Since 2001, this quota was 15,000 families. According to the 2005 data, the overall number of ethnic Kazakhs from abroad who have resettled in Kazakhstan since 1991 was 433,000. These people face their own problems, one of which, interestingly enough, is lack of proficiency in Russian, which is used by many urban Kazakhs in Kazakhstan as their main communication medium.

FREEDOM

The country's Constitution provides for freedom of religion, and the various religious communities worship largely without government interference.

MOVING THE NATIONAL CAPITAL

Apparently, the spatial pattern of two major ethnic groups in Kazakhstan, Russians and Kazakhs, has been a source of concern for Kazakhstan's political establishment. This concern was boosted by some Russian nationalists' assertions that northern Kazakhstan is essentially an extension of Siberia and should be reunited with Russia. Alexander Solzhenitsyn, who, as a prisoner of Stalin's gulags, spent quite some time in Kazakhstan, wrote in his 1990 brochure *How We Ought to Arrange Things in Russia,* "[Kazakhstan's] current enormous territory was drawn up by the communists without much thinking . . . : wherever roaming flocks grazed once a year that's their Kazakhstan. However, where to draw borders was not meant, in the first instance, to be important since all nations were expected to merge into one in a short while. A shrewd man, Lenin called the

(© Ilya Lipkovich)

A farmer's market in Almaty. One can easily associate the farm products on display with specific ethnic groups: Russians sell cottage cheese; Kazakhs sell kumis, a fermented dairy product traditionally made from mare's milk; and Koreans sell marinated vegetables.

issue of borders supremely unimportant. (That's why they adjoined Karabakh to Azerbaijan—who cares where! It was more important at that time to please Turkey, a close friend of the Soviets.) Also, prior to 1936, Kazakhstan was considered an autonomous republic within the Russian Federation, only then was it promoted to union republic status. And it is made up of southern Siberia, the southern piedmont of the Urals, and a desert vastness in the center which has been transformed and built up by Russians, prisoners, and deported peoples. Today, Kazakhs account for much less than half of all people in this much inflated Kazakhstan. Their cluster, their constant and well rooted part—indeed, for the most part, settled by Kazakhs—forms a southern arc of oblasts extending from the extreme east almost to the Caspian Sea. If this arch alone chooses to break away from us they are most welcome." In May 1996, a group of respected Kazakh writers attacked the

Komsomolskaya Pravda daily for publishing an interview with Solzhenitsyn in which he similarly called for northern Kazakhstan to be incorporated into Russia. They demanded that the newspaper be banned in Kazakhstan, accusing it of violating their country's territorial integrity—a charge backed by the Kazakh prosecutor-general, who described Solzhenitsyn's statement as a "gross intervention into the internal affairs of an independent state." The *Komsomolskaya Pravda* was eventually forced to publish an apology.

Solzhenitsyn's opinion on Kazakhstan and its border with Russia comes to mind in conjunction with the issue of the 1997 transfer of the Kazakhstan's capital from Almaty to Astana. The official justifications for this move, such as seismic threats and the exhaustion of land available for Almaty's expansion, are not taken seriously by most analysts. The most plausible reason for moving the capital was

to create a national center for the Kazakh nation in the area which was deemed most vulnerable because of a combination of its ethnic makeup and geopolitical neighborhood, a combination that could make this area prone to secession. Indeed in 1989, in east Kazakhstan, Karaganda, Akmolinsk/Tselinograd, and Kostanay oblasts, the share of ethnic Russians ranged from 44 percent to 65 percent, and the share of ethnic Kazakhs ranged from 17 percent to 28 percent. For example, in Tselinograd Oblast, there were 448,000 ethnic Russians (44.7%), 123,700 ethnic Germans (12.3%); 94,400 Ukrainians (9.4%); and 28,700 Belarusians. As for ethnic Kazakhs, there were 224,800 or 22.4 percent. The transfer of Kazakhstan's capital to that very region was thought of as a way to stimulate migration of ethnic Kazakhs to northern Kazakhstan, where the population was being depleted by intensive outmigration of non-Kazakhs in the 1990s.

(© Leon I. Yacher)

Modern buildings in Astana surround the Baiterek monument, the landmark which represents Astana as the capital of Kazakhstan. It is said that the hand prints of the President have been permanently located on top. The Presidential Palace can be seen in the far background.

Astana's Population and Ethnic Makeup

	1989	1999	Percentage Share in Total Population		1999 as a % of 1989
			1999	1989	
Total	281,252	319,324	100	100	113.5
Kazakhs	49,798	133,585	41.8	17.7	268.3
Russians	152,147	129,480	40.5	54.1	85.1
Ukrainians	26,054	18,070	5.7	9.3	69.4
Germans	18,913	9,591	3	6.7	50.7
Tatars	9,339	8,286	2.6	3.3	88.7
Belarusians	8,220	5,761	1.8	2.9	70.1
Poles	2,762	2,537	0.8	1	91.9
Koreans	1,329	2,028	0.6	0.5	152.6
Ingush	1,889	1,822	0.6	0.7	96.5
Others	10,801	8,164	3.8	2.6	132.3

Source: Kratkie itogi perepisi naseleniya v Respublike Kazakhstan. Almaty 1999: p. 113.

A major mosque in Astana.

(© Leon I. Yacher)

The generation of a replacement migration counter flow from the south to the north was able to destroy the dichotomy "Russian north versus Kazakh south" and create a more balanced ethnic geography.

Following the independence of Kazakhstan, the city of Tselinograd was renamed Aqmola (1992), and just two years later, when the capital transfer plan was ready, it was renamed Astana. The word Astana in Kazakh means capital. Astana became a very ambitious and by all accounts successful project. The original plans for the new Astana were drawn up by the late Japanese architect Kisho Kurokawa. Old buildings that remained from the Soviet era are now being removed and replaced with totally new structures, resulting in significant construction work throughout the city. President Nazarbayev pays particular attention to Astana's architecture: most of the recently completed structures are accredited to internationally acclaimed architects and designers such as Kisho Kurokawa or Norman Foster.

This is how population size and its ethnic makeup changed within just the ten years from 1989 to 1999, proving that, at least in geopolitical terms, the transfer of the capital is a success.

THE NAZARBAYEV RULE

Nursultan Nazarbayev has served as the President of Kazakhstan since the fall of the Soviet Union and the nation's independence in 1991. Nazarbayev was born into a peasant family in 1940 in Almaty Oblast and is therefore a Kazakh from Great Jus. In 1967 he graduated from a ferrous metallurgy college in Karaganda. Since the early 1960s he worked at the Karaganda Metallurgy Combine in several positions, from unskilled laborer to blast furnace operator. In 1969 he was first promoted to lead a Young Communist League cell and began to climb the ranks of the Communist *nomenklatura*. In 1977, he became secretary of the Karaganda regional committee of the CPK; in 1979, a secretary of the CPK's Central Committee; in 1984, Chairman of the Council of Ministers of Kazakhstan; and in 1989, he replaced Gennady Kolbin as the First Secretary of the CPK. It is from that leadership post that Nazarbayev was elected President.

As president of a largely Muslim country Nazarbayev has made an effort to highlight his Muslim heritage by traveling on a hajj to Mecca, supporting mosque renovations, and building new mosques.

There is no doubt that given Kazakhstan's economic success—in no small part attributed to Nazarbayev's leadership, he is popular in Kazakhstan. Nazarbayev's regime is less rigid politically and in particular economically than other ruling regimes in Central Asia, with the possible exception of Kyrgyzstan's. This, coupled with the fact that Kazakhstan is by far the wealthiest state in Central Asia (with a per capita GDP PPP at $11,000 in 2007, according to the CIA Factbook), with a sizable and growing middle class, and with Kazakhstanis' conspicuous absence in Russian cities as menial laborers (contrasting with large numbers of those from all other Central Asian countries but Turkmenistan), there is probably more room for personal advancement in Kazakhstan than elsewhere in Central Asia.

At the same time, Nazarbayev's regime is far from a Western-style democracy. An April 1995 referendum extended his term until 2000. This relieved Nazarbayev from the necessity to launch his election campaign at the end of 1995 or in the beginning of 1996. Instead he did that after moving the capital to Astana and after getting rid of some powerful opponents, like Akezhan Kazhegeldin, Kazakhstan's Prime Minister from 1994 to 1997, who was forced into exile. Nazarbayev won what became "early" elections of January 1999 and he won again in December 2005. The Organization for Security and Cooperation in Europe criticized the last presidential election as falling short of international democratic standards. On May 18, 2007, the Parliament of Kazakhstan approved a constitutional amendment that would allow Nazarbayev to seek re-election as many times as he wishes. This amendment applies specifically and only to Nazarbayev: the original constitution's proscribed maximum of two presidential terms will still apply to all future presidents of Kazakhstan. However, to many (but not all) in Kazakhstan, Nazarbayev is a symbol of order and stability. And as recently shown by Nancy Lubin and Arustan Joldasov, people of Kazakhstan express preference for order over Western-style democracy. When asked what political system would be the best for Kazakhstan in order that the country be able to solve its problems, 55 percent of the Kazakhstani probed in a 2007 representative national survey chose the answer "Any system that will bring order to the country," and only 20 percent chose "Western-style democracy." Although back in 1993, the ratio of those same options was even more skewed in favor of order (62% to 17%), there is hardly a reason to expect the triumph of democracy in Kazakhstan any time soon.

If anything, Kazakhstan's regime evinces most similarity to that of Belarus. This statement may surprise all those who draw knowledge about the post-Soviet space exclusively from what Western media has to say about it. However, in both cases, Kazakhstan *and* Belarus, one sees that political stability and economic success (well above those achieved by other post-Soviet republics) are ensured under extraordinary and seemingly unlimited presidential powers. In both cases vote rigging and squelching the opposition and independent voices in the media have taken place, yet no large-scale violence has occurred and neither has a "color revolution." In both cases, the national leader elicits diametrically opposing insiders' reactions, either admiration or hatred, with very little in between. In both cases, dignitaries who fell out of favor find rescue in the West, as did the former Belarus central banker Tamara Vinnikova and the former

vice director of Kazakhstan's National Security Agency (and Nazarbayev's former son-in-law) Rakhat Aliyev. The reason one of these nations, Belarus, is castigated by Western politicians while the other, Kazakhstan, is not, has little if anything to do with facts on the ground. Rather, the difference is entirely in the "eye of the beholder." In the opinion of this author, the habit of viewing Belarus and Kazakhstan through different lenses has two semi-independent roots. One is what Edward Said called a "subtle and persistent Euro-centric prejudice" against Islamic peoples and their culture. Although in the West, Russia and by association Belarus are also seen as "others," there are, as Dickie Wallace showed so persuasively, various gradations of otherness (Wallace 2009): after all, Belarusians are European "others" while Kazakhs are not. For that reason, the level of democratic expectations vis-à-vis Kazakhstan has always been lower. So while there is a Belarus Democracy Act adopted by the U.S. Congress in 2004 and readopted in 2006, there is no such thing as a Kazakhstan (and likewise any other -stan) Democracy Act. Moreover, in 2006, when Belarus was still "Europe's last dictatorship" and contacts with Lukashenka were still taboo, President Nazarbayev of Kazakhstan was given an opportunity to meet both President George W. Bush (in Washington) and Vice President Dick Cheney (in Astana). These proponents for global democracy did not say a word about well-documented departures from democracy in Kazakhstan. The second reason behind the markedly different optics applied to two remarkably similar political regimes is geopolitical: Kazakhstan is oil-rich and is perceived as an independent player on the world scene, while Belarus does not have oil and, until recently, it was perceived—inaccurately—as being under Russia's thumb. Moreover, Belarus is an important transit country for Russian hydrocarbons en route to Europe, and it is located in a region where virtually every other political regime responded favorably to Western overtures. The failure of Belarus to so respond, until recently, helps explain the great degree of Western irritation with it.

In contrast, Kazakhstan has managed to conduct a well-balanced foreign policy from the beginning without playing off Russia and America or America and China and, even more importantly, without being pressured by any of these countries. Just like Belarus and Ukraine, Kazakhstan disposed of its nuclear nation status in 1992, and the warheads stationed in Kazakhstan were transported to Russia with America actually footing the bill. Kazakhstan is

seemingly one of Russia's closest allies, being a co-signer of the Collective Security Treaty (along with Russia, Armenia, Belarus, Kyrgyzstan, and Tajikistan), a member of the Custom Union (with Russia and Belarus), and a member of the Shanghai Cooperation Organization (with Russia, China, Kyrgyzstan, Tajikistan, and Uzbekistan). The country has demarcated its border with all its neighbors, including China and Russia, and has friendly relations with the United States, with American investments worth $15 billion as of the autumn of 2006.

BORAT AND KAZAKHSTAN

No essay about Kazakhstan would be complete without mention of "Borat." Indeed, Sacha Baron Cohen's 2006 satirical movie was named *Cultural Learnings of America for Make Benefit Glorious Nation of Kazakhstan*. Although the movie was actually about America, not Kazakhstan, with opening and closing village scenes shot in Romania, and the very word "Kazakhstan" used as a stand-in for geographic names few Americans have heard of, there were reasons for the citizens of Kazakhstan to be deeply offended. Indeed, some vented

HEALTH/WELFARE

A government health reform program aims to increase the amount of GDP spent on health care from 2.5 percent in 2005 to 4 percent by 2010. A compulsory health insurance system has been in the planning stages for several years.

their indignation to Western media, for example, in an October 16, 2006 National Public Radio broadcast. It is all the more reassuring that Roman Vassilenko, press attaché of Kazakhstan's Embassy to the United States, revealed in his statement that "Borat is a double-edged sword. It's a problem but also a great opportunity, and it's up to us whether we make use of it or squander it, which is exactly what would happen if we just sit there and ignore the movie or browbeat the comedian into making concessions or come across as not having a sense of humor." *Forbes Magazine* wrote in this regard that the Kazakh diplomat's take is as levelheaded as that from any American PR executive with a black belt: "Borat is frustrating, but people are smart enough to understand that it's a fictional movie about a fictional character. He represents 'Boratistan,' not Kazakhstan—but he helps us get our story out."

Kyrgyzstan

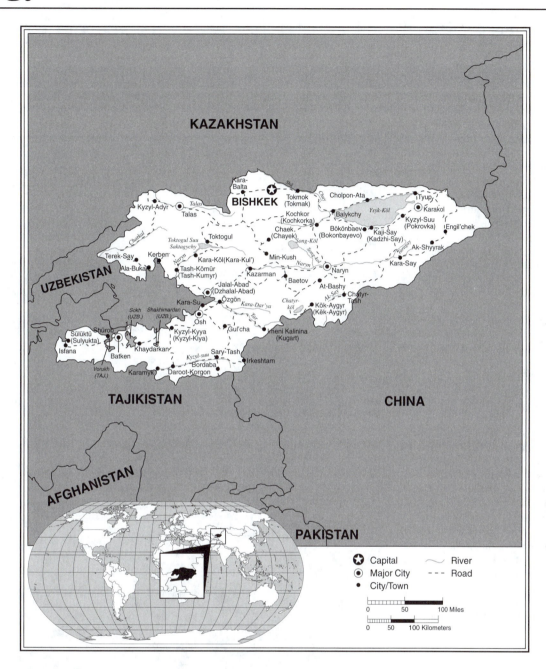

Kyrgyzstan Statistics

Area in Square Miles (Kilometers):
   76,641 (199,900)
Capital (Population): Bishkek
   (1,250,000)

Population

Total (millions): 5.5
Per Square Km: 27
Percent Urban: 35

Rate of Natural Increase (%): 1.6
Ethnic Makeup: Kyrgyz 64.9%; Uzbek
   13.8%; Russian 12.5%; Dungan 1.1%;
   Ukrainian 1%; Uygur 1%; other 5.7%
   (1999 census)

Health

Life Expectancy at Birth: 64 (male);
   72 (female)

Infant Mortality Rate: 31/1,000
Total Fertility Rate: 2.8

Economy/Well-Being

*Currency ($US equivalent on 14 October
   2009):* 43 som = $1
2008 GNI PPP Per Capita: $ 2,130
*2005 Percent Living on Less than US$2/
   Day:* 52

Kyrgyzstan Country Report

The Central Asian republic of Kyrgyzstan is a mountainous, landlocked country with a land area of 199,900 sq km and a population of 5,482,000 (2009 estimate). The capital city is Bishkek. From 1926 to 1991, the city bore the name Frunze (after a Bolshevik military leader who was born there). Kyrgyzstan is located within parts of two huge mountain systems, the Tian Shan (in much of Kyrgyzstan) and the Alai (in its southwest); all of the country is more than 500 m above sea level, one half of its land area is within the 1,000–3,000 m range, and one-third lies more than 3,000 meters above sea level. With such an abundance of rugged terrain, only 8 percent of the land is cultivated. Not surprisingly, the country is settled very unevenly, with most people living either in or near the northern foothills of the Tian Shan in the Chu Valley, which is where Bishkek is located, or in the Fergana Valley, the peripheral part of which belongs to Kyrgyzstan, where, the second-largest city of the republic, the city of Osh, is located. Until the most recent "rehabilitation" of the Bishkek-Osh highway, the two parts of the country were poorly connected. Climatically, all altitudinal zones are represented in Kyrgyzstan, from the subtropics of the Fergana Valley—where in the foothills of the Tian Shan survive some of the world's largest natural walnut forests—to high-elevation tundra and icecap. One of Kyrgyzstan's jewels is Lake Issyk-Kul, the second-largest lake in the world in the alpine zone after Titicaca. One more peculiarity of Kyrgyzstan, rooted in the demarcation of its borders during the early Soviet period, is the existence of a single Kyrgyz exclave surrounded by Uzbekistan's territory in the Fergana Valley. Conversely, four Uzbek and two Tajik enclaves are completely surrounded by Kyrgyzstan.

The area that is modern Kyrgyzstan was incorporated into the Russian Empire in two stages, the northern part during 1855–1864 and the southern part in 1876. It was only as late as 1926 that the ethnonym Kyrgyz became solely associated with the one ethnic group it identifies today. From 1918 to 1924, the whole of today's Kyrgyzstan was part of the so-called Turkestan Autonomous Soviet Socialist Republic within the Russian Federation. An autonomous republic was the second tier of ethnic autonomy in the U.S.S.R., and all such entities were subordinated to union republics. Within the TASSR, the Kyrgyz were called Kara-Kirgiz—to distinguish them from Kirgiz, which was the name Russians then attached to modern Kazakhs. In 1924, when the Kara-Kirgiz Autonomous Republic was formed, it was renamed the Kirgiz Republic; in 1936, it was promoted to union republic status.

Kyrgyzstan has two official languages, Kyrgyz and Russian, although Kyrgyz account for 68.9 percent of the population and Russians for only 9.1 percent. This is a legacy of the Soviet period, when Russian was widely used in Bishkek even among

(© Simon Garbutt)

A manaschi (storyteller) in Kyrgyzstan.

the Kyrgyz themselves and as a language of inter-ethnic communication. Like other titular ethnicities of Central Asia, the Kyrgyz are Muslims. Other ethnicities of Kyrgyzstan include Uzbeks (14.4%) and many smaller groups, such as Dungans (Chinese-speaking Muslims), Russian-speaking Ukrainians, and even Germans whose ancestors were exiled to Kyrgyzstan from the civil divisions of Russia (e.g., from the Autonomous Republic of the Volga Germans, disbanded in 1941). As many as 101,300 ethnic Germans resided in Kyrgyzstan in 1989, but due to emigration to Germany, only 21,000 remained ten years later.

ETHNIC KYRGYZ

The Kyrgyz are of mixed origin. They belong to an Asian race (a legacy of Mongolian domination) but speak a Turkic language. Unlike the other titular ethnicities of Central Asia, Kyrgyz descend from tribes who moved to Central Asia after 840 A.D. from the upper Yenisei area of Eastern Siberia. The closest linguistic relatives of Kyrgyz are Khakasians living in Russia's Krasnoyarsk krai, quite distant from modern Kyrgyzstan, and the Altai people of Russia. However, throughout the long time spent in Central Asia, the Kyrgyz language became closer to other Turkic languages in the region, especially Kazakh.

Perhaps the world's best-known Kyrgyz was Chingiz Aitmatov (1928–2008), a writer whose several short stories (the most famous being, "Jamila," written in 1958, which Louis Aragon has described as the world's most beautiful love story) and novels were translated into many languages, including English.

Researchers of folklore are familiar with the *Epic of Manas,* a traditional epic poem of the Kyrgyz people. Manas is the name of the epic's hero, who consolidated the Kyrgyz into one community. The poem consists of half a million lines, telling the story of Manas, his descendants and his followers. Battles against Chinese, Kalmak, and other enemies form a central theme in the epic. Although the epic is mentioned as early as the fifteenth century, it existed only in oral form until 1856 when Chokan Valikhanov, an ethnic Kazakh (see the essay about Kazakhstan), first wrote down some of its chapters and translated them into Russian. Parts of this classic centerpiece of Kyrgyz literature are often recited at Kyrgyz festivities by the so-called Manaschi, the poem's narrators (see the photo on p. 186).

Prior to the Soviet period, there was no industry in Kyrgyzstan; most of the indigenous population engaged in nomadic sheep herding. Kyrgyz society was at a

pre-feudal stage. In 1916, the Kyrgyz participated in a large-scale uprising against the Russian administration. There were two immediate reasons for the unrest. First, prior to World War I, the indigenous people of Central Asia were not subject to universal draft, but with the beginning of the war they were recruited in growing numbers to dig trenches and do other primitive construction work near the frontlines. Second, many colonists from Russia settled in the northernmost part of Kyrgyzstan and occupied a disproportionate amount of land. The uprising was brutally suppressed. As a result, the number of casualties among the Kyrgyz exceeded 150,000, about one-sixth of their total population; and hundreds of thousands fled to China, where there is still a significant Kyrgyz population in Xinjiang. The northern part of the country became deserted. About half a million Kyrgyz returned from China after the Communist Revolution in Russia. Only in 2008 did the government of Kyrgyzstan begin to pay close attention to the memory of the 1916 tragedy: it decided to commemorate its victims on the first Friday of each August.

KYRGYZSTAN AS A SOVIET REPUBLIC

It would not be an exaggeration to say that sedentary society and the entire Kyrgyz economy are creations of the Soviet period. In 1917, equality of men and women was mandated by law, and in 1921 polygamy and kalym (ransom for a bride) were outlawed. During the 1920s, mandatory elementary education was introduced and literacy improved dramatically; book and newspaper printing in Russian and Kyrgyz commenced. In 1926, it was mandated that the Kyrgyz language dispose of Arabic script and adopt Latin script instead, which simplified learning the written language for most illiterate Kyrgyz. In 1940, the Cyrillic script replaced Latin. Whereas in 1897, only 3.1 percent of Kyrgyz were literate, by 1939, 79.8 percent were Bol'shaya Sovetskaya Encyklopedia (Great or Big Soviet Encyclopedia).

The first local resource to be extracted commercially in Kyrgyzstan was coal. By 1940, the coal mines of Kyrgyzstan accounted for 88 percent of the entire output of Central Asia (minus Kazakhstan). Subsequently, production of antimony and mercury developed.

The resource base of Kyrgyz economy set the country apart from other Central Asian countries. First, Kyrgyz agriculture was not dominated by cotton as was true elsewhere except northern Kazakhstan. Rather, the main specialization of Kyrgyz

agriculture was animal husbandry. Second, much of the rest of Central Asia is flat with a deficit of fresh water, but Kyrgyzstan (and also Tajikistan) displays the opposite, that is, an abundance of water and not much flat land. Of particular importance is the watershed of the Naryn River. The confluence of the Naryn and Kara-Darya rivers form the Syr Darya, one of two rivers that once, before they were diverted into irrigation canals, sustained the level of the Aral Sea. The confluence itself is in Uzbekistan, but the flow of both of Syr Darya's tributaries is controlled by Kyrgyzstan. During the Soviet period, a cascade of hydroelectric stations was built on the Naryn, a mountainous river ideally suited for that purpose. The largest of these stations is the Toktogul station, created in 1976 after 14 years of construction on a dam erected to flood the Kementub Valley. Twenty-six communities were displaced and the main road through the region was re-routed. The reservoir's hydroelectric plant has an installed capacity of 1.2 gigawatts. The dam height is 215 m. This is the major source of power in the republic.

The industrial base of Kyrgyzstan's capital and largest city, developed during the Soviet period, is dominated by machine-building, textiles, and footwear. In the 1960s, significant deposits of gold were discovered in the northern Tian Shan, but their commercial exploitation prior to the breakup of the Soviet Union was limited. Despite tremendous economic progress over the Soviet period Kyrgyzstan remained one of the least urbanized republics of the U.S.S.R. with only 39 percent of urbanites in its population, according to the last Soviet census of 1989.

In 1989, Kyrgyz accounted for only 52 percent of the total population due to a significant migration of Russians and other non-indigenous but for the most part Russian-speaking groups. The migrants and their descendants were largely employed in industry. The share of Kyrgyz in the population of the capital city was especially low, although growing: just 12.3 percent in 1970 and 22.3 percent in 1989. Ethnic Russians, on the other hand, accounted for 66.1 percent of Bishkek's population in 1970 and for 55.8 percent in 1989. In the late 1800s Bishkek was home to only 7,000 people, but by the Soviet Union's end the city's population had grown to 611,000. Even in the second-largest city of Kyrgyzstan, Osh—the major center of Kyrgyzstan's south—ethnic Kyrgyz had not become the largest ethnic community by the end of the Soviet period. There, Uzbeks, not Russians, formed the largest group.

By the end of the Soviet Union's existence, Gorbachev's Perestroika removed

(© Leon I. Yacher)

View of the city of Osh.

constraints on free speech and that, among other things, opened the floodgate of interethnic grievances that had been accumulating for decades. An ethnic minority rooted in a neighboring republic yet gaining a prominent foothold in a city just across the border was a common irritant in ethnic tensions because of a peculiar division of labor: representatives of a titular nationality usually dominated the corridors of power, but minority positions were strong in some of the more lucrative sectors of the economy. (Many parallels can be drawn, including Chinese versus Javanese in Indonesia; Armenians versus Azeri in Azerbaijan; etc.). Indeed, in 1990, Kyrgyz constituted 66.6 percent of the executive committee of the Osh Regional Soviet of People's Deputies, Russians 13.7 percent, and Uzbeks 5.8 percent. At the same time, however, Uzbeks accounted for 71.4 percent of all those working in the retail system of Osh, a powerful, if informal, source of income in the Soviet Union with its chronic shortage of consumer goods.

It should be pointed out, though, that the Fergana Valley in general, not just its Osh periphery, offered a nourishing environment for conflict because of a combination of high population growth and density, insufficient jobs, and above all, poverty. The 1990 conflict between Uzbeks and Kyrgyz was sparked by the allocation of land to Kyrgyz peasants for housing construction within an Uzbek-dominated collective farm not far from the city of Osh. The ensuing tussle between Uzbeks and Kyrgyz ended with a Kyrgyz "victory," but Uzbeks retaliated by forcibly evicting all Kyrgyz from the nearby town of Ozgon, where Kyrgyz had accounted for only 6 percent of the population, most of the rest being Uzbeks. This, in turn, led to Uzbek pogroms in the largest nearby city, Osh. Most murders were committed by strangulation with wire or rope; torture and beating; assault and battery using axes, stones, and other hard objects; and guns. There were cases where the victim was burned to make identification impossible. Rape was commonplace on both sides, as were various forms of humiliation and torture, such as parading women naked through the streets. Officially, the overall number of deaths amounted to about 300, but locals later claimed the

death toll was grossly underestimated. From 30,000 to 35,000 people were involved in the conflict. At the same time, in Kyrgyzstan's capital, mass meetings of Kyrgyz demanded that Russians be evicted from the city so Kyrgyz could get vacated apartments.

Yet, in the March 1991 referendum on the continuation of the U.S.S.R., 88.7 percent of Kyrgyzstan's voters opted for the retention of the Soviet Union. Just as in other Central Asian republics, no proindependence movement had developed in Kyrgyzstan, for the local elite were well integrated into the Soviet elite; and there was a sense of the tremendous economic benefits that Kyrgyzstan had extracted from its membership in the union. In addition, the degree of national consolidation among the Kyrgyz people was not great, though it had progressed significantly during the last three Soviet decades.

POST-SOVIET DEVELOPMENTS

Predictably, along with drastic economic decline—due to the termination of supplies from other republics in the immediate

188

aftermath of the Soviet Union's breakup—came intense outmigration by ethnic Russians and other Russian-speaking groups. In 1995, Kyrgyzstan's GDP was only 55 percent of that of 1991. The following dynamics of the ethnic Russian population of Kyrgyzstan speaks for itself (% of Russians in the total population is given in parentheses): 623,600 (30.2%) in 1959; 855,900 (29.2%) in 1970; 911,700 (25.9%) in 1979; 916,600 (21.5%) in 1989; and 603,200 (12.5%) in 1999. Conversely, the share of Kyrgyz jumped from 52 percent in 1989 to 69 percent in 2007. One-third of all remaining Russians live in Bishkek. In recent years their outflow has slowed, first as a result of the depletion of the most mobile groups and second as a result of the resumption of economic growth. Also, the local political elite became alarmed by the excessive loss of the qualified cadre. In part to prevent further outmigration, Kyrgyzstan made Russian an official language along with Kyrgyz and promoted ethnic Russian officials to high posts. For example, from December 2007 to October 2009 Kyrgyzstan's Prime Minister was an ethnic Russian (Igor Chudinov, appointed in 2007), the only ethnic Russian occupying such a post in all the former Soviet republics except, of course, Russia itself.

Kyrgyzstan became the only Central Asian republic whose ruling Communist Party elite did not ensure the post-communist retention of the top Soviet-era statesman. Everywhere else in Central Asia, the first secretary of the Communist Party prior to independence became the first national president after the breakup of the Soviet Union. Moreover, in three Central Asian republics former first secretaries are still in that position eighteen years after the Soviet breakup, and in a fourth one (Turkmenistan) the only reason this is not the case was natural death of the leader. In Kyrgyzstan, however, First Secretary Abasamat Masaliev, who led the country from 1985–91, could not muster majority support during the October 25, 1990, presidential election in the still-Soviet Kyrgyz parliament; and so a non-establishment candidate, Askar Akayev, a Saint Petersburg- and Moscow-educated scientist (in the area of optical physics) was elected president.

This was a clear indication of weakness and/or the inadequate consolidation of the Kyrgyz national elite; otherwise Moscow would not have been able to promote its own choice, which at the time happened to be Akayev. Fifteen years later (in 2005), the same weakness, most probably, led to his overthrow in the so-called Tulip Revolution. (An alternative reason, the democratization of Kyrgyz society, that some Western political commentators dabbled with for some time, does not seem likely, and post-2005 developments confirm that.) To be sure,

Akayev did not meet popular expectations as living conditions, which had worsened after 1991, did not improve much thereafter, particularly in the southern part of Kyrgyzstan.

Furthermore, several elections were marred by fraud—regrettably, a norm all over Central Asia and beyond—and Akayev was accused of authoritarianism, also scarcely a uniquely Kyrgyz feature. What came to be unique was that in late March 2005 Akayev was literally chased out of his office by the demonstrators who then proceeded to loot retail outlets in downtown Bishkek for several days. In no other country of Central Asia would this be even remotely possible despite the fact that reasons for popular discontent there are no less serious than in Kyrgyzstan. Akayev's fate is a testimony to his inability to create and maintain a deeply entrenched security apparatus, which rather spells *insufficient* authoritarianism, at least by Central Asian standards, for better or worse. Following his ouster Akayev first surfaced in nearby Kazakhstan and then in Moscow, where he is residing now.

Kurmanbek Bakiyev was elected president instead. Unlike all previous Kyrgyz leaders, Bakiyev hails from the south. He does not possess Akayev's academic credentials, increasingly an oddity amongst the post-Soviet officialdom, but rather is a bureaucrat of Soviet vintage. Most Western commentators have disposed of their earlier idealism regarding the Tulip Revolution. When, in July 2009, Bakiyev was reelected for his second five year-term, *The Economist* wrote in an article, characteristically titled "Tulips Squashed," that

> once elected, Mr Bakiyev ignored his promises to limit presidential powers and to hand more authority to parliament. But he complied with the local custom of taking care of kith and kin by making one brother head of the state security service and another ambassador to Germany. For many Kyrgyz, it seems as if the disliked Akayev clan was being replaced by the Bakiyev one. Though salaries and pensions were boosted and government spending increased, corruption has become more endemic and lawlessness has increased. In April Sanjarbek Kadyraliev became the fourth parliamentarian to be assassinated since the revolution. Independent journalists are frequently targeted, including Almaz Tashiev, who died on July 12th after being beaten up by policemen.

But, quite unexpectedly, Bakiyev has proven to be quite skillful at playing off Russia and America. He seems to know just what his country is worth in the fight against international terrorism. Kyrgyzstan is the only country to host both a Russian and an American military base. The American base, opened in 2001, leases part of the Manas airfield, 23 km northwest of Bishkek; the rest of the airfield is used as an international civilian airport, the gateway to Bishkek. Until recently 1,700 servicemen were stationed at the base, mostly Americans but a few from other Western countries (e.g., France and Spain). It is the largest American military presence in Central Asia outside Afghanistan. When in 2005 a similar base was closed in neighboring Uzbekistan, the Manas base became a key element in the military campaign in Afghanistan. Between 500 and 2,500 servicemen fly daily to Afghanistan from Manas. The base is also crucial for aerial refueling.

In poverty-stricken Kyrgyzstan, the presence of even a relatively small number of American troops can have an enormous impact. The base employs more than 500 locals, paying them up to 10 times the average monthly wage of about $100. The base is pumping about $156,000 a day into the local economy and last year accounted for 5 percent of Kyrgyzstan's entire gross domestic product.

On December 6, 2006, U.S. serviceman Zachary Hatfield fatally shot Alexander Ivanov, a Kyrgyz civilian, at a truck checkpoint at the base. According to a statement from the base, the airman "used deadly force in response to a threat at an entry control checkpoint." The killing drew widespread condemnation from Kyrgyz authorities and they quickly demanded that Hatfield's immunity from local prosecution be revoked. In the meantime, U.S. authorities agreed to have Hatfield remain in Kyrgyzstan until the matter was resolved. However, a few days later it was revealed that Hatfield was evacuated from Kyrgyzstan despite initial promises.

In February 2008, Kyrgyz authorities announced that the Manas airbase would be closed. It is unlikely, however, that the above incident played any role in that decision. Rather, intensive lobbying from Russia did. A Russian airbase was opened in

2003 in the town of Kant, just 30 km from the Manas airfield. Bakiyev was promised more than $2 billion in aid and loans by Russia in February 2008, after which he declared the American base at Manas would be closed within six months. Articles appeared in Russian media outlets with titles like "Russia evicted Americans from Kyrgyzstan." However, in 2009, Bakiyev reversed his decision in exchange for more than a tripling of the annual rent of the Manas airfield paid by Americans to $60 million and more than $100 million in investment. Russia's response has been swift. The latest word is that the government in Moscow is negotiating to set up a second Russian military base in Kyrgyzstan. That would bring in even more rent.

Now, neighboring Uzbekistan is grumbling on account of that second Russian base, and the relationships between the two neighbors are tense. This is a legacy not only of the 1990 Osh massacre. There are three other contentious issues. First, as many as 58 stretches of the mutual border are under dispute. Second, Uzbekistan is Kyrgyzstan's sole natural gas supplier, and just from 2002 to 2009, the price of gas rose 500 percent, from $42 to $240 per 1,000 m^3. The price is determined every year during difficult negotiations in which the natural gas price faces off against Kyrgyzstan's crucial role in regulating water discharge downstream from its Toktogul reservoir on the Naryn River. This is the third issue between the two countries. Uzbekistan wants water in this reservoir to accumulate in winter and be released in spring and summer when irrigation needs are at their highest. However, if that is the case, Kyrgyzstan runs the risk of low water levels behind the dam (which was actually the case in 2009) that results in electrical rationing throughout the country with outages lasting hours every day.

Kyrgyzstan is a low-income country whose 2008 per capita GDP (PPP) slightly exceeded $2,000. Since the breakup of the Soviet Union, the country has experienced willy-nilly deindustrialization, and as a consequence currently 48 percent of the labor force is engaged in agriculture. As for industry, its 2002 output was still only 38 percent that of 1991. Gold accounts for 40 percent of Kyrgyzstan's exports and 13 percent of its GDP. In 2003, Kyrgyzstan produced 23 tons of gold, making it the third largest gold producer in the former Soviet Union, after Russia and Uzbekistan.

DEVELOPMENT

As for industry, its 2002 output was still only 38 percent that of 1991. Gold accounts for 40 percent of Kyrgyzstan's exports and 13 percent of its GDP. In 2003, Kyrgyzstan produced 23 tons of gold, making it the third largest gold producer in the former Soviet Union, after Russia and Uzbekistan.

The biggest source of gold is the Kumtor mine, situated in the Tian Shan Mountains at an altitude of over 4,000 m. The Kumtor Gold Company is a joint venture between Canadian Cameco (33.33%) and Kyrgyzaltyn, a Kyrgyz business. The joint venture is a major earner for the country, contributing 6 percent to Kyrgyzstan's GDP.

Another major export is electricity. It is ironic that the country has to ration the electricity supply to its citizens in order to increase export earnings. Other export products include mercury, wool, meat, tobacco, and cotton. Russia is the major trading partner of Kyrgyzstan. Curiously, the second major importer is Switzerland (2007), a fact probably related to Kyrgyzstan's export of gold. In 1998, Kyrgyzstan became the first post-Soviet country other than the Baltic States to join the World Trade Organization.

In 1997, the American University of Central Asia opened in Bishkek. It currently enrolls 1,305 students from 19 countries: Afghanistan, Belarus, China, France, Germany, Great Britain, Iran, Israel, Kazakhstan, Kyrgyzstan, Pakistan, Russia, South Korea, Sweden, Tajikistan, Turkey, Turkmenistan, the United States, and Uzbekistan. It offers courses in English and Russian.

HEALTH/WELFARE

Clean drinking water has been a diminishing problem. The government is working hard to ensure pure and safe drinking water to its citizens, which in return has reduced the risk of water borne diseases.

Turkmenistan

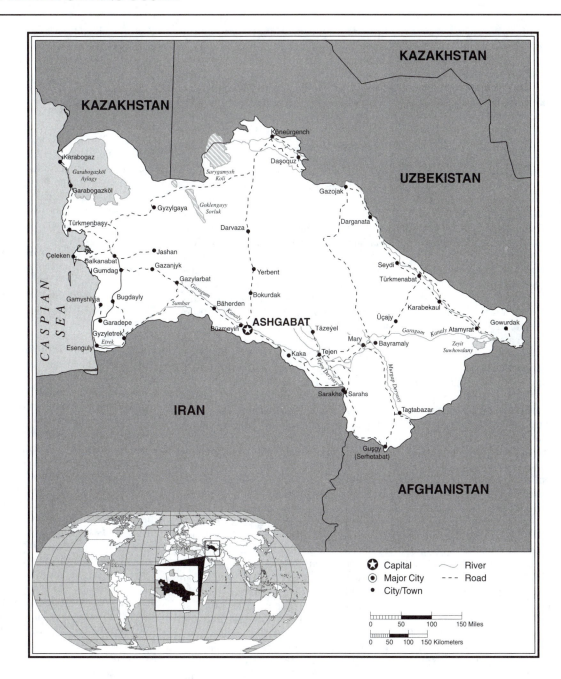

Turkmenistan Statistics

Area in Square Miles (Kilometers):
188,455 (488,100)
Capital (Population): Ashgabat
(est. 1,000,000)

Population

Total (millions): 5.1
Per Square Km: 10
Percent Urban: 47

Rate of Natural Increase (%): 1.4
Ethnic Makeup: Turkmen 85%;
Uzbek 5%; Russian 4%; other 6%
(2003)

Health

Life Expectancy at Birth: 61 (male);
69 (female)

Infant Mortality Rate: 51/1,000
Total Fertility Rate: 2.5

Economy/Well-Being

*Currency ($US equivalent on 14 October
2009):* 14232 manat = $1
2008 GNI PPP Per Capita: $6,210
*2005 Percent Living on Less than US$2/
Day:* 50

Turkmenistan Country Report

The Republic of Turkmenistan has a land area of 488,100 sq km and a population of 5.1 million people (2009 estimate). Although the resulting population density (10 people per sq km) is somewhat higher than that of its northwestern neighbor Kazakhstan, Turkmenistan is located in the most arid part of Central Asia. The Karakum (Garagum) desert, stretching 880 km from east to west and 450 km from north to south, occupies 80 percent of the country's entire land area. However, in the west, south, and southeast this flat (though with sand hills) basin floor is edged by mountains. The climate of Turkmenistan is dry, temperate, and continental, although winters are not as cold as they are in central Kazakhstan. The mean January temperature is $-5°C$ ($23°F$) and the mean July temperature is $28°C$ ($82.4°F$). Only in the Atrek River valley (in the extreme southwest) is the climate dry subtropical with a mean January temperature above freezing at $4°C$ ($39.2°F$). Precipitation ranges from 60 to 150 mm a year, but in the foothills of the mountains it is between 200 and 300 mm. The highest (3,139 m) point in Turkmenistan is in the extreme southeast in the Kugitang Ridge. In the south are the foothills of Kopet Dag (most of this mountain range lies across the border in Iran), where Turkmenistan's capital, Ashgabat, is located. Further east along the southern border lie the Badkhyz and—east of the Murghab River—Karabil uplands. All these zones of contact between desert and mountain are extremely picturesque and home to rare and endemic species, including mammals, birds, and reptiles. Among the endemics are Central Asian gazelle, Turkmen mountain sheep, onager, striped hyena, caracal, leopard, short-toed eagle, and golden eagle. Some of these natural borderlands feature huge scenic caverns, some of them with lakes such as the warm Baherden underground lake some 70 km northwest of Ashgabat. Located in a spacious cave 100 m below the surface, it is up to 15 m deep and has turquoise-colored $37°C$ ($98°F$) water with only a faint odor of hydrogen sulfide; this is one of the most impressive natural wonders this author has ever visited. The Badkhyz upland features a unique pistachio savanna and scores of endemic plant and animal species. In the west, between the Karakum and the Caspian coast, lie the Balkhan (Balkan) hills. The Turkmen coast of the Caspian Sea is sandy and low lying. Connected to the northernmost stretch of the Caspian coast is the Kara-Bogaz-Gol (literally, "mighty straight lake"), a shallow inundated depression forming a bay of the Caspian Sea with a surface area of about 18,000 km^2. It is separated from the Caspian, which lies immediately to the west, by a narrow ridge with an opening in the rock through which the Caspian waters flow. Kara-Bogaz-Gol Bay is one of the saltiest bodies of water in the world; its water salinity amounts to 270–300 g/l.

In a desert environment, access to fresh water is a necessity for the support of a human population. Such access was afforded historically by rivers and alluvial fans in the foothills of the mountains. Turkmenistan has few rivers and all of them are exotic, meaning that recharge takes place in headwaters that form outside the country. One of the two major rivers of Central Asia's closed drainage basin, the Amu Darya, accounts for 94 percent of Turkmenistan's total runoff. Turkmenabat, the country's second-largest city (formerly known as Chardzhou), is on the Amu Darya. Turkmenistan withdraws 28 percent of this river's annual flow. The major withdrawal goes into the Karakum Canal, a 1,300 km long artery carrying water for irrigation. From 10 to 12 cubic km of water is diverted annually from the Amu Darya into this canal; as a consequence the river no longer reached the Aral Sea by the late 1980s. (Actually, in winter a bit of the Amu Darya's water does still occasionally reach the Aral Sea.) Other rivers of Turkmenistan include the Murghab (Murgap) flowing from Afghanistan, the Tejen flowing from Iran and Afghanistan, and the Atrek flowing from Iran. While the Atrek empties into the Caspian Sea, the waters of the Murghab and Tejen are withdrawn to sustain oases, and both rivers now only reach the Karakum Canal. One of the oases sustained by the Murghab, where the city of Mary (pronounced with y under stress)—the fourth-largest city of Turkmenistan—is now located, is the ancient oasis best known as Merv, a UNESCO's world heritage site. Historically a Persian settlement, in the twelfth century, Merv was one of the world's largest and most cosmopolitan cities. Situated amidst an oasis straddling the Silk Road, Merv rivaled Damascus, Baghdad, and Cairo as a trading center. Today, Merv is Central Asia's most extensive archeological site.

Turkmenistan possesses some of the world's largest proven reserves of natural gas and a substantial amount of oil. Also, Kara-Bogaz-Gol is a source of natural salts especially mirabilite—from which sodium sulphate, used in the manufacture of detergents, is made.

TURKMEN ETHNICITY

Ethnic Turkmen are descendants of the Oghuz, a Turkic tribe that migrated to the area in the eleventh century and assimilated local Iranian speakers. The subsequent invasion of the Mongol hordes under Genghis Khan infused the area with more Turkic-speakers (who were the numerically dominant element of Genghis Khan's army) and apparently affected the phenotypic diversity of the Turkmen, who exhibit a continuum from northern Mongoloid to Mediterranean Europoid (Caucasian) types. On average, the Mongoloid types are not nearly as dominant as among Kazakhs and slightly less so than among Uzbeks. The closest linguistic relatives of Turkmen are Anatolian Turks (of Turkey) and Azerbaijani. All three languages are mutually intelligible. The ending "men" in the word "Turkmen" is a kind of an intensifier so that the actual meaning of the word is pure Turk or most Turk-like of the Turks (Curtis 1996). There is an alternative meaning of "men" coming from "iman," in which case a Turkmen is a believing Turk. It is this second meaning that was adopted by the Turkmenistan's late leader, Saparmurat Niyazov, in his book *Ruhnama*, recently mandatory reading for all citizens of Turkmenistan. Islam came to the area with the Arab conquest of the eighth century. Arabs converted local Iranians, who subsequently converted immigrant Turks. The Turkmen adopted a nomadic life style which they retained even longer than Kazakhs and much longer than Uzbeks. Prior to the 1920s almost no Turkmen lived in a town or a city. Aside from Turkmenistan, Turkmen live in neighboring Iran and Afghanistan. Magtymguly Pyragy (1733–1813), whom Turkmen regard as their classical poet and a spiritual leader, actually was born in and lived all his life in northeastern Iran.

According to Glenn Curtis,

> much of what we know about the Turkmen from the 16th to the 19th centuries comes from Uzbek and Persian chronicles that record Turkmen raids and involvement in the political affairs of their sedentary neighbors. Beginning in the sixteenth century, most of the Turkmen tribes were divided among two Uzbek

principalities: the Khanate (or emirate) of Khiva (centered along the lower Amu Darya in Khorazm) and the Khanate of Bukhoro (Bukhara). Uzbek khans and princes of both khanates customarily enlisted Turkmen military support in their intra- and inter-khanate struggles and in campaigns against the Persians. Consequently, many Turkmen tribes migrated closer to the urban centers of the khanates, which came to depend heavily upon the Turkmen for their military forces. The height of Turkmen influence in the affairs of their sedentary neighbors came in the eighteenth century, when on several occasions (1743, 1767–70), [they] invaded and controlled Khorazm.

Despite the uncritical usage of the ethnonym Uzbek, which was conferred on most Turkic-speaking people of Uzbekistan at about the same time the ethnonym Turkmen was applied to the Turkic-speakers of Turkmenistan, that is, in the early 1920s—and hence there were no "Uzbek chronicles" from the sixteenth to the nineteenth centuries, the above quote makes an accurate distinction between sedentary Turks (Uzbeks) and nomadic Turks (Turkmen). Another distinction was that the Turkmen lifestyle was inseparable from horsemanship and frequent raids on nearby sedentary communities. Turkmen horses have been renowned as cavalry mounts and racehorses for hundreds of years. Also, until recently, tribal identity for a Turkmen was stronger than a sense of belonging to a Turkmen nation. The sense of tribal identity was, and to some extent still is, the case everywhere in Central Asia; but in Turkmenistan it seems to have been even more tenacious than among the other titular nationalities of the region. There are five regional communities or tribes among Turkmen, the largest of which are Yomud (Yomut) and Teke. Yomuds migrated and camped with their horses, camels, and sheep within the northern and eastern parts of the country and eventually settled along the Amu Darya; Teke are from the southern foothills. Native to Turkmenistan's south is the Akhal-Teke horse breed, known around the world for its great stamina and courage.

TURKMENISTAN IN THE RUSSIAN EMPIRE

Russians made their first attempt to establish an outpost on the eastern coast of the Caspian Sea in 1717, when Prince Bekovich-Cherkassky founded a fort in preparation for a planned conquest of Khiva. However, his reconnaissance expedition failed, and Bekovich-Cherkassky was ambushed by Yomud horsemen and beheaded. In 1869, at the same spot, General Stoletov founded another fort, which developed into the Caspian port town of Krasnovodsk (now Turkmenbashi). It served as the center of Russia's new Zakaspiiskaya (Trans-Caspian) Oblast until Ashgabat became its center in 1882. The 1869 foundation of Krasnovodsk and the 1985 seizure of Kushka (later renamed Gusgy and then Serhetabat, the town's current name) from the Afghans by Russian forces[1] mark the beginning and the end of the incorporation of Turkmenistan into the Russian Empire.

The Turkmen mounted tough resistance to the Russian army in the north, where 20,000 Yomud horsemen continued to fight long after the sedentary residents (Uzbeks) of the Khiva Khanate conceded defeat in 1873. It appears that the Khan of Khiva himself (who submitted to Russia) paid tribute to Turkmen raiders; consequently, winning the khan's loyalty did not mean winning that of the Turkmen horsemen. In the south, the toughest battle was for Geok Tepe, a fortress that the Teke tribe built to defend the Akhal-Teke oasis from the advancing Russians. In December 1880, its 25,000 defenders were attacked by 6,000 Russians under General Mikhail Skobelev. The siege of Geok Tepe lasted twenty-three days, after which the city was taken by storm.

Following the defeat of the Teke stronghold, Russians founded the city of Ashgabat (first called Askhabad, then for a long time Ashkhabad) at a Teke nomad camp located along a major caravan route. In 1880, in order to secure the conquest of Central Asia's southern part, Russians began to build a railway line from a landing pier near Krasnovodsk to Ashgabat, Merv (Mary), and Chardzhou (now Turkmenabat). The railway line reached Ashgabat by 1885 and Chardzhou by 1886. By 1990, the Merv–Kushka rail line was commissioned, giving the borderline fortress of Kushka an outlet. As an administrative center with a railway station Ashgabat began to grow. It had a rectangular lattice of streets lined by one-story adobe residences with orchards. In 1901, the city had a population of 36,500, including 11,200 Persians, 10,700 Russians, and 11,200 Armenians. Little commercial production was created in Turkmenistan before the Communist revolution. However, an oilfield was discovered in 1877 on the Cheleken Peninsula, which protrudes into the Caspian Sea. To be sure, Cheleken oil had been known since ancient times. Even before the time of Christ, oil from Cheleken was exported to areas that are now Iran, Afghanistan, and other countries of the Middle East. Oil was being produced from shallow pits and transported in goat skins in the eighteenth century. As of 1900, oil was being extracted on Cheleken by 23 companies, including the Branobel Company, established in 1876 across the Caspian in Baku. *Branobel* stands for Bratya (brothers) Nobel—Ludwig and Robert; Alfred Nobel, the inventor of dynamite and founder of the Nobel Prize, also owned some shares in the company. In those days, one of the factors restraining the growth of oil extraction was a deficit of donkeys. Prior to pipelines (the first of which, designed by Vladimir Shukhov for the Branobel Company, was built in 1879) crude oil was transported from oil wells to refineries in barrels loaded onto small carts pulled by donkeys. Reportedly Robert Nobel wrote to his brother Alfred that, while demand for oil soars, donkeys cannot multiply fast enough to match this demand.

Apparently some small portion of the proceeds from oil extraction was used to develop irrigation and expand cotton production east of the Caspian coast. From 1890 to 1915, production of cotton in Turkmen oases grew 15 times! So, Turkmenistan did participate in Russia's pre-World-War-I economic spurt.

Bolsheviks first seized power in Ashgabat in December 1917, but the White Guards and Socialists rebelled in July 1918 and overthrew the Bolsheviks, who did not regain power until 1920.

The Basmachi (Basmach = bandit, robber) emerged as a popular movement in response to the Bolshevik revolution. The Basmachi were particularly active in the Khiva Khanate between 1918 and 1922. (Centered on the historical Khiva oasis, the area of this entity is now divided between Uzbekistan and Turkmenistan. The largest Turkmen city in the oasis is Dashoguz. Formerly known as Tashauz, it is the third-largest city in Turkmenistan.)

Basmachi received substantial backing from anti-Bolshevik Russian forces and from the British military forces under General Wilfred Malleson. Turkmenistan became the only area in Central Asia directly occupied by British troops (Abazov 2005).

The British occupation lasted from August 1918 to September 1919.

At its peak the Basmachi movement represented a powerful military force of between 20,000 and 30,000 people. The movement lasted until 1929–1932. In Turkmenistan, the most prominent Basmachi group was led by Junaid-Khan (Dzhunaid-Khan), a talented military commander who established his base in the Khiva Khanate in early 1918. He received substantial assistance from the Russian White Guards and the British and Turkish governments and established capable military forces. In 1924, Basmachi led by Junaid-Khan organized a large uprising in the Khiva Khanate, but they were defeated. The movement never recovered from this defeat, but occasionally it managed to launch major offensive attacks in various parts of Central Asia from Afghanistan and Iran. In 1925, Junaid-Khan announced that he ended his opposition to the Soviet authorities and opted to remain in Turkmenistan. However, after a series of clashes with the Red army, Junaid-Khan and his close supporters were defeated again and retreated to Afghanistan. After this major setback the movement lost most of its spiritual leadership and gradually degenerated into small, unconnected groups of fighters. They continued their regular assaults against the Soviet administration and major transportation and communication lines, killing supporters of the Soviet regime and destroying crops. In order to contain the Basmachi's military supply, Soviet authorities imposed an "iron curtain" on the borders with Afghanistan and Iran in the late 1920s. At the same time a redistribution of land and improvement in living standards combined with repression against Basmachi sympathizers undermined the movement's support base in the region (Abazov 2005).

Junaid-Khan died in 1938 in Afghanistan at the age of 81 and was the only significant leader of the Basmachi movement not captured by the Red Army.

SOVIET TURKMENISTAN

Soon after the Red Army reclaimed Turkmenistan (or Turkmenia, as it was called in the U.S.S.R.), it became part of the Turkestan Autonomous Soviet Republic; its

capital was Tashkent. However, based on the 1924 national delimitation, it became the Turkmen Soviet Socialist Republic with its capital in Poltoratsk—as Ashgabat was called from 1919 to 1927 (in commemoration of a Russian communist leader who first made his name in Bukhara and in 1918 was executed by the White Guards in Turkmenistan). Thus, Turkmenistan became a full member of the Soviet Union, unlike Kazakhstan, Kyrgyzstan, and Tajikistan, which were to wait years until this status was conferred upon them. Turkmenistan was composed of the former Trans-Caspian Oblast, the southern part of the former Khiva Khanate, and the westernmost part of the former Bukhara Emirate.

Toward the end of the 1920s, Turkmenistan saw the collectivization of agriculture, the construction of irrigation canals, and significant expansion of cropland sown to cotton. In the late 1930s, it became the second-largest producer of cotton in the Soviet Union after Uzbekistan. Oil extraction resumed on the Cheleken Peninsula, and in 1931 a new oil field was discovered south of the city of Nebitdag (now Balkanabat), also in Turkmenistan's west. The Stalinist repressions of the 1930s took a terrible toll on Muslim clerics, a fairly significant part of the new intelligentsia, and thousands of the wealthier cattle owners. In the summer of 1935 a famous ride from Ashkhabad to Moscow took place to demonstrate to Joseph Stalin the formidable strength of Akhal-Teke stallions in the hopes that he would grant permission for their continued breeding. The campaign was a success. Twenty-eight riders, riding purebred Akhal-Tekes, the related Yomud breed, and Anglo-Teke crosses, covered a distance of 4,330 km (2,600 miles) with a broad range of terrain, including a severe three day 360 km (215 miles) test under the scorching sun of the Karakum desert. From the desert, which though stressful was familiar terrain, they rode through steppes, mosquito infested swamps, and rugged stony footing, through heavy rains and densely forested areas. Eighty-four days later they arrived in Moscow. The purebred Akhal-Tekes, notably two stallions, Arab and Alsakar, arrived in significantly better condition than the Anglo-Teke crosses, impressive evidence of the superiority of the purebred Akhal-Teke in terms of hardiness and endurance.

Shortly after World War II, in 1948, the city of Ashgabat, located within the Akhal oasis in the Teke-dominated part of Turkmenistan, suffered a terrible earthquake. Most brick buildings did not survive the quake, concrete structures were destroyed, and trains derailed. In those years, the

Soviet government did not include exact death tolls in its reports. Estimates contained in the Russian-language literature on the Ashgabat earthquake range from 3,000 to 156,000. Most probably at least one half of the entire population of the city or 60,000 people died. There were casualties outside the city, as well as across the border in Iran. Ashgabat was completely rebuilt after the earthquake.

Five years after the end of World War II, the most publicized endeavor in Turkmenistan became the construction of the Main Turkmen Canal from Takhiatash on the Amu Darya (in Karakalpak Autonomous Republic of Uzbekistan, 5 km from the Turkmen border) to Krasnovodsk. Construction, which engaged thousands of people, including prisoners of labor camps, was terminated soon after Stalin's death in 1953. In its stead, in 1954 a new construction project of epic proportions was launched, that of the Karakum Canal. Compared with the earlier endeavor, the Karakum Canal began much farther upstream, near the Bosaga Village in Turkmenistan's southeastern corner. This navigable irrigation canal had reached Mary on the Murghab River, 400 km from the canal's starting point, by 1959. A second, 138-km long stretch, from the Murghab to the Tejen River, was completed in 1960 and included the large (800 million cubic meters) Khauzkhan reservoir. The third, 260 km long stretch to Ashgabat was finished in 1962. The fourth stretch, completed in 1967, was short, extending just to Geog-Tepe (where the Turkmen's toughest resistance to Russian colonization had once been suppressed), but it included three reservoirs. Later, the canal was lengthened another 270 km to the Atrek River, with a branch to Nebitdag (now Balkanabat). The entire project was completed in 1988. Needless to say, the creation of an artificial 1,300 km long navigable river in a desert affected all components of the environment along the canal's path. Considering the huge amount of water used on irrigation and wasted through evaporation from this, the world's longest canal, it is little wonder the Aral Sea is irretrievably lost. However, the output of cotton from Turkmenistan tripled between 1961 and 1975. In addition to cotton, another product of Turkmen crop farming gained fame—melons from around Chardzhou. For most people outside Turkmenistan these melons were more a sweet legend than something they could get to actually eat, although Muscovites and other people along the Moscow to Turkmenistan railway line used to buy them from the train conductors.

In the 1960s, significant new oilfields were discovered in western Turkmenistan, bringing the total number of fields in the entire Western Turkmen Oil-and-Gas Province to more than 100. Two oil refineries were built in the republic, one in Krasnovodsk (1943; subsequently modernized and expanded) and the other in Chardzhou (1989). Given Turkmenistan's current global ranking in proven reserves of natural gas (by some accounts it has one-third of the proven reserves in the entire world) it is surprising how recently these fields were first tapped. Commercial extraction of natural gas in Turkmenistan began in 1967 in the easternmost part of the republic at the Achak field not far from Chardzhou (now Turkmenabat). The country's largest natural gas deposit, Dauletabad-Donmez, was only discovered in 1982, and commercial extraction began one year later. All electric stations in Turkmenistan, including the biggest one in Mary, work on natural gas. In 1967, the Central Asia–Center pipeline was commissioned with feeders from Turkmenistan. In summary, cotton, oil, and natural gas defined the economic profile of Soviet Turkmenistan.

Although ruling from the Akhal-Teke oasis from 1960 to 1985, the appointed leaders of Soviet Turkmenistan did not hearken from the Teke tribe. Balysh Ovezov, First Secretary of the Turkmenistan's Communist Party from 1960 to 1969, was a Yomud; and Muhamed-Nazar Gapurov, First Secretary from 1969 to 1985, was reportedly a Tat, a small Iranian-speaking minority living in and around the city of Chardzhou. Although Gapurov had to resign under Gorbachev as part of his massive sweep of Party leaders, no large-scale scandals were ever associated with his name, and he continued as an honorable (all-Union rank) retiree until the breakup of the Soviet Union. He died peacefully at his dacha in 1999. Gapurov's replacement, ensconced in Ashgabat in 1985, was Saparmurat Niyazov. In retrospect, it is ironic that Niyazov, who would proceed to establish one of the cruelest and most eccentric dictatorships and personality cults in the post-Soviet space and possibly in the whole world, was a Gorbachev appointee.

Shikhrat Kadyrov, the author of the "Russian-Turkmen Historical Dictionary," published in Norway, researched how interpersonal relationships took shape within the communist elite in the Turkmen SSR. "The ethnic Turkmen faction of these elite," writes Kadyrov, "consisted of the representatives of tribal fraternities, which were yet to merge into a nation. Russian Bolsheviks, who assumed the role of the arbiters in the intertribal disputes and were interested in a more or less fair resolution of these disputes in order to sustain their puppet state, were an alien component in opposition to which the idea of national solidarity was maturing." The path to this national solidarity was not straight but lay through the formation of the so-called Euro-Turkmen generation. These were people born of mixed marriages between college-educated Turkmen fathers and non-Turkmen (usually Russian but also Armenian and Jewish) mothers. Kadyrov cites a revealing paragraph from the speech of Mikhail Kalinin, one of the first Soviet leaders, at a 1924 meeting devoted to the establishment of the Turkmen SSR. "Can you imagine a Turkmen who graduated from a university and returned to his aoul (rural village) only to marry an illiterate peasant girl? Do you think that he will preserve the culture he gained during his college years? I have no doubt that in 2 to 5 years all his university knowledge and the culture he absorbed in the capital city will peel off of him as a snake's skin in autumn. This will happen because the setting will bring him back into the primitive state. For only when the cultural level of the woman has risen along with that of the man, only then will culture, however slowly, progress."

The members of the new Turkmen elite got the message. As early as 1927, N. Karadzhayeva, herself a member of that elite, stated: "Many comrades have abandoned their wives in the aouls, moved to the city and started courting European women" (Kadyrov 2001). Indeed, practically all Soviet Turkmen leaders married Russian, Jewish, Armenian, or Tatar women. And because divorces had to be sanctioned by the Party, the wife's "poor education" or "inadequate culture" became an entirely acceptable reason for divorce if the soon-to-be ex-wife was a Turkmen. Children of mixed marriages assumed Turkmen as their internal passport nationality, which ensured career advantages within the republic, but they barely spoke the Turkmen language and were not steeped in local culture. Over the course of 70 years a Euro-Turkmen sub-population grew that was jokingly referred to as "yarym-popolam," which means half-Turkmen-and-half-Russian. This population formed a community from which most of the political elite were recruited in Ashgabat.

Having married Muza Alexeyevna Melnikova, a woman from a Russian-Jewish family who became the mother of two of his children, Niyazov paid tribute to the above tradition, and he appointed to high government posts many Euro-Turkmen when the Soviet Union was still in existence and for some time thereafter. Born in 1940 in a Teke village 10 km from Ashgabat, Niyazov became an orphan at the age of eight. In 1942, his father was killed on the eastern front of World War II, and in 1948 his mother was killed by the Ashgabat earthquake along with his brother and sister. Niyazov was placed in an orphanage, then moved to the family of his uncle, a state farm director, and then (in 1954) was again placed in an orphanage, this time in the city of Ashgabat, where he finished high school with honors. Shokrat Kadyrov writes that for a Turkmen to place a child in an orphanage was quite unusual; a child who lost both parents by custom was adopted by relatives. Putting a child in an orphanage was rather linked to a desire to avail a boy (almost never a girl) of a Russian-Soviet upbringing so he could subsequently enter a career in government, education, or health care. It may well be that Niyazov showed some leadership qualities at an early age. Right after finishing high school in 1957, Niyazov was included in the Turkmen SSR's Moscow college freshman quota, which means that he was not required to pass the usual entrance exams. Thus he became a student at the Moscow Energy Institute (or by some accounts, at the Moscow Institute of Water Management). After failing three courses, he was expelled at the end of his very first semester. He returned to Ashgabat and obtained a job as a clerk in the geologists' trade union. In 1960, he was again included in the Turkmen SSR college freshman quota and this time sent to Leningrad Polytechnic University. He attended evening classes while working as a molding operator at one of Leningrad's defense factories. In 1962, he joined the Communist Party (unlike for office workers and intellectuals, there was no new membership quota for industrial workers) and in 1967 he graduated with a diploma as an engineer-physicist. Close to graduation he married a Leningrad girl but did not stay in Leningrad. He returned to Turkmenistan, and in 1967 he got a job as foreman in the controlling and measuring department of the Bezmeyin Thermal Electric Station, 25 km northwest of Ashgabat, and quickly was elected leader of the station's Communist Party cell. In 1970, Niyazov started his career as a full-time Party bureaucrat and steadily advanced in the ranks. His first position was that of an associate of the Industry and Transportation Department of the Central Committee of the Communist Party of Turkmenistan. In 1975, he became deputy chair of that department and in 1979 its chair. In the meantime he graduated—extramurally (that is, without attending classes, just attending exam sessions and writing papers)—from the Communist Party Higher Learning School in Tashkent. In 1980 he was appointed

First Secretary of the Ashgabat City Committee of the Communist Party—a high *nomenklatura* post, meaning he had been thoroughly vetted and cleared at all levels. The fact that Niyazov was transferred in 1984 to Moscow, where he worked as an associate of the Organization Department of the Central Committee of the Communist Party, clearly shows that he was being cherry-picked for a big-time position. Yegor Ligachev, who at that time was close to Gorbachev, headed the department, and it was he who, in March 1985, arranged Niyazov's appointment as Chairman of the Council of Ministers of Turkmenistan. In December of the same year he became the First Secretary of the Turkmenistan's Communist Party, the number one post in the republic. His candidacy was formally proposed by Gorbachev, who had been General Secretary since March 1985.

Although it is unlikely that Niyazov was at all inclined to play the tribal card, the political culture of Turkmenistan played it for him as it were. Niyazov's appointment terminated a hiatus in Teke leadership—a hiatus which Shokrat Kadyrov describes as being longer than Brezhnev's entire stagnation period in the Soviet Union at large. For some 25 years, the leader of the republic had not been from the Teke tribe. Within a couple of years, Niyazov cleansed the Turkmen Republic *nomenklatura* of 1,380 people: 31 members of the Turkmen Central Committee and of 80 members of the Turkmen Supreme Soviet were expelled. (Each of these bodies had, most probably, no more than 100–120 members.) Most replacements turned out to be from the Teke tribal elite. Thus tribalism rode the wave of Gorbachev's democratization. In the words of Kadyrov, Turkmenistan became an Akhal-Teke sultanate. In 1990, Niyazov became a member of the Central Committee of the Soviet Communist Party's Politburo. Such an honor was by no means bestowed on all first secretaries of the republics. In October 1990, Niyazov combined his supreme leadership post in the Turkmen SSR's Communist Party with chairmanship of the republic's Supreme Soviet (parliament). From there it was just one more step to becoming president of Turkmenistan, a post he acquired on October 27, 1990, as part of the "sovereignty parade" sweeping all the Soviet republics at a time when the fate of the Soviet Union was still unclear.

POST-SOVIET TURKMENISTAN: THE NIYAZOV RULE

Niyazov's dictatorship was consummated *after* the Soviet Union dissolved in December 1991. While *prior* to that he had to abide by certain checks-and-balances (e.g., maintaining relatively autonomous centers of power, such as the Central Committee, the Council of Ministers, and the Supreme Soviet, and having Moscow-based entities, to which these centers were subordinated, endorse their decisions) any constraint to his behavior as an oriental despot disappeared thereafter. Rigging elections, amending a constitution to allow for extra terms as president, replacing a presidential election with a murky and unconstitutional referendum, handpicking members of the parliament, repressing the opposition, and muzzling the media—all of these things occurred in many corners of post-Soviet space; Uzbekistan, Kazakhstan, and Belarus being only the most obvious examples. Niyazov's rule, however, included these and so much more. Obviously much blame can be put on Niyazov himself, but he was surely only part of the problem. Apparently the *need* for multiple centers of power and the tradition of co-opting at least a part of the political elite—not to mention a popular commitment to civility and trust, as the sociocultural basis for democratic forms of governance—were less established in Turkmen *society* than anywhere else in the former Soviet Union. Indeed, even though one and the same political form, to all outward appearances, was imposed on the entire multi-ethnic Soviet Union, different political cultures took shape. Whatever might be said of Soviet Russia, the core of the U.S.S.R., it was hardly a North Korea, whereas Niyazov's Turkmenistan became in many ways its replica.

While Turkmenistan did join the Commonwealth of Independent States, Niyazov quickly reduced the country's membership to the status of "associate member." Political neutrality was set as the foundation of the Turkmen foreign policy, but this neutrality became a means of insulating the political system of Turkmenistan. No foreign observer could attend a court trial, and no Red Cross representative could visit a prison in that country. Beginning in the mid-1990s, Niyazov's personality cult was impossible to ignore. References to his last name disappeared in Turkmenistan, and he was referred to as either Turkmenbashi (the father of all Turkmen) the Great or as Akbar Serdar (The Great Leader). He became also known as the Son of the Best Turkmen Mother, Gurbansoltan-edje, and, after February 19, 2005, (his birthday) as the Impersonated Symbol of Turkmenistan. He personally endorsed the first page of every newspaper, which had to bear his portrait. His underlings had to kiss his hands on televised occasions, which were carried only by state TV. Satellite and cable television were banned, and public access to the internet severely restricted.

On September 12, 2001, Niyazov finished writing his book *Ruhnama,* whose title literally means "the book of the soul." A spiritual message to the Turkmen people, it was endorsed as a national program of spiritual development. The month of September was renamed *Ruhnama,* September 12 became a national holiday, and each Saturday was to be devoted to studying the book. Niyazov subsequently renamed all of the other months, each to commemorate him as Turkmenbashi, his relatives, or some other outstanding Turkmen. In September 2004, the second volume of *Ruhnama* was published. This book included elements of moral guidance, its author's biography, and a historical narrative of dubious quality—essentially a poorly crafted myth. "The Turkmen people have a great history which goes back to the prophet Noah," declares the first section of Volume One. "Prophet Noah gave Turkmen land to his son Yafes and his descendants. Allah made Turkmen prolific and their numbers greatly increased. God gave them two special qualities: spiritual richness and courage. As a light for their road, God also strengthened their spiritual and mental capacity with the ability to recognize realities behind the events. After that, He gave his servants the following general name: Turk Iman. Turk means core, iman means light. The Turkmen name came to this world in this way. Allah by his sacred command sent the Prophet Noah scriptures, including holy orders. The Prophet Noah distributed these to the people of his time. The essence of these pages was indeed beautiful ethics. There were sayings such as honor-honesty to young men; virtue to girls; intellect, sagacity, dignity to old men and women; nobility to brides." In summer 2005, the first book of Niyazov's poems, titled *The Spring of My Inspiration,* was published, and in a few months a second book of poems, *My Kin,* was released. All of these books were included in school curricula at all grade levels. *Ruhnama* replaced such traditional subjects as philosophy. There is a huge mechanical replica of the book in Ashgabat. Each evening at 8:00 P.M., it opens and recorded passages from the book are played with accompanying video. There is also a solar-driven, 12-meter, gold-plated statue of Niyazov that casts his countenance on much of downtown Ashgabat.

From 1998–2000, secondary school programs in Turkmenistan were limited to nine years of instruction and college programs to two years. The Academy of Sciences, a structure that oversaw major research institutions in every Soviet republic, was disbanded and the practice of awarding doctoral degrees terminated—this specific ruling was probably for the better as the quality of those degrees was suspect anyway. Every word or suggestion uttered by Niyazov immediately morphed into a program of activities which applied to everyone. Thus, after Niyazov mentioned that Turkmen do not care about circus, opera, and ballet, all those genres and corresponding institutions were eliminated. In 2003, on a visit to Ashgabat University, Niyazov expressed his displeasure with the long hair and beards worn by some male students. Afterward, all male Turkmen were supposed to shave and wear short hair, and every foreigner with a beard was expected to ask permission to keep it upon entering the country. In 2004, Niyazov launched a campaign against gold teeth. In 2005, he banned using musical recordings during public celebrations; only live music was allowed. In 2006, he said that the elderly should be supported by their children, not by the state; and immediately all those who had children older than 18 were denied their retirement pensions. Needless to say, no political opposition was tolerated. Most members of the Euro-Turkmen community initially co-opted by Niyazov were expelled from power. Moreover, his own non-Turkmen wife had become estranged from him in the early 1990s and moved to Moscow, as had their half-Turkmen children, who subsequently moved to Vienna (son) and London (daughter). As to the Euro-Turkmen removed from power, the fate of Boris Shikhmuradov deserves mentioning.

Shikhmuradov was born into a mixed Turkmen (father) and Armenian (mother) family in 1949. He finished a Russian-language school in Ashgabat, and then graduated from Moscow State University with a diploma in journalism. He subsequently worked for one of the two Soviet press agencies (*Novosti*) and also graduated from the Moscow-based Diplomatic Academy. There were indeed few people in Turkmenistan's political elite with his qualifications. From 1993 to 1999 he worked as a deputy prime minister and in that position attracted foreign investors to Turkmenistan; but in 2000 he was demoted to the minister of foreign affairs, a function he had overseen in his previous position. In the same year, he was pushed a couple notches lower and became the special representative of the president in charge of issues of the Caspian Sea and Afghanistan. In 2001, he became Turkmenistan's ambassador to China. In that position he openly challenged Niyazov and announced that he was organizing and heading the Popular Democratic Movement of Turkmenistan.

(© Leon I. Yacher)

The presidential palace in Ashgabat.

On November 25, 2002, according to the Turkmenistan media, an abortive attempt on Niyazov's life took place. Reportedly the path of his motorcade was blocked by a truck and someone fired into his car. Niyazov, who emerged unscathed, immediately went on TV to announce that he had uncovered a plot to unseat him and denounced Shikhmuradov and other well-known figures as the plotters. Dozens of people, including many prominent government bureaucrats from among the Euro-Turkmen, were thrown behind bars. Still abroad, Shikhmuradov denied his role in the plot and in the alleged attempt on Niyazov's life. However, his relatives, including his younger brother, were promptly arrested. On December 25, 2002, Shikhmuradov arrived in Ashgabat and surrendered to authorities. Before doing so, he announced that his intent was to preclude persecution of his relatives. Four days later, Turkmen State TV broadcasted Shikhmuradov's confession. The stilted syntax and phraseology of his speech and its content led many observers to believe that his confession had been extracted by torture. "While living in Russia, we indulged in using narcotics and, being agitated by them, we recruited mercenaries so they could commit a terrorist act. Our task was to destabilize the situation in Turkmenistan, to undermine the constitutional order, and to make an assault on the president's life." This and other statements in his televised speech were read by

Shikhmuradov from a sheet of paper. On December 30, 2002, the Supreme Court of Turkmenistan sentenced Shikhmuradov to a 25-year prison term, but on the same day the verdict was changed to life in prison.

In 2005, a new wave of purges shook Turkmenistan. This time some of Niyazov's closest associates, his possible successors, were jailed, including vice premiers Redjep Saparov and Yelly Kurbanmuradov, who oversaw the oil and gas industries, as well as the heads of three state corporations operating in this key area of the Turkmen economy. Oil and gas, of course, had buttressed Niyazov's eccentric rule throughout his time in power. Revenues from these lucrative industries also explain why Ashgabat experienced a more significant makeover than any other national capital in post-Soviet space with the exception of the entirely new capital of Astana, Kazakhstan. Today, Ashgabat features a grandiose presidential palace with a massive golden dome-shaped roof, large squares with elaborate fountains, and several newly-built and extensive mosques, including the Ertugrul Gazi Mosque, named after the founder of the Ottoman Empire; Azadi Mosque, resembling the Blue Mosque in Istanbul; the Khezrety Omar Mosque; and the Iranian Mosque. Ashgabat is also home to the Arch of Neutrality. Seventy-five meters in height, the arch was erected to commemorate Turkmenistan's official policy of neutrality. Sundry monuments and museums

(© Jim Fitzgerald)

Ertugrul Gazi Mosque in Ashgabat.

glorify the Turkmen nation and reaffirm its old cultural affinities with Turkey and the Islamic world—and by implication its detachment from Russia.

In 2003, Niyazov suddenly cancelled a dual citizenship agreement co-signed by Boris Yeltsin of Russia in 1993. The agreement had allowed most ethnic Russians and members of other groups with ethnic homelands in Russia but residing in Turkmenistan to obtain Russian citizenship, a kind of insurance policy for them. Niyazov gave holders of dual citizenship only two months to choose either Russian or Turkmen citizenship and created a wave of outmigration, with many fleeing in panic. In 1989, ethnic Russians had accounted for 9.5 percent of Turkmenistan's population. According to Turkmenistan's census of 1995, Russians by then accounted for just 6.7 percent of the population, which still means about 300,000 persons. If Russia had been more welcoming to migrants from other republics and rendered serious financial support, many more would have left, particularly as access to employment became drastically limited in Turkmenistan when all Soviet-period diplomas issued outside the Turkmen SSR were nullified.

As for ethnic Turkmen, their percentage share in Turkmenistan's population has been steadily growing due to a high rate of natural increase as well as the outmigration of Russians and Russian-speakers since the mid-1970s. In 1995, Turkmen accounted for 76.7 percent of the total population, up from 72 percent in 1989 and 61 percent in 1959. During the last 30 years of the Soviet Union's existence the Turkmen population

of Turkmenistan increased 2.7 times. A tradition of large families with more than five children continued in Turkmenistan until, and even after, the end of the Soviet Union; therefore, although the birth rate is now declining, the total fertility rate currently stands at 3.07.

One of Niyazov's last strange initiatives was his 2005 decision to close all hospitals in Turkmenistan that were located outside the capital city. Accordingly, only diagnostic centers were left in five provincial centers; they decided whether or not to hospitalize a patient and send him or her to Ashgabat.

TURKMENISTAN AFTER NIYAZOV

In December 2006, President Niyazov died of cardiac arrest. In fact, he had suffered from several diseases. In 1997, he underwent coronary bypass surgery in Munich, Germany. Physicians from Munich gave him a checkup in September 2004 and were summoned to Ashgabat in early October 2006 to attend to his worsening state of health. They were apparently unable to help. Immediately after Niyazov's death something that would have been considered a coup d'etat in any other country took place. According to Turkmenistan's constitution, if the current president could no longer be in charge, the speaker of the parliament would take over the reins of power until a new presidential election could be held. Thus Mr. Ovezgeldy Atayev, the speaker, should have become the acting president. However, within hours (some say minutes)

of Niyazov's death, the General Prosecutor of Turkmenistan filed a criminal case against Atayev, and he was taken into custody without his immunity to prosecution as a member of the parliament being formally annulled. Niyazov died during the night of December 20–21, and most high-ranking Turkmen officials learned about Atayev's sudden arrest on the morning of December 21. At the same time they learned that Gurbanguly Berdymuhamedov, health minister since 1997 and vice premier since 2001, had been appointed acting president. According to official information, Berdymuhamedov, running against five presidential candidates named and endorsed by the parliament, won 89 percent of the vote in the February 2007 presidential election. Due to his uncanny resemblance to Niyazov, most news agencies speculated that Berdymuhamedov was Niyazov's illegitimate son, but this has never been confirmed.

Berdymuhamedov was born in 1957. He graduated from the Turkmen Medical State Institute in 1979 and entered a career in dentistry. He subsequently took postgraduate studies in Moscow and in 1990 he received his MD in dentistry. Like Niyazov, Berdymuhamedov is from the Teke tribe. The newly-elected president of Turkmenistan made his first state visit, in mid-April 2007, to Saudi Arabia and toured the holy sites in Medina. At the end of April he made his second state visit to Russia. Since late 2007, some of the most odious restrictions initiated by Niyazov have been removed. Thus in December 2007, the ban on the circulation of foreign periodicals in Turkmenistan was lifted; in January 2008, the ban on opera and circus ended; and in July 2008, the traditional names of the months were reinstated. In May 2008, the Arch of Neutrality and the golden statue of Niyazov were moved from the city center to one of the peripheral districts of Ashgabat. Turkmenbashi's name was deleted from the Turkmen national anthem and from the oath of loyalty to Turkmenistan with which students begin their school day. Study of *Ruhnama* continues, but it is no longer mandatory and pervasive. Instead, Turkmen now must purchase a book by Berdymuhamedov titled *To the New Heights of Progress.* While Berdymuhamedov does not come across as imperious to the extent his alleged father was, much depends on his entourage and Turkmen society at large, for it has long been noticed that the followers make leaders, not the reverse.

Turkmenistan's major resource is natural gas. It ranks fourth in the world for proven natural gas reserves after Russia, Iran, and Qatar. Geopolitics and the Soviet legacy impose implacable limitations on Turkmenistan. Restricted by an existing network of pipelines inherited from the Soviet Union, Turkmenistan can export its gas only to or through Russia, that is, through the Central Asia–Center pipeline. Export via Iran, a possible alternative route, has not found willing investors in light of America's and the EU's nuclear argument with Iran. Although a 200 km long pipeline from southwestern Turkmenistan to Iran was actually commissioned in 1997, it will not become a major outlet for Turkmenistan unless Iran becomes a transit country for Turkmen gas going to Europe, a situation unlikely in the immediate future. A route via Afghanistan is even less of a possibility due to well warranted security concerns. A route across the Caspian Sea and Azerbaijan may seem like a real possibility, but Azerbaijan has abundant natural gas from its own Shah Deniz field, enough to fill the Baku-Tbilisi-Erzurum (Turkey) pipeline without Turkmenistan's contribution. Besides, there is a lot of bad blood between Azerbaijan and Turkmenistan over three contested oil fields in the Caspian Sea which both countries believe lie beneath their territorial waters. As a result, Turkmenistan currently has no alternative but to sell most of its gas to Russia's natural gas monopoly Gazprom, which resells it to other countries, mostly to Ukraine.

Due to the current global economic crisis, consumption of natural gas has declined in Western Europe and even more so in Ukraine. However, Turkmenistan insists that previous agreements on the amount and price of gas sold to Russia remain valid. Unable to reach an agreement on a proportional reduction of gas purchases, the Russians decided in April 2009 to unilaterally reduce withdrawals of gas from the Central Asia–Center pipeline. It informed Turkmenistan, Uzbekistan, and Kazakhstan—the three suppliers of natural gas transported through that pipeline—only one day prior to implementing the decision. The two other Central Asian countries reduced the amount of gas they pumped into the pipeline accordingly, but Turkmenistan did not. As a result of

increased pressure in the pipe, an explosion destroyed a stretch of the pipeline within Turkmenistan, and exports of Turkmen gas to Russia came to a halt entirely. This led to mutual accusations. Although the pipeline has since been repaired, Turkmen natural gas sales to Russia have not resumed at this writing, and Turkmenistan has already been deprived of roughly $3 billion.

However, excessive dependency of Turkmenistan on Russian market will be overcome once the pipeline to China via Uzbekistan and Kazakhstan is completed. This pipeline was inaugurated on December 14, 2009. Presidents of Turkmenistan, China, Kazakhstan, and Uzbekistan participated in the ceremony. Turkmenistan committed itself to delivering 40 billion cubic meters of natural gas per year over a period of 30 years. Turkmenistan's participation in the planned Nabucco pipeline (the extension of the existing Baku–Erzurum pipeline to Bulgaria, Romania, Hungary, and Austria) may well become the next step in the implementation of the grand idea of selling Turkmen natural gas to customers outside Russia.

Prior to the Central Asia–Center pipeline explosion, which will inevitably produce economic decline in Turkmenistan, the country's per capita GDP (PPP) was $5,500 (in 2008), which means that Turkmenistan is the second highest ranking country in Central Asia (after Kazakhstan) on this development indicator, but it is a distant second. A double-digit rate of economic growth has been reported since 1999; however, it is unlikely that the living standard of Turkmenistan's population is

on par with that of countries with a similar (or even lower) per capita GDP. It is hard to be certain, though, since Turkmenistan is one of the most repressive and secretive places on earth, sharing this dubious title with North Korea and Myanmar (Burma). As much as 48 percent of its labor force is in agriculture, which is indicative of a poor Third World country. According to official data, the citizens of Turkmenistan do not pay for natural gas, electricity, water, and salt. It is unlikely that many of Turkmenistan's citizens would have access to these basic services and commodities if they were not free.

NOTE

1. The British call this Russian capture of Kushka the Panjdeh incident and/or Panjdeh scare; it nearly brought about a war between Russia and Great Britain, as the latter got apprehensive that soon India might fall victim to a Russian invasion.

Tajikistan

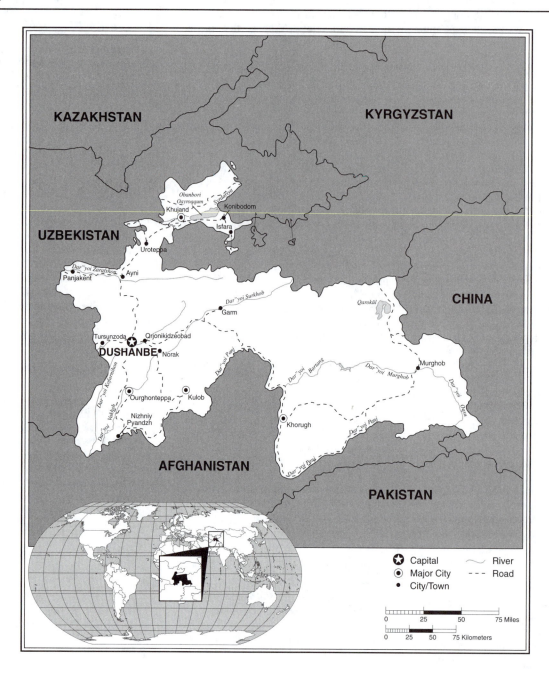

Tajikistan Statistics

Area in Square Miles (Kilometers):
55,251 (143,100)
Capital (Population): Dushanbe
(679,400)

Population

Total (millions): 7.5
Per Square Km: 52
Percent Urban: 26

Rate of Natural Increase (%): 2.3
Ethnic Makeup: Tajik 79.9%; Uzbek 15.3%;
Russian 1.1%; Kyrgyz 1.1%; other 2.6%
(2000 census)

Health

Life Expectancy at Birth: 64 (male);
69 (female)

Infant Mortality Rate: 65/1,000
Total Fertility Rate: 3.4

Economy/Well-Being

*Currency ($US equivalent on 14 October
2009):* 4.4 somoni = $1
2008 GNI PPP Per Capita: $1,860
*2005 Percent Living on Less than US$2/
Day:* 51

Tajikistan Country Report

The smallest country of Central Asia, Tajikistan has a land area of 143,100 sq km and a population of 7.5 million people. Like Kyrgyzstan, Tajikistan is a mountainous country; only 10 percent of its land area lies below 1,000 m above sea level. The mountain ranges of Tajikistan are higher, steeper, and less negotiable than those of Kyrgyzstan, and there are fewer elevated plateaus. As Andrei Rakitnikov, a Russian researcher of Central Asia, used to say, in the mountains a Kyrgyz travels on horseback, whereas a Tajik travels on foot (Rom 1987: 302). As many as 947 rivers (with at least 10 km length) flow across Tajikistan, many through picturesque gorges and canyons. Both of Central Asia's longest rivers, the Amu Darya and the Syr Darya, pass through Tajikistan; and the Amu Darya begins here as the confluence of the Vakhsh and Pyandzh rivers, as does an important tributary of the Amu Darya, the Zeravshan. The mountains of Tajikistan boast some of Asia's largest glaciers, such as the Fedchenko glacier (in the Pamir range)—the world's longest (77 km) glacier outside of the polar regions—and the Zeravshan glacier. There are many lakes

both freshwater and saline. The mountains of Tajikistan belong in the Pamir and Tian Shan systems. Centrally located within Tajikistan are the Gissar (Hissar) and Alai ridges, where many peaks exceed 5,000 m. The capital city of Dushanbe is located in the Gissar Valley in the foothills of the Gissar Ridge. Some of the world's highest mountains are located in the southeastern part of Tajikistan (Pamir literally means the roof of the world), including the highest (7,495 m) peak in what used to be the Soviet Union. Currently known as Ismoil Somoni Peak, from 1932–1962 it was referred to as Stalin Peak, and from 1962–1998 as Communism Peak. In this part of the country, even the lowest pass, Kamaloyak, lies at 4,340 m. In the northwestern part of the republic the Fan Mountains are known for their striking beauty and, among mountain climbers, for their challenging heights. In the north, Tajikistan controls part of the fertile Fergana Valley, most of which is in Uzbekistan. Until the very recent (2006) construction of the Anzob tunnel, no year-round reliable road linked Dushanbe to northern Tajikistan. The easiest way required a detour through Uzbekistan.

TAJIK ETHNICITY

Ethnic Tajiks currently account for almost 80 percent of the population. While Tajik is first and foremost an ethnonym, that is, the name of the majority ethnic group that lives in Tajikistan (as well as in northern Afghanistan), the meaning of the word Tajik has shifted over the course of history. According to Mehmud Qeshqeri (an early medieval Turkic scholar from Kashgar, now in Xinjiang, China), "Tajik" originally derived from Taj, a name borne by an Arabic tribe that resided close to Persia. Persians originally assigned the name to all Arabs. In the eyes of fellow countrymen, every Persian who accepted Islam turned, as it were, into an Arab. However, this meaning of the term is long lost. The Turks who first invaded Persian-dominated Central Asia in the tenth century began to use the word Tajik in reference to Muslims, but as virtually all the Muslims whom Turks initially ran into in Central Asia were Iranian-speakers, "a Tajik" became a speaker of an Iranian language. After the Turks themselves became Muslims, the term Tajik began to indicate any resident of Central Asia who was not a Turk (Lambton 1991). In Central Asia, Iranian speakers adopted this name to differentiate themselves from the steadily expanding Turkic element of the population. In Central Asia, Turkic and Iranian-speakers have coexisted for roughly 1,000 years, with Iranian being the language of high culture and government. Iranian-speakers or Tajiks founded the glorious cities of Samarkand and Bukhara, where such authors of the classics of Tajik literature as Rudaki and scientists as Avicenna (Ibn Sina) lived and worked. These and other Persian-speaking classical writers, such as Omar Khayyam, Rumi, Hafiz, and Saadi are celebrated as their own by both modern-day Iranians and Central Asian Tajiks. Gradually, however, the Turkic-speaking component of the Central Asian population became numerically dominant. Turks advancing from the north steadily pushed Tajik-speakers to the mountains where Tajikistan is currently located. However, most of the educated people "left behind" in the urban places of Central Asia north of Pamir continued to cling to Tajik. These urban clusters became surrounded by predominantly Turkic-speaking populations, which gradually took over the cities as well, allowing Chingiz Aitmatov to call Samarkand and Bukhara "conspicuous embodiments of the Turko-Tajik cultural synthesis." In terms of their

(© Thomas Voekler)

The Fan Mountains of Tajikistan are southeast of Panjakent.

racial type, Tajiks are barely distinguishable from Uzbeks, although Tajiks' faces are, on average, more Europoid (Caucasian) and less Mongoloid (Asian) than Uzbeks' and look much less Asian than Kazakhs and Kyrgyz.

Prior to the incorporation of Central Asia into the Russian Empire, the territory of today's Tajikistan was part of the Bukhara Emirate (1740–1920), an entity with both Turkic and Iranian-speaking populations. However, by 1924–1925 national delimitation saw some major Tajik-speaking cities, notably Bukhara and Samarkand—whose history is permeated by Tajik-speaking architects, statesmen, writers, and clerics, assigned to Uzbekistan, an outcome which has become forever a Tajik grievance. Even today, large groups of people in these cities speak Tajik but identify themselves as Uzbeks. Their ancestors first "became" Uzbeks in the 1930s to avoid discrimination. However, since Uzbek was a newly constructed national idea making its way through a thicket of local and regional identities, this was apparently not much of a sacrifice on the part of those Tajik-speakers. By some accounts, President Islam Karimov of Uzbekistan who was born and raised in Samarkand has divulged that prior to 1952, when he was 14 years old, he had not known any language but Tajik. He reportedly mentioned this in Dushanbe, Tajikistan, during the celebration of the 1,100th anniversary of the Samanid dynasty, which ruled what the Tajik elite now believe to be the first Tajik state. Ironically, that state was centered in Bukhara, a city located beyond Tajikistan's border, in Uzbekistan. According to census data, only 5 percent of Uzbekistan's population is Tajik today. Many more (possibly up to 20%) speak Tajik, but it is hard to say how many of these people actually identify themselves as Tajik.

Linguistically, Iran, much of northern and central Afghanistan (including Kabul), and Tajikistan are parts of the same dialect net. Officially, their languages bear different names—Tajik in Tajikistan, Dari in Afghanistan (the language of at least 30% of Afghanistan's population), and Farsi in Iran—but they are mutually intelligible and are more like a continuum of dialects of the same language. That more Arabic words have been assimilated by Farsi and more Russian words by Tajik does not alter this qualification. Some smaller groups of Tajik speakers also live in northern Pakistan and in the westernmost part of the Chinese province of Xinjiang. While Perso-Arabic script is used in Afghanistan and Iran, the official language of Tajikistan was first (in

1928) transliterated into Latin script and then (in 1939) Cyrillic. Unlike Iranians living in Iran, most Tajiks are Sunni, not Shia, Muslims.

A product of the Soviet national delimitation policy of the early 1920s, Tajikistan was definitely not advantaged by ideologues. As the national home for Soviet Tajiks, not only was it established within only part of the area where Tajik had been historically spoken, but it also was founded in the least accessible part and was replete with natural obstacles to development. When founded in 1925, Tajikistan became just an autonomous republic (the second tier of ethnic subdivision in the Soviet Union) within Uzbekistan; only four years later did it become a separate union republic. Interestingly, while people of Tajik background could be found in the ruling elite of Uzbekistan, so could Uzbeks in Tajikistan. For much of Soviet Tajikistan's existence, power was in the hands of Leninabad-based clans. Situated on the Syr Darya at the mouth of the Fergana Valley in the republic's north, Leninabad was formerly Khujand, the name the city reclaimed in an independent Tajikistan. It is a city in which (and around which) there is a large Uzbek community. Uzbeks accounted for 23.5 percent of Tajikistan's population, according to the last Soviet census of 1989. However, unlike Tajik-speakers in Uzbekistan who identified themselves as Uzbeks, Uzbek-speakers in Tajikistan did not identify themselves as Tajiks—an indication they belonged to a politically stronger, more assertive, and secure group. Even so, the 2000 census of Tajikistan revealed that Uzbeks accounted for less than 17 percent of Tajikistan's population, although no sizable outmigration of Uzbeks was recorded and their natural increase was about as high as that of Tajiks. This indicates that pressures to homogenize have affected Uzbeks in Tajikistan.

SOVIET TAJIKISTAN

Available descriptions of Soviet Tajikistan's economic base, which was created during the Soviet period, are unusually reserved for Soviet economic geography texts, which used to emanate requisite optimism. In the late 1980s, more than half of the gross domestic product was created in agriculture, which employed 45 percent of the labor force. Despite the overwhelmingly mountainous terrain, the structure of agricultural output in Tajikistan was more like that of lowland Uzbekistan than similarly mountainous Kyrgyzstan. That is, the main product was cotton, to which much of the limited flat cropland was

devoted. Meat production, even though 2.4 million sheep were officially owned by Tajik collective and state farms, was not particularly important. Tajikistan produced just 14 percent of Central Asian cotton (about 900,000 tons of raw cotton a year), but half of the highest quality, fine-fiber cotton. Only about one-tenth of all cotton produced, however, was processed within Tajikistan. Another important local specialization—silk—was represented by both cultivation and primary processing (obtaining filaments from the undamaged cocoons) in Khujand, then known as Leninabad; the final stages of production took place outside the republic. Other agricultural products exported from Tajikistan included fruits.

To irrigate almost 600,000 hectares of cropland and simultaneously to generate electricity, Tajikistan's rivers were tapped. A series of five hydroelectric stations was built on the Vakhsh River. The largest was the Nurek station. Its construction began in 1961; and it was commissioned in 1972. It has been operating at full capacity since 1979. At 300 m (984 ft) its dam is currently the tallest in the world. A still larger dam was to be built at the sixth and most upstream station, the Rogun station, but its construction, launched in 1976, was subsequently halted because of lack of funding and a previously underestimated seismic threat. Utilizing the cheap hydroelectric power, the Tajik Aluminum Smelter was built in 1975 in Tursunzoda, 60 km west of Dushanbe, near the border with Uzbekistan. It is the third largest aluminum plant in the world, with a capacity of 517,000 tons of aluminum a year. The closest it has ever come to full capacity was in 1990, when 450,000 tons of aluminum was produced. The plant directly employs 12,000 to 14,000 people, and indirectly supports a community of 100,000. Another industrial enterprise

DEVELOPMENT

In 2005 the President of the Government of Tajikistan (GoT) initiated the long-term National Development Strategy for the period up to 2015 (NDS).

depending on the same source of energy is the Yavan Electrochemical Plant, commissioned in 1978 and later expanded. This plant produces caustic soda and chlorine-based bleaches. In the north of Tajikistan, in the city of Isfara (at the southwestern edge of the Fergana Valley in the foothills of the Turkestan ridge), a unique hydro-metallurgy plant exists, founded in 1947 and subsequently modernized and expanded. It was one of the principal processors of strontium and other rare metals for the Soviet defense industry. The enterprise also mastered the extraction of platinum from the exhaust of oil refineries as well as the extraction of gold and silver (zinc cementers) from poly-metallic concentrates. Gold, uranium, and antimony mining developed during the last two Soviet decades.

Near Nurek, Tajikistan, construction of the Okno (Window), a fully automated optical tracking station used for the identification of satellites, began in 1980. The station was to detect and track satellites orbiting the Earth at altitudes between 2,000 and 40,000 km. Development of the Okno system had begun in the late 1960s. By late 1971, test trials of prototypes of its optical systems were being conducted at the Byurukan Observatory in Armenia. Preliminary design work for the operational system was completed in 1976 and approved in 1977. Construction of the Okno site was not finished at the time the Soviet Union broke up.

Inferior roads and the difficulty of road construction in Tajikistan have created acute development contrasts between the southwestern (Dushanbe and Vakhsh Valley) and northern (Leninabad/Khujand) economic growth poles, on the one hand, and the rest of the republic, on the other; these have never been overcome. Tajikistan remained the poorest Soviet republic with a per capita GDP amounting to just $1,800 in 1990 and average personal consumption at $1,580. Considering that in the Russian Federation the 1990 ratio of personal consumption ($3,220) to per capita GDP ($5,870) was 0.54, while in Tajikistan it amounted to 0.87, it is clear that financial transfers from the federal center were vital in order to sustain the republic's modest level of living.

The last Tajik Communist Party leaders—Jabar Rasulov (1961–1982), Rahmon Nabiyev (1982–1986), and Kakhar Makhamov (1985–1991)—were all from around Leninabad, and two of the three had built their professional careers in agriculture. Nabiyev, who after his ouster as the First Secretary under Gorbachev resurfaced in the Supreme Soviet of the republic, became the first president of independent Tajikistan.

POST-SOVIET TAJIKISTAN

All Soviet republics experienced economic decline after the breakup of the Soviet Union, but this general formula conceals enormous contrasts. The Tajik economy took a dive from an already low level, and unlike in Uzbekistan and Kazakhstan, prospects for import substitution and even for sustaining the most lucrative industries without considerable external assistance looked dim. One of the world's largest aluminum refineries had no domestic bauxite or any other source of aluminum ore. The large silk factory had nobody (for a while) interested in buying the semi-finished product it once manufactured. In 1992, Tajikistan's GDP was 30 percent lower than in 1991; Tajikistan had become one of the world's poorest countries. By 1995, the Tajik economy, additionally crippled by civil war (see the following), was only 41 percent of its former (1991) self and in 1999—despite the war's end two years earlier—just 38 percent. Thereafter, economic growth resumed and from 2004 to 2008 exceeded 8 percent a year, but even at its highest point, (2008) per capita GDP was below $1,700 (adjusted by PPP). More than 60 percent of the population lives on less than $1.25 a day, the World Bank's poverty threshold.

Presently, Tajikistan offers three major goods for which there is demand on the international market: aluminum, cotton, and labor. Aluminum production resumed at the plant in Tursunzoda. Obtained by the Russian aluminum giant Rusal and assuming a new name, Talco, it had by 2006 almost reached its highest ever (1990) level of output. But the enterprise consumes 40 percent of the country's electricity production. Tajikistan has the lowest per capita electricity usage rates among the former Soviet countries, enough for only a few hours a day of electricity in the winter. And this in a country which is the world's third largest producer of hydroelectric power after Russia and the United States. The production of cotton is now at about 40 percent of its late Soviet level (336,000 tons) and it accounts for 70 percent of freshwater consumption in the republic. It is obvious that both industrial and agricultural growth demand further tapping of freshwater resources, including first and foremost building new hydroelectric stations and associated reservoirs on Tajikistan's rivers. This, however, requires external investment, but few international investors have been enthusiastic. When in 2005, Tajikistan signed a treaty with Russia to resume the construction of the Rogun hydroelectric station, the sixth station of the Vakhsh cascade, neighboring Uzbekistan protested, as increased evaporation and diversion of water from the Vakhsh River would diminish the already overextended discharge of the Amu Darya (of which Vakhsh is a tributary) even more and would thus adversely affect Uzbekistan. Uzbek sources engaged in scaremongering, particularly after the August 2009 catastrophe at the Sayano-Shushenskaya hydroelectric station on the Yenisei River in Russia. According to those sources, the Rogun station was to be of the same kind. Situated on a rift between two geologic sub-plates, it too would be prone to destruction by a powerful earthquake; indeed the Rogun station, once complete, might even increase the likelihood of an earthquake. Therefore, they asserted that it would be unwise to resume the implementation of the 20 year old project. In January 2009, during an official visit to Uzbekistan, President Medvedev of Russia said that "hydroelectric stations in the Central Asian region must be built in such a way that the interests of neighboring countries are taken into account. If there is no common consensus," added Medevedev, "Russia will refrain from participation in such projects." Tajikistan interpreted this statement as Russia reneging on its commitment to finish the Rogun station. Russia's charge d'affaires in Dushanbe was handed an official note of protest, and the Tajik ambassador in Moscow received a strong rebuke from the Russian side. Commenting on this exchange, which for a while clouded otherwise overtly friendly mutual relations, news agencies reported that Russian-Tajik negotiations on Rogun were still under way, with Russia demanding at least a 51 percent share in the prospective station in exchange for its investment.

In the absence of employment opportunities at home and with local jobs furnishing only an average $15 a month and retirement pensions $4.40 a month in 2003 (Medvedev 2005), many Tajiks who had been reluctant to pursue employment outside the republic during the Soviet period changed their minds and began to seek work abroad, primarily in Russia and Kazakhstan. According to the World Bank's 2007 estimate, foreign remittances from Tajik workers abroad account for 36.2 percent of the country's GDP (Ratha, Mohapatra 2007). The Russian daily *Kommersant* reported that in 2008 migrant workers transmitted to Tajikistan $2 billion, a sum which exceeds the republic's annual budget (Reutov 2009). There are wide ranging estimates of Tajik labor in Russia, but anecdotal evidence suggests that most trash collectors in Moscow and some other Russian cities are Tajiks. Hundreds of thousands work on construction sites. According to some sources, close to 1.5 million Tajiks live in Russia legally and illegally and from 700,000 to 800,000 of them hold jobs. In 2007, a leader of one Tajik NGO in Russia, the "People's League–Tajiks," Karomat Sharipov, made the peculiar public statement that if Russia offers tolerable legal rights and working conditions to citizens of Tajikistan, "no more than 10% of Tajikistan's population would remain in that country; those would be only people engaged in the army, the KGB, the administration of the president, and his entourage." However, Russian news media are awash in reports of both mistreatment of Tajiks in Russia and Tajiks being the most frequent violators of Russian laws. When in February 2004 in Saint Petersburg a group of skinheads killed a 9-year old Tajik girl, Khursheda Sultanova, the liberal Russian media sounded the alarm; but at the same time sundry reports surfaced with suggestions that her father, who was with the girl when the assault occurred, was a drug dealer (as if it would make killing his daughter less of a crime). One source reports that 2,000 offenses against Tajiks were committed in Russia in 2008, but Tajiks themselves committed almost 7,000 crimes. Selling drugs is indeed an activity in which some Tajiks are involved and of which many more are suspected. Tajiks have a long historical link with drug trafficking because of the production of opium in Afghanistan and because the 2,000-km long Tajik-Afghan border in the Pamir Mountains separating the Badakhshan province of Tajikistan from the eponymous province of Afghanistan reportedly has never been entirely sealed, even during Soviet times. Yury Kazarin, a doctor of philology from Ekaterinburg, Russia (the third-largest Russian city located in the Urals), claims that for the children of Tajik migrants in his region, "smoking marijuana is as routine as is drinking kefir and eating ice cream for Russian children." Yet side by side with this sweeping generalization he offers a more likely explanation of the bad feelings toward Tajiks. "Tajiks who work without breaks and days off for the entire week and earn 100 rubles [a little more than $3—G.I] a day undermined all their Russian competitors [for jobs], who would never work for such a pittance. I wanted to buy land and build a house in the village of Kashino. Local villagers begged me not to bring Tajik construction workers lest locals have no jobs." The article in question is appended by the following editorial note: "According to data from the local Tajik diaspora, 150,000 Tajiks live in Sverdlovsk Oblast [a civil subdivision whose capital is the city of Ekaterinburg—G.I.], but only about 2,000 stay legally and only 500 are legally employed" (Reutov 2009).

THE CIVIL WAR AND ITS AFTERMATH

The large Tajik diaspora in Russia is not the product of poverty and unemployment alone. From 1991 to 1997, Tajikistan experienced a civil war with a death toll estimated between 20,000–50,000 according to some American sources and 100,000–150,000 according to some Russian sources. Various publications offer detailed accounts of the events that triggered the war in late 1991 and of the ensuing hostilities which lasted until June 27, 1997, a date that since 1998 has been celebrated annually as a Day of National Reconciliation.

Two major strands dominate analyses of the Tajik civil war: religion and regionalism. According to the first line of reasoning, a union of religious and democratic forces was set against former Soviet apparatchiks. According to the second line of reasoning, the Khujand clan was set against people from other areas, especially from around Garm (on the Surkhob River, a tributary of the Vakhsh in the central part of the country) and from Badakhshan (Tajikistan's least developed eastern or Pamir part).

Indeed, under Gorbachev's Perestroika a religious revival commenced in all of Central Asia. In Tajikistan, this revival saw the ascent of Sayid Abdulloh Nuri as the most respected Sunni Muslim leader. Because political thaws arrived in Central Asian republics after much delay (if at all), Nuri was imprisoned for 18 months from 1986 to 1988, that is, during Gorbachev's Perestroika, for producing and spreading religious propaganda. This, however, only gained him popularity. During the ensuing civil war Nuri was a military leader representing one emerging center of power in Tajikistan—the one rooted in the central part of the country, specifically in a Tavildara area (east of Garm), which in the 1990s had hosted many foreign Islamists, including those affiliated with the Islamic Movement of Uzbekistan (IMU; see the essay on Uzbekistan). In 1993, Nuri founded the Party of Islamic Revival of Tajikistan over which he presided until his death. Not only was this his central Tajikistan power base, but also Badakhshan lying further east, which had been historically underrepresented in the corridors of power in Dushanbe. Even Kulab (Kulob) and Kurgan Tyube (Qurghonteppa), medium-sized urban centers to the south of Dushanbe with more significant (by Tajikistan's standards) economic bases compared to central and eastern Tajikistan had been

political pariahs throughout the entire Soviet period, because all power had rested in the hands of the Tajik-Uzbek clan centered in Leninabad.

President Nabiyev of Tajikistan, representing the Leninabad region, became one of the civil war's casualties. On September 7, 1992, while trying to escape Dushanbe, which had fallen under the control of the Islamist forces, Nabiyev was apprehended and forced to resign his post. He subsequently died in April 1993 in his home in Khujand (formerly Leninabad) under mysterious circumstances. Nabiyev was succeeded as president by Emomalii Rahmonov, the former chairman of a collective farm near Kulab, a city in southern Tajikistan. Rahmonov became the leader of the Popular Front, a force opposing the union of Islamists, headed by Nuri, and secular liberals. The Popular Front was an affiliation of ex-Soviet bureaucrats alarmed by the rise of the Islamic movement and the possibility of yielding power to it. Though consisting of Soviet bureaucrats, this group was committed to undermining the Leninabad/Khujand stronghold of power in the republic. After the resignation of President Nabiyev, the Supreme Soviet or the parliament of Tajikistan (subsequently renamed Majlis, which is what parliaments are usually called in Turkic-speaking countries and in Iran) eliminated the office of president but elected Rahmonov as its chairman. In 1994, the post of president was reinstated and Rahmonov was popularly elected with 58 percent of the vote.

The Islamic-Democratic opposition was supported by IMU militants and by scores of Afghan fighters from the ethnic-Tajik–dominated Northern Alliance (led by Ahmad Shah Massoud), but the Popular Front enlisted even more powerful supporters, such as Uzbek troops and the 201st Motorized Rifle Division with 12,000 former Soviet troops who had been stationed in Tajikistan since the Soviet Union's 1988 withdrawal from Afghanistan. Initially the division maintained its neutrality, but after Russia reclaimed this military force as its own, it supported the Popular Front. The most disastrous part of the war, during which the Popular Front was pitted against the Islamic-Democratic opposition, lasted from 1992 to 1994. During that time deadly violence erupted twice in Dushanbe. After the end of August 1992, former members of the Soviet political establishment were targeted. However, when in early December 1992 Dushanbe fell to the Popular Front, people from Garm and Badakhshan fell victim to pogroms. According to the Office of the UN High Commissioner for Refugees (UNHCR), approximately 100,000 refugees fled into northern Afghanistan in an exodus that started in late 1992. As late as the end of 1995 nearly 20,000 refugees remained in three main camps maintained by the UN in northern Afghanistan. By the end of the 1990s, the overwhelming majority had returned to Tajikistan. Apparently, despite speaking the same language as most people of northern Afghanistan, Tajik refugees did not like what they saw there. Two Tajik authors related that a little girl born in a refugee camp in Afghanistan heard from her grandfather that "Tajikistan is the best place in the world for it has electricity, asphalt roads, and nice large schools" (Rutland 1998).

In part due to the considerable resistance of the opposition forces and their broad base of support in at least two regions of Tajikistan and in part due to the conciliatory stand taken by Nuri, the warring parties began to actively negotiate in 1994 under the auspices of the United Nations and with Russia's mediation. The eventual reconciliation agreement signed in Moscow by Rahmonov and Nuri on June 27, 1997, (with President Yeltsin of Russia present at the signing ceremony), envisioned the assignment of one-third of government posts to the Islamic-Democratic opposition and integration of thousands of Islamic militants into the security structures of Tajikistan; an unconditional amnesty was offered to most of those militants. However, the top positions remained in the hands of the Popular Front, subsequently renamed the People's Democratic Party of Tajikistan (PDPT). After the signing of the General Agreement in June 1997, Rahmonov performed the hajj, a religious pilgrimage to Mecca. He subsequently consolidated his power so much that when, in November 1999, he faced his second presidential elections, he received 96.9 percent of the vote and thus secured a further seven years as head of government. This election, as well as his third (2006) presidential election, in which he received 79.3 percent of the popular vote, were not considered fair by international observers. As early as 1999, the Party of Islamic Revival (PIR), the third-largest political group—after Rahmonov's PDPT and the Communist Party—had accused Rahmonov of multiple violations of the 1997 reconciliation agreement. In June 2003, the PIR issued a statement protesting a constitutional amendment that would allow Rahmonov to be elected president for a third time. After that, tensions between the PIR and Rahmonov came to a head. In October 2003, Nuri's deputy, Shamsuddin Shamsuddinov, was arrested and charged with organizing a criminal group, illegally crossing the border, and polygamy. While

the PIR participated in the 2005 parliamentary elections, it managed to win just two of the 218 parliamentary seats. The PIR did not recognize the election results and accused Rahmonov's entourage of vote rigging. However, Rahmonov tried to position himself as a traditionalist: in 2007 he purged the Russian ending "ov" from his name and became Rahmon.

Rahmon's old civil war adversary, Nuri, fell ill with cancer in 2004. He was visited by Rahmon, who allocated 150,000 Euros for Nuri's treatment in Germany, but the Sunni leader died in August 2006 at the age of sixty. Nuri's legacy is controversial. Most sources underscore his positive role in national reconciliation in Tajikistan, but he had a darker side. According to an article in *The New York Times,* declassified United States government documents show that in July 1996 Nuri contacted Iranian foreign intelligence officials in Taloqan, Afghanistan, in an attempt to forge an alliance between the government of Iran, Nuri's followers, and Al Qaeda leader Osama bin Laden to attack the United States. While Iranian officials offered to meet with Nuri and bin Laden in Jalalabad, Afghanistan, bin Laden reportedly refused on the grounds that the security risk was too high.

Although overtly the opposing sides in the Tajik civil war were religious versus secular and one region versus another, it seems that on a deeper level the major cause of the war was a general sense of disorientation and despair rooted in

poverty. The general malaise was exacerbated by the collapse of a long lasting power structure, severed economic ties, diminished external assistance, and rapid population growth. The traditional religion was, for awhile, the only institutionalized force once the Soviet Union was gone; the local Soviet's power was not worth much without aid and meticulous guidance from Moscow. So, popular protests against the desperate state of the country took the shape of a religious war. However, it is unlikely that many in Tajikistan seriously opted for an Islamic state as the master key to the existential problems every family and the entire country faced. The realization that popular aspirations pointed elsewhere probably informed Nuri's position; he wanted an Islamic state but was committed to peaceful means for obtaining it.

Over the course of the twentieth century, Tajikistan's population grew six-fold! It had the fastest growing population in the Soviet Union. During the post-Soviet period, some decline in birth rate occurred; however, it was still between 32 and 34 per 1,000 population from 2000 to 2006, and had declined only to 27 per 1,000 in 2008. Even though slowing, natural increase is still 1.9 percent a year. Curiously, in 1989, urbanites accounted for 33 percent of Tajikistan's population, but the 2000 census revealed only 26 percent were urbanites—a substantial deurbanization. In 1989, there had been 388,500 ethnic Russians in the republic (7.6% of the population), for the most part in Dushanbe; the 2000 census showed only

(© Jeremy E. Meyer)

A bazaar in Panjakent, Tajikistan, near the border with Uzbekistan.

68,200. Most of those who had remained after the collapse of the Soviet Union fled during the early phase of the civil war. As was pointed out above, the share of ethnic Uzbeks declined from 23.5 percent to almost 17 percent, whereas ethnic Tajiks expanded their majority from 62.3 percent in 1989 to 79.9 percent in 2000.

The Republic of Tajikistan tries its best to navigate the stormy waters of international politics. In Central Asia, it has uneasy relations with its stronger and more assertive neighbor, Uzbekistan. One bone of contention is Tajikistan's desire to tap into rivers whose downstream sections are in water-deficient Uzbekistan. Another and perhaps more lasting source of irritation is the differing interpretations of a common history, particularly as far as Samarkand and Bukhara are concerned. In the summer of 2009, for example, Tajik news media bemoaned and castigated plans to close the Samarkand-based museum of Sadreddin Aini (1878–1954), a well-known Tajik writer.

No longer is Tajikistan insulated from its close ethno-linguistic relatives in Iran and Afghanistan. These countries and Tajikistan have exchanged governmental delegations and researchers on many occasions. For example, twelve Tajik scholars attended the 6th International Congress of World Teachers of Persian Language and Literature, held at the University of Tehran in January 2009, and Iran has already become the fifth largest trading partner of Tajikistan.

After the civil war ended, Russia's military presence increased in Tajikistan. Russia, which already had 25,000 armed troops in Tajikistan, tentatively agreed

in April 1999 to the establishment of a military base whose nominal goal was boosting stability in Tajikistan. The Russian and Tajik defense ministers signed a treaty on April 16, 1999, that granted Russia's military the right to establish a base on Tajik territory and to quarter troops from the 201st Motorized Rifle Division there for the next 10 years. The agreement provided for the construction of more permanent headquarters for the 6,000–7,000 troops from the 201st Division already deployed there. Until 2004, the bulk of Russia's troops in Tajikistan were stationed in Dushanbe, Qurghonteppa (close to the Uzbek border), and Kulab (near the Afghan border). A newly built Russian base, in Giprozem at the southern outskirts of Dushanbe, opened in 2004. Some observers in Dushanbe see the Russian military presence as a deterrent to Uzbek influence. Until 2003, Russians guarded the entire Tajik-Afghan border; now Tajik border guards are deployed there, but a Russian rapid response team is supposed to aid Tajik border guards in difficult situations. The optical satellite tracking station whose construction was not completed by the time of the Soviet Union's demise has finally been completed and is now the property of the Russian Federation.

Russians get edgy when Americans court the political leadership of Tajikistan. Thus several times during the summer of 2009, Russian media disseminated an unconfirmed rumor that Tajikistan and the United States were negotiating a potential American base in Tajikistan, a rumor that Tajikistan's government and the U.S. embassy in Dushanbe subsequently disavowed. However, slowly but

surely an American presence is growing in Tajikistan. When the U.S. military launched its offensive in Afghanistan against the Taliban in October 2001, Tajikistan granted the U.S. military flyover rights. Subsequently the U.S. government funded a $36 million bridge over the Pyanzh River, which connects Sher Khan, Afghanistan, with Nizhniy Pyanzh, Tajikistan. The bridge was commissioned in August 2007. Trade between Tajikistan and Afghanistan tripled the first year the bridge was in operation.

HEALTH/WELFARE

Since independence, steady reductions in the state health budget have further eroded the salaries of medical professionals and the availability of care. For that reason, health planners have considered privatization of the national health system an urgent priority. In the mid-1990s, however, little progress had been made toward that goal.

FREEDOM

Despite legislation protecting freedom of speech and the press in Tajikistan, in practice freedom of expression is limited. The government of Tajikistan continues to employ a variety of tactics to limit political content in the media.

The Scythians

ALEXANDER BLOK

"Pan-Mongolism—though the word is strange,
  My ear acclaims its gongs."

—Vladimir Solovyov

You are the millions, we are multitude
And multitude and multitude.
Come, fight! Yea, we are Scythians,
Yea, Asians, a squint-eyed, greedy brood.

For you: the centuries; for us: one hour.
Like slaves, obeying and abhorred,
We were the shield between the breeds
Of Europe and the raging Mongol horde.

For centuries your ancient hammers forged
And drowned the thunder of far hates.
You heard like wild fantastic tales
Old Lisbon's and Messina's sudden fates.

Yea, so to love as our hot blood can love
Long since you ceased to love; the taste
You have forgotten, of a love
That burns like fire and like the fire lays waste.

All things we love: clear numbers' burning chill,
The ecstasies that secret bloom.
All things we know: the Gallic light
And the parturient Germanic gloom.

And we remember all: Parisian hells,
The breath of Venice's lagoons,
Far fragrance of green lemon groves,
And dim Cologne's cathedral-splintered moons.

And flesh we love, its color and its taste,
Its deathy odor, heavy, raw.
And is it our guilt if your bones
May crack beneath our powerful supple paw?

It is our wont to seize wild colts at play:
They rear and impotently shake
Wild manes—we crush their mighty croups.
And shrewish women slaves we tame—or break.

Come unto us, from the black ways of war,
Come to our peaceful arms and rest.
Comrades, while it is not too late,
Sheathe the old sword. May brotherhood be blest.

If not, we have not anything to lose.
We also know old perfidies.
By sick descendants you will be
Accursed for centuries and centuries.

To welcome pretty Europe, we shall spread
And scatter in the tangled space
Of our wide thickets. We shall turn
To you our alien Asiatic face.

For centuries your eyes were toward the East.
Our pearls you hoarded in your chests,
And mockingly you bode the day
When you could aim your cannon at our breasts.

The time has come! Disaster beats its wings.
With every day the insults grow.
The hour will strike, and without ruth
Your proud and powerless Paestums be laid low.

Oh pause, old world, while life still beats in you.
Oh weary one, oh worn, oh wise!
Halt here, as once did Œdipus
Before the Sphinx's enigmatic eyes.

Yea, Russia is a Sphinx. Exulting, grieving,
And sweating blood, she cannot sate
Her eyes that gaze and gaze and gaze
At you with stone-lipped love for you, and hate.

Go, all of you, to Ural fastnesses,
We clear the battle-ground for war;
Cold number shaping guns of steel
Where the fierce Mongol hordes in frenzy pour.

But we, we shall no longer be your shield.
But, careless of the battle-cries,
Shall watch the deadly duel seethe,
Aloof, with indurate and narrow eyes.

GLOBAL STUDIES

We shall not move when the ferocious Hun
Despoils the corpse and leaves it bare,
Burns towns, herds cattle in the church,
And smell of white flesh roasting fills the air.

For the last time, old world, we bid you come,
Feast brotherly within our walls.
To share our peace and glowing toil
Once only the barbarian lyre calls.

From *Modern Russian Poetry: An Anthology,* edited and translated by Babette Deutsch and Avrahm Yarmolinsky, 1921. Published in 1921 by Harcourt, Brace, and Company.

The Kremlin Begs to Differ

DIMITRI K. SIMES AND PAUL J. SAUNDERS

Twenty years after the fall of the Berlin wall, Russia remains as Sir Winston Churchill described it: "a riddle, wrapped in a mystery, inside an enigma." Russia's complexity has contributed to an American debate in which policy preferences too often shape analysis rather than analysis driving policy. It's not a sound basis for decisions when key American interests and goals are at stake.

One doesn't need to be a Russian domestic radical or a foreign Russophobe to see major flaws in the way Russia is ruled. The country's president, Dmitry Medvedev, has catalogued its problems: "an inefficient economy, semi-Soviet social sphere, fragile democracy, negative demographic trends and unstable [North] Caucasus," not to mention "endemic corruption" defended by "influential groups of corrupt officials and do-nothing 'entrepreneurs'" who want to "squeeze the profits from the remnants of Soviet industry and squander the natural resources that belong to us all."

Russia's problems are fundamental to its political system, which, while officially democratic, is perhaps best understood as popularly supported semiauthoritarian state capitalism. Russia is clearly not a Western-style democracy, though its citizens enjoy considerable freedom of personal expression, with the level of liberty inversely proportional to the potential impact of criticism. The state dominates "strategic sectors" of the economy like energy and defense, but political-business clans have retained much space to pursue their parochial interests, including through the state's administrative machinery. As throughout its history, Russia is dominated by a ruling class: originally aristocrats, then Communist Party *nomenklatura*, and now a combination of senior bureaucrats and business leaders, including former Soviet managers, ruthless-yet-effective younger entrepreneurs, and outright criminals who took advantage of the decay, collapse, and anarchy of the 1980s and 1990s.

The question now is how long Russia's current political arrangements can hold. Corruption is deeply embedded and pervasive, affecting state and private enterprises along with the media and the courts, severely limiting Russia's modernization and sustainable economic growth. And with so much power concentrated at the top of the system, recent murmurs of a growing rift between President Medvedev and Prime Minister Vladimir Putin raise serious concerns about stability.

Moscow's arbitrary rule affects the people and state of Russia most of all, but it also presents a challenge to the United States. Russia is vital to American interests, and if the Obama administration has any illusions about the nature of Russian politics or, alternatively, surrenders to the long-standing temptation to act as a self-appointed nursemaid, it could severely damage our ability to work pragmatically with Russia to advance important U.S. goals.

It is difficult to overstate the role of corruption in Russia, which in many ways is the glue that holds together the disparate groups dominating Russia's current political system. The Russian state is organically linked to Russian companies, both overtly—through stock ownership and officials' simultaneous service on corporate boards—and covertly, through family ties and secret deals. At the upper levels, Russia's corruption takes the form of private stakes in state firms and profound conflicts of interest; at lower levels, simple bribery is more common. And the scope of corruption is expanding: according to official statistics, Russia's bureaucracy has doubled in size in the last ten years.

Those at the top have a relatively free hand to enrich themselves through insider dealing. Thus, notwithstanding Russia's extensive privatization, it is often difficult to distinguish between government-owned companies and large private conglomerates. Officials are deeply involved with both and the government often acts to protect both, though it is not always clear whether government actions are a result of state or private interests.

Because of corruption, Russia's political system is simultaneously very resistant to change and remarkably fragile. Extensive overseas portfolios and property held by Russian officials and oligarchs are a clear indicator of their own limited confidence in Russia's stability. There is an exuberant Russian presence in London, New York, and on Mediterranean beaches that is totally out of line with the size of Russia's economy. The reluctance of major Russian firms to make long-term capital investments in their own country is further evidence of this mindset. If those who hold power still feel the need to hedge their bets, it is all the more true of international investors.

In addition, the uninhibited power of huge government-owned or government-connected firms to act against their competitors discourages both foreign investment and the development of small- and medium-size businesses in Russia.

Though Gazprom has legitimately wanted Ukraine to pay its bills on time and in full, the resulting disputes took on a very different character when Russian officials became involved. In the end, the repeated crises damaged Moscow's reputation as a reliable energy supplier and a responsible European power. The Alfa Group's struggle with BP over control of their TNK-BP joint venture is another example in which well-connected Russian moguls allied with the state to apply pressure to their business partners. What would be routine commercial disagreements in other countries can rapidly trigger state intervention in Russia, usually to the disadvantage of foreign firms or others outside the system who lack effective protections or recourse.

Corruption and insider dealing can have tragic consequences in Russia, as they did in an August explosion at the Sayano-Shushenskaya dam in Siberia, when over seventy people were killed due to inadequate maintenance. Putin himself described as "irresponsible and criminal" an apparent maintenance contract with a fraudulent firm set up by top managers. Beyond limiting investments in safety and maintenance, however, irresponsibility and corruption have also strongly discouraged investment in other key areas. Russian firms happily squeeze out foreign investors but don't themselves put money into new equipment, training, or research and development. Despite recent increases, state investments in education, health, and science and technology are also inadequate for sustainable economic growth and to diversify beyond energy exports.

Here it is useful to compare Russia to China. China is less free than Russia according to Freedom House, and has a number of similar problems, but is considerably more attractive to foreign investors. The huge scale of China's market is a major inducement, but Beijing's greater willingness to accept international rules and its much more strategic approach to cultivating foreign investors—whose presence China's leaders view as essential to meeting their development goals but energy-rich Moscow has seen as easily replaceable—also make a big difference.

I t will not be possible to modernize Russia without a genuine effort to eliminate corruption—and this includes at the top. Corrupt conduct is not simply tolerated, but a way of life with profound political implications. Any opening in the political system that would allow corruption to be exposed could potentially decimate Russia's elites—and they know it. Everyone who is involved in corruption whether directly or indirectly (by failing to act on knowledge of corrupt acts) may be in legal jeopardy and is therefore a stakeholder in the current system. Russia's weak media and biddable court system and parliament prevent corrupt elite actions from coming to light and ensure few if any consequences.

The media face direct state interference and engage in constant self-censorship as well. The central government effectively controls television news broadcasting, with the exception of minor cable channels, and blocks critical reporting that questions official policy. Moscow newspapers enjoy wider freedom to debate policy, and papers like *Novaya Gazeta* and *Nezavisimaya Gazeta* regularly challenge government decisions. But

they reach only a small audience, and regional papers are usually more cautious, especially in dealing with local officials. The Internet is quite free, though also limited in reach because of its low penetration in Russia. Media that cross the government face harassment through tax inspections and lease problems, to name only a few of the potential consequences. More chilling are the fates of leading investigative journalists like Anna Politkovskaya, Yuri Shchekochikhin and others who have been murdered or died in suspicious circumstances after pursuing major corruption investigations or challenging senior officials. All of this creates a climate of intimidation.

What Dmitry Medvedev has frequently described as the easy manipulation of the judicial system makes impotent yet another potential check on elite power and corruption. Interference in the courts by the government and private companies is a common practice rather than an exception, and the weak judiciary helps to maintain Russia's political status quo by derailing serious efforts to change it. This creates an environment in which the Russian leadership lacks important information and independent analysis when making key decisions. Moscow's control mechanism blocks this critical input and feedback.

T he subservient political system merely adds another layer of top-down control. Russia's formal political system is built around the concept of the "power vertical" introduced by then-President Vladimir Putin and still in effect today. In brief, the power vertical concentrates power at the top by subordinating the country's entire government apparatus to its leaders and ensuring that any and all decisions are their prerogative. While the president and prime minister do not make every policy choice, they retain the right to make (or retroactively question) any particular decision.

Governors and other regional leaders are nominated locally but selected by the presidential administration in consultation with the prime minister's office; mayors are elected through processes that are easily manipulated, as demonstrated by the recent election in Sochi, where opposition candidates were marginalized or disqualified. Pro-government candidates at all levels routinely enjoy considerable advantages, including ready access to television, ease in getting permits for rallies and campaign donors recruited by the regime.

Local leaders—especially those in Russia's ethnically based regions, such as Chechnya's President Ramzan Kadyrov—have what resembles feudal autonomy so long as they remain outwardly loyal to the federal government and can maintain control of their domains. Kadyrov in particular is a striking case, a former rebel turned minidictator who embraces polygamy and honor killings. Yet with his brutality Kadyrov has, at least until recently, maintained stability in Chechnya—allowing Moscow to withdraw Russian troops and remove the issue as a domestic political irritant.

Moscow Mayor Yuri Luzhkov is another interesting example. Luzhkov was attacked as corrupt by the pro-Kremlin media when he and former Prime Minister Yevgeny Primakov led a key challenge to Putin's ascension to power. Luzhkov subsequently failed to endear himself to the federal government's officials as the leader of a political-business clan that did not share the capital city's spoils sufficiently with federal bureaucrats. But Putin came to appreciate Luzhkov after his firm response to protests in the wake of a 2005 decision to reduce social benefits to retirees, veterans, the handicapped and, remarkably, even the police by providing regular benefit payments instead of free services. The Moscow mayor proved that he knew how to run his city when the chips were down. All this goes to show the conduct top officials are willing to tolerate from regional leaders so long as they deliver what counts.

While in theory a separate branch of government, Russia's legislature is in reality subordinate to its top leaders. The executive branch decides which parties hold what number of seats in the State Duma and the Federation Council and can ensure the passage of virtually any legislation. Members of parliament are permitted to lobby for their constituents by trying to secure federal-budget funds or other benefits, but have only a marginal role in policymaking.

The political parties themselves are created and destroyed from above rather than from below. There are only four parties represented in the State Duma, and three of them are creations of the Russian government. United Russia was established explicitly to serve as Russia's ruling party and a vehicle to bring Putin to power. After remaining aloof as president, Putin now leads the party, which holds a supermajority sufficient to amend the constitution. Whenever it matters, United Russia can count on the votes of the Liberal Democratic Party (LDPR), which according to insider accounts was established in part by the Soviet KGB to serve as a nationalist pseudo-opposition.

The presidential administration openly served as the principal architect of Just Russia, a center-left party assembled as a social-democratic alternative to United Russia. Just Russia repackaged and expanded a previous government-inspired opposition party, Rodina (Motherland), itself created as a nationalist and populist alternative to the Communists, but ultimately destroyed when the party and its leader Dmitry Rogozin proved too successful for their own good. The regime's demonization and subsequent rehabilitation of Rogozin as Moscow's ambassador to NATO once he was no longer a threat illustrates Russia's political hardball.

Only the Communist Party—which traces its origins to the Soviet period—appears to be a genuine mass party with a degree of independence. Despite this, the Communists are well aware of the limits of their power and consequently they remain unambitious: they know that the party depends on the government for its national and local registration, for access to television and for relatively easy fund-raising. The Communist Party is not a toy of the Russian government, but neither is it an engine of regime change.

Parties outside the parliament have even less impact. Russia's democrats have failed to capture the public imagination—and not only because of government pressure and limits on their activities. Most pro-Western reformers were never able to successfully demonstrate that they represented the interests of ordinary people or to establish patriotic credentials with a population that remained proud and suspicious of the West. Radical Russian democrats ridiculed fear of NATO enlargement when most opposed it, dismissed concerns over U.S. missile defenses that others stoked, and in some cases supported Georgia's perspectives when Russia and Georgia were at war. More recently, democratic opposition politician and former Prime Minister Mikhail Kasyanov supported the expulsion of Russia's delegation from the Parliamentary Assembly of the Council of Europe—a position with which very few of his fellow citizens could identify. Russian parties that appear insufficiently patriotic are marginalized, while those that embrace nationalism—like Eduard Limonov's rabidly xenophobic and anti-Western National Bolshevik Party—have greater appeal.

Despite its unattractiveness to outsiders, Russia's system of control from above and corruption throughout produces little discontent. Many feel a degree of comfort with strong leaders and a degree of discomfort with democratic freedoms. So long as Russia's citizens reap real benefits from the current arrangement, most see little need to question it.

At its deepest, this is a matter of history. There is less demand for an alternative in part because neither of Russia's two experiments with democracy was stable or successful in Russian eyes. The first experiment, between February and October of 1917, led rapidly to a "dual power" arrangement between Alexander Kerensky's Provisional Government and the Communist-dominated Soviets that degenerated into revolution, collapse and totalitarianism. The second, which began with Gorbachev's perestroika, fell apart as Yeltsin turned democratization to his own purposes and allied himself with separatist elites in order to unseat Gorbachev, splintering the USSR in the process. Yeltsin's aggressive pursuit of radical economic reforms despite popular opposition led him to rely increasingly on revitalized security services, the oligarchs and oligarch-controlled media for political backing. After this, most Russians welcomed Putin's imposition of order.

Russia's present political system is also at least partially attributable to Vladimir Putin and his allies in the security services, who have displayed both a ruthless instinct to establish control and a suspicion of anything they do not control. As president, Putin demolished the national political pretensions of the oligarchs and out-of-control governors (which helped bring an end to the semianarchy of the 1990s), but he could not bring himself to encourage civil society or free markets or to establish alternative centers of influence. From this perspective, Russia's current semiauthoritarian system is not entirely the product of a deliberate process but also the result of a vigorous effort to rein in previous abuses unaccompanied by anything else. It is authoritarianism by default.

At the same time, Russia's democratic leaders have failed to unite and failed to excite. In private conversations, many of Russia's post-Soviet democrats acknowledge that their own limited appeal was a major factor in their failure to win continued representation in the Duma in 2003. Electoral manipulation and skewed media coverage made the task much harder, but the democratic parties themselves clearly also fell short.

Without strong and unified public pressure for change, and with few mechanisms to voice those concerns should they arise, a near-term move toward further openness can come only from the top. No polls thus far show widespread public dissatisfaction with how Russia is governed, nor do any suggest that democracy is a priority for a majority or even a sizable minority of average Russians. More important for most is the fact that real incomes in Russia doubled during Vladimir Putin's two terms as president and poverty dropped by half. Wages and pensions were paid on time, and grew faster than inflation. GDP rose by 70 percent, though Russia has since been hit hard by the current crisis.

Yet even the economic downturn has had a muted political impact. Unlike during the 1998 financial meltdown, Russia today holds considerable gold and hard-currency reserves that it has been able to spend to address emerging problems or potential sources of upheaval and to protect private interests, including by bailing out many of the country's business leaders. Revealingly, though Putin publicly humiliated metals magnate Oleg Deripaska and chastised the local leaders of Pikalevo, a one-factory town near Saint Petersburg, for failing to pay their employees, he ultimately resolved the dispute to Deripaska's advantage by providing state funds to help the plants at the center of the crisis. While the former oligarchs have been shut out of high politics, they remain quite capable of advancing their concrete interests. And Russia's government can thus far afford to satisfy both the economic elite and the public.

This relative wealth and public apathy produce a degree of political stability in Russia. And because those at the top of the power vertical clearly understand that maintaining everyone's well-being is a necessary precondition of avoiding upheaval, they tend to prefer maintaining calm by tolerating the financial-industrial barons and ensuring a decent standard of living for the rest of the population. So absent a continued and more serious financial crisis that drains Russia's reserves, the status quo is likely to hold.

That is, unless the talk of competition between Putin and Medvedev has some real heft. A genuine struggle could tear both the corrupt elite and the power vertical apart, with unpredictable consequences.

Since most power rests at the top, any uncertainty there shakes the entire system. How much influence and ambition Medvedev and Putin possess, and the intentions of each, will powerfully shape how Russia evolves.

But the Medvedev-Putin dynamic is less than clear. Medvedev, the one most challenging to the status quo, is sending mixed signals about his intentions while Putin appears somewhat ambivalent about protecting his primacy. So far, Medvedev seems to be more talk than action. Though he speaks graphically about Russia's challenges and failings, he seems unprepared to act on these sentiments.

This may mean that he enjoys less than full authority, as many have suspected. As one senior official who knows both well put it privately, Putin has a "higher potential" to persuade Medvedev when the two differ. Medvedev has so far been cautious and pragmatic, openly stating that he has no plans to replace the government, therefore minimizing the chances for any fight with Putin over the fate of particular ministers or, of course, over Putin's own role. Revealingly, he has also avoided making significant changes in his own senior Kremlin staff. Khrushchev and Gorbachev learned Stalin's lesson that "cadres decide everything" and moved very quickly to bring in their own people, like Gorbachev's liberal adviser Aleksandr Yakovlev. Medvedev has talked about personnel policy but has done little.

However, judging Medvedev on the basis of his current conduct may be unwise. Russia's still-new president has clearly grown, impressing his foreign counterparts with his confidence and command of the issues. Like many others who have met him, we also noted Medvedev's evolution before and after he became president. Moreover, his behavior in Russia makes sense—he is in no position to challenge Putin directly, and alienating the prime minister prematurely would not advance his career or his agenda.

Thus it should be no surprise that Medvedev has signaled plans to move slowly in attempting to introduce change. He justifies this by referring to mistakes of the past rather than his current constraints: "not everyone is satisfied with the pace at which we are moving," he wrote, adding that he will "disappoint the supporters of permanent revolution" because "hasty and ill-considered political reforms have led to tragic consequences more than once in our history."

Yet despite Medvedev's careful politics, he is clearly trying to establish both his authority and identity, demonstrating a degree of political courage and independence. After Putin declared "we're people of the same blood, with the same political views," the Russian president commented that "we'll have a test to see whether we have the same blood type," an obvious effort to define himself distinctly from his mentor and senior partner. While Putin remains publicly unperturbed, Medvedev's growing assertiveness has clearly not gone unnoticed by the prime minister's supporters. As one close Putin associate put it to us, "at a minimum Medvedev is allowing his ambitious advisers to play a very dangerous game. Vladimir Vladimirovich's [Putin's] patience is not unlimited."

Still, Medvedev's advisers seem optimistic about his prospects and apparently do not fear open retaliation. Igor Yurgens, who leads the Institute of Contemporary Development, which Medvedev chairs, has openly suggested that Putin has outlived his usefulness and should not run again for the presidency, lest he become a new Leonid Brezhnev, the ailing Soviet leader who presided over the country's stagnation in the 1970s and early 1980s. The fact that Medvedev occasionally differs with Putin creates political space that did not previously exist. Medvedev must know this and, at a minimum, is allowing his advisers to criticize Putin and Putin's team while signaling in his own public statements that the two have different views. According to Yurgens, there is now a full-blown "clash of interests" between "conservatives and statists on one side and liberals on the other."

Medvedev's constituency—"the liberals"—seems built around Russia's educated, urban middle and upper classes. It includes some tamed oligarchs who made peace with Putin, but who still resent having been cut down to size, as well as

Westernized elites and professionals. Many were educated or have a presence overseas and see integration into the global economy as among Russia's important national interests. They also see themselves as part of a transnational elite—a few are a part of that coterie that have property and bank accounts overseas—and would suffer both financially and psychologically from Russian self-isolation.

Putin, for his part, enjoys considerable authority because of the strong support of the so-called "*siloviki*" (former KGB men and military types) in the security ministries, his economically driven popular legitimacy, his reputation for machismo and decisiveness, and a widespread sense of Russia's renewed stability and global influence. He has also grown. Appointed by Yeltsin's inner circle, including the unsavory tycoon Boris Berezovsky, who now lives in exile in London, Putin rapidly demonstrated that he was not the tool they sought and became a genuine leader. In the eyes of most Russians, he restored order, prosperity and dignity to their lives and their country, even if many outsiders believe he benefited from high energy prices and U.S. and Western distraction in Iraq and Afghanistan. Relatively few Russians are concerned about the gradual loss of freedom during his rule; Putin gave them what they wanted in exchange for what they didn't think they needed.

Importantly, most Russians believe that Putin cares about their country and cares what its people think. While no longer president, Putin continues to convey the image of a "good czar"—a national leader with clear power, charisma and a certain mystique. When polled, a majority or plurality of Russians regularly state that it is Putin who rules the country, with a smaller group saying that Putin and Medvedev share power, and only a slim share arguing that Medvedev alone is in charge. Strikingly, an August poll by the respected Moscow-based Levada Center found that some 52 percent of Russians credited Putin with leading Russia through the crisis relatively unharmed, compared to just 11 percent who praised Medvedev. When assigning responsibility for the economic hard times, 36 percent blamed "the government," 23 percent blamed Medvedev and only 17 percent blamed Putin.

Many Russian liberals recognize Putin's power and see no path toward reform without him. Yevgeny Gontmakher, who is also affiliated with the pro-Medvedev Institute of Contemporary Development, wrote that Russia needs "modernization with the prime minister" because "we do not have another person capable of somehow influencing the situation." Both Medvedev and his advisers also seem to fear moving too quickly and impulsively, reluctant to be crushed under the wheels of history like Kerensky and Gorbachev by unleashing a process that would develop its own momentum, not only bringing very different people to power (in today's case, possibly virulent left-leaning nationalists), but also creating considerably more upheaval than planned, and risking Russia's collapse or disintegration. Gontmakher writes that Russia has two options: "a ruthless mutiny" that he believes would not succeed or "some kind of modernization from above."

Much hinges on the relationship between Medvedev and Putin, which is perhaps Russia's most carefully kept secret. Medvedev says that they meet only once per week, a fact his partisans share to dispel the notion that the president receives regular guidance from Putin. Both frequently cite Russia's constitutional division of labor, which puts the president firmly in charge of foreign and security policy and leaves the economy and social issues to the prime minister. However, each routinely acts to blur the lines; Medvedev summons ministers who report to Putin to issue public instructions on the economy, while Putin often takes a visible role on security and foreign-policy issues, such as last year's war with Georgia, the decision to apply to the WTO as a customs union with Belarus and Kazakhstan (announced just days after Medvedev's advisers said Moscow would continue with its previous approach), and high-profile foreign trips, like a 2009 visit to Poland around the seventieth anniversary of the Molotov-Ribbentrop Pact and the German (and Soviet) invasion. Nevertheless, without knowing what understandings may exist between them, it is difficult to be certain what behavior and statements to ascribe to a good cop/bad cop routine, to personal ambitions or to real policy differences.

Both Medvedev's intentions and Putin's potential responses are unclear. After all, it was Putin who spent eight years systematically eliminating or weakening any potential rivals in Russian politics only to create just such a rival when he left the presidency to become prime minister. Medvedev's supporters clearly hope that the prime minister will be prepared to fade away after receiving appropriate assurances. Putin's confidants deride this as "daydreaming," arguing that their man is reenergized and sees a continuing mission for himself in Russian politics.

The big unknown is whether Putin has a sufficient lust for power to fight back if Medvedev appears successful and loyal at the same time. If so, Putin seems unlikely to be as inept as the Soviet Union's so-called "anti-party group" that tried to remove Khrushchev in the 1950s or the anti-Gorbachev coup plotters of 1991.

Some Russian pundits suggest this will all come to a head with Medvedev and Putin running against one another for president in 2012. This seems unlikely. Russian politics have so far been decided before rather than during its elections and this probably will not change in the next three years. There is also no clear institutional base for Medvedev in an electoral competition against Putin. Putin chairs the United Russia party, and among the other parties in the Duma, only Just Russia could conceivably be an appropriate home for Medvedev. But Just Russia remains loyal to Putin and is both fairly weak and less supportive than United Russia of the Western-style reforms that Medvedev's camp seems to want. Thus, for Medvedev to build an institutional base, he would more likely have to be a divider rather than a uniter, splitting apart Russia's elite, its government and the United Russia party—something many will resist and many others will fear. The steady approach of 2012 and the preelection decisions it will force only fuel the tension in the Medvedev-Putin relationship—and add to Russia's uncertainty. And while Putin remains dominant thus far, his power has never been seriously challenged.

Political and economic liberalization in Russia would advance American interests and improve U.S.-Russian cooperation. However, real political conflict or a stalemate in Russia will likely be a problem for the United States as well. This is not an argument for stability for the sake of convenience in U.S. policy; it is simply a statement of fact. Political competition in Russia creates pressures to take harder lines in defining Moscow's positions and goals and can even lead to dangerous Russian actions. U.S. officials, members of Congress and others will take note of these attitudes or actions and react. At the same time, domestic uncertainty in Russia only increases the difficulty that outsiders have in understanding its government decision making and predicting Moscow's conduct—which in turn undermines American policy.

This situation means that taking sides in Russia's internal political debates could come at a great cost. This is a central lesson of the 1990s, when American support for Boris Yeltsin—who many thought was a pro-Western democrat despite early signs to the contrary—in fact persuaded most Russians that Washington was more interested in ensuring that Moscow's leaders remained weak and compliant than in helping the country's citizens or preserving democracy.

Another reason to avoid taking sides is the murky relationship between domestic reforms and foreign policy in Russia. Russia's past is replete with rulers who supported both ambitious internal reforms and aggressive foreign policies, ranging from Czar Alexander II (who freed the serfs but pursued wars in the Balkans and the Caucasus) to Soviet leader Nikita Khrushchev (who led the post-Stalin thaw but provoked the Cuban missile crisis). It also includes a few pragmatic autocrats like Alexander III, who reversed many of his father Alexander II's reforms but was careful to avoid reckless foreign pursuits.

With all the uncertainty about Russia, it may be helpful to focus on what the country is not. First and foremost, Russia is not a country governed by a messianic ideology and is neither intrinsically antidemocratic nor anti-Western. Secondly, however, Russia is not a nation of altruistic do-gooders upon whose support the United States can rely when its interests and priorities differ from Washington's. Here it is useful to recall Churchill's entire quote: "I cannot forecast to you the action of Russia. It is a riddle, wrapped in a mystery, inside an enigma; but perhaps there is a key. That key is Russian national interest."

Russia's domestic situation will be an obstacle to cooperation in some areas, but not in others. For example, Russia does not particularly seem to care how its foreign partners run their countries. Moscow has worked quite successfully with democracies such as Germany and Italy, demonstrating that Russia does not have a problem with democratic governments as such. In fact, Russia's leaders seem to have gotten along much better with German Chancellor Angela Merkel than Belarusian strongman President Alexander Lukashenko, with whom they have frequent public spats. Conversely, contrary to Georgian President Mikheil Saakashvili's assertions, Russia's problem with Georgia is not its democracy but its hostile conduct. When Saakashvili acted on his sometimes-authoritarian instincts in Georgia, he received no credit for it in Moscow.

Similarly, Russia's internal politics will not prevent its leaders from cooperating pragmatically with the United States on issues like arms control and nonproliferation when they view such efforts as promoting their interests. Nor will it prohibit Russia from becoming an American partner on some issues at some times, or from viewing partnership with the United States as sufficiently serving its national interests to influence other calculations.

Ultimately, of course, Russia's domestic practices present the greatest obstacles to Russia itself, which severely limits foreign investment, modernization and the country's integration into the international system without making real changes. This will in turn affect not only Moscow's hard power but also its soft power. So while Vice President Joseph Biden's recent blunt assessment of Russia's decline may have been exceedingly undiplomatic—and mistaken in its conclusion that Moscow would have no choice but to cooperate with Washington—it was not fundamentally incorrect if Russia does not alter its course. While Medvedev has promised to do this, he admits that little has happened even in the wake of Russia's dismal performance in the global financial crisis.

Russia has been a difficult interlocutor since its independence nearly two decades ago and is unlikely to become an easier one anytime soon. But for all of its faults—and they are many—Russia is not inherently an American foe. Russia's leaders may be ruthless, but they do not need foreign enemies. With care and determination, the United States can work with Moscow to advance important national interests.

DIMITRI K. SIMES is the president of the Nixon Center. **PAUL J. SAUNDERS** is executive director of the Nixon Center.

From *The National Interest*, October 28, 2009. Copyright © 2009 by National Interest. Reprinted by permission.

The Bells

How Harvard helped preserve a Russian legacy.

ELIF BATUMAN

Father Roman is the head bell ringer at the Danilov Monastery, the working residence of the Moscow Patriarch and the seat of the Holy Synod. But the first time he rang the monastery's original bells was in 2003, at Harvard's Lowell House, where they had hung since the early years of Stalin's Terror, when bell ringing was banned throughout Soviet Russia. I met Father Roman this past June, when he returned to Harvard to help commemorate the bells' repatriation. He was presented to me rather abruptly by one of the Lowell House masters. "This is Father Roman," she announced, and, turning, I beheld a towering figure in his early thirties, well over six feet tall, with enormous hands and a flowing chestnut beard, wearing a long black habit and a *skufya,* the black pointed cap of a Russian Orthodox monk. "And *this,*" she continued, gesturing toward an easel, "is a picture of Father Roman." A large photograph showed a gray-eyed man in a belfry, beard blowing in the wind, against a leaden sky crossed by the ropes of Russian church bells. When I asked Father Roman whether I might meet with him later in the day, he produced from the folds of his habit a slim black Nokia telephone, whose ring tone, I learned, was a recording of Russian church bells.

The eighteen Danilov bells, which range in weight from twenty-two pounds to thirteen tons, were among the last to ring in the Soviet Union. Cast between 1682 and 1907, they are one of only a few intact sets of pre-Revolutionary Russian bells in existence. Some of them tolled at the burial of Nikolai Gogol. In 1929, the Danilov Monastery was converted into an orphanage; during the nineteen-thirties, it also housed the children of "enemies of the people". Most of the monks were shot in 1937, during the Great Purge, but the bells escaped the country. In 1930, an American philanthropist purchased the entire set at scrap-metal prices—about eighteen thousand dollars—and donated it to Harvard. The bells hung in Lowell House, where they rang on Sundays—as well as after home football victories, and victories over Yale, regardless of venue—for more than seventy years.

In 1983, Danilov became the first monastery to be officially returned to the Russian Orthodox Church. The belfry was restored, and equipped with fifteen bells, salvaged from various defunct churches. But these bells were mismatched, and some had been severely damaged. In 1988, Ronald Reagan expressed openness to the idea of returning the bells during a visit to the Danilov Monastery, where a photograph of Ronald and Nancy Reagan hangs to this day. But many things had to change, both in Russia and at Harvard, before the return could take place.

Ten years ago, as an undergraduate living at Lowell, I knew nothing about the bells' provenance, or the hopes and fortunes they embodied. I had a vague sense that they were old, and a more vivid sense of their being rung every Sunday at one in the afternoon—an hour at which I particularly liked to be undisturbed by loud clanging noises. The bells were the purview of a student group called the Klappermeisters, who took a keen interest in house traditions. For fifteen minutes on end, they hammered out peculiarly tuneless variations on melodies like "Twinkle, Twinkle, Little Star."

In Russian history and culture, church bells occupy a mysteriously important position. Their tolling, Father Roman told me, has been known to bring miserly or hard-hearted people to repentance, and to dissuade would-be murderers and suicides. In "Crime and Punishment," Raskolnikov falls into a guilt-induced fever, hearing the ringing of Sunday church bells; he gives himself away by returning to the scene of the crime and compulsively ringing the murder victim's doorbell. In "War and Peace," the Kremlin's bells ring during Napoleon's invasion, disquieting the Grande Armée. Considered in Russian folklore to be animate beings, bells exercise a profound power over humankind—a power that lay dead or dormant for most of the twentieth century. The return of the Danilov bells represents the hope that this power might finally be recovered.

In the Russian Orthodox faith, bells are widely considered to be "aural icons," symbolizing the trumpet calls blown on Mt. Sinai and the sounding of the trumpet for the raising of the dead before the Last Judgment. Just as painted icons are not intended to be mimetic representations of a spiritual object but magical windows into the world of the spiritual, a Russian bell is not a musical instrument but, as Father Roman puts it, "an icon of the voice of God." A Russian bell, he said, must sound rich, deep, sonorous, and clear, for how can the voice of God be otherwise? It must be loud, because God is omnipotent. Above all, Russian bells must never be tuned to either a major or a minor chord. "The voice of a bell is understood as just that," he said. "Not a note, not a chord, but a *voice*." Whereas Western European bells are tuned on a lathe to produce familiar major and minor chords, a Russian bell is prized for its individual, untuned voice, produced by an overlay of numerous partial frequencies, with only approximate relations to traditional pitches—a feature that gave the Lowell Klappermeisters' performances the denatured effect of music played on a touchtone telephone. Where Western European bells play melodies, Russian bell ringing consists of rhythmic layered peals.

The afternoon I met Father Roman, I climbed the hundred and twenty-eight steps to the top of the Lowell bell tower. In the belfry were Father Roman, five of his Danilov colleagues, and Konstantin Mishurovsky, one of the bell ringers at the Kremlin. The window openings in the belfry were covered by netting. Outside, the distant Boston skyline glittered against an overcast sky. On the floor lay tiny feathers and other traces of falcon life, along with a few discarded fluorescent-orange foam earplugs of a type worn by some bell ringers—to the contempt of other bell ringers. Mishurovsky told me that earplugs prevent ringers from achieving the requisite delicate touch. When I asked him about hearing loss, he sighed, and said, "People always ask me this question. And to all of them I answer the same thing: *Ehhh?*"

At Lowell House, the largest bell was known as the Mother Earth bell. The Russians call it simply *Bolshoy* ("big") or *Blagovestnik* ("bearer of good tidings"), because the biggest bell is always used to play the *blagovest*—deep, regular tolls that announce the beginning of a service. The three largest bells are decorated with inscriptions in Old Church Slavonic, depictions of saints, or imperial medallions; the oldest bell in the collection—purportedly cast by a famous master, Fyodor Matorin, in 1682—also has cherubs, and what resembles a face of Bacchus, open-mouthed and surrounded by grapes. Unlike Western church bells, which move when they are rung, Russian bells remain stationary, and are rung by means of ropes attached to the clappers. To ring the vestibule-size Mother Earth bell in the Lowell belfry (every belfry is set up differently), one or two people stood inside the bell on an elevated platform, swinging the thirteen-hundred-pound clapper on a rope. The remaining bells were rung from a lower platform: the bass bells controlled by foot pedals; the tenors and altos, by a crisscross web of ropes. The "trills"—the smallest bells, some no bigger than a top hat—were attached to reinlike ropes held in the right hand and rung with a flicking wrist motion, producing an eerie birdlike jangle.

Shadowy, covered with verdigris, the bells seemed to move slightly even in repose, as if under water. It was easy to grasp why such bells might be treated as living things. Russian bells are given names like Swan (for producing a swanlike cry), Bear (for rumbling or unwieldiness), or Sheep (for a rattling or uneven tone). They ring with their "tongues," hang by their "ears," and have shoulders, waists, crowns, and skirts. They are considered to be capable of suffering, and even of a certain metallic obstinacy. When Peter the Great ordered that the Troitse-Sergiyeva bells be made into cannons, each bell resisted him in its own fashion: one disappeared and only later was discovered at the bottom of a pond; two proved impossible to remove from their hooks, then suddenly crashed to the ground. A fourth bell was beaten for a long time but didn't shatter, and buzzed reproachfully for three days afterward.

The ringing began. The deafening clangor seemed to be produced by something midway between metal and a living thing, and one felt physically assailed by an impossibly large and powerful force. At one point, I reached into my bag for a bottle of water; it was vibrating so vigorously that I mistook it for my cell phone. Between two and four ringers worked the ropes and pedals at any given time, while the others stood to one side, giving advice about who should stand where.

When Mishurovsky rang the bells, his normally skeptical expression was replaced by one of smiling beatitude, and he swayed and waved his arms. ("I hope you noticed how I move when I'm ringing," he told me later, rather sternly. "Like flying an airplane.") According to Mishurovsky, weather conditions affect the sound of bells, and they are heard to best advantage during a light rain, so it was lucky for us that a cold drizzle began to spray into the belfry through the screens. After a few minutes, I climbed a little ladder to the upper platform, to seek cover beneath the Mother Earth bell. The verdigris had worn off at the two points in the bell where the clapper struck, leaving them a shiny, burnished bronze. The graffiti scratched inside the bell included Russian names spelled with pre-Revolutionary orthography.

Mishurovsky and Father Roman climbed up to the platform. Roman caught my eye and pointed solicitously at his ears: they were going to ring the bell. Against the backdrop of the rhythmic din of the other bells, the two men began to swing the clapper slowly back and forth. Each time, the arc expanded, until finally the clapper touched the lip of the bell, producing a plangent, unforeseeable boom that made my whole body vibrate. They continued to swing the massive pendulum at the same rhythm, but varying the movement, so that sometimes the clapper hit both sides of the bell and sometimes only one side. When the peal came to an end and the bells had fallen silent, a hum continued to animate the air. Mishurovsky motioned to me to touch the big bell. The sensation resembled an electric shock.

One of the first mentions of bells in Russia occurs in the *Novgorod Chronicle,* which records the seizure, by the Prince of Polotsk, of the bells of Novgorod's St. Sophia Cathedral, in 1066. At the end of the fourteenth century, Russian metal foundries began forging both bells and cannons, inaugurating an uneasy partnership between the two great noisemakers: during times of war, the former were in constant danger of being melted down into the latter. But bells were also honored objects of state. Ivan the Terrible used to climb the belfry at four o'clock every morning in order to personally ring the Matins. Tardy monks were imprisoned for eight days.

The most famous item of Russian bell lore involves the matter of Ivan's heirs. Ivan had three surviving sons: Ivan Ivanovich, the eldest, and favorite, whom Ivan accidentally killed during a domestic quarrel; the mentally retarded Fyodor, who loved bells so much that he was known as Fyodor the Bell Ringer; and the epileptic infant Dmitri. Although Ivan was officially succeeded by Fyodor, the de-facto ruler was Fyodor's brother-in-law, Boris Godunov, an ambitious boyar who had his eye on the throne. To safeguard his own interests, Godunov banished Dmitri to the town of Uglich—where, a few years later, the young boy was found stabbed to death. When the Uglich bell sounded the death knell, a riot broke out: surely the boy had been murdered by Godunov's agents. (According to the official report, the wounds were self-inflicted during an epileptic seizure.) Godunov was so enraged that he ordered the bell's ears and tongue to be torn off. The mute and mutilated bell was then literally banished to Tobolsk, in Siberia, where the governor registered it as Uglich's "first inanimate exile."

Defying orders that the bronze guest remain silent forever, the people of Tobolsk outfitted it with a clapper and installed it in a local belfry. Because the bell had defended the Tsarevitch Dmitri, and was surely fond of children, a legend arose that, if you washed the clapper and collected the water in a special container, it became an elixir for curing children's diseases. This and other stories about bells standing up for the little man perhaps helped inspire Alexander Herzen, the great pre-Revolutionary intellectual populist, to name his influential clandestine journal *Kolokol:* "The Bell."

Both the tremendous popular attachment to Russian church bells and their systematic destruction by a radical regime have a historical precedent in France. During the French Revolution, bells were melted into coins and cannons, churches were converted into "Temples of Reason," and bell ringing was banned—suddenly bringing to light the bell's deep cultural significance. The French historian Alain Corbin, in his study "Village Bells," cites a report by an administrator in Somme: "There is no disputing the fact that what hit the people hardest in the Revolution was being deprived of their bells." The public outcry was such that French authorities were ultimately unable to enforce the anti-bell legislation. Belfries were eventually restored and augmented in a campanological golden age that lasted until the end of the nineteenth century.

In Russia, things happened differently. The Soviets took the destruction of bells, as they took many things, with deadly seriousness. Stalin's "Great Break" of the late nineteen-twenties aimed for a total transformation of every aspect of life. A staggered workweek was implemented, abolishing the concept of Sundays. Moscow's Church of the Nativity became a holding pen for circus lions, the Cathedral of Christ the Saviour became the world's largest open-air swimming pool, and rural churches were turned into carpentry or plumbing collectives. Bell ringing was prohibited by law, and the destruction of thousands of tons of consecrated bronze that hung in belfries across the Soviet Union was regulated by the All-Russian Central Executive Committee. "They fell with a roar and a thud, digging holes some five feet into the ground," the novelist Boris Pilnyak wrote of the bells in Uglich. "The whole town was full of the moaning of these ancient bells." Photographs from the period show proletarians swinging mallets in a sea of shattered bronze; teen-agers sit inside toppled bells, cigarettes dangling from their mouths. The smashed bells were melted into industrial

material, realizing the state's favorite transformative metaphor: the recasting of metal. One peasant from Orlov later recalled that, the year the village dismantled its belfry, "a new tractor, its spurs glittering in the sun, drove down the length of the village street, making real to us boys the strange transformation from bell to machine."

Whereas France's bells were ultimately allowed to outlive their relevance, as industrialization and urbanization dissolved the village unit, the Soviets eliminated bells at a time when they were still beloved features of the cultural landscape. In twentieth-century Russian poetry, the bell appears as the hub of a timeless, rounded universe. "The sky's like a bell—the moon is its tongue," the popular poet Sergei Esenin wrote; and when Osip Mandelstam wanted to describe meaninglessness, he wrote, "Someone has pulled the bells / out of the blurred tower." If bells in Revolutionary France were the voice of the village, bells in Revolutionary Russia were still the voice of God.

The bells' history at Harvard begins with Charles R. Crane, who devoted a family fortune made in plumbing and fixtures to the pursuit of a career in international philanthropy and political activism. His beneficiaries ranged from Syrian revolutionaries to Macedonian Slavophiles and the Czech painter Alphonse Mucha (who subsequently used Crane's daughter, Josephine, as the model for Slavia on the hundred-koruna bill). A novel based on his adventures depicted Crane as an eccentric plutocrat who, according to an article in the *Times,* in 1909, employed "not only his colossal American fortune, but his brilliant American business ability for organizing a revolution throughout the world on strictly business principles." Crane traversed Albania on horseback and Iraq by car, and made some twenty-five trips to Russia; in 1917, he was on the same Petrograd-bound steamer as Leon Trotsky. He first learned about the Danilov bells in the late nineteen-twenties, through his friend the Byzantologist Thomas Whittemore, who is best known today for his restoration of the mosaics of Istanbul's Hagia Sophia, which date to the age of Justinian.

Luis Campos, a former Klappermeister who now teaches history at Drew University, is completing a book for Harvard University Press on the purchase and return of the bells, highlighting the escapades of Crane and Whittemore. According to Campos, Crane in his youth deplored religious traditions. His earliest positive reaction to any religious object appears to have occurred when he saw Tibetan prayer wheels: the young businessman was greatly impressed by the idea, as Campos puts it, of "a machine that could do your praying for you." In later life, Crane became a devoted patron of Russian Orthodox art. Icons and pictures of cathedrals hung in his house, and he eventually bankrolled an entire Orthodox church choir in New York. When Whittemore heard about the dismantling of the Danilov belfry, he contacted Crane, who gave him carte blanche to purchase the bells. Crane donated the bells to Harvard in 1930, during the construction of Lowell House. A projected clock tower was hastily redesigned as a belfry capable of accommodating up to thirty-five tons of Russian bells.

Shortly after the bells arrived in Cambridge, a young Russian named Konstantin Saradzhev appeared at Lowell House. Whittemore had appointed the young man, who was described as "Moscow's most famous bell ringer," to oversee the bells' installation. Saradzhev spoke no English, and his entire luggage consisted of four pairs of socks and two handkerchiefs. At his request, he was taken immediately to see the bells, which were not as he had expected. "You have one bell that does not belong in the set," he announced. "And you should have seventeen others." A cable was sent to Addis Ababa, where Whittemore was travelling, to attend the coronation of Haile Selassie. Whittemore replied that Saradzhev was mistaken. Nonetheless, the offending bell was exiled across the river to the Harvard Business School.

During his time at Harvard, Saradzhev irritated the Lowell residents with incessant tapping on the bells. One day, the university's president, Abbott Lawrence Lowell, found him scraping the bells with a file, apparently trying to alter their pitch. Lowell told him to stop, and he became upset. Saradzhev, who turned out to be epileptic, began suffering recurrent seizures. He was eventually sent to the infirmary, where, one morning, his sheets were found covered in dark stains. According to a Lowell House tutor, Saradzhev admitted that he had been drinking ink—an antidote, he said, to poison that was being secretly administered to him. This was the last straw for President Lowell, who sent him back to Moscow.

The bells' installation was completed by a Russian émigré who was found by Crane's Russian Orthodox choir and who claimed to have rung the Danilov bells before the Revolution. The official public concert that Easter was a huge failure. "At once the horrid truth became apparent," the Lowell tutor wrote. "This . . . was no carillon . . . on which each note could be played independently with some semblance of a tune." "GIFT CHIMES PROVE WHITE ELEPHANT," a local headline read. In subsequent decades, however, Lowell House became rather proud of its cacophonous white elephant. Some students were devoted enough to start ringing the bells on Sundays and after football games, and the Klappermeisters were formed.

As for Konstantin Saradzhev, he was forgotten altogether—until the nineteen-seventies, when Anastasia Tsvetaeva, the sister of the poet Marina Tsvetaeva, published a memoir about his life and fate. As it turns out, Saradzhev really *was* Moscow's most famous bell ringer, known not just for his ringing but also for his superhuman aural acuity: between two adjacent whole tones, he perceived not just one half tone but a half tone flanked on either side by a hundred and twenty-one flats and a hundred and twenty-one sharps.

When Saradzhev was seven years old, the sound of a particularly powerful church bell caused him to lose consciousness, and he was captivated for life. Although he was a skilled pianist, he always referred to the piano as "that well-tempered nitwit": a piano can produce only twelve tones per octave, whereas Saradzhev perceived one thousand seven hundred and one. This sensitivity perhaps explains Saradzhev's intense delight in Russian bells, which are unparalleled in their microtonal complexity. Each bell sounds a unique cloud of untempered frequencies, producing intervals unplayable on any twelve-tone keyboard.

By such acoustic fingerprints, Saradzhev could distinguish all four thousand of Moscow's church bells. He described his hearing as "true pitch" (by contrast with perfect pitch). The capacity for true pitch, he said, lay dormant in all humans, and would someday be awakened. But in the meantime he was, like a superhero, cruelly isolated by his own powers. He spent most of his time working on a theory of the future of music that was incomprehensible to anyone who couldn't hear a thousand or more distinct microtones in an octave.

Still, the "bell symphonies," which Saradzhev played in place of traditional church peals, were so popular that, even in the dead of winter, crowds filled the courtyard of the Church of St. Maron the Hermit to hear them. It is difficult to guess exactly how the symphonies sounded—Saradzhev struggled all his life with the problem of musical notation—but the effect, according to Tsvetaeva, was explosive: the sky seemed to collapse under a massive weight of thunder; the square filled with the metallic cries of gigantic bronze birds; the trills clattered overhead like swallows. Many listeners characterized Saradzhev's ringing as "the music of the spheres."

To his great frustration, Saradzhev was never able to play his symphonies as he imagined them. This is because Russian bells were invariably church bells, and had to be played according to church rules. Saradzhev had devised a new layout that would make it possible for the bells to be rung in more combinations, but no church would let him implement it. Professors at the Moscow Conservatory, including the composer Reinhold Glière, petitioned the People's Commissariat of Enlightenment to give Saradzhev a set of bells for secular use. The petition was denied.

When, in 1930, Saradzhev accepted a contract to work on installing the bells at Lowell House, he thought his dream had come true. Finally, some Americans, recognizing the greatness of his bell symphonies, were going to build him a symphonic belfry in America, equipped with thirty-four Russian bells: the seventeen Danilov bells, plus seventeen others of his own choosing. One can imagine Saradzhev's feelings when he got to Harvard. Nobody was interested in the more than a hundred symphonies he had composed—which were unplayable now, anyway, since the seventeen bells he had painstakingly chosen were nowhere to be seen.

Saradzhev is said to have returned to Moscow one night in 1931, unannounced, bringing his father a raccoon-fur hat with earflaps. America appeared not to have made much of an impression on him. "There's nothing there but Americans," he reportedly said, when pressed for details. Little is known of his later years. He is believed to have died in 1942, in a mental asylum in Moscow.

A long period of impasse followed the monastery's conversation with Ronald Reagan about the return of the Danilov bells. The university administration, pointing out that Harvard was the legal owner of the bells, objected that their removal would be expensive and possibly harmful to the Lowell buildings. The Lowell House master William Bossert expressed reservations about the project, which he suggested would cost tens of millions of dollars and result in a student-housing crisis.

In 1998, Bossert and his wife were succeeded by Diana Eck and her partner, Dorothy Austin, who were particularly attuned to the importance of spiritual objects. (Eck, who is a professor of comparative religion, has written several books on Hinduism and religious pluralism in America; Austin is an Episcopalian priest at Harvard's Memorial Church.) Another development took place in 2000, during a lunchtime conversation at the Danilov Monastery, where the American Ambassador to Russia and a United States Under-Secretary of Agriculture were working with the monastery on a food-aid-distribution project, and sampling some dishes made from soy protein (recalled as inventive but quite bland). Upon learning that both the Ambassador and the Under-Secretary were Harvard alumni, the monks raised the subject of the bells' return. The diplomats relayed to Harvard a suggestion that the monastery swap its replacement set of bells for the ones at Lowell House and also arrange for transport. In 2003, Harvard received a letter from Archimandrite Alexei, the father superior of Danilov Monastery. Alluding to the Book of Ecclesiastes, Alexei wrote that a new time had come in his country—"a time to gather stones together" which had been scattered in past decades.

In December, 2003, in the middle of a snowstorm, a Russian delegation, including Father Roman and Archimandrite Alexei, arrived at Lowell House. In the bell tower, the Orthodox members of the group crossed themselves, and Father Alexei kissed his hand and touched an image of the Holy Trinity on one of the bells. Father Roman tapped some of the bells, and tested the ropes. Then he began to play. "This was the moment when we realized that these were truly their bells," Eck recalled. "He brought a sound from those bells that we had never heard before." Soon afterward, Harvard commissioned a study, which determined the bells' removal to be expensive but feasible.

In 2004, a group of Lowell bell ringers travelled to Russia for master classes in bell ringing, conducted by Father Roman and bell ringers from the Kremlin. The same year, Patriarch Alexei II, who was the head of the Russian Orthodox Church until his death, last December, issued a worldwide appeal for funding. The appeal was answered by the Link of Times, a nonprofit organization founded by the oil-and-metals tycoon Viktor Vekselberg with the aim of repatriating historically significant Russian works of art. Earlier that year, Vekselberg had purchased the entire Forbes family Fabergé collection, including nine imperial Easter eggs, valued at more than ninety million dollars. The removal and transportation of the Lowell House bells, and their replacement by an entirely new set, was comparatively inexpensive, costing around ten million dollars. In 2006, Eck and other Harvard representatives made a tour of several Russian bell foundries to commission the set of replacement bells.

During an exchange of gifts at Harvard last spring, the Harvard project manager presented Archimandrite Alexei with a handful of Liberty Bell-shaped lollipops. It was a wry allusion to an unspoken and potentially divisive asymmetry in the bells' significance: for Harvard, the

bells were a curio or a fad; for the Danilov Monastery, they are a sacred patrimony. Beaming, Father Alexei kissed the lollipops and made the sign of the cross. Throughout the exchange, neither he nor his colleagues betrayed any sign of displeasure at the decades-long use of their "singing icons" to commemorate the defeats of the Yale football team. Meanwhile, it was the university's implicit stance that Harvard was trading a precious object for another of equal or greater value: a twenty-seven-ton aural monument to the university's public-spiritedness.

Even within Russia, it has been difficult to recapture the significance once accorded bells. Alain Corbin, writing about France, characterizes bell ringing as one of the historical phenomena that, "even though close to us in time, bear witness to a paradoxical distance." Bells, "once the occasion of so many obscure and forgotten passions," afford the modern historian a key to "a world we have just lost." In Russia, the challenge of unlocking this lost world faces not only historians but also a new generation of bell ringers and casters. Konstantin Mishurovsky, like Father Roman, is in his thirties, and grew up in a bell-deprived Soviet Union. Mishurovsky's mother gave him a recording of bells when he was a child. Mishurovsky told me that he first took up bell ringing to play in church services. In the nineteen-nineties, while he was studying biology at Moscow State University, he found himself at a research station on the White Sea, in a Karelian village called Kovda. Kovda had managed to preserve one of its church bells and, in honor of the village's five-hundredth anniversary, to acquire three more. Mishurovsky installed the bells and rang them at the anniversary celebrations. He soon realized that his true vocation was bell ringing, but he took the precaution of earning his biology degree. "Now I am so grateful to my studies," he said. "They gave me the scientific method."

Mishurovsky uses the scientific method to research bell acoustics, a field fraught with mystifications: apparently, many of the pitches that we hear when a bell is rung are "sonic mirages," which exist only in the listener's mind and do not appear in a spectrographic analysis. His work is geared toward modelling the sounds of bygone bells. He told me about one particularly valuable resource: a single 1963 Mosfilm recording of Rostov cathedral bells, as played by some of the last living bell ringers to have been trained in Imperial Russia. (The recording was made for a film adaptation of "War and Peace," and then used in virtually all the classic bell-ringing scenes of Soviet cinema, including the end of Andrei Tarkovsky's "Andrei Rublev.") After Moscow's Cathedral of Christ the Saviour was rebuilt, in the nineties, Mishurovsky performed on the church's new bells for the first time after closely studying the Mosfilm recordings. The bells were cast by the manufacturer AMO-ZIL, which, in Soviet times, had produced armored cars for Stalin, the Politburo's limousines, and the Red Army's cargo trucks. Just as bells were once transformed into tractors, so truck factories are now being "refunctionalized" as bell foundries.

The replicas of the Danilov bells made for Lowell House were cast by Vera, one of the first post-Soviet bell foundries, in Voronezh. When I met the director and owner, Valery Anisimov, he was wearing a ponytail, a black shirt, a white suit, and white pointy-toed shoes. He said that the dream of building a bell factory came to him in 1988, the thousand-year anniversary of Christianity in Russia. As an engineer in a press-equipment factory, he had no idea how bells were made. At the Lenin State Library, he found exactly one book on the subject, written in 1906 by a Yaroslavl bell-caster—Saradzhev's favorite book, according to Tsvetaeva. Russian bells, like Western European bells, are cast using the "lost wax" method. An inner core is made of a brick-like substance—the Yaroslavl caster recommends a blend of mud, flax, manure, horsehair, and straw—providing the shape of the bell's interior. This core is covered by a layer of wax in the shape of the bell, and then by a layer of clay. The wax is then melted, leaving a clay mold. Finally, the mold is filled with bronze. It wasn't easy, Anisimov said, to follow the 1906 instructions, using the facilities he had rented in a nineteen-eighties Soviet factory, and the first bells didn't come out very well. Nonetheless, demand was so high, as the Russian Orthodox Church reestablished itself around the country, that they were bought up immediately. In 1991, Anisimov decided to open his own factory, on the site of a moribund collective farm. Problems arose in securing the property, however. Anisimov was indicted on charges of "gangsterism" and spent three months in jail before being acquitted. Today, Vera is the world's largest bell foundry, producing two hundred tons of bells each year.

Meanwhile, Russian bells continue to generate a folklore reflective of the times. In today's health- and environment-obsessed climate, much is made of the medieval belief that bell ringing prevented the spread of plagues and epidemics. The psychiatrist Andrei Gnezdilov, the founder of Russia's first hospice for the terminally ill, uses "bell therapy" as a pain-management technique; various rhythms and frequencies, he says, bring relief to about a third of cancer patients.

The curative and purifying properties of church bells are elaborated on numerous Russian Orthodox Web sites. Mishurovsky's mentor, who was the official bell ringer of Patriarch Alexei II, has said that bells can kill flu viruses, and that he hasn't had a cold since 1984. According to an oft-quoted ecologist, Fotiy Shipunov, formerly of the Soviet Academy of Sciences, bells produce ultrasonic waves that can curdle the proteins of jaundice-causing virus cells in a matter of seconds. Bell ringing has been used to combat mental retardation in children and to encourage the growth of flax plants. The physicist Anatoly Okhatrin claims that bells activate a field of super-lightweight particles called microleptons, which purify the air on a molecular level. Other Internet legends include accounts of bells warding off hurricanes and dispelling radiation.

In many of these stories, it is perhaps really the body politic, and not the human body, that is being cleansed of the accumulated ills of the Soviet era. Shipunov had spoken of an "amplifier" that can pick up the sound of bells from long-lost churches, even from a certain monastery near Kaluga, currently used as a cattle yard. The magical promise of bells is that we might be able to pluck the past from the very airwaves around us.

Last July, workers removed the entire northern face of the Lowell House bell tower. A hundred-and-fifty-foot crane gently lowered the old bells onto two flatbed trucks, where they briefly sat side by side with some of Anisimov's replicas. Father Roman watched the proceedings perched on a nearby fire escape. A few weeks later, he helped install the new bells in Lowell House. The old bells, meanwhile, had embarked on a monthlong transatlantic voyage to St. Petersburg, where they were transferred to trucks and conveyed to Moscow via Novgorod and Tver. In each city, they were greeted by local authorities, clerics, and throngs of worshippers.

On September 12th, Patriarch Alexei II blessed the returned bells at Danilov, in a ceremony attended by Viktor Vekselberg, President Dmitri Medvedev, a Harvard delegation, and thousands of Orthodox believers. Father Roman rang the bells from a temporary ground-level platform. It took several months to install the bells in the monastery's rather eye-catching salmon-colored belfry, where their first tolls rang out on March 17th. (September 12th and March 17th are consecrated to St. Daniel, who founded the monastery more than seven hundred years ago.) Patriarch Kirill I, who succeeded to the Patriarchate after the death of Alexei II, spoke of the bells' return as marking the transition between "an epoch of destruction and an epoch of creation," and expressed the hope that it would mark a new period in Russian-American relations as well. "It is not without the Divine Providence that the bells found themselves in one of the most important intellectual centers of the United States," Kirill observed. He said that the bells' odyssey could "help the scientists and students of Harvard University to also think about God and preserve the Christian faith in their hearts."

Toward the end of my stay in Cambridge, in June, I made a trip to a gray waterfront warehouse in South Boston, where the seventeen replacement bells were in the care of a firm that stores, transports, and installs art objects. The head of the firm led me through multiple security doors and climate-controlled hallways filled with crates, canvases, and statues. Ten of the bells were lined up along a long corridor, beside two massive bronze statues that had once sat in Boston Common; standing vigil around the corner was a ten-foot statue of Mary Baker Eddy, the founder of Christian Science. We moved on to a cement-floored storeroom, where the new Mother Earth bell was enthroned on a wooden platform, surrounded by smaller bells, like a gigantic hen surrounded by chicks. Under fluorescent lights, the bas-relief faces of saints stood out on the pale bronze with luminous clarity. On the Mother Earth bell, the reliefs copied from the old bell had been supplemented by the coats of arms of Lowell House and Harvard University, the logo of the Link of Times Foundation, and an image of the Danilov Monastery. Because of differences in the shape of the mold, the new Mother Earth bell weighs two tons more than the original. Nonetheless, the sounds of the two bells are almost indistinguishable.

From *The New Yorker*, April 27, 2009, pp. 22–28. Copyright © 2009 by Elif Batuman. Reprinted by permission of Elif Batuman.

To Die for Tallinn

PAT BUCHANAN

All week, young toughs in Moscow have besieged the Estonian embassy to harass Ambassador Marina Kaljurand. Her bodyguards had to use a mace-like spray to drive back the thugs, who call Estonia a "fascist country." Estonian diplomats and their families are being pulled out of Moscow and sent home.

Relations between the countries are about to rupture, if the Kremlin does not reign in the bully-boys.

Behind this nasty quarrel is the decision by Estonia to move the giant statue of a Red Army soldier, and the remains of Soviet soldiers, from the center of its capital, Tallinn, to a military cemetery. In Tallinn, patriots and nationalists have clashed with citizens of Russian ancestry over the perceived insult to Mother Russia and the "liberators" of Estonia from the Nazis.

Both points of view in this quarrel are understandable.

To Russians, who lost millions of their grandfathers, fathers and uncles in the Great Patriotic War, the Red Army liberated Europe from Nazism, and their sacrifices ought to be honored. And the Estonians are a pack of ingrates.

To Estonians, the Red Army did not liberate anyone. Having won their independence from the Russian Empire in World War I, they were raped by Russia—to whom they had been ceded as part of the Hitler-Stalin Pact. In June 1940, the Red Army stormed into the three Baltic republics, butchered the elites and shipped scores of thousands off to Stalin's labor camps never to be seen again.

While the Soviets were expelled from the Baltic republics by the Germans in 1941, they returned in 1944 and held the Baltic peoples in captivity until the Evil Empire collapsed. It was only then that Estonia regained her independence and freedom.

Why should Estonians honor a Red Army that brutalized them and, after driving out the Germans, re-enslaved them for half a century?

Why should this issue be of interest to America?

If President Putin decides the Estonians need a lesson, and sends troops to teach it, the United States, under NATO, would have to treat Russian intervention in Estonia as an attack upon the United States, and declare war on behalf of Estonia.

So we come face to face with the idiocy of having moved NATO onto Russia's front porch, and having given war guarantees to three little nations with historic animosities toward a nuclear power that has the ability to inflict 1,000 times the destruction upon us as Iran.

Latvia, too, is now a member of NATO. And Latvia, too, has a quarrel with Moscow over its treatment of the descendants of those Russians whom Stalin moved into Latvia to alter its ethnic character. Their children and grandchildren have grown up in Latvia, and know no other home, though they are unwelcome to ethnic Latvians.

Settling these quarrels is essential to peace in Europe. But the notion that Russian intervention in a Baltic republic should be met by a U.S. declaration of war, or any attack upon a nation with thousands of atomic weapons, is the definition of insanity.

Nor are these the only quarrels we have with Putin's Russia that could explode into full-blown crises. Washington has persuaded the Czech Republic and Poland, two former Warsaw Pact countries, to accept radars and missiles for a U.S. anti-missile system.

We say the missile defense system is directed at Iran. Russians see it as a piece of the move eastward of NATO and targeted at them. Can we blame them for so thinking, when we responded to their pullout of troops from Central and Eastern Europe by bringing Central and Eastern Europe into a U.S.-led alliance?

If the Russia-baiters in this capital have their way, Ukraine and Georgia will also be brought into NATO. That would commit us to go to war with Russia over control of the Crimean peninsula and the Russian-speaking Donbass of eastern Ukraine, and over the birthplace of Stalin and who should control South Ossetia and Abkhazia.

"Moscow would not dare intervene in the Baltic republics!" comes the retort. Perhaps not. But the Russians are now fiercely nationalistic and anti-American. And it is always a mistake for a great power to cede to a minor power the ability to draw it into a great war. Just as it is always a mistake to hand out war guarantees one cannot honor.

In March 1939, Britain gave a war guarantee to Polish colonels who had not requested it, a guarantee Britain had no way of fulfilling. The war that followed cost Britain her empire and Poland 50 years of freedom.

In August 1914, King Albert of Belgium informed King George V that the Kaiser's troops had crossed his border. He invoked a treaty assuring Belgian neutrality that the British had signed—in 1839!

So, Britain declared war, and 700,000 Brits perished in the Great War that hurled the West onto its present path of self-destruction.

And the march of folly continues on.

From *Creators.com* by Pat Buchanan, May 4, 2007. Copyright © 2007 by Creators Syndicate. Reprinted by permission.

McCain, Obama and Russia

Stephen F. Cohen

Despite its diminished status following the Soviet breakup in 1991, Russia alone possesses weapons that can destroy the United States, a military-industrial complex nearly America's equal in exporting arms, vast quantities of questionably secured nuclear materials sought by terrorists and the planet's largest oil and natural gas reserves. It also remains the world's largest territorial country, pivotally situated in the West and the East, at the crossroads of colliding civilizations, with strategic capabilities from Europe, Iran and other Middle East nations to North Korea, China, India, Afghanistan and even Latin America. All things considered, our national security may depend more on Russia than Russia's does on us.

And yet U.S.-Russian relations are worse today than they have been in twenty years. The relationship includes almost as many serious conflicts as it did during the cold war—among them, Kosovo, Iran, the former Soviet republics of Ukraine and Georgia, Venezuela, NATO expansion, missile defense, access to oil and the Kremlin's internal politics—and less actual cooperation, particularly in essential matters involving nuclear weapons. Indeed, a growing number of observers on both sides think the relationship is verging on a new cold war, including another arms race.

Even the current cold peace could be more dangerous than its predecessor, for three reasons: First, its front line is not in Berlin or the Third World but on Russia's own borders, where U.S. and NATO military power is increasingly ensconced. Second, lethal dangers inherent in Moscow's impaired controls over its vast stockpiles of materials of mass destruction and thousands of missiles on hair-trigger alert, a legacy of the state's disintegration in the 1990s, exceed any such threats in the past. And third, also unlike before, there is no effective domestic opposition to hawkish policies in Washington or Moscow, only influential proponents and cheerleaders.

How did it come to this? Less than twenty years ago, in 1989–90, the Soviet Russian and American leaders, Mikhail Gorbachev and George H.W. Bush, completing a process begun by Gorbachev and President Reagan, agreed to end the cold war, with "no winners and no losers," as even Condoleezza Rice once wrote, and begin a new era of "genuine cooperation." In the U.S. policy elite and media, the nearly unanimous answer is that Russian President Vladimir Putin's antidemocratic domestic policies and "neo-imperialism" destroyed that historic opportunity.

You don't have to be a Putin apologist to understand that this is not an adequate explanation. During the last eight years, Putin's foreign policies have been largely a reaction to Washington's winner-take-all approach to Moscow since the early 1990s, which resulted from a revised U.S. view of how the cold war ended [see Cohen, "*The New American Cold War,*" The Nation, July 10, 2006]. In that new triumphalist narrative, America "won" the forty-year conflict and post-Soviet Russia was a defeated nation analogous to post-World War II Germany and Japan—a nation without full sovereignty at home or autonomous national interests abroad.

The policy implication of that bipartisan triumphalism, which persists today, has been clear, certainly to Moscow. It meant that the United States had the right to oversee Russia's post-Communist political and economic development, as it tried to do directly in the 1990s, while demanding that Moscow yield to U.S. international interests. It meant Washington could break strategic promises to Moscow, as when the Clinton Administration began NATO's eastward expansion, and disregard extraordinary Kremlin overtures, as when the Bush Administration unilaterally withdrew from the ABM Treaty and granted NATO membership to countries even closer to Russia—despite Putin's crucial assistance to the U.S. war effort in Afghanistan after September 11. It even meant America was entitled to Russia's traditional sphere of security and energy supplies, from the Baltics, Ukraine and Georgia to Central Asia and the Caspian.

Such U.S. behavior was bound to produce a Russian backlash. It came under Putin, but it would have been the reaction of any strong Kremlin leader, regardless of soaring world oil prices. And it can no longer be otherwise. Those U.S. policies—widely viewed in Moscow as an "encirclement" designed to keep Russia weak and to control its resources—have helped revive an assertive Russian nationalism, destroy the once strong pro-American lobby and inspire widespread charges that concessions to Washington are "appeasement," even "capitulationism." The Kremlin may have overreacted, but the cause and effect threatening a new cold war are clear.

Because the first steps in this direction were taken in Washington, so must be initiatives to reverse it. Three are essential and urgent: a U.S. diplomacy that treats Russia as a sovereign great power with commensurate national interests; an end to NATO expansion before it reaches Ukraine, which would risk something worse than cold war; and a full resumption of negotiations to sharply reduce and fully secure all nuclear stockpiles and to prevent the impending arms race, which requires ending or agreeing on U.S. plans for a missile defense system in Europe. My recent discussions with members of Moscow's policy elite suggest, whether Russia's real leader is its new President Dmitri Medvedev or Prime Minister Vladimir Putin, that there may still be time for such initiatives to elicit Kremlin responses that would enhance rather than further endanger our national security.

American presidential campaigns are supposed to discuss such vital issues, but neither Senator McCain nor Senator Obama has done so. Instead, in varying degrees, both have promised to be "tougher" on the Kremlin than George W. Bush has allegedly been and to continue the encirclement of Russia and the hectoring "democracy promotion" there, which have only undermined U.S. security and Russian democracy since the 1990s.

To be fair, no influential actors in American politics, including the media, have asked the candidates about any of these crucial issues. They should do so now before another chance is lost, in Washington and in Moscow.

STEPHEN F. COHEN, professor of Russian studies at New York University, is the author (with Katrina vanden Heuvel) of *Voices of Glasnost: Conversations With Gorbachev's Reformers, Failed Crusade: America and the Tragedy of Post-Communist Russia* (both Norton) and, most recently, *Soviet Fates and Lost Alternatives: From Stalinism to the New Cold War* (Columbia).

Why Russia Is So Russian

ANDREW C. KUCHINS

Certain eternal questions—"Whither Russia?" "What to do?" and "Who is to blame?"—have long vexed observers of Russia, as well as Russians themselves. They continue to do so. Russian elites have a ready answer to the question of blame (anybody but Russians), but no consensus prevails on the first two queries, either inside or outside Russia. All nations and peoples have their idiosyncratic features, yet there is something special about Russia that perplexes policy makers and analysts in capitals from Beijing to Tashkent to Berlin. Efforts to penetrate the forces driving Russian behavior are so frustrating that sometimes one can only conclude that the Russian national interest is to bewilder the rest of the world.

It would be highly presumptuous, even perilous, for me to suggest I have any convincing answers to the eternal questions. But I believe certain common themes, core features, and driving factors evident through czarist, Soviet, and post-Soviet Russian history can help elucidate a query that underlies these eternal questions. That is: Why is Russia so . . . Russian?

The Imperial State

Looking at a map of the world, one cannot help but be impressed by Russia's sheer vastness. From the sixteenth century until the beginning of the First World War, the Russian government was in a nearly constant process of state expansion. During the 150 years from the beginning of the sixteenth century through the middle of the seventeenth, Russia added territory every year, on average, equivalent to the Netherlands. No state in world history has expanded so persistently or held onto land more tenaciously than has Russia.

Russia grew as a multi-national and multi-cultural empire at roughly the same time that Western European empires came on the world stage. But at least one important difference distinguished the Russian and the Western European empires of the time. The European empires—Spanish, Portuguese, British, French, Dutch, and so forth—were multinational, but the colonies were overseas and thus clearly separated from their ruling capitals. Russia, on the other hand, was a continental empire, and there was not such a clear differentiation between ruling core and colonial periphery. While the Western European states developed national identities apart from their colonial possessions, Russia never did. In fact, many historians have argued

that Russia never was a nation-state, but rather developed as an empire from the very beginning.

Historically, as one traveled east from Western Europe, regions became progressively poorer and rule more autocratic. Russian governments, in order to compete with a succession of real and potential adversaries from the West, invested much authority in the central sovereign, the czar, and later in the head of the Communist Party of the Soviet Union. Both of these allocated relatively large resources to the military. For much of Russian history such resources were extracted from a chiefly agrarian population that struggled just to maintain near-subsistence standards of living in climatically and geologically adverse conditions.

Whether this militarization was undertaken for offensive or defensive purposes has been a matter of considerable historical debate, but the phenomenon of a centralized and militarized state and society has distinguished Russia for hundreds of years. In addition, Russian expansion created a self-perpetuating dynamic. Russia continually conquered and acquired territory populated by non-Russian ethnic and nationalist groups that formed a contiguous belt or frontier in regions where political loyalty to Moscow was dubious. This encouraged a real or imagined state of permanent insecurity for the core state, whose response was repression and pushing the boundaries yet further out to create bigger buffer zones.

The geography of Eurasia was conducive for the Russians—as it had been for the Golden Horde, Tamerlane, and others before—to create a huge continental empire. The Russians also developed a deeply ingrained sense of territorial security that, while not entirely unique, placed great strains on the central government, creating the need for a large state bureaucracy and military. Commercial trade, economic growth, and technological development consistently lagged in Russia, compared to its European neighbors. Yet its vast natural resources, large territory and population, and ability to mobilize a large army made Russia a formidable player in European politics beginning with the rule of Peter the Great.

After Peter the Great moved the Russian empire's capital to his newly built city near the Baltic Sea, St. Petersburg, Russia continued to expand in the Baltic region thanks to the defeat of Charles XII and Sweden; to the west with the partitioning of Poland; and to the south at the expense of the Ottoman Empire.

In the nineteenth century the expansion continued to the south in the Caucasus and to the southwest in Central Asia.

The historian Edward Keenan has suggested that Moscow showed pragmatic opportunism, that it was not inherently bent on expansion but did take advantage of opportunities as they emerged. In other words, Russia expanded because it could, in an international context of anarchy, in which conflict and wars were commonplace. Keenan argued that Russia expanded more or less in the manner of normal major powers of the period.

The historian George Vernadsky further argued that the peculiar geography of Eurasia makes it a natural setting for an unusually dynamic national grouping, in this case the Russians, to extend its domination as far as possible for security reasons. "The fundamental urge which directed the Russian people eastward lies deep in history," he wrote. "It was not 'imperialism,' nor was it the consequence of the petty political ambitions of Russian statesmen. It was in geography, which lies at the basis of all history." (Pyotr Chaadaev had written in this vein in 1837, in *The Vindication of a Madman,* declaring "Russia, it is a geographical fact.") Keenan's views mesh well with Vernadsky's geographical determinism by suggesting that Russia's expansionism could be explained as normal behavior in an unusual geography.

Manifest Destiny

Richard Pipes and others would suggest, however, that the Russians, and later the Soviets, adapted an ideology—be it "Moscow as the Third Rome" or Marxism-Leninism—that inspired an extraordinary imperial appetite. Keenan's explanation places more emphasis on the external environment that a state faces, while Pipes argues there is something fundamentally different about Russia that leads it to be inherently aggressive and expansionist.

Upon reflection these views—those of Keenan and Vernadsky versus those of Pipes—may not be so contradictory. The geography of Eurasia does present a truly Darwinian dilemma, given its susceptibility to invasion, so it is not surprising that the imperatives deriving from a territorial view of security drove a peculiarly militarized kind of economic development in both czarist Russia and the Soviet Union. At the same time, if such extensive territories are to be dominated for such a long period, a powerful national myth is required. The Russians developed this under the czars, and after a brief "time of troubles" following World War I, the Soviet Union adopted a different but also powerful myth.

In the fifteenth century a potent national myth emerged to fuel a messianic vision for the Russian state and people. It began with the notion of Moscow as the "Third Rome," or the historical protector and purveyor of Orthodox Christianity. The first Rome had fallen centuries before, and in 1453 the "second Rome," Constantinople, fell. In 1472 the Russian Czar Ivan III married Sophia Palaeologina, the niece of Byzantium's last emperor, Constantine, and the marriage supported the idea that Russia was Byzantium's historical successor.

Keenan argues this myth was actually suggested by the post-Renaissance West, as part of efforts to mobilize the Russians to act as a bulwark against the Ottomans. In 1520 the monk Philotheus supposedly wrote a letter to the czar which stated: "And now, I say unto them: Take care and take heed, pious czar; all the empires of Christendom are united in thine, the two Romes have fallen and the third exists and there will not be a fourth."

In 1547 Russian rulers officially adopted the title of "czar," which was derived from the Roman Caesar; this emphasized that the succession of Christian capitals was matched by a succession of rulers. Iver Neumann has argued that, while the doctrine of the Third Rome anointed Russia as the divine successor, the temporal borders of that state were never clearly identified. This provided a powerful religious justification for the expansion of the Third Rome at the expense of territories supposedly abandoned by God in favor of Moscow.

This kind of religious imagery has reemerged on numerous occasions throughout Russian history, often invoking "Holy Russia," with Christlike qualities, as the suffering savior of the world. Russia was imbued with a historical mission; this was the crux of the "Russian idea." The Russian philosopher Nikolai Berdyaev wrote that the Russians and the Jews have "the most vigorous messianic consciousness of all peoples." Berdyaev attributed this messianism in Russians to the unique combination of Western and Eastern qualities that make up the Russian character:

> The Russian people is not purely European and it is not purely Asiatic. Russia is a complete section of the world—a colossal East-West. It unites two worlds, and within the Russian soul two principles are always engaged in strife—the Eastern and the Western.

Enduring Insecurity

While the identity of the state as Marxist-Leninist was important throughout the Soviet period, Soviet leaders' perceptions of security were dominated by the more traditional dilemmas of geography and power. The Soviet Union that emerged in the 1920s from seven debilitating years of World War I and the Russian Civil War was economically devastated and physically smaller than its czarist predecessor state, a situation somewhat analogous to the Russian Federation after the demise of the Soviet Union in 1991. Joseph Stalin was especially sensitive to the impact of Soviet economic and technological backwardness on military power as the European security situation grew increasingly perilous in the 1930s. In 1930 Stalin warned that if the Soviet Union did not rapidly industrialize it would be overrun once again, as Russia had been numerous times in its history:

> To slow down the tempo [of industrialization] means to lag behind. And those who lag behind are beaten. The history of Old Russia shows . . . that because of her backwardness she was constantly being defeated. By the Mongol Khans, by the Polish-Lithuanian gentry, by the Anglo-French

Capitalists. . . . Beaten because of backwardness—military, cultural, political, industrial, and agricultural backwardness. . . . We are behind the leading countries by fifty-to-one-hundred years. We must make up this distance in ten years. Either we do it or we go under.

The experience of the 1930s and World War II strengthened Stalin's territorial view of international security, and his obsession with territorial security fueled the cruel symbiosis of Soviet domestic and foreign policies from the 1930s onward. An internal regime of unprecedented terror in the 1930s was justified by the supposed prevalence and influence of capitalist spies and saboteurs who conspired to destroy the Soviet regime, just as the capitalist powers had tried to "choke the baby in its crib" with the allied intervention in 1918.

Moscow was a deeply surreal place in the 1930s, when a series of show trials condemned to death many of the leaders of the Bolshevik Revolution who were falsely accused of, among other things, espionage. Vladimir Lenin had referred to czarist Russia as "the prisonhouse of nations," but the purges and the creation of the gulag system in the Soviet Union under Stalin were far more brutal than the oppression under czarism.

With the defeat of the Nazis in May 1945, Stalin stood triumphant as no Russian leader had since Alexander I after the defeat of Napoleon in 1812. During the war the terror subsided and the leadership made ideological concessions to appeal to patriotic Russian nationalism. For the first time in its history (anticipating the imminent defeat of Japan), the Soviet Union was not under threat from any major power. The dean of U.S. Soviet specialists, George Kennan, wrote in May 1945 from Moscow, where he served in the American embassy:

By the time the war in the Far East is over Russia will find herself, for the first time in her history, without a single great power rival on the Eurasian landmass. She will also find herself in physical control of vast new areas of this landmass: some of them areas to which Russian power had never before been extended. These new areas (although their exact frontiers are deliberately kept vague) will probably contain well over 100 million souls—most of them in the European sector. Those are developments of enormous import in the development of the Russian state.

It was not an unconscious slip on Kennan's part that he referred repeatedly to "Russia" rather than the "Soviet Union." He strongly believed that traditional Russian nationalist goals and concerns were the most informative guide for understanding Soviet foreign policy under Stalin.

The breakup of the wartime alliance and the onset of the cold war marked a seminal event that defined the core structure of international relations until the collapse of the Soviet Union. As with the eventual defeat of the Golden Horde in the fifteenth century and the defeat of Napoleon in the nineteenth century, Russia during World War II had made huge sacrifices to "save the West" from a continental hegemon—this time, Hitler. Stalin believed the West should pay its debt by allowing the Soviet Union to expand its domination to extensive territories of Eastern and Central Europe, including a neutralized Germany. The Soviet victory over Germany was the event that cemented the legitimacy of the Soviet regime, and the regime did its utmost, through propaganda and education, to ensure that the citizens did not forget.

After that, the development of a Soviet identity was increasingly defined by the nation's position in a superpower confrontation with the United States. Moscow and Washington each maintained a series of alliance relationships in Europe and Asia and they held each other in a balance of nuclear terror. This balance of power remained critical even as Soviet leader Leonid Brezhnev sought détente with the United States and Western Europe. Détente was defined as a relaxation in tension; it resulted in important arms control agreements, including the 1972 Anti-Ballistic Missile Treaty and the SALT I and II treaties. The Brezhnev administration's achievement of nuclear parity with the United States helped consolidate the Soviet international identity as a superpower of equal standing with the United States.

Yet even at the peak of its powers, the Soviet Union, as Robert Legvold noted in 1977, was like a "deformed giant . . . mighty in its military resources and exhilarated by its strength, but backward in other respects." Paradoxically, the Soviet Union of the 1980s was simultaneously a global superpower and a third world country. In 1989, one Soviet official described his country as "Upper Volta with rockets."

The notion of the Soviet Union's relative decline in comparison with the West is essential to understanding the motivations for economic, social, and political reform during the perestroika years. It is rather difficult to imagine that the Soviet leaders would have embarked on reform if their economy had been growing rapidly while the West stagnated. Indeed, the catalyst for perestroika was Mikhail Gorbachev's perception that the Soviet Union in the 1980s was in a "pre-crisis situation." The Soviet president's motivations were somewhat similar to Peter the Great's at the end of the seventeenth century.

The New Thinking

As the perestroika years proceeded into the late 1980s, it became increasingly clear that the new thinkers' attacks on the traditional Soviet approach to foreign policy were aimed primarily at reducing the impact of the military on Soviet security and foreign policy, and then by logical extension ending the military's lock on domestic economic resources. The Gorbachev team profitably used diplomatic success with the West to help support efforts to redefine Soviet security demands and so to justify reductions in defense spending.

However, Gorbachev's failure consistently to implement an effective set of economic reforms left him without the promised benefits of increased economic production and subsequent trade with and investment from the West. In fact, the very half-hearted measures toward economic reform effectively destroyed the previously inefficient but functioning system. They helped reduce Soviet markets to chaos and the state to near bankruptcy.

Russian history before the collapse of communism reveals, in short, core continuities that formed organizing principles of domestic and foreign policy from Muscovite to czarist to Soviet Russia. These deep grooves include highly centralized and unaccountable political authority, weak and often virtually nonexistent institutions of private property and rule of law, and a great power mentality that is deeply militarized as well as colored by messianism and xenophobia. Russia's experience of either being in or preparing for war for most of its history, coupled with its unique geography, engendered a highly territorially based sense of security. This in turn drove Russia's impulse to dominate near neighbors and to expand "buffer zones" against presumed and potential enemies.

The revolution of 1991, led by Boris Yeltsin and his team of reformers, marked the most concentrated and successful effort in Russia's more than thousand-year history to break out of its traditional patrimonial and imperial paradigm. In retrospect, it seems remarkable that this effort has been as successful as it has been, given the disastrous conditions at the starting point. The new Russia was effectively bankrupt. With the demise of the institutional glue of the Communist Party of the Soviet Union, state power and authority were gravely weakened.

And probably most devastating was the inheritance of an economic system and infrastructure that for 70 years had been based on nonmarket principles, resulting in perhaps the greatest misallocation of resources in history. The late economist Gregory Grossman captured the magnitude of this legacy in the 1980s when he described the Soviet economy as "negative value-added," and suggested that, from an economic point of view, Russia would make better use of its resources by simply shutting down its entire misdeveloped industrial structure.

Yeltsin and his team were exuberantly optimistic, too optimistic as it turned out, about the future of Russian foreign policy and Russia's place in the world order. Immediately after being elected president in June 1991, Yeltsin came to the United States and formulated his vision of Russian-American relations based on common interests—for example, creating a "common political and economic system in the Northeastern hemisphere in which the United States and Russia would play a leading role." These dreams of deep partnership evaporated all too quickly, as the reformers' economic and foreign policy plans came under increasing political attacks. In the spring of 1992, the Yeltsin administration's "shock therapy" approach was already in rapid retreat, even before much shock had been administered.

Putin's Thermidor

Since Vladimir Putin became president in 2000, the more traditional themes that marked the continuity between Russian czarist and Soviet foreign policy have gradually come to predominate. First is the virtual obsession with Russia's status as a great power that both deserves respect and is entitled to "privileged relations" with its neighbors. Certainly the traditional sense of territorial security drives Moscow's approach to its neighbors, and control of or hegemony over vast swaths of territory remains a key factor in Russia's sense of what constitutes a great power. In addition, the recentralization of political and economic power has been cast as a security imperative. (The militarization of the Russian economy, a core link between

czarist and Soviet domestic orders and their highly security-driven foreign policies, has, so far, not taken place.)

Russia's extraordinary economic recovery since 1999 has fueled its transformation from a reluctant follower in the 1990s to an obstructionist power with aspirations to revise the world order from unipolar to multipolar. The country's recent economic setbacks, a consequence of global recession and declining oil prices, have not altered this perspective as yet.

Indeed, recovery from the economic crisis of the 1990s is only part of Moscow's rather Darwinist outlook on the increasing tilt in the global economic balance of power toward large emerging-market economies and hydrocarbon producers—two categories in which Russia figures prominently. Thirty years ago when the Group of Seven was formed to manage the global economy, its member countries constituted more than 60 percent of the world economy; today those countries are no longer so dominant and account for only 40 percent.

During the 1990s, President Bill Clinton and President Yeltsin believed that they shared the same Western values, which could be described as "market democracy." Today, the Russian leadership no longer subscribes to those values. The country's capitalism persists but is increasingly becoming state capitalism, and political freedom in the Western sense has been curtailed.

Given these changes in its perspective, Moscow was bound to reevaluate its interests in the international system. Then-President Putin did so starkly in his famous February 10, 2007, speech at the *Wehrkunde* Security Conference in Munich. He made essentially two points: first, that the United States was behaving irresponsibly in managing global affairs, and second, that the international system of American hegemony was evaporating and being replaced by genuine multipolarity. Most commentary focused on the first point and missed the importance of the second, which Putin summarized:

> The combined GDP measured in purchasing power parity of countries such as India and China is already greater than that of the United States. And a similar calculation with the GDP of the BRIC countries (Brazil, Russia, India, and China) surpasses the cumulative GDP of the EU. And according to experts this gap will only increase in the future. There is no reason to doubt that the economic potential of the new centers of global economic growth will inevitably be converted into political influence and will strengthen multipolarity.

Putin and his colleagues elaborated on this theme in a series of important speeches in 2007, and the call for a "new international architecture" of global governance became a campaign theme of the Russian parliamentary and presidential electoral cycle of 2007–08.

Russians are right to point out that institutions of global governance are anachronistic and often ineffective. However, Russians' own capacity to contribute to a solution is less obvious because of their emotionally charged view of the past 20 years. The Kremlin considers many changes to the international system since the late 1980s as illegitimate, because Russia was too weak to assert its positions. In its narrative the West—mainly the United States—took advantage of Russian weakness through NATO enlargement in 1997, the bombing of Yugoslavia in 1999, the abandonment of the Anti-Ballistic Missile Treaty in 2001, the endorsement of regime change (the "color revolutions") on Russia's borders in 2003 and 2004, the promotion of missile defense, and the recognition of Kosovo in 2008. The Russian elite sees these Western moves as detrimental to Russia's national interests.

Russian leaders see themselves as "realists," and describe their foreign policy as pragmatic and driven by national interests. When they discuss international relations, they rarely talk of "public goods" or "norms," and they receive U.S. and European references to them with cynicism or, more often, with defensive hostility about "double standards." They view U.S. efforts to promote American "values" as hypocritical justifications for the promotion of U.S. interests—and, ultimately, U.S. influence and hegemony.

Russian leaders frame their country's international cooperation in terms of realpolitik bargains.

In place of norms and public goods, Russian leaders and political analysts frame their country's international cooperation in terms of realpolitik bargains and "tradeoffs" of interests. If the United States wants Russia to take a stronger position in isolating Iran, Washington is expected to compensate Moscow by ending NATO enlargement or, as was announced recently, halting missile defense plans for Central Europe.

One of Russia's oft-repeated grievances is the U.S. betrayal of a supposed "gentleman's agreement" between George H.W. Bush and Gorbachev in 1990 to allow the unification of Germany as long as NATO would not deploy new bases on the territory of former Warsaw Pact countries. U.S. officials contest the Russian interpretation, thus illustrating the problem with such unwritten exchanges.

The Medvedev Doctrine

The Russian government holds one norm dear, that of national sovereignty—but it applies it very selectively. Russian policy is itself rife with double standards when it comes to the sovereignty of countries like Georgia and Ukraine. President Dmitri Medvedev made this eminently clear in his September 2008 remarks on Russian television presenting the five principles that would guide his country's foreign policy:

First, Russia will comply in full with all of the provisions of international law regarding relations between civilized countries.

Second, Russia believes in the need for a multipolar world and considers that domination by one country is unacceptable, no matter which country this may be.

Third, we are naturally interested in developing full and friendly relations with all countries—with Europe, Asia, the United States, Africa, with all countries in the world. These relations will be as close as our partners are ready for.

Fourth, I see protecting the lives and dignity of Russian citizens, wherever they may be, as an indisputable priority for our country, and this is one of our foreign policy priorities.

Fifth, I think that, like any other country, Russia pays special attention to particular regions, regions in which it has privileged interests. We will build special relations with the countries in these regions, friendly relations for the long-term period.

Western analysts interpreted Medvedev's speech as aiming at a diminished role for NATO and the Organization for Security and Cooperation in Europe. But his formulation, which analysts have dubbed "the Medvedev Doctrine," is also a striking contrast with the idealistic universalism that marked Gorbachev's "new political thinking" of the late Soviet period. And it bears a strong resemblance to traditional "realist" balance-of-power thinking.

A security framework that would allow for Russia's special spheres of interest sounds straight out of the nineteenth century playbook of great powers.

A European security framework that would allow for Russia's "privileged relations" with neighbors and special spheres of interest sounds straight out of the nineteenth century playbook of great powers, including the American Monroe Doctrine, which justified the United States' repeated violations of the sovereignty of its neighbors. Such anachronistic notions are nonstarters in twenty-first century Europe, where the trend is toward common and cooperative security institutions.

A contradiction exists between Russia's domestic goals for economic growth and its insistence on hyper-sovereignty.

An even more fundamental contradiction exists between Russia's domestic goals for economic growth and its increasingly belligerent insistence on hyper-sovereignty. Russia is more integrated today in the global economy than it has ever been. However, as its ambitious strategic economic goals for 2020 make clear, the best-case growth scenarios for Russia require much deeper integration with the West—first and foremost Europe, but also the United States and Japan.

These partners are far more important to Russia for trade, investment, technology, and management transfer than are the Commonwealth of Independent States, China, Iran, Venezuela, and the rest of the world. Yet, despite Russia's deepening economic integration and the imperative for more such integration, Moscow's political ties with the West have been worsening in recent years. This is an unsustainable contradiction that runs counter to Russia's national interests.

ANDREW C. KUCHINS is a senior fellow and director of the Russia and Eurasia program at the Center for Strategic and International Studies. His most recent book, coauthored with Anders Aslund, is *The Russia Balance Sheet* (Peterson Institute for International Economics and CSIS, 2009), from which this essay is adapted.

Fear Comes to the Russian Heartland

It may be too late for Putin to avoid trouble in hundreds of rust-belt 'monotowns.'

Owen Matthews and Anna Nemtsova

Traces of the boom years linger in Magnitogorsk, a steel town of nearly 500,000 straddling the Ural River, about 900 miles east of Moscow. Boutiques, beauty salons and sushi bars remain open along grimy downtown avenues named for Lenin, Marx and Engels. A vast poster of Prime Minister Vladimir Putin—until recently the symbol of the nation's prosperity and stability—covers the façade of a brand-new shopping mall, overshadowed by the Magnitogorsk Metallurgical Plant. But prosperity and stability are vanishing fast in Magnitogorsk. World steel prices have fallen by half since last summer. Four of the city's eight giant steel smelters have been shut down, their workers placed on indefinite leave. In Magnitogorsk and other hard-hit cities across the Russian heartland, Putin is becoming a symbol of something else: the Kremlin's failed promises.

While the global economy was pushing oil prices to new highs, Putin could do no wrong. Many Russians only shrugged as he crushed democratic opposition and strangled the independent media. Now serious unrest seems imminent despite those autocratic moves. "We are expecting mass unemployment and mass riots," says Gennady Gudkov, a former KGB colonel and current chairman of the Duma's Security Committee. "There will be not enough police to stop people's protests by force."

The biggest threat lies in cities like Magnitogorsk—company towns that are dominated by a single industry. Russia has an unusually high number of them: a quarter of the population is estimated to live and work in one of roughly 1,500 "monotowns," as they are called. They're relics of the old Soviet-era command economy, dating back to Stalin's efforts to create an industrialized nation overnight. They are especially vulnerable in this frightening global downturn. "Our plant has been the main employer in this city for 80 years," says MMK (the steel plant's Russian initials) spokeswoman Elena Azovtseva. "When this plant stops, everything here stops."

The fear is that the slowdown may have only begun. "Every second ruble earned in this city is made at the metal plant," says Boris Shtelter, a consultant for small businesses in Magnitogorsk. "The plant also provides most of the heating, electricity and water supplies for the city." Workers in hundreds of one-company towns are protesting job losses, inflation, government corruption and unpaid wages. The anger is spreading. In December, riot police had to be flown to Vladivostok, on Russia's Pacific coast, to crush demonstrations against new taxes on imported cars. Within the past two weeks, thousands have taken to the streets in Moscow, Vladivostok, Novosibirsk, Volgograd and Ulan-Ude. In Vladivostok they stood, shouting anti-Putin slogans, in temperatures of close to zero Fahrenheit.

The Kremlin evidently sees worse trouble ahead. In December it shelved plans to retire 280,000 Army officers (part of a sweeping military reform). A long-expected reduction in the number of Interior Ministry troops was also abruptly canceled. At the same time, the Interior Ministry set up a special command center in Moscow, packed with surveillance equipment designed to deal with street unrest. The Duma, on Kremlin instructions, added seven new articles to the criminal code. One expands the definition of high treason and espionage to include advisory "and other" assistance to foreign and international organizations; another makes "participating in mass disorders" such as the one in Vladivostok a "crime against the state." More sinister still, defendants accused of the new crimes can only be tried by a special court of three judges, not by a jury—a system reminiscent of the Stalin-era troika courts that sent millions to the Gulag. "This is a very dangerous development," says Lyudmila Alexeyeva, a veteran human-rights activist and head of the independent Moscow Helsinki Group. "It returns the Russian justice system to the norms of the 1920s."

Putin isn't relying solely on brute force. The government has handed out more than $13.5 billion in soft loans to Russian companies since October. Last month Putin unveiled an additional $7.6 billion bailout plan aimed at nearly 300 of the biggest monotowns. Major beneficiaries of both handouts include oligarchs like MMK's owner, Victor Rashnikov, and Russia's current No. 1 billionaire, Oleg Deripaska, owner of the AvtoVaz auto plant. But those efforts are fatally mired in corruption, says Gudkov. "We already hear angry towns complaining that bureaucrats are demanding kickbacks in exchange for aid," he says. "In a country where everybody steals, the government should not be pumping cash into the monotowns."

Instead, he suggests, the funds would be more wisely spent on tax cuts and cheaper credit.

The question may be academic. Moscow's rescue plans depend on money saved up from the boom years, but there's too little left to go around. The ruble has lost more than 25 percent of its value since August, while foreign-currency reserves have plunged from more than $700 billion to less than $400 billion. The country's 2009 budget would balance if oil were $70 a barrel, but the current price is closer to half that. Worse, the government is now reallocating some $40 billion of the bailout money to save hundreds of banks from failure—and even then, the Central Bank says up to half the country's private banks will be gone in a year's time. Though most small depositors' savings are government guaranteed, many businesses are set to suffer, as will confidence in the banking system in general.

It may be too late anyway to avoid real trouble in the monotowns. The government is planning to raise utility prices for all homeowners this year. A similar increase was the cause of the last serious round of social unrest in 2005. This past November the only two industries in Verkhny Ufalei, a town of 70,000 not far from Magnitogorsk, stopped paying salaries. Workers who protested were told they were lucky to have jobs at all. "All it takes is one spark for the entire Urals to catch fire," says Magnitogorsk labor activist Sergei Vasilyev. Yuri Babylyev, an activist with the Magnitogorsk City Organization of Workers, reports that the authorities' indifference to the plight of laid-off workers is fueling resentment and driving more and more onto the streets. "The plant's managers treat us like domestic animals, like slaves," he says. "Their corruption is pushing us toward massive public protests." In a recent poll by the Moscow-based Levada Center, 20 percent said they were ready to join public protests in the coming months. "The time has come for us to act," says Vasilyev. "Our employers force us to sign papers saying that we quit our jobs voluntarily. The courts are corrupt, and nobody defends workers' rights. Putin cares for oligarchs and their pockets more than for us."

Economists have warned for decades that Russia's system of company towns was dangerously vulnerable to market fluctuations. Even so, the Kremlin failed to help them diversify while times were flush. Some residents worried that the boom wouldn't last. Bulat Stolyarov, director of the Institute of Regional Policy in Moscow, says leaders in several monotowns, including Chelyabinsk, Krasnoyarsk and Magnitogorsk, sought permission from the central government to use budget surpluses to create citywide stabilization funds—but Moscow refused.

Sometimes even ordinary people could sense how fragile their prosperity was. Vitaly Polishuk, a window-frame maker in Magnitogorsk, says he consciously took comfort from the sight of the steel mills' smoke-belching chimneys. "We could see our breadwinner breathing," he recalls. Now half the chimneys in town have fallen idle. "When the smoke disappears, fear comes," says Polishuk. It's coming to millions across Russia who face unemployment and poverty. And it's coming to the Kremlin, which fears that all its troops and rescue funds won't be enough to control angry people who believed in their leaders' promises of wealth, national greatness and political stability.

Pride and Power

RICHARD PIPES

Russia is obsessed with being recognized as a "Great Power." She has felt as one since the 17th century, after having conquered Siberia, but especially since her victory in World War II over Germany and the success in sending the first human into space. It costs nothing to defer to her claims to such exalted status, to show her respect, to listen to her wishes. From this point of view, the recent remarks about Russia by Vice President Joe Biden in an interview with this newspaper were both gratuitous and harmful. "Russia has to make some very difficult calculated decisions," he said. "They have a shrinking population base, they have a withering economy, they have a banking sector that is not likely to be able to withstand the next 15 years."

These remarks are not inaccurate but stating them publicly serves no purpose other than to humiliate Russia. The trends the vice president described will likely make Russia more open to cooperating with the West, Mr. Biden suggested. It is significant that when our secretary of state tried promptly to repair the damage which Mr. Biden's words had caused, *Izvestiia,* a leading Russian daily, proudly announced in a headline, "Hillary Clinton acknowledges Russia as a Great Power."

Russia's influence on world affairs derives not from her economic power or cultural authority but her unique geopolitical location. She is not only the world's largest state with the world's longest frontier, but she dominates the Eurasian land mass, touching directly on three major regions: Europe, the Middle East and the Far East. This situation enables her to exploit to her advantage crises that occur in the most populous and strategic areas of the globe. For this reason, she is and will remain a major player in world politics.

Opinion polls indicate that most Russians regret the passing of the Soviet Union and feel nostalgia for Stalin. Of course, they miss not the repression of human rights which occurred under Communism nor the miserable standards of living but the status of their country as a force to be reckoned with: a country to be respected and feared. Under present conditions, the easiest way for them to achieve this objective is to say "no" to the one undeniable superpower, the United States. This accounts for their refusal to deal more effectively with Iran, for example, or their outrage at America's proposal to install rocket defenses in Poland and the Czech Republic. Their media delight in reporting any negative news about the United States, especially the dollar, which they predict will soon be worthless (even as their central bank holds $120 billion or 30% of its reserves in dollar-denominated U.S. securities).

One unfortunate consequence of the obsession with "great power" status is that it leads Russians to neglect the internal conditions in their country. And here there is much to be done. To begin with: the economy. The Russian aggression against Georgia has cost it dearly in terms of capital flight. Due to the decline in the global prices of energy, which constitute around 70% of Russian exports, exports in the first half of 2009 have fallen by 47%. The stock market, which suffered a disastrous decline in 2008, has recovered, and the ruble has held steady, but the hard currency reserves are melting and the future does not look promising: The latest statistics indicate that Russia's GDP this year will fall by 7%. It is this that has prompted President Dmitry Medvedev recently to demand that Russia carry out a major restructuring of her economy and end her heavy reliance on energy exports. "Russia needs to move forward," he told a gathering of parliamentary party leaders, "and this movement so far does not exist. We are marking time and this was clearly demonstrated by the crisis . . . as soon as the crisis occurred, we collapsed. And we collapsed more than many other countries."

One of the major obstacles to conducting business in Russia is the all-pervasive corruption. Because the government plays such an immense role in the country's economy, controlling some of its most important sectors, little can be done without bribing officials. A recent survey by Russia's Ministry of the Interior revealed, without any apparent embarrassment, that the average amount of a bribe this year has nearly tripled compared to the previous year, amounting to more than 27,000 rubles or nearly $1,000. To make matters worse, businesses cannot rely on courts to settle their claims and disputes, and in extreme cases resort to arbitration.

The political situation may appear to a foreigner inculcated with Western values as incomprehensible. Democratic institutions, while not totally suppressed, play little role in the conduct of affairs defined by the leading ideologist of the regime as "sovereign democracy." Indeed, President Medvedev has publicly declared his opposition to "parliamentary democracy" on the grounds that it would destroy Russia.

A single party, One Russia, virtually monopolizes power, assisted by the Communists and a couple of minor affiliates. Parliamentary bodies duly pass all bills presented to them by the government. Television, the main source of news for the vast country, is monopolized by the state. One lonely radio station and a few low-circulation newspapers are allowed freedom of expression in order to silence dissident intellectuals. And yet, the population at large seems not to mind this political arrangement—an acquiescence which runs contrary to the Western belief that all people crave the right to choose and direct their government.

The solution of the puzzle lies in the fact that during their 1,000-year old history of statehood, the Russians have virtually never been given the opportunity to elect their government or to influence its actions. As a result of this experience, they have become thoroughly depoliticized. They do not see what positive influence the government can have on their lives: They believe that they have to fend for themselves. Yes, they will gladly accept social services if offered, as they had been under the Soviet government, but they do not expect them. They hardly feel themselves to be citizens of a great state, but confine their loyalties to their immediate families and friends and the locality which they inhabit. From opinion polls it emerges that they believe democracy everywhere to be a sham, that all governments are run by crooks who use their power to enrich themselves. What they demand of the authorities is that they maintain order: when asked what is more important to them—"order" or "freedom"—the inhabitants of the province of Voronezh overwhelmingly expressed preference for "order." Indeed, they identify political freedom, i.e., democracy, with anarchy and crime. Which explains why the population at large, except for the well-educated, urban minority, expresses no dismay at the repression of its political rights.

One aspect of the "great power" syndrome is imperialism. In 1991, Russia lost her empire, the last remaining in the world, as all her colonies, previously disguised as "union republics" separated themselves to form sovereign states. This imperial collapse was a traumatic experience to which most Russians still cannot adjust themselves. The reason for this lies in their history. England, France, Spain and the other European imperial powers formed their empires overseas and did so after creating national states: As a result, they never confused their imperial possessions with the mother country. Hence, the departure of the colonies was for them relatively easy to bear. Not so in the case of Russia. Here, the conquest of the empire occurred concurrently with the formation of the nation-state: Furthermore, there was no ocean to separate the colonies. As a result, the loss of empire caused confusion in the Russians' sense of national identity. They have great difficulty acknowledging that the Ukraine, the cradle of their state, is now a sovereign republic and fantasize about the day when it will reunite with Mother Russia. They find it only slightly less difficult to acknowledge the sovereign status of Georgia, a small state that has been Russian for over two centuries. The imperial complex underpins much of Russia's foreign policy.

These imperial ambitions have received fresh expression from a bill which President Medvedev has submitted in mid-August to parliament. It would revise the existing Law of Defense which authorizes the Russian military to act only in response to foreign aggression. The new law would allow them to act also "to return or prevent aggression against another state"

and "to protect citizens of the Russian Federation abroad." It is easy to see how incidents could be provoked under this law that would allow Russian forces to intervene outside their borders.

How does one deal with such a difficult yet weighty neighbor, a neighbor who can cause no end of mischief if it becomes truly obstreperous? It seems to me that foreign powers ought to treat Russia on two distinct levels: one, which takes into consideration her sensitivities; the other, which responds to her aggressiveness.

We are right in objecting strenuously to Russia treating her one-time colonial possessions not as sovereign countries but dependencies lying in her "privileged zone of influence." Even so, we should be aware of their sensitivity to introducing Western military forces so close to her borders. The Russian government and the majority of its citizens regard NATO as a hostile alliance. One should, therefore, be exceedingly careful in avoiding any measures that would convey the impression that we are trying militarily to "encircle" the Russian Federation. After all, we Americans, with our Monroe Doctrine and violent reaction to Russian military penetration into Cuba or any other region of the American continent, should well understand Moscow's reaction to NATO initiatives along its borders.

This said, a line must be drawn between gentle manners and the hard realities of politics. We should not acquiesce in Russia treating the countries of her "near abroad" as satellites and we acted correctly in protesting last year's invasion of Georgia. We should not allow Moscow a veto over the projected installation of our anti-rocket defenses in Poland and the Czech Republic, done with the consent of their governments and meant to protect us against a future Iranian threat. These interceptors and radar systems present not the slightest threat to Russia, as confirmed publicly by Russian general Vladimir Dvorkin, an officer with long service in his country's strategic forces. The only reason Moscow objects to them is that it considers Poland and the Czech Republic to lie within its "sphere of influence."

Today's Russians are disoriented: they do not quite know who they are and where they belong. They are not European: This is attested to by Russian citizens who, when asked. "Do you feel European?" by a majority of 56% to 12% respond "practically never." Since they are clearly not Asian either, they find themselves in a psychological limbo, isolated from the rest of the world and uncertain what model to adopt for themselves. They try to make up for this confusion with tough talk and tough actions. For this reason, it is incumbent on the Western powers patiently to convince Russians that they belong to the West and should adopt Western institutions and values: democracy, multi-party system, rule of law, freedom of speech and press, respect for private property. This will be a painful process, especially if the Russian government refuses to cooperate. But, in the long run, it is the only way to curb Russia's aggressiveness and integrate her into the global community.

RICHARD PIPES is Frank B. Baird Jr. professor of history, emeritus, at Harvard University. In 1981 and 1982 he served as Director of East European and Soviet Affairs in President Reagan's National Security Council.

Speech at the 43rd Munich Conference on Security Policy

VLADIMIR PUTIN

Thank you very much dear Madam Federal Chancellor, Mr Teltschik, ladies and gentlemen!

I am truly grateful to be invited to such a representative conference that has assembled politicians, military officials, entrepreneurs and experts from more than 40 nations.

This conference's structure allows me to avoid excessive politeness and the need to speak in roundabout, pleasant but empty diplomatic terms. This conference's format will allow me to say what I really think about international security problems. And if my comments seem unduly polemical, pointed or inexact to our colleagues, then I would ask you not to get angry with me. After all, this is only a conference. And I hope that after the first two or three minutes of my speech Mr Teltschik will not turn on the red light over there.

Therefore. It is well known that international security comprises much more than issues relating to military and political stability. It involves the stability of the global economy, overcoming poverty, economic security and developing a dialogue between civilisations.

This universal, indivisible character of security is expressed as the basic principle that "security for one is security for all." As Franklin D. Roosevelt said during the first few days that the Second World War was breaking out: "When peace has been broken anywhere, the peace of all countries everywhere is in danger."

These words remain topical today. Incidentally, the theme of our conference—global crises, global responsibility—exemplifies this.

Only two decades ago the world was ideologically and economically divided and it was the huge strategic potential of two superpowers that ensured global security.

This global stand-off pushed the sharpest economic and social problems to the margins of the international community's and the world's agenda. And, just like any war, the Cold War left us with live ammunition, figuratively speaking. I am referring to ideological stereotypes, double standards and other typical aspects of Cold War bloc thinking.

The unipolar world that had been proposed after the Cold War did not take place either.

The history of humanity certainly has gone through unipolar periods and seen aspirations to world supremacy. And what hasn't happened in world history?

However, what is a unipolar world? However one might embellish this term, at the end of the day it refers to one type of situation, namely one centre of authority, one centre of force, one centre of decision-making.

It is a world in which there is one master, one sovereign. And at the end of the day this is pernicious not only for all those within this system, but also for the sovereign itself because it destroys itself from within.

And this certainly has nothing in common with democracy. Because, as you know, democracy is the power of the majority in light of the interests and opinions of the minority.

Incidentally, Russia—we—are constantly being taught about democracy. But for some reason those who teach us do not want to learn themselves.

I consider that the unipolar model is not only unacceptable but also impossible in today's world. And this is not only because if there was individual leadership in today's—and precisely in today's—world, then the military, political and economic resources would not suffice. What is even more important is that the model itself is flawed because at its basis there is and can be no moral foundations for modern civilisation.

Along with this, what is happening in today's world—and we just started to discuss this—is a tentative to introduce precisely this concept into international affairs, the concept of a unipolar world.

And with what results?

Unilateral and frequently illegitimate actions have not resolved any problems. Moreover, they have caused new human tragedies and created new centres of tension. Judge for yourselves: wars as well as local and regional conflicts have not diminished. Mr Teltschik mentioned this very gently. And no less people perish in these conflicts—even more are dying than before. Significantly more, significantly more!

Today we are witnessing an almost uncontained hyper use of force—military force—in international relations, force that is plunging the world into an abyss of permanent conflicts. As a result we do not have sufficient strength to find a comprehensive solution to any one of these conflicts. Finding a political settlement also becomes impossible.

We are seeing a greater and greater disdain for the basic principles of international law. And independent legal norms are, as a matter of fact, coming increasingly closer to one state's legal system. One state and, of course, first and foremost the United States, has overstepped its national borders in every way. This is visible in the economic, political, cultural and educational policies it imposes on other nations. Well, who likes this? Who is happy about this?

In international relations we increasingly see the desire to resolve a given question according to so-called issues of political expediency, based on the current political climate.

And of course this is extremely dangerous. It results in the fact that no one feels safe. I want to emphasise this—no one feels safe! Because no one can feel that international law is like a stone wall that will protect them. Of course such a policy stimulates an arms race.

The force's dominance inevitably encourages a number of countries to acquire weapons of mass destruction. Moreover, significantly new threats—though they were also well-known before—have appeared, and today threats such as terrorism have taken on a global character.

I am convinced that we have reached that decisive moment when we must seriously think about the architecture of global security.

And we must proceed by searching for a reasonable balance between the interests of all participants in the international dialogue. Especially since the international landscape is so varied and changes so quickly—changes in light of the dynamic development in a whole number of countries and regions.

Madam Federal Chancellor already mentioned this. The combined GDP measured in purchasing power parity of countries such as India and China is already greater than that of the United States. And a similar calculation with the GDP of the BRIC countries—Brazil, Russia, India and China—surpasses the cumulative GDP of the EU. And according to experts this gap will only increase in the future.

There is no reason to doubt that the economic potential of the new centres of global economic growth will inevitably be converted into political influence and will strengthen multipolarity.

In connection with this the role of multilateral diplomacy is significantly increasing. The need for principles such as openness, transparency and predictability in politics is uncontested and the use of force should be a really exceptional measure, comparable to using the death penalty in the judicial systems of certain states.

However, today we are witnessing the opposite tendency, namely a situation in which countries that forbid the death penalty even for murderers and other, dangerous criminals are airily participating in military operations that are difficult to consider legitimate. And as a matter of fact, these conflicts are killing people—hundreds and thousands of civilians!

But at the same time the question arises of whether we should be indifferent and aloof to various internal conflicts inside countries, to authoritarian regimes, to tyrants, and to the proliferation of weapons of mass destruction? As a matter of fact, this was also at the centre of the question that our dear colleague Mr Lieberman asked the Federal Chancellor. If I correctly understood your question (addressing Mr Lieberman), then of course it is a serious one! Can we be indifferent observers in view of what is happening? I will try to answer your question as well: of course not.

But do we have the means to counter these threats? Certainly we do. It is sufficient to look at recent history. Did not our country have a peaceful transition to democracy? Indeed, we witnessed a peaceful transformation of the Soviet regime—a peaceful transformation! And what a regime! With what a number of weapons, including nuclear weapons! Why should we start bombing and shooting now at every available opportunity? Is it the case when without the threat of mutual destruction we do not have enough political culture, respect for democratic values and for the law?

I am convinced that the only mechanism that can make decisions about using military force as a last resort is the Charter of the United Nations. And in connection with this, either I did not understand what our colleague, the Italian Defence Minister, just said or what he said was inexact. In any case, I understood that the use of force can only be legitimate when the decision is taken by NATO, the EU, or the UN. If he really does think so, then we have different points of view. Or I didn't hear correctly. The use of force can only be considered legitimate if the decision is sanctioned by the UN. And we do not need to substitute NATO or the EU for the UN. When the UN will truly unite the forces of the international community and can really react to events in various countries, when we will leave behind this disdain for international law, then the situation will be able to change. Otherwise the situation will simply result in a dead end, and the number of serious mistakes will be multiplied. Along with this, it is necessary to make sure that international law have a universal character both in the conception and application of its norms.

And one must not forget that democratic political actions necessarily go along with discussion and a laborious decision-making process.

Dear ladies and gentlemen!

The potential danger of the destabilisation of international relations is connected with obvious stagnation in the disarmament issue.

Russia supports the renewal of dialogue on this important question.

It is important to conserve the international legal framework relating to weapons destruction and therefore ensure continuity in the process of reducing nuclear weapons.

Together with the United States of America we agreed to reduce our nuclear strategic missile capabilities to up to 1700–2000 nuclear warheads by 31 December 2012. Russia intends to strictly fulfill the obligations it has taken on. We hope that our partners will also act in a transparent way and will refrain from laying aside a couple of hundred superfluous nuclear warheads for a rainy day. And if today the new American Defence

Minister declares that the United States will not hide these superfluous weapons in a warehouse or, as one might say, under a pillow or under the blanket, then I suggest that we all rise and greet this declaration standing. It would be a very important declaration.

Russia strictly adheres to and intends to further adhere to the Treaty on the Non-Proliferation of Nuclear Weapons as well as the multilateral supervision regime for missile technologies. The principles incorporated in these documents are universal ones.

In connection with this I would like to recall that in the 1980s the USSR and the United States signed an agreement on destroying a whole range of small- and medium-range missiles but these documents do not have a universal character.

Today many other countries have these missiles, including the Democratic People's Republic of Korea, the Republic of Korea, India, Iran, Pakistan and Israel. Many countries are working on these systems and plan to incorporate them as part of their weapons arsenals. And only the United States and Russia bear the responsibility to not create such weapons systems.

It is obvious that in these conditions we must think about ensuring our own security.

At the same time, it is impossible to sanction the appearance of new, destabilising high-tech weapons. Needless to say it refers to measures to prevent a new area of confrontation, especially in outer space. Star wars is no longer a fantasy—it is a reality. In the middle of the 1980s our American partners were already able to intercept their own satellite.

In Russia's opinion, the militarisation of outer space could have unpredictable consequences for the international community, and provoke nothing less than the beginning of a nuclear era. And we have come forward more than once with initiatives designed to prevent the use of weapons in outer space.

Today I would like to tell you that we have prepared a project for an agreement on the prevention of deploying weapons in outer space. And in the near future it will be sent to our partners as an official proposal. Let's work on this together.

Plans to expand certain elements of the anti-missile defence system to Europe cannot help but disturb us. Who needs the next step of what would be, in this case, an inevitable arms race? I deeply doubt that Europeans themselves do.

Missile weapons with a range of about five to eight thousand kilometres that really pose a threat to Europe do not exist in any of the so-called problem countries. And in the near future and prospects, this will not happen and is not even foreseeable. And any hypothetical launch of, for example, a North Korean rocket to American territory through western Europe obviously contradicts the laws of ballistics. As we say in Russia, it would be like using the right hand to reach the left ear.

And here in Germany I cannot help but mention the pitiable condition of the Treaty on Conventional Armed Forces in Europe.

The Adapted Treaty on Conventional Armed Forces in Europe was signed in 1999. It took into account a new geopolitical reality, namely the elimination of the Warsaw bloc. Seven years have passed and only four states have ratified this document, including the Russian Federation.

NATO countries openly declared that they will not ratify this treaty, including the provisions on flank restrictions (on deploying a certain number of armed forces in the flank zones), until Russia removed its military bases from Georgia and Moldova. Our army is leaving Georgia, even according to an accelerated schedule. We resolved the problems we had with our Georgian colleagues, as everybody knows. There are still 1,500 servicemen in Moldova that are carrying out peacekeeping operations and protecting warehouses with ammunition left over from Soviet times. We constantly discuss this issue with Mr Solana and he knows our position. We are ready to further work in this direction.

But what is happening at the same time? Simultaneously the so-called flexible frontline American bases with up to five thousand men in each. It turns out that NATO has put its frontline forces on our borders, and we continue to strictly fulfill the treaty obligations and do not react to these actions at all.

I think it is obvious that NATO expansion does not have any relation with the modernisation of the Alliance itself or with ensuring security in Europe. On the contrary, it represents a serious provocation that reduces the level of mutual trust. And we have the right to ask: against whom is this expansion intended? And what happened to the assurances our western partners made after the dissolution of the Warsaw Pact? Where are those declarations today? No one even remembers them. But I will allow myself to remind this audience what was said. I would like to quote the speech of NATO General Secretary Mr Woerner in Brussels on 17 May 1990. He said at the time that: "the fact that we are ready not to place a NATO army outside of German territory gives the Soviet Union a firm security guarantee." Where are these guarantees?

The stones and concrete blocks of the Berlin Wall have long been distributed as souvenirs. But we should not forget that the fall of the Berlin Wall was possible thanks to a historic choice—one that was also made by our people, the people of Russia—a choice in favour of democracy, freedom, openness and a sincere partnership with all the members of the big European family.

And now they are trying to impose new dividing lines and walls on us—these walls may be virtual but they are nevertheless dividing, ones that cut through our continent. And is it possible that we will once again require many years and decades, as well as several generations of politicians, to dissemble and dismantle these new walls?

Dear ladies and gentlemen!

We are unequivocally in favour of strengthening the regime of non-proliferation. The present international legal principles allow us to develop technologies to manufacture nuclear fuel for peaceful purposes. And many countries with all good reasons want to create their own nuclear energy as a basis for their energy independence. But we also understand that these technologies can be quickly transformed into nuclear weapons.

This creates serious international tensions. The situation surrounding the Iranian nuclear programme acts as a clear example. And if the international community does not find a reasonable solution for resolving this conflict of interests, the world will continue to suffer similar, destabilising crises because there are more threshold countries than simply Iran. We both know this. We are going to constantly fight against the threat of the proliferation of weapons of mass destruction.

Last year Russia put forward the initiative to establish international centres for the enrichment of uranium. We are open to the possibility that such centres not only be created in Russia, but also in other countries where there is a legitimate basis for using civil nuclear energy. Countries that want to develop their nuclear energy could guarantee that they will receive fuel through direct participation in these centres. And the centres would, of course, operate under strict IAEA supervision.

The latest initiatives put forward by American President George W. Bush are in conformity with the Russian proposals. I consider that Russia and the USA are objectively and equally interested in strengthening the regime of the non-proliferation of weapons of mass destruction and their deployment. It is precisely our countries, with leading nuclear and missile capabilities, that must act as leaders in developing new, stricter non-proliferation measures. Russia is ready for such work. We are engaged in consultations with our American friends.

In general, we should talk about establishing a whole system of political incentives and economic stimuli whereby it would not be in a states' interests to establish their own capabilities in the nuclear fuel cycle but they would still have the opportunity to develop nuclear energy and strengthen their energy capabilities.

In connection with this I shall talk about international energy cooperation in more detail. Madam Federal Chancellor also spoke about this briefly—she mentioned, touched on this theme. In the energy sector Russia intends to create uniform market principles and transparent conditions for all. It is obvious that energy prices must be determined by the market instead of being the subject of political speculation, economic pressure or blackmail.

We are open to cooperation. Foreign companies participate in all our major energy projects. According to different estimates, up to 26 percent of the oil extraction in Russia—and please think about this figure—up to 26 percent of the oil extraction in Russia is done by foreign capital. Try, try to find me a similar example where Russian business participates extensively in key economic sectors in western countries. Such examples do not exist! There are no such examples.

I would also recall the parity of foreign investments in Russia and those Russia makes abroad. The parity is about fifteen to one. And here you have an obvious example of the openness and stability of the Russian economy.

Economic security is the sector in which all must adhere to uniform principles. We are ready to compete fairly.

For that reason more and more opportunities are appearing in the Russian economy. Experts and our western partners are objectively evaluating these changes. As such, Russia's OECD sovereign credit rating improved and Russia passed from the fourth to the third group. And today in Munich I would like to use this occasion to thank our German colleagues for their help in the above decision.

Furthermore. As you know, the process of Russia joining the WTO has reached its final stages. I would point out that during long, difficult talks we heard words about freedom of speech, free trade, and equal possibilities more than once but, for some reason, exclusively in reference to the Russian market.

And there is still one more important theme that directly affects global security. Today many talk about the struggle against poverty. What is actually happening in this sphere? On the one hand, financial resources are allocated for programmes to help the world's poorest countries—and at times substantial financial resources. But to be honest—and many here also know this—linked with the development of that same donor country's companies. And on the other hand, developed countries simultaneously keep their agricultural subsidies and limit some countries' access to high-tech products.

And let's say things as they are—one hand distributes charitable help and the other hand not only preserves economic backwardness but also reaps the profits thereof. The increasing social tension in depressed regions inevitably results in the growth of radicalism, extremism, and feeds terrorism and local conflicts. And if all this happens in, shall we say, a region such as the Middle East where there is increasingly the sense that the world at large is unfair, then there is the risk of global destabilisation.

It is obvious that the world's leading countries should see this threat. And that they should therefore build a more democratic, fairer system of global economic relations, a system that would give everyone the chance and the possibility to develop.

Dear ladies and gentlemen, speaking at the Conference on Security Policy, it is impossible not to mention the activities of the Organisation for Security and Cooperation in Europe (OSCE). As is well-known, this organisation was created to examine all—I shall emphasise this—all aspects of security: military, political, economic, humanitarian and, especially, the relations between these spheres.

What do we see happening today? We see that this balance is clearly destroyed. People are trying to transform the OSCE into a vulgar instrument designed to promote the foreign policy interests of one or a group of countries. And this task is also being accomplished by the OSCE's bureaucratic apparatus which is absolutely not connected with the state founders in any way. Decision-making procedures and the involvement of so-called non-governmental organisations are tailored for this task. These organisations are formally independent but they are purposefully financed and therefore under control.

According to the founding documents, in the humanitarian sphere the OSCE is designed to assist country members in observing international human rights norms at their request. This is an important task. We support this. But this does not mean interfering in the internal affairs of other countries, and especially not imposing a regime that determines how these states should live and develop.

It is obvious that such interference does not promote the development of democratic states at all. On the contrary, it makes them dependent and, as a consequence, politically and economically unstable.

We expect that the OSCE be guided by its primary tasks and build relations with sovereign states based on respect, trust and transparency.

Dear ladies and gentlemen!

In conclusion I would like to note the following. We very often—and personally, I very often—hear appeals by our partners, including our European partners, to the effect that Russia should play an increasingly active role in world affairs.

In connection with this I would allow myself to make one small remark. It is hardly necessary to incite us to do so. Russia is a country with a history that spans more than a thousand years and has practically always used the privilege to carry out an independent foreign policy.

We are not going to change this tradition today. At the same time, we are well aware of how the world has changed and we have a realistic sense of our own opportunities and potential. And of course we would like to interact with responsible and independent partners with whom we could work together in constructing a fair and democratic world order that would ensure security and prosperity not only for a select few, but for all.

Thank you for your attention.

Speech by Vladimir Putin, delivered at the 43rd Munich Conference on Security Policy, February 10, 2007. (Public Domain)

The Incredible Shrinking People

Russians are dying out, with dire consequences.

You do not need to travel far to find evidence of Russia's demographic problems. Just 250km west of Moscow, in the Smolensk region, it is glaringly obvious. Turn off the main road in the village of Semlevo and you will see rusting gates and derelict buildings. Continue for a short distance on what just about resembles a road and you will see a new cowshed.

A third of the people working here are from Tajikistan. Marina, a 38-year old Russian milkmaid, is married to one of them. Talib, she says, may be strict, but he does not drink or beat her up. She prefers him to her first, Russian, husband who drank himself to death at the age of 47, leaving her with three children. Her 19-year old son is unemployed and drinks heavily.

Talib came to Russia so that he can feed his other family in Tajikistan, and Marina does not mind Talib having another wife. But local residents resent the Tajik and Uzbek migrants because "they are prepared to work for a pittance and take our jobs." Yet finding sober local working men in the village is difficult, says Sergei Pertsev, the farm manager. "Here, everyone has fallen ill with alcohol." To make sure the farm functions properly Mr Pertsev keeps people in reserve, to fill in for those who go on a binge. It used to be mainly men who drank, but now women do too.

Semlevo's collective farm was built in 1962. Since then the village's population has dropped to a third of its former level. Thirty years ago the village school had 500 pupils. This year only one girl entered the first grade. Some people have left the village, others have died of drink. Those who remain drink heavily. Still, by local standards, Semlevo, with its 900 residents, is a thriving metropolis. Some nearby villages have just two or three people left.

Tatyana Nefedova, a geographer and specialist on Russian agriculture, calls these deserted areas "Russia's black hole." In the European part of the country alone they account for one-third of the land mass. Urbanisation has drawn people from villages into larger cities and to the vast industrial building sites in the east and north of the country. Active life is concentrated in a radius of 35–40km from the centre of these large cities. Russia has only 168 cities with a population over 100,000 and their number is dropping. The average distance between large cities is 185km. According to Ms Nefedova, this means that a stretch of 100km between them is a social and economic desert. In villages closer to large cities, especially Moscow and St Petersburg, or in the south of the country, things are better.

But for Mr Pertsev, the idea that Russia is "rising from its knees" seems like a bad joke. Two years ago he lost his son, who was in the army. The young man was killed by a drunk driver who crashed into a column of soldiers in the dark. "The only people who live well in this country are those who make decisions, sit on an [oil] pipe or those who guard it," says Mr Pertsev.

The sorry state of villages like Semlevo is the result of "negative social selection," says Ms Nefedova: the most active and able people have migrated to large towns. Few people have stayed behind, and most of those are unable to work. In Semlevo there is only one farmer who keeps his own sheep and chickens. Most houses there have no running water, plumbing or gas heating. Still, Semlevo's old collective farm is considered lucky: it was recently bought by a businesswoman from Moscow. Most other collective farms in this district are dead.

Russia's demography befits a country at war. The population of 142m is shrinking by 700,000 people a year. By 2050 it could be down to 100m. The death rate is double the average for developed countries. The life expectancy of Russian males, at just 60 years, is one of the lowest in the world. Only half of Russian boys now aged 16 can expect to live to 60, much the same as at the end of the 19th century.

No Babies to Kiss

"If this trend continues, the survival of the nation will be under threat," Mr Putin said in his first state-of-the-nation address in 2000. Six years later he offered an increase in child support and a bonus for second babies. Since then the birth rate has started to climb, the number of deaths has declined and life expectancy has edged up a little (see Figure 1, next page). Mr Putin rejoiced: "We have overcome the trend of rising deaths and falling births. . . . In the next three to four years we can stabilise the population figures."

But demographers say there are few grounds for optimism, and Russia's goal of increasing the population to 145m is unattainable. Anatoly Vishnevsky, Russia's leading demographer, says an increase in the number of births in a single year does not reverse a trend. People may respond to financial incentives by

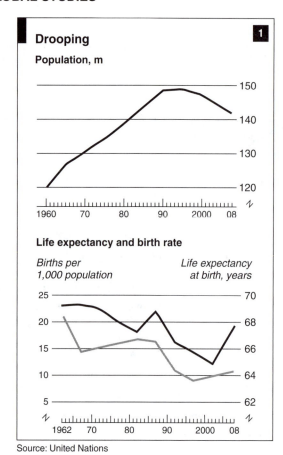

Drooping [1]

Population, m

150
140
130
120

1960 70 80 90 2000 08

Life expectancy and birth rate

Births per 1,000 population

Life expectancy at birth, years

25 — 70
20 — 68
15 — 66
10 — 64
5 — 62

1962 70 80 90 2000 08

Source: United Nations

changing their timing rather than having more babies overall. An extra $100 a month is helpful to people with low incomes and rural or Muslim families, who have more children anyway, but is unlikely to persuade middle-class families to produce more babies.

The main reason for the recent rise is that there was an uptick in the birth rate in the 1980s and the people born during that period are now having children themselves. But the next generation to reach child-bearing age is much smaller. "What we are going to see over the next few years is a rapid decline in the number of births," says Mr Vishnevsky. At the same time the death rate is likely to go up, not because people will die any earlier but because the generation which is getting to the end of its life now is larger than the one born during the war.

Russia's demographic crisis is one of the main constraints on the country's economy. Although Russia's population has been ageing, over the past decade the country has enjoyed a "demographic dividend" because the age structure was in its favour. This dividend has now been exhausted and the population of working age will decline by about 1m a year, increasing the social burden on those that remain. Over the next seven years Russia's labour force will shrink by 8m, and by 2025 it may be 18m–19m down on the present figure of 90m.

What makes a shrinking population dangerous for a country that has always defined itself by its external borders is the loss of energy it entails, Mr Vishnevsky argues. The Soviet Union did not just try to exploit the resources of its vast and inhospitable land, it tried to populate it. Now large swaths of land in Siberia and the far east are emptying out as people move to central Russia. The population density in the country's far east is 1.1 people per square kilometre. On the other side of the border with China it is nearly 140 times that figure.

The decline in Russia's population is often linked to the collapse of the Soviet Union. In fact, the pattern was set back in the mid-1960s when the number of births fell below replacement level and life expectancy started to shrink. In 1964, after several years of post-Stalin thaw, life expectancy for men was 65.1 years, only slightly lower than in the West. But by 1980 the gap with the West had widened to more than eight years and is now 15 years.

Russia's health problems, says Mr Vishnevsky, were partly a legacy of the cold war. By the middle of the 20th century the developed world had learnt to control infections that killed large numbers of people. The next targets were illnesses caused by lifestyle, such as heart attacks, pollution and respiratory diseases. But whereas the West invested heavily in health-care systems and better lifestyles, Russia was putting its financial and human capital into the arms race and industrialisation.

If life expectancy in Russia had improved at the same pace as in the West, the country would have had an extra 14.2m people between 1966 and 2000, adding 10% to the population. The Soviet Union's spending on health care was less than a quarter of the American figure. The Communist Party elite was well looked after, but ordinary people were less fortunate.

Crucially, the paternalistic Soviet system, which survives in today's Russia, was geared towards fighting epidemics and infections rather than to empowering people to look after their own health. Even now Russian doctors treat patients and their relatives like imbeciles. Officially the state guarantees free care for all, but half the patients offer gifts and money to doctors and the other half often have to forgo necessary treatment.

Boosting the country's health-care spending was one of the "national projects" Mr Medvedev was in charge of before becoming president. Last year its funding doubled to 143 billion rubles ($5.8 billion). That is still below Western standards, but the main problem is that it is not well spent. For example, the government more than doubled general practitioners' pay. But as Sergei Shishkin of the Independent Institute for Social Policy argues, the pay increase was not linked to performance and created a sense of injustice among specialists.

The Curse of the Bottle

Russian history, particularly in the 20th century, has encouraged the view that life is cheap. But there is also a strong self-destructive streak in the national character. Drinking yourself to death is one of the most widely used methods of suicide.

Alexander Nemtsov, a senior researcher at the Institute of Psychiatry, points to a clear correlation between the death rate and the consumption of alcohol in Russia. A short anti-alcohol campaign conducted by Mikhail Gorbachev in the late 1980s extended life expectancy by three years. Mr Nemtsov estimates that nearly 30% of all male deaths and 17% of female deaths are directly or indirectly caused by excess alcohol consumption and that over 400,000 people a year die needlessly from drink-related causes, ranging from heart disease to accidents, suicides and murders.

The average Russian gets through 15.2 litres of pure alcohol a year, twice as much as is thought to be compatible with good health. The problem lies not just with how much but also with what is drunk: moonshine and "dual-purpose" liquids, such as perfume and windscreen wash, make up a significant proportion of alcohol consumption, according to Russia's chief physician, Gennady Onishchenko. Tens of thousands a year die of alcohol poisoning, against a few hundred in America. In large cities the fear of losing a job, and growing car ownership, is keeping people soberer.

The most obvious reason why Russians drink so much is the low price and easy availability of alcohol. Consumption increased dramatically in the 1960s when the state hugely boosted production. People started drinking not just on special occasions but during the week and at work too.

Vodka is one of a very few Russian products that seem relatively immune to inflation. Between 1990 and 2005, for example, the food-price index increased almost four times faster than the alcohol-price index. A cheap bottle of vodka in Russia costs the same as two cans of beer or two litres of milk. The easiest way to curb consumption would be to make hard spirits much more expensive and less accessible, as many Nordic countries have done. But as with many other things in Russia, corruption gets in the way: two-thirds of hard liquor is produced illegally and sold untaxed.

Occasionally the government raises the alarm about alcohol poisoning, but it does little to curb drinking. Instead it has declared war on Georgian wine and mineral water, which it claims is not fit for consumption. But life expectancy in Georgia remains 12 years higher than in Russia.

Unlike drinking, AIDS is a relatively new problem for Russia. The first case of HIV was recorded in 1987, but it took a long time for the country to take notice. By 1997 the number of cases had grown to 7,000. Now the official figure is over 430,000, the largest in Europe. The real number could be double that, according to the World Health Organisation. Most victims are under the age of 30. Some two-thirds are drug-takers, but the epidemic is now spreading to the general public.

The government seems to have woken up to the danger and has increased spending. But Vadim Pokrovsky, head of the federal AIDS centre, says Russia still lacks adequate prevention measures.

About 28,000 people have already died of AIDS-related illnesses, but the real number could be masked by a co-infection of HIV and tuberculosis which kills people after two or three months. The incidence of TB is the highest in Europe. Last year 24,000 people died of the disease, almost 40 times as many as in America, not least because most TB hospitals are crumbling and some lack sewerage or running water.

The only solution to Russia's demographic problems appears to be immigration, as in the village of Semlevo, but the Russian public is hostile to it.

Wanted but Not Welcomed

Most low-skilled migrant workers in Russia come from Central Asia. In the east of the country they are mainly Chinese. The precise figures are impossible to pin down because the vast majority of immigrants over the past decade have been illegal. Until recently they were treated much like serfs. They could not apply for work permits but had to rely on their employers, who would often impound their passports and refuse to pay them for their work. Thousands of corrupt police officers grew fat on the proceeds.

In the past couple of years the rules have become more accommodating and migrant workers can now apply for their own work permits and sign contracts with their employers. But for tax reasons only a quarter of immigrants do so. The new law has increased the number of legal migrants to more than 2m, but the real figure is thought to be five times that.

Many migrants are scared to venture outside on their own for fear of running into police or skinheads. Employers much prefer illegal immigrants to local workers because they are cheaper and will put up with worse conditions. But a xenophobic public habitually vents its anger on the immigrants, even though they are estimated to generate 8% of Russia's GDP.

Support for extremist organisations such as the Movement Against Illegal Immigration has risen sharply in the past few years. More than half the population supports the slogan "Russia for the Russians" and almost 40% feel "irritation, discomfort or fear" towards migrants from Central Asia and Azerbaijan. Hate crimes are on the rise. SOVA, a body that monitors racism, last year counted 667 racists attacks, including 86 racially motivated murders.

At a recent pep talk with the editors of Russia's leading media, Mr Putin urged them to guard against xenophobia. Russia's new day of unity is traditionally marked by ultra-nationalist marches, and the youth wing of Mr Putin's own United Russia party campaigns against immigrants. Dmitry Rogozin, a nationalist politician who built his campaign for parliament in 2003 on anti-immigrant rhetoric, is now Russia's ambassador to NATO. On his office wall hangs a portrait of Stalin.

NATO and Russia: Partnership or Peril?

The Western alliance has no reason to fear its members' 'defecting' to Moscow, and it has every reason to engage with the Russians on common security concerns.

DMITRI TRENIN

Among the many anniversaries of world events marked in 2009, NATO's founding 60 years ago stands out. At the start of the cold war, the North Atlantic Treaty Organization was built to provide for the external as well as internal security of its member states. By pooling West European defense efforts, it put an end to intra-European wars. By permanently involving the United States and Canada with Western Europe, it created a security community spanning the North Atlantic: the modern world's first zone of stable peace.

Since the end of the Western-Soviet confrontation, NATO has not withered away—it has evolved, alongside the European Union, into a premier pillar of European security. The transatlantic link has withstood both the loss of a common adversary and divisions among the allies. The alliance has demonstrated, in the Balkans, a resolve for military action on its periphery and, with its involvement in Afghanistan, a capability to project power into the heart of another continent. Meanwhile, NATO membership has expanded to almost double its level at the end of the cold war.

Twenty years after the fall of the Berlin Wall, however, one major piece of unfinished post–cold war business remains: fitting the former Soviet lands into a pan-European security framework. The heart of the issue is Russia's absence from the European and Euro-Atlantic security structures. Moscow's one-time favorite among intergovernmental bodies, the Organization for Security and Cooperation in Europe (OSCE), has ceased to be an ongoing dialogue platform and has failed to live up to its title. This institutional deficiency affects not only Russia, but also its neighbors, such as Ukraine and Georgia. The brief war in the Caucasus in August 2008 and the tensions it produced in Crimea—which continue to linger and may produce another crisis in the future—point to the reality and potential severity of the problem.

There is no simple way to resolve this conundrum. Russia's membership in the NATO alliance, sought by Moscow in the 1990s and again explored in the early part of this decade, is not a realistic proposition for the foreseeable future, if ever. Above all, it is not realistic at this stage to expect Russia to join a U.S.-led alliance such as NATO is today, and it is even less realistic to

anticipate some sort of NATO coleadership by the two nuclear superpowers. Even if one of these highly unlikely conditions were met, Russia's hypothetical accession would needlessly exacerbate Russia's own, and the West's, relations with China, much to the detriment of global stability and security.

Thus, since no shortcut is possible, the West and Russia need to embark on a long, tortuous, and potentially rocky path toward creating a security community in Europe that would include both NATO members and nonmembers. Russian President Dmitri Medvedev's idea of revamping European security, which he first announced before the Georgia war but has amplified since, is useful not so much because he calls for a new, legally binding treaty on security, but because it represents a de facto invitation to an ongoing dialogue. NATO needs to see the importance and urgency of the situation, seize the opportunity, and generate forward-looking ideas of its own.

Talk to Each Other

In the 12 years since the NATO-Russia Founding Act, which called for cooperative relations between NATO and Moscow, and the 7 years since establishment of the NATO-Russia Council (NRC), which created an official diplomatic vehicle for cooperation, the relationship between the alliance and its biggest neighbor has not lived up to the expectations of 1997 or 2002. The NRC, instead of becoming the instrument of Western-Russian security interaction, has turned into a mostly technical workshop—useful, but extremely narrow in scope. The major contentious issues in European security, such as Kosovo, the South Caucasus, all the "frozen conflicts" in former Soviet republics, and ballistic missile defense, have not been constructively discussed and dealt with in the NRC context. This needs to change if NATO means to avoid a new crisis down the road.

At minimum, the NRC is the place to engage the Russians in serious discussions, both formal and informal, on issues of common concern. It is important that the Russians do not feel that a common front of Western allies is ganging up on them: This was the fatal flaw of the NRC's predecessor, the Permanent Joint Council. Making the Russians feel that they are equals among equals in the joint body would do more for Europe's security than making sure that all allies speak with one voice during NRC sessions. The Western alliance has no reason to fear its members' "defecting" to Moscow, and it has every reason to engage with the Russians on common security concerns.

To be sure, sharp disagreements occasionally will occur. That is why it is important that the NRC become an all-weather operation. This has not been the case so far. In the wake of NATO's 1999 Kosovo campaign, Russia suspended the work of the Permanent Joint Council. Nine years later, NATO followed suit by severing contacts with Russia as punishment for Moscow's "peace enforcement" in Georgia. The latter action created an absurd situation. NATO representatives were talking more *about* Russia among themselves, but no one was talking *to* the Russians. Isolated diplomatically, Dmitry Rogozin, the Russian ambassador to NATO, took his case to journalists. He had to break a wall in his office to accommodate ever more members of the press who attended his lively media conferences.

For Russia and NATO, keeping in touch and hearing each other out are essential, but the key task is to lay down elements of confidence in their badly, even dangerously frayed relationship. The urgency of the task is not apparent to all. NATO's current focus is very much on Afghanistan, as a decade ago it was on the Balkans. Seen from Brussels, Russia has almost receded over the horizon. Yet, east of Berlin, the NATO alliance continues to be perceived as "being about Russia." This is the view shared in Moscow and Minsk, Tallinn and Tbilisi, even if views of NATO itself differ greatly in those capitals. Recognizing this situation and managing it will be crucial for the security of Europe's east. Ukraine and Georgia are the cases in point.

The Hard Cases

The real problem with Ukraine's bid for NATO membership is not so much Russia's opposition—nor is it, of course, NATO's spurious threat to Russia's security. The actual issue is Ukraine itself. If that nation of 46 million were overwhelmingly pro-NATO, no force in the world, and certainly not the Kremlin, could prevent it from acceding to the alliance, provided it met the relevant criteria. However, since in Ukraine, too, NATO is largely "about Russia," the Ukrainian population faces a dilemma that it cannot resolve for the time being.

Perhaps one-quarter to one-third of Ukrainians—like most Poles, Balts, and Romanians—believe that their country needs NATO as a security hedge against Russia, the historical hegemon in the region. However, just over one-half of Ukrainians—*unlike* the Poles, Lithuanians, and so forth—view Russia as part of an extended family. To regard Russia as a notional adversary, which is fine for the majority of Estonians, is unnatural for people in Odessa and Kharkiv and perverse for the Russian-populated Sevastopol. To force such a stark choice on a nation so split on the issue would be to court disaster. In particular, it would reignite Crimean separatism and make Russian interference virtually unavoidable.

Since a near-totality of Ukrainians do not want to be part of Russia, while a majority do not want to part with Russia,

the best way to handle the Ukrainian security issue is along the lines of progressive integration with, and ultimately into, Europe. The brunt of this task needs to be carried by the Ukrainians themselves. Fortunately, a broad consensus exists in Ukrainian society that Ukraine belongs in Europe. Less fortunately, since the Orange Revolution of 2004, Ukraine's elites have not done nearly enough to implement reforms that can actually make Ukraine a modern European country. Ukrainians, of course, will need all the help they can get. The European Union's Eastern Partnership Program, inaugurated in May 2009, is a useful, albeit modest, step toward providing help.

As for NATO membership, it would be wise to take up this issue only when a comfortable majority in Ukraine, including in Crimea, favors such a step. If Moscow feels confident that the issue of NATO membership is de facto left to the people of Ukraine to decide and not rushed by pro-Western nationalists, it can be expected to proceed with termination of the Black Sea Fleet's presence in Crimea when the current lease expires in 2017.

Georgia's NATO problem is different from Ukraine's. Most Georgians want NATO precisely because of a keenly perceived threat from Russia. They are in the same league as the Central and Eastern Europeans who fear Russia—only their fear is more acute, and their anger less well contained. And unlike the new NATO members, Georgia is not protected by a formal Western security guarantee. However, admission of Georgia into the alliance according to Georgia's internationally recognized borders would put NATO in danger of a direct conflict with Russia, which no longer recognizes those borders. Conflicts within Georgia, or what used to be Georgia, now create a potential for conflict between Georgia and Moscow's two formal allies in the region, the breakaway provinces of Abkhazia and South Ossetia, which host Russian regular forces and border guards.

All the Western references to Georgia's territorial integrity notwithstanding, Russia will not go back on its recognition of Abkhazia and South Ossetia. However, one must remember that Abkhazia did not secede from Georgia in 2008, when Moscow recognized it. It seceded a decade and a half before. South Ossetia, which also broke away from Tbilisi in the early 1990s, is a more awkward case. The option of a peaceful reintegration was seriously undermined by Georgian President Mikheil Saakashvili's ill-fated attempt in 2004 to bring the territory back by police force. Georgia probably lost this option entirely as a result of its reckless attack on the territories in 2008. But, unlike Abkhazia, South Ossetia does not appear capable of evolving into a viable state. The status quo is likely to continue for quite some time.

This situation is fraught with constant dangers. Russia's military deployments in Abkhazia and South Ossetia—and Moscow's angry reaction to NATO's May 2009 Partnership for Peace exercise in Georgia—make the South Caucasus a sore spot in NATO-Russia relations. The termination of a United Nations mission in Abkhazia and of an OSCE mission in South Ossetia, caused by Russia's insistence that both missions recognize the authority of the two new states, leaves EU monitors deployed on the Georgian side as the only third-party observer in the zones of conflict.

Hence NATO needs to support conflict prevention in the area. Predictability, transparency, and consultation with

Moscow can help avoid misunderstanding, as was evidenced by U.S. efforts to explain to Moscow the mission of its military trainers who returned to Georgia in September 2009, a year after the war. Military disengagement in the zones of conflict; confidence- and security-building measures; protection of minorities and general support for human rights in the region: All of these need to be constantly discussed within the NATO-Russia Council, even though the final resolution of Abkhazia and South Ossetia's status is certainly years away.

Practical ways could and should be found around the apparently incompatible legal positions of the West and Russia on the status of the two territories. Provided the political will exists, diplomats can be very resourceful. It was possible in 1971, for example, to conclude a critically important agreement that defused tensions around West Berlin—even when the Soviet Union and three Western powers could not agree on whether the title of the document should refer to "Berlin" or "West Berlin." In the end, it was just called a Quadripartite Agreement. On the other hand, lack of attention to conflict prevention in the South Caucasus might lead to new violence, perhaps even international hostilities.

Expand the Agenda

The NRC's agenda needs to be expanded to include other items as well. One is strategic missile defense, which is both a contentious issue on the U.S.-Russian agenda and a potential area of bilateral collaboration. For some time, NATO and Russia have been successfully cooperating on theater missile defenses. It is in the interest of the alliance, as well as Western-Russian relations, that topics related to both theater and strategic missile defense be brought together under the auspices of the NRC. Russia objected strongly to U.S. plans for deploying missile defenses in Poland and the Czech Republic. Washington took an important step in September when it decided to scrap those plans.

U.S. President Barack Obama's visit to Moscow in July 2009 resulted in a vague and general statement on the subject, which nevertheless opened the door for missile defense cooperation. Depending on progress in U.S.-Russian strategic arms talks, missile defense could become the flagship project of NATO-Russian cooperation. If successful, this cooperation would start the long trek away from nuclear deterrence and mutually assured destruction as the foundations of Russian-Western security relations, and toward something that could become a security community extending across the entire Euro-Atlantic area: the old Vancouver-to-Vladivostok formula.

Another issue to come under NRC review is conventional arms control in Europe. As things now stand, the NATO countries have not ratified the recently revised Conventional Forces in Europe (CFE) treaty, due to Russia's noncompliance with its separate commitment to withdraw forces from Moldova (Transnistria) and Georgia. Russia, for its part, has suspended its participation in the original 1990 document, which was based on the idea that the Warsaw Pact still existed. Moscow is particularly incensed by the treaty's flank limitations, which constrain Russian forces in the south of Russia, including the North Caucasus, in terms of numbers and movements. Moscow sees these limits as both discriminatory in principle and constraining in practice. The original CFE treaty also leaves out the Balkans and the Baltic states—a clear flaw.

Negotiated as a cold war confidence-building measure, the CFE treaty served as a kind of reinsurance policy, a material basis for classic military stability on the continent. Despite its many inadequacies, it compensated somewhat for the asymmetry represented by most of the region's countries belonging to one alliance, NATO, while others are outside it *and* divided in their attitude toward it. At present, no one benefits from the NATO countries' failing to ratify the revised treaty and Russia's countering by suspending it. Revisiting this subject, and opening a serious discussion of ways out of the present impasse, would be a useful step in confidence building.

Help with Afghanistan?

If NATO and Russia could maintain a modicum of mutual confidence when it comes to Europe, Western-Russian cooperation beyond Europe would in turn be facilitated. For example, cooperative missile defenses, if they were conceived and constructed, would protect Europe from missile risks from the greater Middle East.

Regarding Afghanistan, no Russian troops will be deployed there, of course. But a further expansion of transit supply routes across Russian territory—beyond what was agreed in the past two years with America, Germany, and France, and amplified at the 2009 U.S.-Russian summit—is a distinct possibility. Antidrug cooperation is another area of genuine mutual interest. NATO rightly sees the illicit drug trade as a major source of funding for the Taliban insurgency in Afghanistan. And the Russians are reeling from proliferation of Afghan-produced drugs inside Russia, where the number of drug addicts has shot up in the past decade.

Russia, which has been providing training, in cooperation with NATO, to Afghan antinarcotics officers, is also a source of training and equipment for the Afghan government's military forces. As more responsibility eventually is shifted to the Afghan government, the importance of this connection could increase. Russia, with its knowledge of and experience in the country, could also contribute to Afghanistan's economic reconstruction. Moscow has a genuine interest in preventing the emergence of an extremist regime in the region.

In Central Asia, Russia is a formal ally of several countries, and it keeps some military presence there (in Tajikistan and Kyrgyzstan). For years, Russia has been seeking NATO's recognition of the Moscow-based Collective Security Treaty Organization (CSTO) as the Atlantic alliance's security partner in Eurasia, from Belarus to Central Asia. Until now, the Western alliance has been reluctant to extend such recognition, in order not to bolster Moscow's geopolitical role in the former Soviet Union. NATO and the United States, moreover, have sought to capitalize on bilateral relations with all the countries concerned and have seen no need for an additional link with the Russia-led organization.

This stance requires a hard look. The United States, despite its own presence and contacts in Central Asia, stands to lose, not to gain, from a geostrategic competition with Russia in the region, or even the appearance of competition. Washington may spend only a little time and energy keeping track of Russia's moves on the historical turf of the "Great Game." In Russian eyes, however, America is reenacting that very game. The United States does not intend to stay in Afghanistan indefinitely; it does not envision bearing all the burden itself. It needs others assuming more responsibility, not less.

As to concerns about Moscow's dominating Central Asia, Russia is no Soviet Union and the CSTO is no Warsaw Pact. Not only self-confident Kazakhstan, but even the tiny and cash-strapped Kyrgyzstan, feels free to deal with the West without first clearing everything with its nominal Russian ally. The 2009 story of the Manas Air Base is revealing: Kyrgyzstan first asked the United States to vacate it, apparently pleasing Moscow, but eventually allowed the Americans to keep it under a slightly different label, while giving Russia an option of establishing a small base of its own in the country. In any case, a regular NATO-CSTO contact is more likely to help the latter become a more efficient structure, and to give its Central Asian members a higher platform and more of a voice. And NATO-CSTO contacts are more likely to strengthen security in this strategically important and vulnerable region than they are to resuscitate Moscow's domination of Central Asia.

At a macro-regional level, Russia—together with China, India, and Iran—belongs to a group of major powers key to the establishment of stability in the area. NATO would be right to engage in a structured dialogue with the Shanghai Cooperation Organization, which includes China and Russia as members and India, Iran, and Pakistan as observers. A NATO link with that organization, alongside the NATO-CSTO link, would help create a security nexus covering almost the entire continent of Eurasia and engaging all its major powers in security collaboration.

Russia, of course, cannot "deliver" Iran: Nobody can. But Moscow should be a significant part of any international effort seeking diplomatic resolution of the Iranian nuclear issue. Russia's abstention from such efforts would likely undermine their effectiveness and contribute to the possibility of military confrontation or a continued standoff with Iran.

NATO countries' cooperation with Russia is highly desirable on a number of other issues, from fighting piracy off the coast of Somalia, where the Russian navy has recently been active, to achieving a settlement in the Middle East, where Russia is a member of the diplomatic Quartet (along with America, the EU, and the UN) that is trying to advance the peace process.

The Strategic Question

Anders Fogh Rasmussen, NATO's new secretary general, has identified relations with Russia as his top priority after Afghanistan. This is an opportunity not to be wasted. One of the alliance's imminent tasks is to develop a new "strategic concept," to replace the one agreed to in 1999. As it sets out to draft its new guiding document, the alliance faces an important question: Does NATO foresee building Europe's security together with Russia, or with an eye on Russia?

Clearly, not everything depends on the policies of the West. Russia, which faces a similar question in reverse, will have to come up with an answer of its own. Most likely, for both the Western alliance and Russia, the preferred option is cooperation. But for each of them, a failure to cooperate raises the need to hedge against the perils of a partner-turned-opponent. In the outcome that nobody wants, Europe might walk back to its future.

To achieve a good outcome, NATO and Russia need to engage with each other seriously. As they start a review of their relationship, nothing should be off the table. NATO has a number of legitimate concerns to share with Russia. None of these should be ignored or given short shrift by the Russian side. Zones of influence, as well as great games, belong to the past. Nations today decide for themselves. Powers of attraction trump those of coercion. Great power means great responsibility. Challenges to Russian security in the early twenty-first century lie in the south, not the west.

Still, the seventh decade of NATO-Russia relations is going to be difficult. Building confidence and a modicum of trust will take a long time. Managing the post-conflict situation in the South Caucasus will not be easy, but it is possible, given a measure of attention to the region. Preventing tensions over Ukraine from leading to a new and incomparably worse crisis in the Black Sea area is an absolute must. Turning missile defense from an issue of contention into one of collaboration would be reversing a tide in Russian-Western security relations. After 40 years of wintry adversity and two decades of delusions and disappointments, Russia and the West both deserve better.

DMITRI TRENIN is director of the Carnegie Moscow Center. His books include *Getting Russia Right* (Carnegie Endowment for International Peace, 2007).

From *Current History,* October 2009, pp. 299–303. Copyright © 2009 by Current History, Inc. Reprinted by permission.

Gorbachev on 1989

On September 23, *Nation* editor Katrina vanden Heuvel and her husband, Stephen F. Cohen, a contributing editor, interviewed former Soviet President Mikhail Gorbachev at his foundation in Moscow. With the twentieth anniversary of the fall of the Berlin Wall approaching, we believed that the leader most responsible for that historic event should be heard, on his own terms, in the United States. As readers will see, the discussion became much more wide-ranging. —The Editors

Katrina vanden Heuvel and Stephen F. Cohen

KVH/SFC: Historic events quickly generate historical myths. In the United States it is said that the fall of the Berlin Wall and the end of a divided Europe was caused by a democratic revolution in Eastern Europe or by American power, or both. What is your response?

MG: Those developments were the result of perestroika in the Soviet Union, where democratic changes had reached the point by March 1989 that for the first time in Russia's history democratic, competitive elections took place. You remember how enthusiastically people participated in those elections for a new Soviet Congress. And as a result thirty-five regional Communist Party secretaries were defeated. By the way, of the deputies elected, 84 percent were Communists, because there were a lot of ordinary people in the party—workers and intellectuals.

On the day after the elections, I met with the Politburo, and said, "I congratulate you!" They were very upset. Several replied, "For what?" I explained, "This is a victory for perestroika. We are touching the lives of people. Things are difficult for them now, but nonetheless they voted for Communists." Suddenly one Politburo member replied, "And what kind of Communists are they!" Those elections were very important. They meant that movement was under way toward democracy, glasnost and pluralism.

Analogous processes were also under way in Eastern and Central Europe. On the day I became Soviet leader, in March 1985, I had a special meeting with the leaders of the Warsaw Pact countries, and told them: "You are independent, and we are independent. You are responsible for your policies, we are responsible for ours. We will not intervene in your affairs, I promise you." And we did not intervene, not once, not even when they later asked us to. Under the influence of perestroika, their societies began to take action. Perestroika was a democratic transformation, which the Soviet Union needed. And my policy of nonintervention in Central and Eastern Europe was crucial. Just imagine, in East Germany alone there were more than 300,000 Soviet troops armed to the teeth—elite troops, specially selected! And yet, a process of change began there, and in the other countries, too. People began to make choices, which was their natural right.

But the problem of a divided Germany remained. The German people perceived the situation as abnormal, and I shared their attitude. Both in West and East Germany new governments were formed and new relations between them established. I think if the East German leader Erich Honecker had not been so stubborn—we all suffer from this illness, including the person you are interviewing—he would have introduced democratic changes. But the East German leaders did not initiate their own perestroika. Thus a struggle broke out in their country.

The Germans are a very capable nation. Even after what they had experienced under Hitler and later, they demonstrated that they could build a new democratic country. If Honecker had taken advantage of his people's capabilities, democratic and economic reforms could have been introduced that might have led to a different outcome.

I saw this myself. On October 7, 1989, I was reviewing a parade in East Germany with Honecker and other representatives of the Warsaw Pact countries. Groups from twenty-eight different regions of East Germany were marching by with torches, slogans on banners, shouts and songs. The former prime minister of Poland, Mieczyslaw Rakowski, asked me if I understood German. "Enough to read what's written on the banners. They're talking about perestroika. They're talking about democracy and change. They're saying, 'Gorbachev, stay in our country!'" Then Rakowski remarked, "If it's true that these are representatives of people from twenty-eight regions of the country, it means the end." I said, "I think you're right."

KVH/SFC: That is, after the Soviet elections in March 1989, the fall of the Berlin Wall was inevitable?

MG: Absolutely!

KVH/SFC: Did you already foresee the outcome?

MG: Everyone claims to have foreseen things. In June 1989 I met with West German Chancellor Helmut Kohl and we then held a press conference. Reporters asked if we had discussed the German question. My answer was, "History gave rise to this problem, and history will resolve it. That is my opinion. If you ask Chancellor Kohl, he will tell you it is a problem for the twenty-first century."

I also met with the East German Communist leaders, and told them again, "This is your affair and you have the responsibility to decide." But I also warned them, "What does experience teach us? He who is late loses." If they had taken the road of reform, of gradual change—if there had been some sort of agreement or treaty between the two parts of Germany, some sort of financial agreement, some confederation, a more gradual reunification would have been possible. But in 1989–90, all Germans, both in the East and the West, were saying, "Do it immediately." They were afraid the opportunity would be missed.

KVH/SFC: A closely related question: when did the cold war actually end? In the United States, there are several answers: in 1989, when the Berlin Wall came down; in 1990–91, after the reunification of Germany; and the most popular, even orthodox, answer, is that the cold war ended only when the Soviet Union ended, in December 1991.

MG: No. If President Ronald Reagan and I had not succeeded in signing disarmament agreements and normalizing our relations in 1985–88, the later developments would have been unimaginable. But what happened between Reagan and me would also have been unimaginable if earlier we had not begun perestroika in the Soviet Union. Without perestroika, the cold war simply would not have ended. But the world could not continue developing as it had, with the stark menace of nuclear war ever present.

Sometimes people ask me why I began perestroika. Were the causes basically domestic or foreign? The domestic reasons were undoubtedly the main ones, but the danger of nuclear war was so serious that it was a no less significant factor. Something had to be done before we destroyed each other. Therefore the big changes that occurred with me and Reagan had tremendous importance. But also that George H.W. Bush, who succeeded Reagan, decided to continue the process. And in December 1989, at our meeting in Malta, Bush and I declared that we were no longer enemies or adversaries.

KVH/SFC: So the cold war ended in December 1989?

MG: I think so.

KVH/SFC: Many people disagree, including some American historians.

MG: Let historians think what they want. But without what I have described, nothing would have resulted. Let me tell you something. George Shultz, Reagan's secretary of state, came to see me two or three years ago. We reminisced for a long time—like old soldiers recalling past battles. I have great respect for Shultz, and I asked him: "Tell me, George, if Reagan had not been president, who could have played his role?" Shultz thought for a while, then said: "At that time there was no one else. Reagan's strength was that he had devoted his whole first term to building up America, to getting rid of all the vacillation that had been sown like seeds. America's spirits had revived. But in order to take these steps toward normalizing relations with the Soviet Union and toward reducing nuclear armaments—there was no one else who could have done that then."

By the way, in 1987, after my first visit to the United States, Vice President Bush accompanied me to the airport, and told me: "Reagan is a conservative. An extreme conservative. All the blockheads and dummies are for him, and when he says that something is necessary, they trust him. But if some Democrat had proposed what Reagan did, with you, they might not have trusted him."

By telling you this, I simply want to give Reagan the credit he deserves. I found dealing with him very difficult. The first time we met, in 1985, after we had talked, my people asked me what I thought of him. "A real dinosaur," I replied. And about me Reagan said, "Gorbachev is a diehard Bolshevik!"

KVH/SFC: A dinosaur and a Bolshevik?

MG: And yet these two people came to historic agreements, because some things must be above ideological convictions. No matter how hard it was for us and no matter how much Reagan and I argued in Geneva in 1985, nevertheless in our appeal to the peoples of the world we wrote: "Nuclear war is inadmissible, and in it there can be no victors." And in 1986, in Reykjavik, we even agreed that nuclear weapons should be abolished. This conception speaks to the maturity of the leaders on both sides, not only Reagan but people in the West generally, who reached the correct conclusion that we had to put an end to the cold war.

KVH/SFC: So Americans who say the cold war ended only with the end of the Soviet Union are wrong?

MG: That's because journalists, politicians and historians in your country concluded that the United States won the cold war, but that is a mistake. If the new Soviet leadership and its new foreign policy had not existed, nothing would have happened.

KVH/SFC: In short, Gorbachev, Reagan and the first President Bush ended the cold war?

MG: Yes, in 1989–90. It was not a single action but a process. Bush and I made the declaration at Malta, but Reagan would have had no less grounds for saying that he played a crucial role, because he, together with us, had a fundamental change of attitude. Therefore we were all victors: we all won the cold war because we put a stop to spending $10 trillion on the cold war, on each side.

KVH/SFC: What was most important—the circumstances at that time or the leaders?

MG: The times work through people in history. I'll tell you something else that is very important about what subsequently happened in your country. When people came to the conclusion that they had won the cold war, they concluded that they didn't need to change. Let others change. That point of view is mistaken, and it undermined what we had envisaged for Europe—mutual collective security for everyone and a new world order. All of that was lost because of this muddled thinking in your country, and which has now made it so difficult to work together. World leadership is now understood to mean that America gives the orders.

KVH/SFC: Is that why today, twenty years after you say the cold war ended, the relationship between our two countries is so bad that President Obama says it has to be "reset"? What went wrong?

MG: Even before the end of the cold war, Reagan, Bush and I argued, but we began to eliminate two entire categories of nuclear weapons. We had gone very far, almost to the point when a return to the past was no longer possible. But everything went wrong because perestroika was undermined and there was a change of Russian leadership and a change from our concept of gradual reform to the idea of a sudden leap. For Russian President Boris Yeltsin, ready-made Western recipes were falling into his hands, schemes that supposedly would lead to instant success. He was an adventurist. The fall of the Soviet Union was the key moment that explains everything that happened afterward, including what we have today. As I said, people in your country became dizzy with imagined success: they saw everything as their victory.

In Yeltsin, Washington ended up with a vassal who thought that because of his anticommunism he would be carried in their arms. Delegations came to Russia one after the other, including President Bill Clinton, but then they stopped coming. It turned out no one needed Yeltsin. But by then half of Russia's industries were in ruins, even 60 percent. It was a country with a noncompetitive economy wide open to the world market, and it became slavishly dependent on imports.

How many things were affected! All our plans for a new Europe and a new architecture of mutual security. It all disappeared. Instead, it was proposed that NATO's jurisdiction be extended to the whole world. But then Russia began to revive. The rain of dollars from higher world oil prices opened up new possibilities. Industrial and social problems began to be solved. And Russia began to speak with a firm voice, but Western leaders got angry about that. They had grown accustomed to having Russia just lie there. They thought they could pull the legs right out from under her whenever they wanted.

The moral of the story—and in the West morals are everything—is this: under my leadership, a country began reforms that opened up the possibility of sustained democracy, of escaping from the threat of nuclear war, and more. That country needed aid and support, but it didn't get any. Instead, when things went bad for us, the United States applauded. Once again, this was a calculated attempt to hold Russia back. I am speaking heatedly, but I am telling you what happened.

KVH/SFC: But now Washington is turning to Moscow for help, most urgently perhaps in Afghanistan. Exactly twenty years ago, you ended the Soviet war in Afghanistan. What lessons did you learn that President Obama should heed in making his decisions about Afghanistan?

MG: One was that problems there could not be solved with the use of force. Such attempts inside someone else's country end badly. But even more, it is not acceptable to impose one's own idea of order on another country without taking into account the opinion of the population of that country. My predecessors tried to build socialism in Afghanistan, where everything was in the hands of tribal and clan leaders, or of religious leaders, and where the central government was very weak. What kind of socialism could that have been? It only spoiled our relationship with a country where we had excellent relations during the previous twenty years.

Even today, I am criticized that it took three years for us to withdraw, but we tried to solve the problem through dialogue—with America, with India, with Iran and with both sides in Afghanistan, and we attended an international conference. We didn't simply hitch up our trousers and run for it, but tried to solve the problem politically, with the idea of making Afghanistan a neutral, peaceful country. By the way, when we were getting ready to pull out our troops and were preparing a treaty of withdrawal, what did the Americans do? They supported the idea of giving religious training to young Afghans—that is, the Taliban. As a result, now they are fighting against them. Today, again, not just America and Russia can be involved in solving this problem. All of Afghanistan's neighbors must be involved. Iran cannot be ignored, and it's ill-advised for America not to be on good terms with Iran.

KVH/SFC: Finally, a question about your intellectual-political biography. One author called you "the man who changed the world." Who or what most changed your own thinking?

MG: Gorbachev never had a guru. I've been involved in politics since 1955, after I finished university, when there was still hunger in my country as a result of World War II. I was formed by those times and by my participation in politics. In addition, I am an intellectually curious person by nature and I understood that many changes were necessary, and that it was necessary to think about them, even if it caused me discomfort. I began to carry out my own inner, spiritual perestroika—a perestroika in my personal views. Along the way, Russian literature and, in fact, all literature, European and American too, had a big influence on me. I was drawn especially to philosophy. And my wife, Raisa, who had read more philosophy than I had, was always there alongside me. I didn't just learn historical facts but tried to put them in a philosophical or conceptual framework.

I began to understand that society needed a new vision—that we must view the world with our eyes open, not just through our personal or private interests. That's how our new thinking of the 1980s began, when we understood that our old viewpoints were not working out. During the nuclear arms race, I was given a gift by an American, a little figure of a goose in flight. I still have it at my dacha. It is a goose that lives in the north of Russia in the summer and in the winter migrates to America. It does that every year regardless of what's happening, on the ground, between you and us. That was the point of this gift and that's why I'm telling you about it.

KVH/SFC: Listening to you, it seems that you became a political heretic in your country.

MG: I think that is true. I want to add that I know America well now, having given speeches to large audiences there regularly. Three years ago I was speaking in the Midwest, and an American asked me this question: "The situation in the United States is developing in a way that alarms us greatly. What would you advise us to do?" I said, "Giving advice, especially to Americans, is not for me." But I did say one general thing:

that it seems to me that America needs its own American perestroika. Not ours. We needed ours, but you need yours. The entire audience stood and clapped for five minutes.

KVH/SFC: And do you think President Obama will be the leader of such an American perestroika?

MG: As far as I know, Americans did not make a mistake in electing him. Barack Obama is capable of leading your society on a very high level and of understanding it better than any political figure I know. He is an educated person with a highly developed capacity for dialogue, and that too is very important. So I congratulate you.

What the World Really Wants

Fareed Zakaria

The Bush administration describes spreading democracy as the lodestar of its foreign policy. It speaks about democracy constantly and has expanded funding for programs associated with it. The administration sees itself as giving voice to the hundreds of millions who are oppressed around the world. And yet the prevailing image of the United States in those lands is not at all as a beacon of liberty. Public sentiment almost everywhere sees the United States as self-interested and arrogant. There is a huge disconnect between what the Bush administration believes it stands for and how it is seen around the world.

Why? Well, consider Vice President Cheney's speech on May 4 in Lithuania, in which he accused Russia of back-pedaling on democracy. Cheney was correct in his specific criticisms. If anything, he was coming a little late to this party. Senators like John McCain and Joe Lieberman have been making this case for more than a year. Russia watchers have been pointing to these trends for longer. But to speak as Cheney did last week misunderstands the reality in that country, and squanders America's ability to have an impact in it.

In Cheney's narrative, Russia was a blooming democracy during the 1990s, but in recent years it has turned into a sinister dictatorship where people live in fear. In castigating Vladimir Putin, Cheney believes that he is speaking for the Russian masses. He fancies himself as Reagan at the Berlin wall. Except he isn't. Had Cheney done his homework and consulted a few opinion polls, which are extensive and reliable in Russia, he would have discovered that Putin has a 75 percent approval rating, about twice that of President Bush.

Most Russians see recent history differently. They remember Russia in the 1990s as a country of instability, lawlessness and banditry. They believe that Boris Yeltsin bankrupted the country, handed its assets over to his cronies and spent most of his time drunk and dysfunctional. Yeltsin's approval ratings by 1994 were below 20 percent and in 1996 he actually went into the single digits for a while. Russians see Putin, on the other hand, as having restored order, revived growth and reasserted national pride.

Why? Well, for the average Russian per capita GDP has gone from $600 to $4,500 during Putin's reign, much, though not all of which, is related to oil prices. The poverty rolls have fallen from 42 million to 26 million. College graduates have increased by 50 percent and a middle class has emerged in Russia's cities. And yet the backsliding that Cheney described is quite true, too. I've been critical of Putin power grabs for years now. But the truth is that even so, Russia today is a strange mixture of freedom and unfreedom. (The country publishes 90,000 books a year, espousing all political views.) Polls in Russia show that people still rate democracy as something they like and value. But in the wake of the 1990s, they value more urgently conditions that will allow them to lead decent civic and economic lives. We went to Iraq with similar blinders, believing that all people thirsted for was the end of Saddam. But when that meant the end of order, stability and civilized life, they were horrified and blamed us. If we had paid attention to this fundamental (and conservative) insight, we might not be in the mess we are in today in Iraq.

Or consider Nigeria. American officials have been debating how to help that country, by ensuring that its elected president, Olusegun Obasanjo, would not run for a third term (which would have required amending election laws). Last week the Nigerian Senate ruled out a third term, and Washington applauded. But in fact this whole drama is largely irrelevant to what is really happening in Nigeria. Over the last 25 years, the country has gone into free fall. Its per capita GDP has collapsed, writes Jeffrey Tayler in the April issue of *The Atlantic,* from $1,000 to $390. It ranks below Haiti and Bangladesh on the Human Development Index. In 2004 the World Bank estimated that 80 percent of Nigeria's oil wealth goes to 1 percent of its people. Sectarian tensions are rising, particularly between Muslims and Christians, and 12 of the country's 36 provinces have imposed Sharia. Violent conflict permeates the country, with 10,000 people dead over the last eight years. In this context, Obasanjo's third term is really not the big issue that will determine Nigeria's future. (Obasanjo has actually presided over a series of important improvements, which will probably collapse in his absence.) But these are the only issues that we talk about, because we're spreading democracy.

The United States should stand for and help promote freedom around the world. But we can do so effectively only if we ally ourselves with the aspirations of the people we are trying to help. For many of them, the great struggle going on in so much of the world today is to end civil strife, corruption, extreme poverty and disease, which destroy not just democracy but society itself. And on those issues, I don't think I've ever heard a speech by Dick Cheney.

Russia's Search for a New Migration Policy

Zhanna Zayonchkovskaya

At the beginning of 2007 Russia took a decisive step towards liberalisation of its migration policy by introducing a new procedure for registering migrants and issuing work permits for foreign workers. In its own way, this step is as significant for the life of the country as the l993 law on freedom of entry and exit, which brought the Iron Curtain crashing down.

The two events are separated by more than a decade; a time in which the orientations of Russia's migration policy and its priorities changed repeatedly. In the first half of the 1990s, attention was focused on forced migration from the conflict zones on post-Soviet territory and the repatriation of ethnic Russians from the former Soviet republics. As for migration from beyond the former USSR borders, it can be said that Russia's gates were wide open during this time, for emigrants and immigrants alike. In the absence of checks on immigration, the phenomenon of illegal migration took root and rapidly proliferated. The country thereupon swerved to the other extreme, with a restrictive policy built around migration control. One of the things that made this development palatable was the mistrust with which foreigners were viewed by a population that had got used to living in a closed country.

The first stage of the migration policy was completed in 2001 when the federal migration service of Russia, an independent state institution created in mid-1992, was disbanded. The second stage got off to a clear start with the adoption, in 2002, of a law on the legal status of foreigners in the Russian Federation and the creation of a new federal immigration service—now a subsidiary organ of the ministry of the interior. A complicated legal and procedural apparatus, reflecting the new approach, set up imposing barriers for foreigners wishing to legally immigrate and work in the country; a paradox, given the country's real need to supplement the population, in the interests of security as well as economic development. In addition, the overly restrictive legal framework forced immigrants into a shadowy extra-legal zone, thereby contributing to a rise in illegal migration and the growth of criminal gangs engaged in migration-related activities.

The need to reorient migration policy so as to create immigration incentives was becoming increasingly urgent. This finally took place at the initiative of President Putin. In a speech before the federal security council on 17 March 2005, he called the stimulation of migratory processes "our priority task", and went on to reiterate its importance in the presidential communication to the federal parliament in May 2006. Since then, it has been reinforced in subsequent legislative and administrative decisions.

The Demographic Context of Immigration

Although migration has always been very significant in the history of Russia, never before has its role in the development of the country been as important as the one it is expected to play in the next few decades.

The immigration problem emerged in close interplay with the demographic crisis. Regarding the future of Russia's population, the verdict of demographic experts—whether those of the United Nations, Russia's official statistics agency Rosstat, or independent experts—is unanimous: the population is expected to shrink drastically. The median scenario in the Rosstat prognosis, based on the 2002 census, has the population of Russia decreasing to 137 million by 2026, assuming slight improvements in the birthrate and mean life expectancy, and an increase in net immigration from the 126,000 registered in 2005 to 415,000 in 2025. If the prognosis is made strictly on the basis of current trends for birthrate, mortality and immigration, the population diminishes to 125,000.[1] The United Nations forecast for the same date has Russia's population at 130 million.[2]

Of course, the Russian population has been declining gradually since 1992, but that process has not affected the part of the population that is of working age.[3] On the contrary, while the overall population declined, the proportion made up by people of working age actually grew noticeably for a time, thanks to a favourable ratio of persons entering the working-age population to those leaving it. In 2006 this temporary phenomenon came to an end, and a phase of rapid natural attrition of the working-age population began. Attrition is relatively modest at this early stage: in 2007, it is expected that approximately 300,000 workers will be lost. However, in the following year already annual attrition will double, and in the decade 2010–2019 the annual figure is predicted to exceed one million (see Figure 1). By

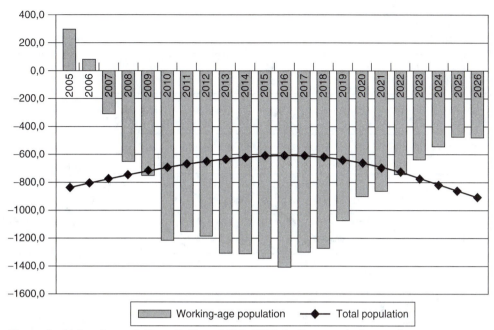

Figure 1   Natural population attrition, 2005–2026, in thousands (assuming zero immigration)

Source: Center for Demography and Human Ecology, Institute for Economic Forecasting, Russian Academy of Sciences

2026, the total attrition of Russia's working-age population from natural causes, if it is not compensated by immigration, will exceed 18 million. If this number is juxtaposed with the number of employed workers in the Russian economy, 69 million, the full seriousness of the situation becomes evident. It also makes it clear that labour will be among the most scarce of resources in Russia, if not the most scarce.

This drastic shrinking of the workforce makes it necessary to compensate by a sizeable increase in immigration, and gives the immigration component of Russia's migration policy strategic significance.

Given that two thirds of immigrants tend to be of working age, more than 25 million immigrants would be required in the next row decades, if it was desired to completely make up the workforce losses due to natural attrition. It should be emphasised that this is *net* immigration, the difference between the arrivals and departures; which means that the real number of immigrants has to be even higher. A massive influx of this type is hardly practical, regardless of how active and liberal migration policy is; but the calculation is indicative of the magnitude of the problem.

The expected additional labour demand in Russia cannot be met by increasing labour productivity alone. This is confirmed by the experience of European countries, in most of which the working population continues to grow, even though their level of economic development is far higher than that of Russia. Furthermore, calculations show that, the higher the economic growth rate, the greater is the marginal increase in demand for labour. Simulations show that, if the Russian gross domestic product (GDP) continues to grow at present rates, five million more workers will be required in 2015 than currently; but if annual growth rates of 7% are used, then that figure increases to seven million.[4] If we further factor in the need to compensate for natural attrition, the figure obtained for the aggregate additional

workforce requirement by 2015 is 14–16 million. Better use of the existing workforce will not compensate for more than a fraction of the losses due to workers approaching retirement age, leaving the basic need for massive immigration unchanged.

The country is already experiencing the effects of an acute labour shortage, even though the workforce has only just begun to shrink. Aleksandr Shokhin, the head of RSPP, the association of Russian businessmen, explains that "In many areas, the workforce simply does not exist, physically. Businesses have to either bus workers in . . . or relocate people from economically depressed regions."[5] (We should note, however, that the labour reserves in economically depressed regions are very limited.)

The depopulation of Russia, including the shrinkage of the working-age population, was anticipated as early as the 1980s. However, in the USSR it was not possible to publish demographic forecasts in a form accessible to the general public; instead, they were restricted to "official use". This state of affairs ruled out the possibility of timely steps being taken to prepare the population for the current situation. In the 1990s there was a prolonged blackout in forecasting. This is why the current situation is such an unexpected, indeed shocking experience, both for the new generation of political leaders, and for the population as a whole. The realisation of what was impending did not take place until the country was on the verge of workforce implosion, the effects of the diminishing population began to be felt in the army and the educational system, and the question of pensions became acute.

The current situation leads to several immediate conclusions:

First, the country requires a decisive reorientation of its migration policy. A sustainable migration policy for Russia is, before anything else, a policy for immigration.

Second, the size of the expected immigration requirements means that Russia in the first half of the twenty-first century

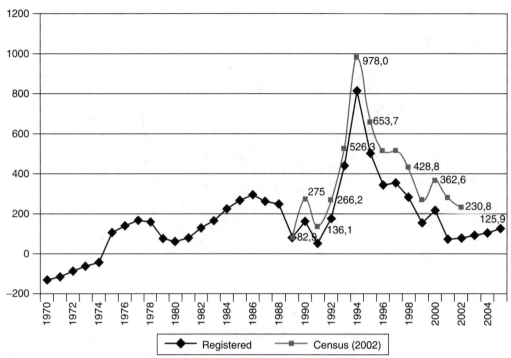

Figure 2   Net migration of Russia, 1970–2005, thousands

could become one of the most attractive countries on Earth for migrants. This development already began in the 1990s; the 2002 census figures showed that Russia had become the third most popular destination for immigrants, after the United States and Germany. From 1989 to 2002, Russia absorbed some eleven million immigrants, an average of 781,000 per year, compared with 865,000 for Germany and 924,000 for the United States. Converting the immigrant statistics to figures per 10,000 inhabitants gives Russia a score of 54, ahead of the United States with 32, but far behind Germany, with 142. In the preceding period of the same duration, 1975–1988, Russia had even more immigrants: thirteen million. Recent experience thus confirms that Russia is capable of absorbing large numbers of immigrants. Nonetheless, even an influx of this magnitude is not up to the task of fully compensating for the natural attrition of the workforce. The Russian economy will thus of necessity continue to function under conditions of labour deficit.

Third, migration will become a matter of vital national interest for Russia in this time. The country's success in dealing with the task of attracting the required volume of immigrants will determine its rate of economic growth, the standard of living in the population, the security of pensions, social stability, balanced regional development, and, finally, the size of the country and its integrity. The outcome will be decided in the course of the next ten years.

Migratory Trends

The analysis of Rosstat is that the migratory population increase in Russia during the 14 years between the 1989 census and that of 2002 was 5.6 million. This is 2.3 times as great as

the increase during the previous period of the same duration. Russia's population increased thanks to 6.8 million migrants from its former partners in the Soviet Union, while it lost 1.3 million emigrants itself. Beginning in 1994, the migratory trend slowed down abruptly (see Figure 2). Even though the sights were raised for immigration after the 2002 census, the trend is unchanged, which is clearly at odds with the country's need for migrants, and shows that the policy being implemented is ineffective.

The migration trends that emerge from the official resident registration statistics probably say more about the obstacles facing immigrants than about actual immigration. Survey results show that the number of immigrants in Russia is in the neighbourhood of seven million people.[6] The diagram shown in Figure 3 makes it clear that a large proportion is made up of labour migrants working unofficially, with one in two possessing neither official registration, nor an employment contract; some 30% are registered, but work without a contract; while 20% have a contract, but work in contravention of quotas. Only slightly more than 10% of all labour migrants have an official position. Thus, between 3.5 and 4.5 million immigrants are quite simply illegal. This is a high figure, even for a large country such as Russia; but it is considerably lower than the estimates issued by the federal migration service, FMS, which does not trouble itself with substantiating them. The service claims that "there are some ten million illegal labour migrants on the territory of our country."[7] The declining migration figures and the enormous imbalance between illegal immigrants and those who have duly registered and obtained official status show the utter bankruptcy of migratory policy.

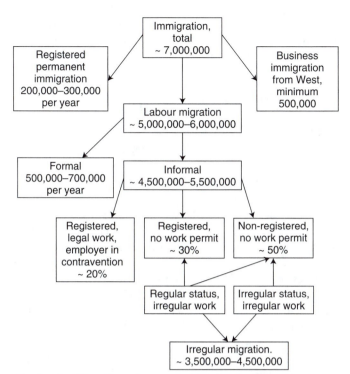

```
                    ┌──────────────────┐
                    │  Immigration,    │
                    │  total           │
                    │  ~ 7,000,000     │
                    └──────────────────┘
  ┌─────────────┐            │              ┌──────────────┐
  │ Registered  │            │              │  Business    │
  │ permanent   │            │              │  immigration │
  │ immigration │            │              │  from West,  │
  │ 200,000–    │            │              │  minimum     │
  │ 300,000     │            ▼              │  500,000     │
  │ per year    │   ┌──────────────────┐   └──────────────┘
  └─────────────┘   │ Labour migration │
                    │ ~ 5,000,000–     │
                    │   6,000,000      │
                    └──────────────────┘
  ┌─────────────┐            │
  │  Formal     │            ▼
  │ 500,000–    │   ┌──────────────────┐
  │ 700,000     │   │  Informal        │
  │ per year    │   │  ~ 4,500,000–    │
  └─────────────┘   │    5,500,000     │
                    └──────────────────┘
 ┌──────────────┐ ┌──────────────┐ ┌──────────────┐
 │ Registered,  │ │ Registered,  │ │ Non-registered,│
 │ legal work,  │ │ no work      │ │ no work      │
 │ employer in  │ │ permit       │ │ permit       │
 │ contravention│ │ ~ 30%        │ │ ~ 50%        │
 │ ~ 20%        │ └──────────────┘ └──────────────┘
 └──────────────┘
        ┌────────────────┐ ┌────────────────┐
        │ Regular status,│ │ Irregular status,│
        │ irregular work │ │ irregular work │
        └────────────────┘ └────────────────┘
              ┌──────────────────────┐
              │ Irregular migration. │
              │ ~ 3,500,000–         │
              │   4,500,000          │
              └──────────────────────┘
```

Figure 3 Migration flows to Russia (estimates) at the beginning of 2007

The Revolution in Migration Legislation

On 15 January 2007, two new federal laws came into force in Russia, laying the foundations for a drastic change in the way immigration is regulated. One is the law on migrant registration for foreign citizens and stateless persons, and the other is the amended law on the legal status of foreign citizens. Changes concerned a simplified procedure for registering foreign citizens at their place of residence, along with employment; the very issues that were the most serious obstacles for legalisation of immigrants.

Registration used to be the first step in the process of obtaining legal status as a foreign resident in Russia. This first step, being so complex, involved and burdened with red tape, was frequently insurmountable. Registration, in its form at that time, was the most powerful deterrent to legal migration, and the main source of corruption in the world of migration. The rotten core of the Russian system of registration was the requirement to give an official address, formally linking registration to the prospective immigrant's place of residence. This made it exceedingly difficult for immigrants to acquire legal status. Before any application could be submitted for registration, the immigrant needed to find a place to live, or at least a paper address that he or she would be allowed to use for registration purposes. The primitive conditions of the residential market, and the weak protections afforded owners in Russia, along with certain other factors, meant that it was far from easy to find a place to live where the owner would agree to let a migrant register.

Studies show that, among migrants who remained unregistered, one in five did so because he or she did not have an

address to give. For the same reason, the officially registered place of residence and the actual residence were frequently not the same; something that hollowed out the usefulness of registration as a tool for tracking migrants and providing a deterrent against crime among them. Registration, even for a short time, meant that the migrant had to collect a variety of official papers and appear in person before the police, who conducted the procedure, accompanied not only by the owner or the responsible individual at the rental address, but also by all permanent residents at that address, whose concurrence was required for the new migrant to be registered.

This cumbersome system meant that immigrants arriving in a new place depended on the presence of relatives or acquaintances prepared not only to register them, but also to accompany them through the whole procedure. Investigations showed that, among registered migrants, only one in ten was at an address other than that of relatives or acquaintances. Few were able to meet the statutory three-day deadline for registration. Again, studies showed that only some 20% were able to accomplish registration within one week; but the same number was unable to complete the procedure even after two full months had gone by. Since proper registration was a mandatory prerequisite for employment, there was, unsurprisingly, a boom in illegal services offering to help foreigners find an official address and obtain the necessary documents; and, in parallel, a boom in exploitative employment practices. This is the historical reality that must be borne in mind, to fully appreciate the changes that have taken place.

The new system greatly simplifies the registration procedure for persons arriving in the country for a temporary stay, which is the case for most migrants. This category is exempt from the requirement to register at the place of residence, with a stamp in the passport. It has been replaced by a system of voluntary residence notification. Under this system, migrants can submit official applications without necessarily indicating their ultimate place of residence; the employment address or the address of the company providing employment services is sufficient. The application can be submitted to the police, but it can also be submitted by mail to the local unit of the migration service. The application stub with a stamp confirming submission of the document constitutes a certificate of legal residence. At the same time, the number of substantiating documents that must be submitted has been reduced to a minimum: a passport and the landing card. For the first time since the passport system was introduced, a procedure has been brought into use that takes migrant registration out of the hands of the police. This is a truly revolutionary development, which promises to reduce corruption. It is also the first time that people arriving in Russia for a temporary stay do not require the registration stamp in their passport.

At the same time as new registration rules were introduced, the process governing migrant employment was transformed just as dramatically. The changes here concerned only non-visa migrants arriving in Russia from one of the countries of the Commonwealth of Independent States (CIS). From Russia's perspective, labour from CIS countries is considered to be preferential, which explains the special treatment. Labour permits are now issued directly to the immigrant, and not, as in the past, to the employer. The employer, in turn, now has the right to hire any legal immigrant having the work permit. This means that workers are no longer tied to one employer.

The next major transformation involved taking the procedure for issuing work permits out of the hands of the employment authorities. Before 2007, the procedure was that permits for hiring immigrants were only issued once the approval of the employment authorities had been obtained; and the authorities had an obligation to protect the interests of the local workers. That procedure has been discontinued; the only obligation that remains is for the migration service to inform the employment authorities about work permits issued to foreign immigrants.

The quota system for issuing work permits has been retained, but the quota criteria have been made more sophisticated. Thus, quotas can be established on the basis of profession, level of qualification and specialisation, country of origin, and other economic and social criteria, taking into account the particular nature of the regional labour market. Furthermore, qualified specialists working in their own domain have been exempted from the quota, another innovation. Accordingly, two quotas were established in 2007: one for immigrants from countries for which no visa requirement exists (six million work permits), and one for visa immigrants arriving for employment purposes (308,800 invitations). The enormous quota for countries without a visa requirement provides an indication of the government's highly laudable intention to regularise the status of the accumulated masses of illegal immigrants from those countries.

The problem remains that, in the past, the quota mechanism in Russia did not prove to be an effective method for regulating the migratory labour flows. Satisfactory methods for calculating quotas to attract foreign workers do not exist. Indeed, the very notion of calculating quotas under the conditions that exist on the Russian labour market today seems dubious: it is a rapidly changing and unpredictable market, heavily dominated by the shadow economy. Furthermore, relations between the players in the labour market need to be fine-tuned.

It is as yet too early to assess the new policy on its merits. In its first few months, it brought to light the primitive state of the migration infrastructure in the country, and the lack of coordination between the various entities involved in the legal processing of migrants. Unfortunately, as is often the case in Russia, the new laws were brought in before the necessary administrative instructions had been drawn up, administrators trained in applying the new rules, information documents for migrants prepared, etc. All of this is being done only now. Nonetheless, the reaction among the migrants themselves has been enthusiastic. According to FMS statistics, some 700,000 migrants registered in the first month of the new system. Naturally, this reaction created chaos, with inevitable queues, cheating and therefore corruption. The regional services responsible for implementing the new rules proved to be badly prepared for the transition, and in many cases showed themselves to be at a loss, far from being prepared for the liberal course of the new migration policy. According to a preparatory survey carried out in the summer of 2006, among FMS professional administrators, one in three opposed the new policy completely or partly, and was against the introduction of the simplified registration regime based on voluntary reporting. Almost 40% expressed their strong preference for the existing rules for employment of foreign workers, based on permits issued to employers.

Where Will the Immigrants Come from?

Naturally, Russia pins its hopes, in the first instance, on the CIS countries, which have traditionally been sources of immigrants. In the post-Soviet era, the largest flow of migrants into Russia has come from Kazakhstan, with more than two million persons, accounting for one third of all immigrants. The other countries of central Asia contributed a similar number of immigrants to Russia, while immigrants from the countries of the Caucasus made up 20%. Only 8% were from Ukraine, and less than 5% from the countries of the Baltic. The majority of the new arrivals—some two thirds of the total—were in fact ethnic Russian returnees. Another 10% is made up of other nationalities that are domiciled within the Russian Federation: Tatars, Bashkirs and others.

To stimulate immigration from the CIS countries, a government programme has been set up for voluntary repatriation according to ethnic criteria. The programme provides for significant material assistance, covering transportation costs for the entire family and its effects, an accelerated procedure for obtaining Russian citizenship, assistance with finding employment, and other forms of assistance. Laudable though it may be, this action comes too late. Those who may have wished to return to Russia have already done; and the vast majority of those that have not made their own way in those countries. In any event, the situation in the other CIS countries has changed. Many of them are developing even faster than Russia, and qualified workers are in demand everywhere.

The potential population of ethnic Russians remaining in the other CIS countries, for whom the programme is primarily intended, is estimated by experts to number no more than four million—patently inadequate for meeting Russia's need for immigrants. The potential candidates from other CIS nationalities and ethnic groups would probably come to six or seven million at the most by the year 2025, according to our estimates. These are primarily Uzbeks, Tajiks and Kyrgyz. The immigrant potential in the Caucasus, contrary to widespread opinion, is almost exhausted. Thus, if the potential from all the CIS countries is taken together, it may cover the demand on the Russian labour market and substantially alleviate the workforce situation in the near term; however, for the period leading to 2025, or even further into the future, it will not be sufficient. Furthermore, to attract just this existing potential, Russia will have to rid itself of the legacy of arbitrariness that has plagued its migration policy. If liberalisation measures are overshadowed by campaigns like the recent roundup of Georgian immigrants, or crackdowns on migrant labour at open markets, our credibility as a country that welcomes immigrants will suffer. Such campaigns not only aggravate the widespread xenophobia in Russia but also create alarm among potential immigrants, who may well think twice about facing such an uncertain reception and the prospect of suddenly finding themselves unwelcome guests in a foreign land.

In any event, to plug the emerging labour deficit, Russia will have to attract immigrants from other countries, too. Clearly, it would be to our advantage to develop such relations with countries which possess cultural affinities to Russia, such as those of eastern Europe. But this would be a forlorn hope. The global market for immigrant labour is marked by competition, which will only grow fiercer. Accordingly, Russia is competing against the United States and the European Union, among others. This is an uneven contest, with Russia in the role of clear outsider. Russia's options are therefore limited. Its advantage lies among the countries of southeastern and southern Asia, and perhaps the Arab countries.

While diversity in the countries of origin of immigrants is of course a desirable goal, it is also an undeniable fact that the Chinese, our immediate neighbours, are by far the most promising candidates. The Chinese, at any rate those who live in the north-eastern provinces, are not deterred by the Russian climate; they are successful in agricultural work in Siberia and Russia's far east. Being neighbours, they are better informed about the Russian labour market, what opportunities exist and what rules must be followed in Russia; this makes it easier for them to find their place among Russians, and, by comparison with other immigrants, greatly improves the likelihood that they will decide on such a move. Another positive influence is the large number of Russians who travel to China, roughly twice the number of Chinese travelling here.

Research surveys of Chinese visitors arriving in Russia for a temporary stay show a fairly constant picture, with no more than 500,000 for the entire territory of the country.[8] Most of these are traders or construction and agricultural workers. Local estimates of the Chinese presence in Russia's far east are very moderate, no more than 25,000–30,000.[9] The 2002 census listed 35,000 Chinese living in Russia. Despite the modest scale of Chinese immigration, the hoary myth of two million Chinese in Russia's far east, based on an old, spurious article in the newspaper *Izvestia*,[10] continues to persist. It has contributed to creating in Russian society a lasting suspicion of Chinese expansion, which is taken to be a *fait accompli*. The alleged Chinese threat has become a favourite among demagogues, who thrill the popular imagination with increasingly ominous depictions of immigrants "flooding" across the border. Thus, *Rossiiskiye vesti* quite recently informed its readers that there are three to four million Chinese in Russia.[11] Among the scaremongers, one of the loudest, and most irresponsible, is surely L. K. Aksenov, with his prediction that "Today we have one Chinese for every 150 Russians, and in 50 years it will be the other way around."[12] In this way, the ground is made fertile for Russian nationalism and racism, preparing a toxic environment of xenophobia for future migrants.

And yet, Russia, with its dwindling population, stands to gain far more than it would lose from its position as neighbour to a demographic titan—a country with a population that is mobile and adaptable under Russian conditions. There is no consensus in society on the question of immigration. Despite official recognition, the suggestion that Russian development depends on immigration invariably arouses hostile reactions. Migration has become the plaything of electoral campaigns, the subject of political speculation, and a sensational theme for the mass media. The anti-immigration hysteria is being fanned, and it is directed primarily at the "invaders from the south and the east."

Thus, the innovations in Russian migration policy have their share of enemies. Of course, the new measures have their faults; in the short time they have been in practice, numerous discrepancies and omissions have come to light. But if migration policy is to develop along the lines charted out by the new laws, and in accordance with the country's own interests, it is important that the new initiative should not be stifled, but rather consolidated and allowed to grow.

Notes

1. *Предположительная численность населения Российской Федерации до 2025 года.* ("Projected population of the Russian Federation to 2025.") Rosstat statistics bulletin. Moscow, 2005, p. 8 and p. 116.

2. 2005 World Population Data Sheet, issued by the Population Reference Bureau (PRB). Retrieved from www.prb.org.

3. The working age in Russia is 16–54 for women and 16–59 for men.

4. *Политика иммиграции и натурализации в России Аналитический доклад.* Eurasia Heritage Foundation, Moscow: 2005, p. 125. ("Immigration and naturalisation policy in Russia").

5. Moskovsky komsomolets, 18 March 2007.

6. In the following, the cited results are based on studies that the author directed at the Migration Research Center.

7. Figure given by Konstantin Romodanovsky, the head of the Russian federal immigration service FMS. Cited by Alliansmedia, see www.allmedia.ru (retrieved 3 November 2006).

8. Gelbras V. G. *Россия в условиях глобальной китайской миграции.* ("Russia and global Chinese migration"). Moscow, 2004, p. 36.

9. Larin, V. L. Chinese Migration in the Far East. A Bridge Across the Amur River. Moscow–Irkursk, 2004, p. 109.

10. *Izvestia,* 2 November and 30 November 1993.

11. Rossiiskiye Vesti, 19 May 2004.

12. *Izvestia,* l9 December 2005.

ZHANNA ZAYONCHKOVSKAYA is the Director of Migration Research Center and Chief of Laboratory of Migration in the Institute for Economic Forecasting (Russian Academy of Science, Moscow).

From *European View,* vol. 5, Spring 2007, pp. 137–145. Copyright © 2007 by European People's Party. Reprinted by permission of Zhanna Zayonchkovskaya.

Behind the Central Asian Curtain: The Limits of Russia's Resurgence

ALEXANDER COOLEY

The August 2008 Russia-Georgia war produced a wave of international media commentary about Russia's resurgence as the dominant actor in Eurasia. Moscow's military campaign, once it had routed the Georgian army and consolidated Russia's control over the breakaway territories of Abkhazia and South Ossetia, drew only a tepid response from a divided West. The Georgia war, coupled with Russia's high-profile rise as a self-styled energy superpower, seemingly marked a new era of Russian confidence and assertiveness.

This resurgence in Russian power seemed to extend beyond Georgia, to Central Asia. During a February 2009 visit to Moscow, Kyrgyz President Kurmanbek Bakiyev announced at a joint news conference with Russian President Dmitri Medvedev that he would close the U.S. military air base at Manas. The air base—near Bishkek, Kyrgyzstan's capital—plays a pivotal role in supplying U.S.-led military operations in Afghanistan.

Bakiyev made the announcement just as Russia unveiled a $2 billion economic package to help Kyrgyzstan cope with the global financial crisis. No official quid pro quo was acknowledged, but most saw Moscow's hand behind Bakiyev's decision to close the base. Once again Russia had used its regional influence to block Western interests—at least for a while—dealing the new U.S. administration an ill-timed blow to its Afghanistan operations.

It is tempting to conclude from selected, dramatic events such as these that Russian influence in Eurasia is, indeed, resurgent. On closer inspection, however, this narrative fails to capture important structural limits on Moscow's actions, as well as unintended consequences that its regional interventions can create. Although Russia will continue to be a vitally important player in Central Asia, the events of the past year suggest that Moscow's efforts to maintain regional hegemony will become increasingly difficult and costly.

Despite Russia's numerous levers of influence, its recent policy in Central Asia has been more reactive than strategic. Moscow's actions seem opportunistic and tactical, and they are frequently driven by a desire to theatrically counter Western influence in the region, rather than to position Russia to better manage the rapid changes currently under way.

As political developments surrounding the Manas Air Base eventually demonstrated, the Central Asian states themselves (the former Soviet republics of Kazakhstan, Kyrgyzstan, Tajikistan, Turkmenistan, and Uzbekistan) are pursuing a multivector diplomacy: They pragmatically cultivate their strong ties to Russia but at the same time strengthen cooperation with China and maintain openness to the West. What some have described as Russia's resurgence in Central Asia now appears more akin to a rearguard action.

Levers of Influence

When analyzing Russia's interventions in Central Asia, it is important to distinguish between Russia's strategy, which in some respects is wanting, and its considerable capacity to project power across the region. Moscow continues to control a number of critical levers, many of which will not be matched or undermined by competing regional powers any time soon.

In terms of security ties, the Central Asian countries remain firmly within the Russian orbit. With the exception of Turkmenistan, all five nations in the region are members of the Russian-dominated Collective Security Treaty Organization. Tajikistan and Kyrgyzstan also host Russian military facilities and troops. All of the Central Asian states retain close and cooperative ties with Russia's internal security and intelligence services, and these ties have been strengthened since Western-supported "color revolutions" swept through Georgia, Ukraine, and Kyrgyzstan.

Economically, Russia remains a leading player in Central Asia and is the region's largest single trading partner. In 2007, before the onset of the global recession, Central Asia's annual trade with Russia totaled over $21 billion, compared to $14 billion with China and $7 billion with Germany. Russian energy companies are particularly active in the region, and Moscow still controls most of the area's energy pipelines and its distribution infrastructure.

Moreover, since the start of this decade, an estimated 3 to 5 million Central Asian migrants have gone to Russia to find work in the country's construction and services sectors. According to estimates by international financial institutions, overseas remittances in 2008—the overwhelming majority of which originated in Russia—constituted 35 percent to 45 percent of Tajikistan's GDP, and 25 percent to 35 percent of Kyrgyzstan's.

Thus, Moscow's ability to restrict the activities and legal status of these migrants represents a significant political lever over the Central Asian sending states, all the more so because the global economic downturn has adversely affected migrant-dominated economic sectors.

Regional public opinion toward Russia remains generally positive, with most Central Asians regarding Russia as a privileged partner rather than a former colonizer. Russia supports Russian-language programming on television and radio and maintains cultural centers and educational institutions in the region. Polls regarding Central Asian attitudes toward other countries consistently find Russia viewed far more favorably than the United States, the European Union, or China. And according to a Gallup survey in November 2008, strong majorities in the region hold favorable views of the Russian language and view speaking it as a desirable professional skill. In sum, Russia continues to maintain an impressive, and unmatched, network of security, economic, and social ties with the Central Asian states.

The Politics of Noninterference

One of Moscow's most important mechanisms of influence in recent years has been to strongly back regional governments against perceived Western-led attempts to democratize them and impose Western-style political and economic conditions. The aftermath of the Western-backed color revolutions (Georgia's Rose Revolution in 2003, Ukraine's Orange Revolution in 2004, and Kyrgyzstan's Tulip Revolution in 2005) provided Russia with a significant political opening in Eurasia. Moscow successfully painted Western political activities in the region as unwanted political interference in domestic affairs and a potential security threat to these regimes. Russian leader Vladimir Putin's notion of "sovereign democracy" found strong support among the Central Asian governments, which have steadily restricted the activities of local civil society, the media, and transnational nongovernmental organizations.

The Central Asian states also have backed Russia's campaign to curtail the election-monitoring activities of the Organization for Security and Cooperation in Europe (OSCE). The OSCE's Office for Democratic Institutions and Human Rights has since the mid-1990s conducted the largest monitoring missions in the post-communist space. At an OSCE summit in 2007, Russia threatened to withhold funds from the organization and proposed a number of reforms that would dilute its election-monitoring work. The proposals included allowing host governments to exercise a veto over the appointment of a monitoring mission's director, limiting observer missions to just 50 people, and mandating that a mission issue its recommendations in direct consultation with—and with the approval of—host governments.

By the time of the Georgia War, Moscow had effectively cast itself as an active supporter of Central Asian states' sovereignty against external, particularly Western, political meddling. In fact, the Georgian government of pro-Western President Mikheil Saakashvili was itself viewed by Central Asian governments as an American political client born of Washington's direct intervention in Georgia's political affairs.

It was quite unexpected, then, when the Georgian War actually undermined Moscow's role as a sovereign guarantor. It did so in two critical ways. First, Moscow's distribution of Russian passports to citizens of the Georgian breakaway territories of Abkhazia and South Ossetia (a distribution that had accelerated before the war), along with Moscow's invocation of a unilateral right to defend militarily the rights of Russian citizens everywhere, sent alarm bells ringing across the region. After all, the Central Asian states, especially Kazakhstan and Kyrgyzstan, still host large Russian and Russian-speaking communities. The prospect of Russia distributing passports in an unregulated fashion clearly clashes with Central Asian governments' sovereign authority.

Second, Moscow's decision to recognize the independence of Abkhazia and South Ossetia heightened concerns among all the Central Asian states about the basic inviolability of post-Soviet political borders. In fact, four of the five Central Asian states are members of the Shanghai Cooperation Organization (SCO), also comprised of China and Russia, which since its founding has focused on maintaining the sanctity of state borders and eradicating the so-called "evil" of separatism in Central Asia.

Thus, when President Medvedev went to the August 2008 SCO summit in Dushanbe, Tajikistan's capital, to secure international recognition for Abkhazia and South Ossetia's independence, the Central Asian states—under China's direction—firmly refused. Their leaders feared that any recognition of separatist territories in the Caucasus could prove destabilizing elsewhere in Eurasia.

As events in the western Chinese province of Xinjiang some months later underscored, China views the preservation of its territorial integrity as its utmost national priority. Under no circumstances would Beijing acquiesce to conferring legitimacy on separatist political entities. In Dushanbe, as a result, Moscow was publicly rebuffed according to the very same doctrine of sovereign noninterference that it had promoted in prior years.

Economic Strains

Meanwhile, the global financial crisis and the subsequent global recession have eroded Russia's standing as the dominant economic player in Central Asia. The Russian economy was one of the most severely affected at the onset of the crisis, as a collapse in energy and commodity prices blew a hole in the government's budget and sent the stock market tumbling by 70 percent. Moscow has burned through over $100 billion of its accumulated hard currency reserves, and many of its firms and banks face mounting external debts.

The crisis also has revealed that many of Russia's high-profile investments and contracts in Central Asia are motivated primarily by politics and may not be economically sustainable. Nowhere is this more evident than in the energy sector—in particular, in Russia's dispute with Turkmenistan regarding gas prices. In 2007, when Russia and Turkmenistan concluded a new deal for Russia to buy and transport Turkmenistan's gas, Moscow seemed to have locked the Central Asian country into its distribution network for the foreseeable future. But the financial crisis wreaked havoc with Russia's energy-based foreign policy.

Not only did energy prices fall sharply on world markets, but global energy demand, including European demand, also declined. By the spring of 2009, Gazprom found itself committed to purchasing unneeded natural gas from Turkmenistan for the princely sum of $300 per 1,000 cubic meters. At that price the Russian energy giant was losing money on its Turkmen contract.

On April 9, 2009, an explosion along the main Turkmen-Russian gas pipeline, which the Turkmen side claims was caused by Russia suddenly reducing Turkmen gas volume, plunged relations between the countries into crisis. Whether Russia deliberately caused the explosion is still unclear, but Turkmenistan's gas trade with Russia was abruptly halted, resulting in Turkmenistan's losing an estimated $1 billion worth of monthly revenues.

Afterwards, Turkmenistan conducted a high-profile search for alternative partners to use as leverage in its bargaining with Moscow. In May, Ashgabat even declared that it would be ready to supply Turkmen gas to the Nabucco pipeline, the costly and high-profile project, opposed by Moscow, that is designed to bring Caspian and Middle Eastern gas to Europe while bypassing Russian territory.

Moscow in the spring of 2009 also had a public economic spat with Tajikistan, this one over allegedly unfulfilled investment commitments. Under pressure from the economic crisis, Tajik President Emomali Rahmon demanded that Russia deliver on promises it had made in 2004, as part of a bilateral basing and security accord, to provide $2 billion worth of investments in Tajik hydroelectric projects and infrastructure development.

Rahmon was particularly upset that the Russian company Rusal had not followed through and invested in the massive Rogun hydroelectric plant, and the Tajik president indicated that he would seek alternative partners to provide the required billions for the project. Some analysts have interpreted recent overtures by Tajikistan to the United States and NATO as an attempt to use Western security cooperation as leverage to force Moscow to make good on its investment commitments.

Hedging Bets

Even if Moscow could commit resources at the level demanded by its Central Asian partners, it would still face two important structural impediments to its attempts to secure regional supremacy.

First, Moscow's diplomacy of influence, especially its energy politics, itself seems to be driving the Central Asian countries to seek alternative partners in order to hedge their bets and leverage their relations with Russia. All of the Central Asian states remain wary of Russia's monopolizing their foreign relations, in terms of both economic and security ties. Meanwhile, it is now clear that Beijing is increasing its role as an alternative source of partnership and economic patronage.

In the Turkmen case, Ashgabat has continued to expand its energy cooperation with China. Additionally, in the summer of 2009, Turkmenistan announced plans to expand its gas supplies

to Iran, and it concluded an oil and gas exploration agreement for its Caspian shore with Germany's energy company RWE.

Turkmenistan's diversification strategy is echoed in Kazakhstan. Astana seeks to gain alternative energy sources and new customers from the Turkmenistan-China pipeline. In July 2009 Kazakhstan announced completion of a 1,300-kilometer segment of a pipeline that traverses Kazakh territory and will carry western Kazakh gas and supply Kazakh industries in the south. Also in July 2009, the third and final segment of the 1,800-kilometer China-Kazakhstan oil pipeline, running from Atyrau in Kazakhstan to Alashankou in China's Xinjiang region, was completed, allowing China direct access to oil production from Kazakhstan's western fields.

Multi-vectorism has also characterized the foreign policy orientations of the other Central Asian states, as they have used security cooperation with the West for their economic benefit. Following Kygyzstan's initial Manas eviction announcement, Uzbekistan moved quickly to conclude an array of supply and transit contracts with the United States and its allies. The most noteworthy, and legally creative, of these is Uzbekistan's commercial contract with Korean Air to supply U.S. forces in Afghanistan through a renovated cargo airport at Navoi. As a commercial contract with a third-party intermediary, the deal avoids any direct U.S. military basing on Uzbek soil. Yet the net result is clearly a rebuilding of the U.S.-Uzbek ties that had been badly disrupted after Tashkent evicted the United States from its base at Karshi-Khanabad in the summer of 2005.

With the spring 2009 signing of a U.S. transit agreement with Tajikistan and an ongoing refueling and supply accord with Turkmenistan, cooperation with the American military is now an important economic, if not military, source of partnership for all of the Central Asian states.

Taking Sides

Along with having to cope with the Central Asian practice of strategic multi-vectorism, Moscow faces a second impediment to regional hegemony: Russia's interventions in Central Asia run the risk of entangling it in local conflicts and disputes. For instance, regarding the politically thorny issue of regional water management, Russia must carefully thread the needle between reassuring Central Asia's most significant power, Uzbekistan, and safeguarding its own hydroelectric investments in Kyrgyzstan and Tajikistan.

On a visit to Uzbekistan in January 2009, President Medvedev stated that Russian investments in major hydroelectric power stations in neighboring Kyrgyzstan and Tajikistan would go ahead only after the interests of all states in the region on the water issue had been adequately addressed. The Russian president's remarks were greeted with shock in Bishkek and Dushanbe. His comments seemed to signal a shift away from Russia's previous position of unequivocally supporting the construction of hydroelectric power projects in these upstream countries.

The dam-construction issue is particularly sensitive for Uzbekistan, which fears that it could lose access to the water resources that it requires for its large-scale irrigation and cotton cultivation. Just a few months before, in the fall of 2008, Uzbekistan had withdrawn from the Eurasian Economic Community in protest against a deal reached by Kazakhstan, Kyrgyzstan, and Tajikistan to provide reciprocal supplies of coal, oil, and water. The water dispute is now the most significant source of potential interstate conflict in the region. Over the winter, there were even reports of Uzbek security forces crossing the border into Kyrgyz villages near the hydroelectric power sites.

In principle, Russia sees itself as playing a constructive role and even brokering a future regional solution to water management disputes. In practice, however, tensions over water are escalating, not receding, as upstream and downstream countries now view the issue in stark zero-sum terms. No evidence exists, moreover, to suggest that the parties involved would necessarily accept Russian arbitration as either neutral or credible.

The Manas Debacle

The evolution of negotiations over the status of the air base in Manas, Kyrgyzstan, which is used by the U.S.-led coalition, illustrates many of Russia's constraints and strategic challenges. In early February 2009, when President Bakiyev announced that the small Central Asian state had taken the decision to close the U.S. air base, the move was widely perceived as an important victory for Russia in rolling back American influence in Central Asia.

Neither Russia nor Kyrgyzstan acknowledged any formal link between a Russian pledge to provide the cash-strapped state with emergency financing and the status of the U.S. base. But by offering Kyrgyzstan a $2 billion package in short-term financing and investments in its Kambarata hydroelectric power project, Moscow had seemingly seized an opportunity presented by Kyrgyzstan's economic problems to leverage its influence and emphasize to Washington that it could control access to Afghanistan's northern transit routes.

Subsequent events, however, revealed that the Central Asian state retained considerable capacity to play the United States and Russia against one another for its own short-term economic advantage. Negotiations between the Kyrgyz and U.S. sides continued throughout the spring of 2009, and a final deal was concluded just a few days after the Kyrgyz government had secured $350 million in payments from Russia. The formal announcement of the new deal—which tripled to $60 million the annual rent paid by the United States, and reclassified the military base as a "transit center"—was delayed until late June, by which time Moscow had little choice but to publicly support the new agreement as Kyrgyzstan's "sovereign right."

In fact, with several of the Central Asian states concluding new transit or logistical support deals with the United States, Moscow itself has abandoned, for now, its attempts to block Washington's access to the Central Asia supply routes. Instead, at a Moscow summit meeting in early July, Medvedev and U.S. President Barack Obama announced a new bilateral accord that would allow the transit of American weapons and lethal cargo through Russian territory and airspace.

The Manas deal, besides provoking a high-profile bidding war with the United States, has created other potential regional headaches for Russia. As an equal partner in the construction of Kyrgyzstan's Kambarata hydroelectric power plant, Moscow now has a major stake in a project plagued by poor governance, and one whose economic viability remains in doubt. Moreover, Russia's 50 percent equity share in the project is likely to increase once the Kyrgyz side, as is expected, falls behind on its financing commitments. Thus Russia has firmly allied itself with Kyrgyzstan against Uzbekistan's vehement objections over Kambarata's construction.

In response to Kyrgyzstan's apparent double-crossing over Manas, Russian Defense Minister Anatoly Serdyukov and Deputy Prime Minister Igor Sechin were dispatched to Bishkek in early July. Shortly thereafter, Moscow announced it would open a second military base in Kyrgyzstan, in the southern Osh district, to be used by the Collective Security Treaty Organization's rapid response forces. Although Kyrgyz officials suggested that the new base is needed in the fight against growing extremist activity in the Fergana Valley, the timing of the measure seems to suggest that the new base represents, at least in part, a concession by Bishkek to Moscow for the Manas fiasco.

However, as with many of its other actions in Central Asia, Moscow's decision to open a second Kyrgyz base seems rushed and potentially fraught with unintended consequences. It remains unclear what, if any, additional regional influence this second base will actually grant Russia. Also, a future military presence in Ferghana risks becoming a magnet and target for local militant actions.

The hasty announcement, moreover, has greatly angered Uzbekistan, which insists it was not consulted on the issue. Tashkent worries about the presence of a Russian-led military on its doorstep in southern Kyrgyzstan. Some Uzbek planners are even concerned that, in the event of an escalating conflict over water or other regional disputes, the new base would allow Russia to support Kyrgyzstan militarily. Establishing a Russian military base in southern Kyrgyzstan is likely to increase Uzbek anxieties regarding Moscow's intentions as much as it will offer Russia an additional lever of influence in the region.

Beijing on the Rise

On top of all this, Russian influence in Central Asia now faces its most significant geopolitical challenge since the former Soviet republics gained independence. And this competition comes not from the West, with which Russia has been so preoccupied, but from the East. For much of this decade, Russia and China have pursued a strategic partnership, including their development of the SCO, in an attempt to limit Western influence in the region and to secure the area from transnational threats. But it is now clear that, as a result of China's recent massive investments in Central Asia, Russia-China relations in the region will be characterized by competition as well as by collaboration.

China's recent moves to secure access to Central Asian energy are particularly noteworthy, as is the impressive scale of these new investments. In May 2009, as Russia-Turkmenistan tensions escalated as a result of the pipeline explosion, China announced that it would lend Turkmengaz $4 billion to develop the South Yoloten gas field. Meanwhile, the current Turkmenistan-China pipeline, scheduled for completion at the end of 2009, will be expanded to provide to China 40 billion cubic meters (bcm) of gas per year, up from a previously planned 30 bcm.

From Gazprom's perspective, this pipeline represents Turkmenistan's first major alternative to the Russian-controlled Soviet-era gas distribution network. While Russia and some Western states have fought behind the scenes over the construction of future gas pipelines to Europe, China has steadily worked to complete construction of the massive 7,000-kilometer project that will send Turkmen gas eastward on a scale approaching the 40 to 50 bcm that Turkmenistan has annually supplied to Russia over recent years. Ashgabat insists it has enough gas reserves to supply all of these standing and new contractual commitments.

Elsewhere in Central Asia, Beijing has used the financial crisis as an opportunity to provide emergency financing to states in exchange for acquiring investments in key energy sectors. In April, Kazakhstan and China announced a $10 billion package that included a $5 billion loan by China Exim Bank to the Development Bank of Kazakhstan and a $5 billion investment by the China National Petroleum Corporation in the state company KazMunaiGas.

Chinese companies have also concluded an array of new investment deals with Tajikistan, including a commitment by Sinohydro Corporation to build hydroelectric power stations at Yovon and Nurobod, along with a transmission line to the town of Panjakent. China's strategic goal is to use these projects to supply cheap electricity to Xinjiang and its energy-intensive industrial base. Further investments in Tajikistan's infrastructure will assist the forging of future linkages. At a time when Russia is stalling on financing the Rogun dam, Chinese investments and commitments in Tajikistan are approaching $3 billion.

Competing Agendas

Differences in Russia's and China's national priorities are also reflected in how each power is now trying to shape the focus and regional agenda of the Shanghai Cooperation Organization. The SCO attracted concern among Western policy makers when it declared, at its 2005 annual summit in Astana, that U.S. military bases had served their purpose in Central Asia and should be put on a timetable for withdrawal. Just days later, the Uzbek government formally served notice to the United States to vacate its Karshi-Khanabad military facility, seemingly demonstrating that the SCO as a security organization had an anti-Western agenda.

Yet, unlike Russia, China now is concerned that the organization will come to be identified as an anti-Western security bloc. Beijing certainly sees value in the security functions of the SCO that target the so-called three transnational "evils" within the Central Asian region—extremism, terrorism, and separatism—especially in the wake of the recent political upheaval and violence involving Uighurs in Xinjiang. But China also sees the SCO as a potential vehicle for playing some future role in Afghan reconstruction, possibly in cooperation with the United States and NATO, and in general China remains reluctant to antagonize the West by identifying the organization as a counterweight to NATO. Even Russia has of late cooled on promoting the SCO as a security tool, preferring instead to promote the Collective Security Treaty Organization, which it effectively directs.

Above all, China sees the SCO as a vehicle for promoting its economic interests in Central Asia and as a regional provider of public goods. Yet these economic aspirations for the organization have been received with little enthusiasm by Russia. Moscow realizes that any meaningful regional economic integration will disproportionately benefit China's more dynamic and competitive economy. Russia prefers to maintain its own set of bilateral ties and political bargains with the Central Asian governments, rather than channel them through a multilateral forum that it does not completely control.

In fact, at the June 2009 SCO summit in Yekaterinburg, Russia blocked Chinese efforts to create an official SCO financial stabilization fund, arguing that legally it could not allocate funds to a multilateral lending institution without legislative authorization. Notwithstanding Moscow's objection, Beijing announced the creation of a $10 billion fund for SCO members, a sum of money well beyond what Moscow or even international financial institutions can allocate for the region's short-term financing needs. Beijing has signaled that it is willing to step in and provide large-scale investment and financial assistance in Central Asia, with or without Moscow's acquiescence.

New Rules of the Game

For nearly two decades, Russia has been able to translate the economic and military ties with Central Asia that it inherited from the Soviet era into networks of influence. Maintaining these mechanisms of influence and regional monopolies, especially in the energy sector, was itself the driving logic behind Russian–Central Asian relations. But recent events have made Moscow's attempts to preserve its exclusive regional control seem no longer feasible or cost-effective.

The dynamics of the Russia–United States–China triangle in Central Asia are now starting to take shape. China is clearly accelerating the promotion of its economic interests. The United States now sees Central Asia primarily as a set of countries to be stabilized, and when necessary economically placated, in the service of the military and reconstruction campaign in neighboring Afghanistan. The active promotion of human rights and democracy in the region is clearly not a priority of the Obama administration.

By contrast, the Russian strategic vision for the region remains unclear. Certainly, the so-called reset of U.S.-Russia relations has alleviated, at least in the short term, some of the sense of geopolitical competition between Russia and the United States. But regardless, Russia must prioritize and formulate a positive agenda for the region, one that is not fixated on constantly countering Western power in Central Asia.

To do this, Moscow may have to abandon its attempts to promote influence over the entire region—a task that is proving increasingly difficult—in favor of more clearly privileging a set of selected client states. Shifting from a strategy of "unite and rule" to "divide and rule" would carry risks, but it would give Russia a sharper focus in the region and a more solid political platform from which to effectively counter the rise of regional competitors.

Meanwhile, all three interested great powers need to be concerned about current economic and social trends in Central Asia. As economic conditions deteriorate amid the global downturn, as remittances from the overseas labor force drop, and as regional militant groups reconstitute their transnational ties, the potential for some sort of state failure in Central Asia is increasing. Such a scenario would not serve the interests of any of the regional powers and would only add to global worries about the instability of the greater Central Asian region.

ALEXANDER COOLEY is an associate professor at Barnard College, Columbia University, and an Open Society fellow. He is the author of *Base Politics: Democratic Change and the U.S. Military Overseas* (Cornell University Press, 2008) and coauthor, with Hendrik Spruyt, of *Contracting States: Sovereign Transfers in International Relations* (Princeton University Press, 2009).

From *Current History,* October 2009, pp. 325–332. Copyright © 2009 by Current History, Inc. Reprinted by permission.

Armenia: Parliament Debates Diplomatic Normalization with Turkey

GAYANE ABRAHAMYAN

Parliamentary debate in Armenia on diplomatic normalization with Turkey opened on October 1 with an emotional opposition attack on the government for supposedly selling out Yerevan's interests. Despite the political maneuvering, the Armenian legislature is widely expected to ratify protocols that open the way for a rapprochement.

The 400-seat gallery at the National Assembly was packed to capacity, with deputies, diplomats, analysts and various public figures on hand to witness the seven hours of give-and-take by MPs on the merits of the protocols. (For background see the Eurasia Insight archive). It is not yet clear how long parliamentary debate will last.

An August 31 agreement between Armenia and Turkey envisaged six weeks for public debate before the protocols were signed by the two countries' presidents, and then passed on to their respective parliaments for ratification "within a reasonable timeframe."

Little doubt exists that Armenia's parliament—controlled by the government coalition triumvirate (Republican Party of Armenia, Prosperous Armenia Party and Country of Law Party)—will vote for ratification. President Serzh Sargsyan, currently on a tour of Diaspora communities to discuss the reconciliation plan, has already agreed to sign the protocols on October 10.

While vociferous, parliamentary opposition to the protocols is minimal. The Heritage Party and the Armenian Revolutionary Federation-Dashnaktsutiun (ARF), which both oppose the documents, hold only 23 seats in the legislature. The governing coalition, by comparison, controls 94 seats. A senior ARF-Dashnaktsutiun member, Vahan Hovhannisian, admitted to EurasiaNet that "the forces are too unequal" to stop the documents' ratification.

Many opposition politicians affiliated with parties that do not have parliamentary representation, as well as representatives of some nongovernmental organizations, have suggested the protocols contain wording that is potentially disadvantageous for Armenia.

To buttress their concerns, critics point to protocol provisions covering the mutual recognition of the Turkish-Armenian state border, territorial integrity and the reaffirmation of border stability and the creation of a subcommittee to study archives and historic documents. The provisions, as seen by some government critics, can do serious damage to Armenia's long-term diplomatic goals. For instance, critics say the protocols can potentially formalize the inclusion of former Armenian cities within Turkish borders, de-legitimize Nagorno-Karabakh's independence bid and call into question the appropriateness of classifying the 1915 Ottoman Turkish slaughter of ethnic Armenians as genocide.

President Sargsyan has charged that such criticism only confuses the public. "Where did you see preconditions in the publicized protocols?" Sargsyan asked at a September 30 meeting with the 1,700-member Public Council, an informal advisory group. "If it were so, it would have been written in the document that it is signed with preconditions."

"Compromises" were necessary, he continued, but "many points" in the protocols also "are in Armenia's interests."

ARF parliamentarian Armen Rustamian countered during the October 1 parliamentary debate that Sargsyan appeared to have been outmaneuvered by Turkish negotiators. "The preconditions are not put in exact words," Rustamian claimed. "When one wants to poison somebody, they do not write 'poison' on the cup."

Fellow ARF member Hovhannisian, a former deputy parliamentary secretary, added that the protocols are the "result of dilettante and cynical diplomacy" that has "disgraced" Armenia's foreign policy.

Taking aim at the ARF's nationalist image, Republican Party of Armenia (RPA) parliamentarian Vardan Ayvazian, a staunch supporter of the Sargsyan administration, called on fellow deputies "not to pretend to be more patriotic than the Republicans," and to take greater care in reading the protocols.

The protocols, he said, "are going to ensure normal relations [with Turkey] and a new economic path for our country."

"We are for this process started by the president and we will take that path [of economic opportunities]," stated fellow RPA MP Eduard Sharmazanov, who also acts as the party's spokesperson. The ARF's Hovhannisian dryly rejoined that the "path will be a very short one."

One political analyst who attended the debates, Gagik Harutyunian, director of the Noravank Foundation, argued that the Armenian parliament "should wait for the decision of the Turkish parliament" before ratifying the documents. Making the first move, he contends, "would be wrong."

Other analysts, however, believe that Armenia should settle for what it has gotten from Turkey. One of the pragmatists, Ruben Melkonian, a Turkish expert in Yerevan, said the protocols are "the best documents at the moment."

Editor's Note—Gayane Abrahamyan is a reporter for ArmeniaNow .com in Yerevan.

Azerbaijan Can Resort to Multiple Options for Its Gas Exports

VLADIMIR SOCOR

Countries and companies along the Nabucco route in Europe (Bulgaria, Romania, Hungary, Austria, Germany) as well as Greece, Italy, and Switzerland are all expressing interest in purchasing Azerbaijani gas. If Turkey continues to block the transit agreement and if the E.U. and the U.S. fail to pull their weight with the AKP government, Azerbaijan can resort to alternative solutions for its gas exports.

Baku is now actively considering other markets and pipeline options. On October 14 Azerbaijan's State Oil Company signed an agreement with Gazprom for an initial volume of 500 million cubic meters in annual deliveries to Russia, starting on January 1, 2010. The small volume is susceptible to further increases at any time, without upper limits other than those of Azerbaijan's own gas surplus. The Russian purchase price is apparently equivalent to European netback prices. Moreover, this agreement apparently precludes the re-export of Azerbaijani gas by Gazprom to third countries at Azerbaijan's expense (EDM, October 15).

Baku is also considering the possibility of starting gas exports to Iran, initially in small volumes, by early 2010. Iran is already importing Turkmen gas for consumption in Iran's northern provinces. Those volumes, however, do not fully meet requirements there. Iran intends to import additional volumes of gas for off-season storage and peak-season consumption. This creates a market for Azerbaijani gas in northern Iran (Trend, October 17).

No third country transit solutions are necessary for Azerbaijani gas to reach Russia or Iran. Nor is the construction of new pipelines necessary. Pipeline connections to Russia and to Iran existing since the Soviet era, now require modernization of lines and compressors. Pipelines in both of these directions add up to approximately 10 billion cubic meters (bcm) in annual capacities. These can accommodate Azerbaijan's annual export surpluses for the next few years, in the event that the Nabucco project falters, or if Turkey's AKP government remains uncooperative on pricing and transit terms for Azerbaijani gas.

Azerbaijan plans to upgrade the Baku-Novo Filya and Gazimahomed-Mozdok pipelines for gas exports to Russia's North Caucasus territories. These Soviet-era pipelines can easily be adapted for use in the reverse-mode. Their combined capacity (after upgrading) would enable Azerbaijan's State Oil Company to deliver up to 7 bcm of gas to Russia annually, according to the company's president Rovnag Abdullayev (APA, October 19).

Similarly, Azerbaijan plans to upgrade the Gazakh-Astara and Gazimahomed-Astara gas pipeline links to Iran. Pending this, Iran's gas storage authority is expressing interest in using gas storage sites on Azerbaijan's side of the common border, with a view to using those gas volumes during winter in northern Iran (Trend, October 17, 19).

During the October 16 session of Azerbaijan's government (EDM, October 21), President Ilham Aliyev clearly alluded to the proposed White Stream pipeline as a possible option for Azerbaijan's gas exports (www.day.az, October 17). White Stream is being proposed by a London-based project company to carry Azerbaijani and Turkmen gas via Georgia and the seabed of the Black Sea to Romania and onward into E.U. territory (an earlier, now-discarded version would have run on the seabed to Ukraine). White Stream is one element in the E.U.'s Southern Corridor concept, designed to increase capacity and security of transportation for Caspian gas to Europe.

Aliyev discussed the White Stream proposal for the first time with the Romanian President Traian Basescu in late September in Bucharest, where the two presidents signed a strategic partnership agreement. In parallel, Azerbaijan's State Oil Company intends to examine the option of gas liquefaction for export via the Black Sea to E.U. territory. According to company president Abdullayev, "we are ready to review these forward-looking proposals in detail" (APA, October 20). Consideration of the Black Sea options suggests that Turkey does not necessarily enjoy a monopoly on gas transportation from the Caspian basin to Europe; and that Turkey can ultimately be circumvented, if the AKP government overplays its hand.

The AKP government's gas conflict with Azerbaijan is two years older than the Turkish-Armenian political normalization, which is now taking its first, uncertain steps. The two processes have no relationship to each other and Baku insists on keeping them separate. Meanwhile, Ankara's price extortion and its delaying tactics on the transit agreement have hurt Azerbaijan financially in two ways: by cutting into Azerbaijan's annual export revenues and by slowing down the development of

Azerbaijan's Shah Deniz field, the main designated source for the Nabucco pipeline.

Baku's October 16 response and its follow-up measures seek to concentrate attention in Brussels and Washington to a festering situation that puts Nabucco and the Southern Corridor at risk. According to Aliyev, at the government session, Ankara's "unacceptable terms proposed to us may lead to a failure of this entire project" (www.day.az, October 16). While Azerbaijan is irreplaceable as a producer as well as a transit country, Turkey is not irreplaceable.

From *Eurasia Daily Monitor,* vol. 6, Issue 194, October 22, 2009. Copyright © 2009 by Jamestown Foundation. Reprinted by permission. www.jamestown.org

Belarus after Its Post-Georgia Elections

Andrew Wilson

A New Paradigm or the Same Old Balancing Act?

The five day war in Georgia has already had far-reaching consequences throughout the East European Neighbourhood. In Belarus the signs are contradictory.

Political prisoners were released in August, and President Lukashenka has so far resisted Russian pressure to recognise Abkhazia and South Ossetia. On the other hand, the political conditions for the parliamentary elections on 28 September were not liberalised in the way the authorities had been hinting was possible before August.

Nevertheless, the EU went ahead with a tentative rapprochement on 13 October, suspending most of its visa bans on Belarusian officials, including Lukashenka himself, for six months, and announcing a study mission.

This article looks at three related questions. How much has Belarus really changed? How should the EU respond? And has EU policy been properly calibrated to the degree of change?

Three Balancing Acts

Lukashenka has long performed two simultaneous balancing acts. In domestic policy he has varied the degree of authoritarianism needed to pre-empt any challenge to his rule while trying to maintain the populist image of a benign 'batka.' In foreign policy he has developed a 'Titoist' game of playing Russia off against the West—or, more exactly, of making periodic and secondary overtures to the West to secure the maximum gains in the primary game with Russia. More recently he has added a third balancing act: maintaining the welfare populism turned consumerism that is his key 'social contract' with the nation, while trying to partially accommodate the growing pressures for nomenklatura privatisation. Despite his one-time job as a chicken farmer and image in some quarters as a brutish hick, Lukashenka has actually been quite successful in parleying his limited resources in these three strategies. The key question about the recent elections therefore must be the following: are they just the latest recalibration of these various balancing acts, or do they mark a more fundamental shift in the nature of the regime?

Luka the Chameleon

Lukashenka survived the Yeltsin era with ease. Relations with Putin were initially more difficult, but the Orange Revolution gave him a second lease of life as Belarus became Russia's laboratory for 'counter-revolutionary technology' in 2005–06. Three powerful pressures have built up in his third term (since 2006), however. First, Russia has recalibrated the price of its support.[1] It has not ended the subsidies regime, but it has made its financial and other support more conditional. The speed with which Russia moved to raise gas prices and win 50% of Beltransgaz as soon as the March 2006 election was out of the way was a real shock in Minsk—as it was intended to be. Lukashenka's links with the likes of Sergei Ivanov and Igor Sechin (through past oil deals) have also proved a double-edged sword under the new Putin-Medvedev tandem.

Second, the EU has belatedly begun to rethink its Belarusian policy. Until the 2004 enlargement, Brussels was under no real pressure from member states to change the isolationist approach it adopted in 1997 (after Lukashenka staged two controversial referenda to enforce a constitutional coup d'état in 1995–6). After enlargement, Poland made the running in setting strategy towards Belarus for the 2006 election; but now realises that its one-shot policy of pushing Aliaksandr Milinkevich as the 'united opposition candidate' was counter-productive, allowing Lukashenka the propaganda gift of a 'Polish plot.' Lithuania was tempted to promote a coloured revolution in Minsk in 2006, which also played into Lukashenka's hands; but Vilnius has also since become more pragmatic. By 2007–08, moreover, Poland and Lithuania were joined by other new member states and some older members such as Sweden in pushing for a strategic review of Belarus policy. The EU was also joined by the United States in rethinking the strategy of placing all bets on a weak and divided opposition. (In fact, after getting its fingers burnt over the Belneftekhim sanctions affair, the United States seemed to swing from being the leading hawk to one of the principal doves). Former Polish President Alexander Kwaśniewski has played a key role in this process with his 'Belarus Task Force.' Though the EU has not stuck to a common line, as a common line has not yet been agreed upon. Germany has followed one strategy, Polish Foreign Minister Sikorski another (criticised by President Kaczyński), yet another line has been taken by Finnish Foreign Minister Stubb as chair of the OSCE.

Third, the balance has shifted within the Belarusian elite from the 'siloviki,' who were essential to Lukashenka surviving the threat of a 'coloured revolution' in 2006, to the 'technocrats' who would like to enrich themselves via nomenklatura privatisation; although it is not yet clear whether this is primarily a political change—due to the purge of pro-Russian siloviki and/or the rise of Viktar Lukashenka—or an economic development. Long-term resource problems are clearly building up, but short-term GDP growth, still strong at 8% in 2007, and the state's fiscal position are currently holding up reasonably well—though Belarus's accounts are not exactly full and transparent. Belarus had relatively little direct exposure to the global financial crisis, but will worry greatly about the indirect effects on its privatisation programme and its search for alternative sources of international loans. In late October, Minsk agreed to another two-stage Russian loan for $2 billion.

Belarus Has 'Siloviki Wars' Too

The internal pressures produced by these multiple balancing acts were already apparent during Belarus's equivalent of Russia's 'siloviki wars' in the summer of 2007. As in Russia during the 2007–08 election cycle, there was a simultaneous clan struggle for power and economic assets. The extraordinary public beating of Zianon Lomat, head of the State Control Committee, in July 2007 coincided with management purges at Belneftekhim in May 2007 and Beltransgaz and the Belarusian Oil Company in July. The fall of KGB chief Stsiapan Sukharenka after the attack on Lomat was the first sign of the waning influence of the strange coalition of interests around Viktar Sheiman, representing certain Russian oligarchs and the domestic oil business as much as a hard line in domestic affairs. This was confirmed by Sheiman's dismissal after the even more bizarre (and still largely unexplained) affair of the July 2008 bombings in Minsk, along with his ally Hennadz Niavyhlas from his position as head of the Presidential Administration. The removal of Sheiman, Lukashenka's long-term number two, was a dramatic and potentially risky step, as he knows where many bodies are buried—both literally, given his role in the 1999–2000 'disappearances', and metaphorically, as he has long been at the centre of the local web of kompromat (not to mention the 'Liozna incident,' the fake attempt on Lukashenka's life apparently staged by Sheiman that helped secure Lukashenka his first election victory in 1994).

The decline of one clan was matched by the rise of another, centred around the President's son Viktar Lukashenka, who has recently built up a strong position in construction and property development. The reshuffles also showed that clan politics mattered more than competence, as the alliance between Viktar Lukashenka and the 'technocrats' pushed their men forward. Both Sukharenka's replacement at the KGB, Yury Zhadobin (moved to head the National Security Council in July 2008) and Zhadobin's successor, Vadzim Zaitsaw, were born in Ukraine and lack direct security experience. Zhadobin previously headed the Presidential Guards Service. Zaitsaw was a protégé of Ihar Rachkowski of the State Border Committee, another ally of Viktar Lukashenka. Niavyhlias was replaced at the Presidential Administration by Uladzimir Makei, who is a long term associate of both Lukashenkas—though it is less clear just who is riding on whose coattails. Makei may be the new regime's eminence gris. He is an arch manipulator, but no liberal.

The net effect of all this protracted game of musical chairs is, however, clear enough. The 'old guard' are down and almost out. Of their number, only Uladzimir Naumaw and Leanid Maltsaw remain at the Ministries of Interior and Defence—assuming Prime Minister Sidorski is now in the 'technocrats' camp. But the new 'technocrats' are just as self-interested a group as the old Sheiman clan. They do not want Belarus to learn from the mistakes of Russia and Ukraine in the 1990s. Quite the opposite. They want to enrich themselves in the same fashion. They want Western support, but they don't want too much Western capital. They want control of key economic assets for themselves; the West is perceived as a useful counterweight to a Russian incursion that comes with too many strings attached. Russian oligarchs like Roman Abramovich are already hovering over the juiciest Belarusian assets.[2] The global economic crisis may force some Russian oligarchs to rein in their ambitions; others are already starting to think they can buy assets up cheap.

Some 'technocrats' may be deluding themselves that they can use Lukashenka the younger as a 'battering ram' to win power, just as the 'young Turks' (would-be equivalents of Russia's shock therapy liberals of the 1990s) tried to do with Lukashenka the elder in 1994.[3] However, President Lukashenka is unlikely to be so easily outmaneuvered; nor is he likely to let his son monopolise power. It is often remarked that the precise moment when Lukashenka began to introduce a 'state ideology' in 2003 was ironically also the moment when many of its key tenets were being ditched. But Lukashenka's long-standing rhetoric against oligarchs and corrupt privatisation will be difficult to abandon completely. The introduction of limited curbs on state welfare in May 2007 (pensioners' health subsidies, free student travel) was a significant milestone; but so was the partial backtracking soon after. One interesting sign is that Lukashenka has been reluctant to sanction the establishment of a ruling party (Belaia Rus) that would bind him more closely to the new elite.

Belarus and the EU

How should the EU react to the prospect of a new Belarus? First and foremost, it should not over-react. The realist case for direct engagement with the regime has become slightly stronger since the war in Georgia. But in terms of the twin-track policy adopted in late 2004 and amplified in the Commission's non-paper or 'shadow Action Plan' addressed to the Belarusian people in December 2006 that set out an unofficial road-map of steps that Belarus could take to improve relations,[4] Belarus has not changed much. Arguably the release of political prisoners before the September 2008 elections was a more important change than the cosmetic improvements made for the elections themselves. Allowing in OSCE observers was an important step, but one which was predicated on maintaining a system that would not allow them to see very much. The presence of fifty to sixty opposition candidates led to unseemly speculation, even bargaining, over just how many opposition victories would be necessary for the elections to 'count'—hence the obvious disappointment when none whatsoever won through. The traditional processes of candidate registration weeded out many significant opponents; the opposition was largely excluded from the election committees (and there were no real observers at the higher levels where the votes were actually counted); a punitive new media law and criminal code were approved in June 2008 (on top of the restrictions, dubbed 'Sukharenka's law,' introduced before the 2006 election); 'active measures' seemed to have been used to split Kazulin's Social Democratic Party, possibly the UDF and others;[5] in June state media soured the atmosphere with a traditional propaganda film 'The Network,' pitching the now traditional message of a foreign-financed and frankly treasonous opposition. And as the OSCE-ODIHR report makes clear,[6] 'transparency' disappeared in the actual counting process.

With little to show in terms of short-term progress, optimistic observers look to the longer term. In time, the 'technocrats' may bring about regime change. Once they become oligarchs, they will develop an interest in the rule of law. But in order to become oligarchs they will abuse the system in the short run. The EU should be pushing for more of a rule of law in the region, not less.

The EU can welcome many of the economic reforms introduced in 2007–08, with more promised by Lukashenka immediately after the September elections: notably a 12% flat income tax, an expanded privatisation programme and hints at easier access for foreign investors.[7] But a business-friendly Belarus will be difficult to construct without political change, particularly a rule of law (the jailing of the lawyer Emmanuel Zeltser) and open tender processes.

Lukashenka meanwhile is selling himself not as a defender of democracy, but as a defender of Belarusian sovereignty; calculating that, after Georgia, this would suit a putative Western realpolitik. The 'technocrats' will of course also want to protect their target assets with a stronger state in the future. But the West should not rush into adopting a new policy on Belarus before it has decided on its general approach to the region as a whole. The EU (and NATO to a lesser extent) must face the reality of a newly competitive environment in the 'Neighbourhood'; but it must also choose between two types of competition. It can accept the reality of spheres of influence politics, and struggle over where to draw the line (Can we shift Belarus? Is Georgia lost? Can Ukraine be brought in?). Or it can choose to compete over the rules instead, and continue to uphold European values against the very different political and business culture that underlies Russia's version of 'neighbourhood policy.'

Some EU politicians have already rushed to declare that the overriding priority after the Georgia war should be to 'counteract Russian influence' in Belarus. Aliaksandr Milinkevich has been reported as saying that securing the country's independence from Russia takes precedence over all other objectives; and that there is therefore 'no alternative to a policy of dialogue' with the current regime.[8]

However the West will be accused of double standards if it accepts Lukashenka's overtures after so little has changed internally in Belarus. On the other hand, the broader Neighbourhood clearly has changed after the war in Georgia. There has been a paradigm shift to which the EU, NATO and the United States must all adjust. But Lukashenka has yet to develop into a true Tito. There is, at the moment, no 'equidistance' in Belarusian

foreign policy (Yugoslavia was 'non-aligned,' though still in the Communist Block). The overture to the West is but one part of a broader strategy of diversifying foreign and energy policy that has involved Belarusian missions to all sorts of energy-rich regimes, including Norway, but also Venezuela, Iran, Azerbaijan, Nigeria and China. The West should not base its policy on the mistaken assumption that Lukashenka is already a full-grown and fully-adept geopolitical balancer; though there is a case for basing its policy on encouraging him to become one.

Policy Implications

What should happen next after the tentative rapprochement between the EU and Belarus in October? The EU's priority should be to prevent both the 'Tito-isation' and the 'Franco-isation' of Lukashenka and Belarus. The latter has been proposed as a scenario when Belarus, like Spain in the twenty years before Franco's death in 1976, pursues economic liberalisation but remains resolute in defending political authoritarianism.[9] The EU should continue to push for political as well as economic change; just as it should make sure it is not simply part of Lukashenka's east-west balancing act.

Belarus is not going to turn into a democracy overnight, so the EU should not be wasting its energy by over-reaching and staking its entire policy on an obviously desirable but currently unachievable future. Instead it should be pressing for greater leverage in areas where it already has a toehold. For example, if the authorities in Minsk have not really moved on elections, but have released political prisoners; the EU should press for further progress in human rights, such as a moratorium on the death penalty, following the ruling by the Constitutional Court in 2004. It is significant that Lukashenka has moved aside the regime's chief thug Dzmitry Pawlichenka (who was in charge of suppressing demonstrations after the 2006 elections, exploiting his previous notoriety as head of Lukashenka's alleged 'disappearance squad').

Whether rightly or wrongly, the EU has lifted most of the visa bans imposed on the Belarusian elite for the next six months. The EU should encourage that elite to distance itself further from the darkest parts of its past by maintaining the remaining visa bans on Pawlichenka and others involved in the 1999–2000 'disappearances' and the violent suppression of protests after the 2006 election.

On sanctions, the United States has moved first as it has got itself into a muddle over Belneftekhim; but there is only a limited case for the EU to move on GSPs. Belarus has yet to move on the linked issue of rights for labour unions, but the broader point is that the EU should also be using its economic leverage more strategically, to secure more than simply economic gains.

Supporting the European Humanities University in Vilnius is an excellent project. The EU should also be looking to move back into supporting education and civil society within Belarus.

RTV1 and the European Radio for Belarus have only a limited audience. The EU should be seeking to support a freer media within Belarus itself. Minsk has only recently, and partially, moved to censor the internet. It has put a lot of energy into its youth strategy and does not want to alienate the internet-savvy younger generation. The EU should be pressing for a much lighter system of web regulation.

A real testing point will be NGOs. It is probably a fair assumption that Lukashenka has paid so much attention to closing genuinely independent NGOs and setting up government-sponsored alternatives in recent years that he will be reluctant to see them come back. But the EU should be prepared to try.

Both Belarus and the EU have an interest in energy diversification. It is in neither party's interest for Gazprom to gain full control over Beltransgaz. The EU could help Belarus explore the feasibility of linking up with the Odesa-Brody pipeline and of reverse energy supply from Lithuania.

The one issue constantly mentioned by ordinary Belarusians is visas. The mooted reduction in visa fees from €60 to €35 would be a step in the right direction. There is no conceivable reason why Belarusians should pay more than Russians for a Schengen visa. But the EU should go further and consider waiving fees for key sectors like students and professionals when and if visa negotiations begin.

Conclusions

The EU is between a rock and a hard place with respect to Belarus. The post-1997 isolation policy has clearly not worked, at least in terms of providing leverage over Minsk. But a new policy is not yet in place, or even on the horizon. The EU has its own balancing acts to perform. On the one hand, there is a case for moving quickly. If Belarus were to recognise Abkhazia and South Ossetia, the window of opportunity for improved relations may close almost as soon as it opened. On the other hand, the EU must situate its Belarus policy within a coherent strategy towards the Neighbourhood as a whole, where the priority should be combating Russian modus operandi, rather than the fact that Russia exercises influence in the neighbourhood; just as it should be combating Lukashenka's modus operandi in Belarus.

Notes

1. The increase in gas price to $100 per 1,000 m^3 in January 2007 cost Belarus an estimated $1.6 billion. The 'stabilisation loan' granted by Russia in December 2007 was for $1.5 billion. See George Dura, "The EU's Limited Response to Belarus' Pseudo 'New Foreign Policy' ", CEPS Policy Brief, no. 151, February 2008, www.shop.ceps.eu/BookDetail.php?item_id=1598

2. See www.charter97.org/index.php?c=ar&i=412&c2=&i2=0&p=1&lngu=en

3. Aleksandr Feduta, *Lukashenko: Politicheskaia biografiia*, (Moscow; Referendum, 2005), pp. 64–66, 72.

4. See "What the European Union Could Bring to Belarus," at www.delblr.ec.europa.eu/page3242.html

5. Lukashenka may have a long game in mind—further disabling the opposition before the 2011 presidential election.

6. See www.osce.org/odihr-elections/item_12_32542.html

7. There are legitimate fears that many 'foreign' investors mobilising through the city of London may in fact be Russian.

8. Ahto Lobjakas, "Belarusian Opposition Backs EU U-Turn on Lukashenka," RFE/RL Belarus, Ukraine and Moldova Report, October 8, 2008.

9. Vitali Silitski, "Why Lukashanka is changing (some) spots," *European Voice*, September 25, 2008, www.europeanvoice.com/article/imported/why-lukashenka-is-changing-some-spots/62434.aspx

Marching through Georgia

Wendell Steavenson

Mikheil Saakashvili, the President of Georgia, keeps late hours. During and after Georgia's five-day war with Russia this August over the breakaway region of South Ossetia, he spent long days receiving Western dignitaries—Condoleezza Rice, Dick Cheney, Joe Biden, Nicolas Sarkozy, Angela Merkel—and he spent his nights rallying foreign journalists to Georgia's side, often until the early hours of the morning. When I went to see him, at his office in Georgia's new, unfinished Chancellery building, in the capital, Tbilisi, I was told to arrive "sometime after midnight." Saakashvili's adviser Daniel Kunin, an American in his late thirties whose salary, until recently, was paid by the United States Agency for International Development, was there to meet me. Kunin is blond, boyish-looking, and usually cheerful, but his eyes were bloodshot. "I'm afraid it's going to be a little wait," he said.

Saakashvili's office is lined with icons of Georgian Orthodox saints and book-shelves containing biographies of Stalin and the Kennedy family. On a side table is a history of the Winter War of 1939–40, in which greatly outnumbered Finnish forces thwarted a Soviet attempt to annex their country. Saakashvili entered the office after 2 A.M., wearing a dark suit and a loosened tie. He is forty years old—he was thirty-six when he became President, after leading Georgia's Rose Revolution as a charismatic young democratic hero who, with a flower in his hand, confronted Georgia's postSoviet leader. He is known to all as Misha. Nearly six feet four, he has a lumbering gait, and his bulk and restless energy lend a curious awkwardness to his movements. "He can't sit still," one Georgian journalist told me. "There used to be a joke about it: people would take bets on how long he would sit down for—ten minutes, twelve minutes."

"There is an A.F.P. story saying that Condi Rice will announce on Thursday certain measures," Saakashvili said to Kunin. Saakashvili has staked his country's security on its close relationship with the United States, and he had been hoping that the Bush Administration would take action against Russia for sending its forces into Georgia.

"Yeah," Kunin said. "I don't know what they are going to say, though."

"They are going to bomb! From Alaska!" Saakashvili said, smiling. (Governor Sarah Palin had been chosen as John McCain's running mate two weeks earlier.) "Or they are going to shoot their mooses!" Saakashvili asked an aide to bring a bottle of Georgian Saperavi wine and said that he hoped my story wouldn't turn out to be "an obituary." In August, Russia's Prime Minister, Vladimir Putin, reportedly told President Nicolas Sarkozy, of France, that he wanted to see Saakashvili hanged "by the balls." (To which Saakashvili replies, "He would not have enough rope.")

The animosity between Saakashvili and Putin made the summer war look intensely personal. "People who have been in the room when Putin and Saakashvili have been together tell me that it has been electric with hatred," Richard Holbrooke, the former Ambassador to the United Nations, told me. "It's about a tiny man from the big country and a big man from the tiny country." (Putin is about five feet six. He is said to have been particularly annoyed to hear his Tbilisi nickname, Lilli-Putin.) When he first became President, Saakashvili said, Putin "was pretty polite, and more and more he would be cynical and aggressive. The Putin I saw in 2006 already behaved like an emperor." Putin, he said, never respected Georgian sovereignty. "Not only Georgia," he added. "He always said Ukraine was not a real country."

The Russians' "arrogance and their nastiness grew in exact proportion with oil prices, and with American problems in Iraq," Saakashvili told me. But as much as Saakashvili talks about Russian high-handedness, the affront is felt on both sides. There is a sense in Russia, Masha Lipman, of the Carnegie Moscow Center, told me, "that Georgia is this tiny country, with a population of four and a half million and a weak economy, who have been provoking them in an arrogant manner because they have a bigger guy"—the United States—"behind them, and that their leader has been emboldened to talk arrogantly in this way." The summer war, Lipman said, was seen as "a standoff against America, with the Russians victorious, as the Americans were unable to defend their client and ally."

Despite the Russian bullying of Georgia, most foreign observers believe that Saakashvili has been needlessly provocative. On August 7th, Saakashvili ordered a strike on Tskhinvali, the capital of South Ossetia, after a week of escalating clashes between Georgian police and South Ossetian separatist forces, who have largely controlled the region since the fall of the Soviet Union. (South Ossetia, in the north-central part of the country, bordering Russia, has been a part of Georgia for centuries, but the Ossetians are ethnically distinct and have their own language.) Saakashvili maintains that the situation had become untenable, saying that he had received intelligence that a Russian invasion had begun, or was about to. He claims that he had no choice but to attack Tskhinvali—that it was either fight or surrender. The Russians deride this version of events, saying that Georgia acted as the aggressor, threatening the lives of Russian citizens in Tskhinvali and of Russian peacekeepers stationed there. In any event, the Russians reacted quickly and decisively, launching air attacks and sending columns of tanks into Georgia. The tanks rolled to a stop within twenty miles of Tbilisi.

Saakashvili seemed to be winning the propaganda war in August and September, as Western media and politicians denounced Russia's incursion and its disproportionate use of force. In recent weeks, however, Western observers (and plenty of Georgian politicians, too) have become more skeptical of Saakashvili's story. In late November, one of his former allies, Erosi Kitsmarishvili, who had briefly been Georgia's Ambassador to Moscow, said during parliamentary hearings that he believed Saakashvili had been planning to invade South Ossetia for some time. When I spoke to Kitsmarishvili, before his appearance in parliament, he complained about the sense of being under siege that he felt Saakashvili had brought on Georgia. "I don't want to live in the new Caucasian Israel," he said. "I'd rather live in the new Caucasian Ireland. They still have their problems in the north, but no one is scared." Saakashvili had dismissed his criticisms ("Erosi has always been all about the money," he told me), and his government denied the charges. Still, even those who think that Russia simply seized on a pretext to invade now look at Saakashvili's decision to attack Tskhinvali—an obviously provocative move, with no clear backup plan—and ask, What was he thinking?

Saakashvili's mind is capacious but digressive. When talking about the summer war, he darts from the history of Soviet policy toward minorities to anecdotes about generals and presidents, and from bits about phone intercepts to citations from the *Washington Post*. In one conversation, he described the beauty of the women in Georgia's mountainous Tusheti region ("They're all blondes"); the layout of Tskhinvali ("It has only three streets: Moscow Street, Lenin Street, and Stalin Street"); and the characteristics of Georgian tribes ("I have two ancestors: South Ossetian—half wild, impulsive sometimes, maybe—but I am also Mingrelian. Mingrelians are all about compromise").

"His mouth is like a machine gun," one of his advisers told me. A Westerner who has worked closely with him said, "After you've had a discussion with him, you need to lie down. You need a drink."

It was past 3 A.M., and Kunin was yawning. Saakashvili looked up and asked him why he had not brought any good news for him that day. Kunin mentioned the latest McCain-Obama polls, which were indicating a tight race. McCain has ardently supported Saakashvili ever since the two men met, in Washington, in the mid-nineties; he has visited Georgia several times.

"He told us if he gets elected he'll make his first official visit to Georgia," Saakashvili said. I suggested they might have to rename George W. Bush Street, the airport highway.

Saakashvili laughed. "No, we'll call Tskhinvali New McCain or something."

Saakashvili's enthusiasm, volubility, and charm have helped him to cultivate American politicians in a way that has not come naturally to former Soviet leaders. (Kunin, whose mother is Madeleine Kunin, a former governor of Vermont, told me, "Really, he's just a typical Western politician.") Vice-President-elect Biden has also been a supporter, and visited Georgia in August. In some ways, the force of Saakashvili's personality has elevated his country's status beyond what its size would usually merit. One Western diplomat formerly stationed in Tbilisi told me that, though he was wary of Saakashvili's impetuousness, he understood his appeal. "Here's a little beacon of light coming from the Caucasus, a leader on the Western path of democracy and market reform who speaks impeccable English," he said. Georgia had come to represent a small spot of good news against the morass of Iraq and Afghanistan. The diplomat said, "We saw in him things we wanted to see."

Positioned on the southern slope of the Caucasus Mountains, which cut across the neck of land between the Caspian Sea and the Black Sea, Georgia lies along the ancient Silk Road, a route that has been partly retraced with oil pipelines. The country has long been a battleground for competing powers, and in the early nineteenth century it acceded to the Russian Empire, which it hoped would protect it against the Persians. But it stubbornly maintained its own identity, with its Georgian Orthodox Church and a language that is not Turkic or Slavic or Indo-European and is written with a distinctly curlicued alphabet.

With the collapse of the Communist bloc, in 1991, Georgia regained its autonomy but was immediately mired in a civil war and two separatist wars—in Abkhazia, on the coast of the Black Sea, and in South Ossetia. In 1992, a triumvirate of militia commanders invited Eduard Shevardnadze, who had been First Secretary of the Georgian Communist Party and Foreign Minister under Soviet President Mikhail Gorbachev, to be their leader. The Georgian forces were defeated in Abkhazia, and a quarter million ethnic Georgians were pushed out. In South Ossetia, tens of thousands fled, and the enclave became a "frozen" conflict zone, with an unrecognized separatist government, policed by mistrustful Russian, Georgian, and Ossetian peacekeepers monitored by a half-dozen or so European military observers.

Shevardnadze was elected President in 1995, and brought stability, but by the time he won reelection, in 2000, Georgia had sunk into an era of economic stagnation and endemic corruption. In winter, there was sometimes less than four hours of electricity a day in Tbilisi. Many government salaries were the equivalent of thirty-five dollars a month. Police would stop cars to demand, often apologetically, two-dollar bribes.

Parliamentary elections in November, 2003, were seen as a referendum on Shevardnadze's rule; Georgians had become openly frustrated with the status quo. Independent exit polls showed the United National Movement, led by Saakashvili, then head of Tbilisi's city council, in front, with twenty-six per cent of the vote; the government party second, with nineteen per cent; and the rest divided among smaller parties. Somehow, though, the official results gave Shevardnadze's party a small plurality. Saakashvili and other opposition politicians led demonstrations in Tbilisi, a model of well-organized, tenacious but peaceful protest. (Ukraine's Orange Revolution followed a year later.) At the climax of three weeks of protests, as Shevardnadze refused to negotiate or call another election, Saakashvili and other protesters pushed through the doors of the parliament, disrupting a speech by Shevardnadze. Saakashvili, holding up a rose in one hand, cried, "Resign immediately!" Shevardnadze was hustled away by his bodyguards—he resigned the following day—and Saakashvili picked up his abandoned glass of tea and drank it. He was elected President in January, 2004, with ninety-six per cent of the vote.

Six months into his first term, Saakashvili told the *Financial Times,* "People compare my style with that of J.F.K., but in terms of substance, I feel much closer to Ataturk or Ben-Gurion or General de Gaulle—people who had to build nation states."

Saakashvili was born in Tbilisi in 1967. His parents divorced before he was three, and he grew up as an only child among five adults—his mother, a university history lecturer; her parents; and her grandparents—in an apartment on a wide avenue in a new suburb of Tbilisi. When I visited his mother recently, we drank tea and ate raspberry cream cake in the same small sitting room where Saakashvili had grown up. She told me that as a boy he had "a lot of attention, and everyone somehow contributed to raising him." They were a family of professionals, but they were not part of the Soviet elite. "I remember my grandfather had a special coupon from some institute he lectured in," Saakashvili told me. "Once a year we could go and receive two Finnish sausages, five or six specialty sprats from Latvia, Finnish chocolate, and sometimes, if we were very lucky, a box of red caviar. It was a very humiliating experience. You had to stand in long lines in some basement of the Communist Party headquarters or whatever, and we were standing there together—these professors together with the drivers of heads of departments of the Communist Party." He had formed a clear world view. "I grew up with the idea, like all my generation, that the West was an absolute paradise and that we lived in an absolute hell. That was the identity of my generation."

Saakashvili wanted to travel outside the Soviet Union. He was fascinated by Western politicians—"I loved Reagan and Thatcher"—and, after serving in the Soviet Army for two years, he began studying international law at the Institute of International Relations, in Kiev. He was there as the Soviet Union fell apart. After graduating, Saakashvili worked for a while as a human-rights officer for the Georgian government and married a Dutch lawyer, Sandra Roelofs, within three months of meeting her, at a summer course on human rights in Strasbourg. Saakashvili told me that Sandra had wanted to go to Somalia on a humanitarian mission: "I told her that Georgia was as bad as Somalia, except there is some hope." Still, options for an ambitious young lawyer in Tbilisi during the early nineties were few, and in 1993 Saakashvili won a scholarship to Columbia Law School.

Sandra and Misha initially lived in Astoria, Queens, in a basement apartment whose landlord "always wanted to turn the lights out." Sandra worked at Cheesy Pizza for four dollars an hour. But within a year or so both had found jobs at New York law firms, and they moved to the Upper West Side. "Perfect, almost posh," Saakashvili said.

The following year, Saakashvili began working toward a Ph.D. in international law at George Washington University. As a dissertation topic, he chose *uti possedetis,* a principle in international law maintaining that newly autonomous states should have the same borders as they had before independence. The point of it, he said, in a clear reference to Abkhazia and South Ossetia, was that "illegal acts cannot produce results."

When he arrived in Washington as a student, he was dismayed to find Lafayette Square, opposite the White House, full of homeless people and rats, and the nearby streets in disrepair. "You get a sense what different governments are," he recounted. "The road was really very bad, worse than roads in Shevardnadze's time. But because local D.C. government was broke—even if it was leading to the White House, who cares? There he sits, the most powerful President in the world, but he cannot fix the road!" He went on, "They call it separation of powers. Some people would call it democracy. I would call it inefficient."

Saakashvili never finished his doctorate. In 1995, Zurab Zhvania, one of the young reformers in Shevardnadze's Citizens' Union party, asked him to return to Georgia and get involved in judicial reform. His dissertation adviser, Thomas Buergenthal, a Holocaust survivor who is now a judge at the International Court of Justice, in The Hague, also encouraged him to return. Saakashvili recalls his saying, "Look, you don't have to write this thesis. Change your country and call it 'The Making of a New Country.' It will be much more interesting."

Saakashvili went back and entered parliament on the governing-party list. In 1999, he ran to represent Vake, a district of Tbilisi built in the nineteen-fifties and sixties, when Soviet architecture could still manage grand, high ceilings and parquet floors for the apparatchik class. "I was running against the very corrupt head of tax inspection," he said. "We thought I would easily win because I am young, modern, et cetera, et cetera. But we did some polls and saw he was ahead by a very large margin. So that was the first real election campaign in Georgia. The first billboards, the first negative ads—the first real political ads on TV. It was really a classical Western political campaign," he said. "And I won by fifty votes."

In 2000, Shevardnadze made Saakashvili his Justice Minister. Soon after, he stormed into a cabinet meeting trailing a TV crew and brandishing photographs of new dachas owned by various ministers. This move—audacious, self-righteous, and media-savvy—became a template for his future. He resigned, after less than a year, formed his own political party, and soon afterward was elected head of Tbilisi's city council.

As a young parliamentarian, Saakashvili made several trips to Washington with Zhvania, the man who had urged him to return to Georgia, and knocked on any door on Capitol Hill that might open. (Saakashvili grasped very early the importance of connections in Washington, and as President he has paid eight hundred thousand dollars to Orion Strategies, a two-man lobbying firm run by Randy Scheunemann, a close adviser to John McCain.) He recalled, "We would have a big map and tell them, 'This is America. This is Russia. This is Georgia. This is the pipeline. This is why Georgia matters.' "

Saakashvili's first months as President were heady. Most of the police force was fired and replaced, and crime went down.

Tax rates were cut, but revenues increased. Civil-service jobs were eliminated, and ministers accused of corruption were arrested on national television. In Tbilisi, one common criticism of Saakashvili is that his impatience to get things done has subverted the institution-building that should be the bedrock of democratic reform. "I've seen him do things that are right out of Giuliani's playbook, and I've seen him do things that are right out of Putin's," Lincoln Mitchell, a professor of international politics at Columbia who worked in Tbilisi for two years for the National Democratic Institute, an American nonprofit, told me. "It's like he has a good angel and a bad angel on his shoulders, and the good angel is telling him to do the right thing and the bad angel is telling him just to do what he needs to do to get it done."

By 2007, Georgia's G.D.P. was growing by more than twelve per cent. There were freshly laid roads, renovated schools and hospitals, new housing developments, and elaborate public fountains (including one that spouts to the theme from "Mission Impossible"). Foreign direct investment was about $1.7 billion last year, and there have been hundreds of privatizations. A foreign businessman who has lived in Georgia for almost a decade described Saakashvili's government as "libertarians," and said, "Milton Friedman would have been proud." He was skeptical about much of the apparent success, though. "Many of the privatizations have been untransparent, and those of the privatizations that have been transparent have met all kinds of obstacles," he said. (For example, some public utilities were sold to foreign-registered companies with obscure ownership structures.)

One adviser described Saakashvili's project as "two hundred Western kids trying to drag four and a half million Georgians into the twenty-first century." Saakashvili made liberal use of European and American advisers like Kunin, received early financial backing from the U.S. State Department and, indirectly, from the Soros Foundation, and included in his government senior officials with foreign passports (his first Foreign Minister was the former French Ambassador to Georgia, his current Minister of Economics is Georgian-American, his Defense Minister, until last week, was Georgian-Israeli), all of which contributed to the Russian sense that Georgia was a Western—and for "Western" the Russian elite tends to read "American"—proxy.

Saakashvili's ministers argue that their postmodern economic experiment was threatening to Moscow because it provided an example of what it was possible to achieve in the region without Russian sponsorship. But the economic reforms failed to help many ordinary Georgians, and the euphoria that followed the Rose Revolution was dissipating.

Levan Ramishvili, who works for the Liberty Institute, a think tank in Tbilisi, noted that unemployment in Georgia was eighteen per cent, and even higher in Tbilisi. Many of the reforms in health and education were poorly implemented or had not been funded. "We spend one of the smallest amounts of G.D.P. per capita on education in all of the post-Soviet countries," he said. Meanwhile, the defense budget had increased to a third of all government expenditure. "There's always another crisis that diverts political attention and resources or economic means away."

In Tbilisi, Saakashvili's government was increasingly seen as high-handed, ignoring common concerns and the rule of law. On November 2, 2007, fifty thousand protesters took to the streets—crowds that rivalled those of the Rose Revolution. They included the economically disaffected, those who believed the government had become corrupt, and others who were outraged by the arrest of Saakashvili's former Defense Minister on corruption charges; in the midst of the investigation, the former minister had gone on television to announce that he was forming an opposition party, and accused Saakashvili of ordering a political murder and of other crimes. (Saakashvili called those charges absurd, and the former minister, after a brief time in prison, retracted them.) A television channel, Imedi, owned by Badri Patarkatsishvili, a millionaire oligarch, covered the demonstrations live and urged the protesters on, broadcasting denunciations of Saakashvili's regime. When Saakashvili sent police to remove about seventy protesters who were camping out on the steps of the parliament building, Imedi broadcast what amounted to an appeal for reinforcements. As hundreds of protesters gathered, riot police cleared the streets with tear gas and rubber bullets. That night, the police stormed and wrecked Imedi's studios, forcing journalists and studio guests to lie face down on the carpet, and shut the station down. Saakashvili imposed ten days of emergency rule and claimed the protests were part of a Russian-backed plot to bring down his government.

Sozar Subari, Georgia's public defender—his job is to be a sort of civil-liberties ombudsman—told me that he had tried to intervene between demonstrators and police, only to be beaten himself: "My staff member said, 'What are you doing? He is the public defender!' and he also got one hit." The laws that Saakashvili had passed included positive reforms, but in practice, Subari said, political pressure on the courts and on judges had increased. "In reality, the prosecutor just moves his eyebrow and the judge knows what to do," he said. "The law has been adjusted for the comfort of the ruling party as if it were the reign of Louis XIV." He outlined several cases of police violence, confiscated properties, and businessmen harassed by tax police. Four years into a five-year tenure, Subari, constitutionally, cannot be fired, but he told me ("And this is how it always works in Georgia") that some months ago he got a call from an old university friend who insinuated that he should resign and take a job with an independent agency at a higher salary. "They could say I was resigning for health reasons," he said.

Saakashvili's American supporters tend to disregard questions about his democratic credentials. One Western diplomat told me that the measures against the demonstrators were not so different from those practiced in European capitals. But the numbers on the streets had shaken Saakashvili, and he resigned and called for early Presidential elections. Holding the election, he told me, was the riskiest thing he had ever done. "We were living in a city that just hated us," he said. In January, 2008, he was reelected, with fifty-three per cent of the vote, but he lost in large parts of Tbilisi.

One night this fall, I had dinner on the terrace of the Hotel Kopala with a member of parliament. He told me that before the Rose Revolution he couldn't have afforded to eat at a restaurant as nice as this; in terms of economic development, he appreciated what Saakashvili had achieved. But he spent most of the meal criticizing Saakashvili's arrogance, imperviousness to advice, and recklessness. He told me that a few weeks after the Rose Revolution he and a friend had bumped into Saakashvili, and the friend said, "You understand now you'll be President of this damned country. And you'll destroy us, because you are crazy!" Saakashvili laughed and said, "But I can always invent something"—he snapped his fingers—"and get out of a mess."

Saakashvili campaigned for President both times on a platform that promised greater economic growth but also made it clear that the separatist regions of Abkhazia and South Ossetia would be a priority of his Presidency. Georgians generally view Abkhazia and South Ossetia as an integral part of their homeland—the regions have long been ethnically mixed—and their return as a pressing patriotic issue.

Saakashvili did manage to regain control of Adjara, an autonomous region on the Black Sea along the Turkish border, in the spring of 2004. Adjara had been ruled since Georgian independence as a personal fiefdom of Aslan Abashidze, a former minor functionary. He decreed that all the houses in the capital, Batumi, should be painted white; tried to inveigle his way into a connection with Washington by offering Hillary Clinton's brothers an exclusive deal on hazelnuts; and pocketed revenue from the port and border trade. Abashidze was unpopular, but he was supported by Moscow. Saakashvili encouraged protests and imposed sanctions, hoping for a sequel to the Rose Revolution. As Abashidze was seeking asylum in Russia, Saakashvili went on television to proclaim, "Aslan has fled. Adjara is free!" Soon after this intoxicating success, he began to devote serious attention to the problem of South Ossetia.

In August, 2004, Saakashvili sent Georgian police into South Ossetia to shut down the market in smuggled cigarettes and gasoline that was flourishing on the border there. This move led to clashes between Georgian peacekeepers and South Ossetian militia. The Americans were alarmed. One Western diplomat told me, "They were stumbling into this mess, and I think there was plenty of fault on both sides." He said that he told Saakashvili to stop the fighting. "He got furious. He said, 'Damn it, we cannot afford to be in this situation,' " and he conceded.

Saakashvili told me that, back then, he had still hoped that he could persuade the South Ossetian leadership that a democratic, economically resurgent Georgia would be a better partner than Russia. "I thought we needed patience," he said. The problem, according to Oksana Antonenko, a senior fellow at the International Institute for Strategic Studies, in London, was that Saakashvili fundamentally underestimated the depth of feeling and hostility on the Ossetian side. "When he realized that he was not going to prevail by sheer charisma and vision for integrating with the West, he started to operate more and more in a pattern where he just didn't want to deal with those people who didn't like him—with the Abkhaz and the South Ossetians—and he created a kind of semblance and image, a Potemkin village of a peace process."

He also may have misjudged Russia's commitment to maintaining its hegemony in the region. "It's impossible to imagine a weak Russia in the Caucasus," Sergey Markedonov, of the Institute for Political and Military Analysis, in Moscow,

said. "It's our tradition. It's not European, I understand, but it's really, really impossible to ignore it."

Saakashvili, meanwhile, strengthened his military ties to the West. He sent two thousand troops to Iraq (they formed the third-largest contingent there, after the American and the British, before being called home in the wake of the August war) and lobbied, with the Bush Administration's support, to be admitted to NATO. Three former Soviet Republics—Estonia, Latvia, and Lithuania—have become NATO members, and even the remote possibility that Georgia might join them has infuriated the Russian government. Masha Lipman told me that in Moscow NATO expansion felt like "an encroaching force advancing closer and closer to what Russia regarded as its legitimate sphere of influence."

In 2006, Georgia arrested four Russian officers on charges of espionage; Russia expelled several thousand Georgian nationals living in Russia, embargoed Georgian exports, and cut air and postal links. A few weeks later, Saakashvili organized elections in South Ossetia's Georgian villages; Dmitri Sanakoev, a former South Ossetian prime minister and militia leader, became head of a new administration, to compete with the region's Russian-backed government. Sanakoev, with money from Tbilisi, built houses, clinics, and sports facilities, but many Ossetians saw him as a traitor and a puppet. Russia increased the amount it paid in pensions to South Ossetians and the number of Russian passports it handed out. There was often gunfire around Ossetian and Georgian checkpoints, and, this July, Sanakoev narrowly escaped an assassination attempt.

The sequence of events that followed is much in dispute. The skirmishes between Ossetian militia and Georgian police in early August appeared at first to be just another flareup. Then, according to the Georgians, the attacks on their positions intensified, and gunshots came close to some Georgian villages; attempts to engage the Ossetians in negotiations failed. "The Russians clearly just refused to talk to us, and the separatists refused to back off, which was very unusual," Saakashvili said. He concluded that this was "preplanned" to provoke Georgian action. (The Russians say that it was the Georgians who didn't want to talk.) Temuri Yakobashvili, the Georgian Minister for Reintegration, watched these events with unease and some confusion. On the afternoon of August 7th, he drove to Tskhinvali for the second time in a week. He found the streets almost deserted. "You know, it was weird," he told me. "In Tskhinvali, we discovered a completely empty city. You just had a bad feeling—there were no people on the streets, no cars on the streets. The city was sort of pregnant for war."

Yakobashvili told me that he met with General Marat Kulakhmetov, commander of the Russian peacekeeping force, who told him that the Ossetians were not under Russian control and suggested that a unilateral Georgian ceasefire might calm things down. At seven o'clock, Saakashvili announced the ceasefire on television, and offered an amnesty to Ossetian fighters and broad autonomy for the region. But, according to the Georgians, there was continued mortar fire from South Ossetian positions. (According to the Times, European military observers reported no sign of shelling in the area until the Georgian artillery began firing.)

Later that evening, Yakobashvili said, Saakashvili received a call that seemed to confirm reports of Russian units moving south through the Roki Tunnel, the only road between Russia and South Ossetia. He described Saakashvili putting down the receiver: "He got pale. I asked him what had happened, and he said, 'They're moving.' " Saakashvili asked Yakobashvili, "What do you think? Are we becoming Israel?" (Russian authorities say traffic through the tunnel that night was either an ordinary troop rotation or supplies for their peacekeepers.) Saakashvili told me that he had tried to call Western allies, but Bush was already in Beijing, along with Putin, for the opening ceremony of the Olympics. At 11:30 P.M., on Saakashvili's order, Georgian artillery opened fire on Tskhinvali. Rockets hit civilian housing blocks, South Ossetian government buildings, and the Russian peacekeeping barracks.

In the next four days, Russian aircraft decimated Georgian battalions; Russian tank columns blocked the main national highway, surrounded the port of Poti and Georgian military bases, and roamed the roads of western Georgia. The Russians presented themselves as the defenders of the Ossetians, but much of the world saw the conflict in terms of the Cold War paradigm of a Russian military threat. T-shirts appeared in Tbilisi with "Stop Russia!" printed on the front and "Prague 1968" on the back.

It is unclear what Saakashvili thought or hoped he could achieve by escalating the conflict so dramatically. The immediate effect was to turn the fight with the South Ossetians into all-out war and to pull in the Russians. On August 11th, Saakashvili ordered that there be no further resistance. Sarkozy intervened to broker a ceasefire, a six-point plan that required all parties to withdraw to prewar positions—although Russian and Ossetian forces still occupy parts of the enclave that had been under Georgian administration, as well as Abkhazia's Kodori Gorge. (The Russians say that they are there at the invitation of the Ossetians and the Abkhaz.) More than four hundred Georgians were killed, more than half of them civilians, along with more than a hundred and fifty Ossetians and, according to Russian military sources, sixty-four Russian soldiers. Two hundred thousand people were displaced; months later, twenty thousand Georgians remain unable to return to their villages in South Ossetia.

Privately, every Western diplomat I spoke with said that the Georgian attack on Tskhinvali was a mistake. They blamed Saakashvili's hubris and questioned the broader policy of provoking Russia. In some ways, he was lucky. The Russian reaction—which Saakashvili insisted to me was "not a reaction but an action"—had the effect of uniting the Americans and the Europeans in condemnation. The long-term consequences of the debacle remain unclear. "I think everyone lost," Richard Holbrooke told me. "It's a Georgian loss because their economy has been heavily damaged and foreign direct investment has dried up, and now this credit crunch is going to have an adverse affect. I don't think Misha's position has been strengthened politically. Russia lost because it was their real objective to topple Saakashvili, and they found that they had alienated the West. And it revealed Russia as being very dangerous to their neighbors. And the United States and NATO lost because

there was nothing we could do for a small democratic country we were supporters of."

In September, Saakashvili came to New York for the annual opening session of the United Nations. He was eager to shore up American support—a billion dollars' worth of direct aid had already been announced. I met Daniel Kunin at Rockefeller Center, where he was waiting for Saakashvili to arrive for an interview with NBC's Brian Williams. Kunin was happy. Transparency International had just released its new "corruption perceptions index," and Georgia's rating had improved from No. 79 to No. 67 in the world, well above Russia, which was No. 147. "It's a gift. I just added it into the speech," Kunin said.

That afternoon, Saakashvili addressed the General Assembly. He did not mention Russia by name, calling it "our neighbor," but he decried the threat to territorial integrity and called for an international investigation into the causes of the war. He professed his commitment to democratic values and promised a redoubled reform effort, a "Second Rose Revolution." After the speech, he went outside to greet a crowd of Georgian protesters holding up "Stop Russia!" placards and chanting "Misha! Misha! Misha!" Then he went back in to attend a reception hosted by President Bush.

After about half an hour, Saakashvili emerged from the reception and stopped at the bottom of an escalator to talk to a Central Asian president. The man shook his hand and left, but Saakashvili lingered. Just then, President Bush came out, accompanied by the American Ambassador to the U.N., Zalmay Khalilzad. Saakashvili approached and, in one fluid motion, put his arm around Bush's shoulder, shook his hand, and warmly greeted Khalilzad. The three of them rode up the escalator, chatting like old friends, until Bush and Khalilzad stepped toward a bank of elevators and disappeared.

Waiting for his car by the side entrance, Saakashvili was ebullient. He flipped through a copy of *Newsweek* with his picture in it and showed off his new watch: "Kenneth Cole. I got it in Miami." (Saakashvili has a penchant for watches; he buys several a year and tends to give them away when he gets bored with them.) He said that he didn't want to stay for the dinner after Bush's reception. "First, they don't feed you well at this thing," he said, "and then they sit you next to Mugabe."

In New York, Saakashvili kept his Secret Service detail busy. When he first came to the city as President, the Secret Service referred to him as "the Energizer Bunny." Saakashvili walks fast: heading to a meeting with *Time* editors one afternoon, he left a trail of staff members along Sixth Avenue. "Is this is a Presidential race?" his protocol chief asked, panting. He eats fast: two-handed, reaching for a second helping even as he is chewing the first. He talks fast, and in New York he talked to everyone from McCain and Palin to Henry Kissinger, George Soros, Ban Ki-moon, and Richard Gere. (The only person he missed was Barack Obama.) He gave interviews to NBC, CNN, the BBC, PBS, and Fox News. By the end of the week, he was joking that the only channels he had not appeared on were the Food Network and Animal Planet.

The morning before his speech at the U.N., Saakashvili spent an hour chatting with Bill Clinton. "It's amazing! I am not American—and that they would perceive me as one of them and that it's interesting to know my opinion," he said. "No European would gossip with me this way."

In the evenings, he went out until two or three in the morning. He likes parties: one night, he and his bodyguards crammed into a corridor at the Chelsea Hotel; another night, he went to watch the first McCain-Obama debate at a book party in the West Village. He visited an architectural exhibition at the Museum of Modern Art to scout out ideas. Saakashvili seemed to want to stay in New York, even after most of his delegation had returned to Tbilisi. Protocol officials scrambled to rearrange flights as he postponed his departure again and again. I asked him how things were at home. "As well as can be expected," he said.

Saakashvili enjoyed a postwar bounce in his approval rating at home, but there have been signs of discontent. Nino Burjanadze, who had been one of the leaders of the Rose Revolution but split from Saakashvili's party just before the May elections, announced that she had forty-three questions for Saakashvili about the events of the summer, and called for an official inquiry. (When I asked Saakashvili about Burjanadze, he said, "Nobody likes her.") One of his closest ministers told me he would support a referendum on early elections, a constitutional mechanism that might allow for Saakashvili's resignation if popular opinion turned against the government. (His current term lasts until January, 2013; constitutionally, he cannot run again.) The minister seemed torn about his and his colleagues' responsibility for the summer defeat. "People will judge," he said. "People will judge depending on how we proceed with the economic recovery and how we handle the political discussion about the war."

One evening after his return from New York, Saakashvili and I sat by the Black Sea, at a government dacha in Adjara. Two days earlier, Angela Merkel, the German Chancellor, had met with the Russian President, Dmitry Medvedev, in St. Petersburg, and discussed a new Baltic pipeline to carry gas to Germany and Western Europe. In the same press conference, Merkel had said discouraging things about Georgia's prospects for NATO membership. I asked Saakashvili if he had been upset.

"I think it wasn't nice," he replied. "She tried to balance, but it's not the right time to balance." He talked about the pressure Russia was putting on Ukraine. He said that he had spoken to Victor Yushchenko, Ukraine's President, the night before, and that Yushchenko had been "very emotional." He mentioned, again, the arrogance of Putin.

Saakashvili didn't think, really, that he had been the loser of that summer's war. "If we thought winning was taking over Tskhinvali—well, it didn't mean much for me anyway. To get another hundred Georgian towns to administer?" He dismissed the loss of the territory that had fallen under separatist control: "So what? These are two districts of Georgia. It's not a setback. We are in a fight, and this is the position of fighting." The larger point was "to get rid of Russian influence," he said. "And the Russians overreacted."

He went on, "If anyone had any illusions that we could get rid of two centuries of Russian influence in one week, that would have been a big mistake, but we are on the way. This is the beginning of the end for them in this region." He added, "But I would like to be successful myself, not one generation from now."

During the American Presidential campaign, Saakashvili was always careful, when asked about his friendship with John McCain, to emphasize his bipartisan support in America and his relationship with Joe Biden. When I spoke to him after the election, he drew parallels between himself and Obama: they had both spent their first nights at Columbia sleeping on the street, having been locked out; and they had both felt the weight of popular expectations as young Presidents. "O.K., I led a revolution for five million people in Georgia, he leads a revolution for six billion. But I know, when everyone loves you, everyone wants to see in you something they want to see for themselves." Still, he worried about "this new trend emerging—telling Obama, warning him and the new people that will be coming into the White House, that Georgia and Ukraine are like this poisonous stock that you should drop, abandon."

He said that he had talked to Obama a few days before, and that although they hadn't gone into specifics, he found him well informed and broadly supportive. But, he acknowledged, an Administration preoccupied with the economic crisis might not consider Georgia a priority. "To give up Georgia is easy, because Georgia doesn't have a big lobby in America or some vested interests. It's all about how idealistic America can be."

Earlier, he had told me, "With Americans, you can be yourself, and they accept you. I've never heard from Americans that I'm hotheaded, bossy. From Europe, I've heard a lot of this. Because, for them, in some cases, in Europe spontaneity looks like a dangerous thing."

The Country That Europe Forgot

F inding the offices of the only English-language paper in a foreign city is hardly a tough assignment—except when the paper is the *Tiraspol Times,* the slickly produced propaganda outlet of the Transdniestrian authorities. This purports to be a proper newspaper, with a website and a print edition. The editor until recently was one Mark Street, assisted by Jason Cooper (bald) and Karen Ryan (blonde). The news, features and analysis are better than anything Moldova produces in English.

Yet this fine organ is remarkably elusive. Its website shows no address, but the domain registration gives 118 October 25th St. That is shared by the infamous Hotel Druzhba and a political party. Nobody at either building has heard of the *Tiraspol Times.* Local journalists cannot recall meeting anyone from the paper. Even a person such as Dmitri Soin, who has featured in its columns, says he has had contact "only by e-mail."

The internet reveals no trace in any previous life of the journalists who supposedly work there. The only discoverable person seems to be its publisher, Mr Des Grant, an Irish local-newspaper owner with an unexplained but lively connection with Transdniestria.

It is all a bit reminiscent of a Transdniestrian disinformation exercise last year that involved the creation of a grandly named think-tank, the International Council for Democratic Institutions and State Sovereignty. This published a report—supposedly by top international lawyers—backing Transdniestria's legal case for independence. But the lawyers named had nothing to do with the report, and the "international council" turned out to be merely a website, with no offices, staff, legal existence, or claim to credibility. (An *Economist* investigation prompted it to add an entertaining entry in its website FAQ entitled: "What is your response to claims that ICDISS does not exist?").

A clue comes from Mr Antufeyev, the state security minister. Asked about his country's impressively energetic English-language propaganda efforts, he smiles broadly and says, "Why should we not use the West's technology, on our side?" A supplementary question about the striking success of the Transdniestrian efforts, compared to those in Moldova, provokes a hearty chuckle, and the cryptic: "We are not lazy."

It is a mistake to see Transdniestria only as part of a tussle between Moldova and Russia. Ukraine, the region's eastern neighbour, is another big factor. Senior Ukrainian politicians and officials have made a lot of money out of trade with the breakaway region.

A typical scam is to import chicken from Odessa, claiming that it is bound for Transdniestria (and therefore duty free).

Once across in Transdniestria the shipment is switched to smaller trucks and brought back into Ukraine. A truck of cheap American chicken meat brings a profit of several thousand euros if it can be resold in Ukraine's protected and inefficient home market.

The cleverest bit of the scam is that re-importation ceases to qualify as a criminal offence if it is done in small enough loads. Another dodge is to import luxury cars as "car parts," by the simple expedient of removing the windscreen wipers, hub caps and other minor items. Once cleared by a cooperative customs official, the car can be "reassembled" at an instant profit of half its sale price.

Ukraine, in consequence, is not terribly enthusiastic about an Odessa-based EU monitoring mission, led by a formidable Hungarian policeman, Ferenc Banfi, which is trying to beef up security on the border with Transdniestria. Mr Banfi, a marmite-loving anglophile, is undaunted. Smuggling certainly continues, he concedes, but it is a lot less blatant. Transdniestrian firms are having to register with Moldovan customs authorities and conduct their trade legally. It may no longer be quite right to call Transdniestria a "black hole." And it is hard to avoid the impression that, whatever the *Tiraspol Times* may say, it is slowly imploding.

Thursday

The misnamed Hotel Druzhba (Friendship) used to be the only place to spend the night in Tiraspol. For connoisseurs of truly dismal Soviet-style rudeness, apathy, squal or and clashing shades of muddy pastel, it is still unmissable. As a place to stay, its noisy, draughty rooms, with their nylon sheets, uneven tiles, flimsy locks and eccentric plumbing, leave a lot to be desired.

Even Dmitri Soin, the chief Transdniestrian cheerleader and director of the magnificently named Che Guevara School of Political Leadership, shows visible relief that his foreign visitor's enthusiasm for authentic local flavours does not stretch to the Druzhba. His youth movement, *Proriv* (Breakthrough), apes the pro-Western flagwavers of the "coloured revolutions" that toppled autocrats in Georgia, Ukraine and elsewhere.

Proriv's headquarters look flashy, with elegantly designed logos (featuring Mr Guevara) and lots of computers. But the aim is not to promote Western-style democracy, rather, its opposite: *Proriv* is much closer to the pro-Putin youth movements of Russia. The accent is on fun, and a positive and loyal portrayal of the motherland. Transdniestria is a modern, vibrant, multi-ethnic society, happily linked to dynamic, prosperous Russia. A much better bet than muddy Moldova.

The real prize is not meeting the pony-tailed, yoga-loving Mr Soin, but his boss, the Transdniestrian security chief Vladimir Antufeyev. How to reach him, in his secrecy-shrouded, fear-inducing ministry? Even the gutsiest independent journalists in Transdniestria do not have a phone number for his office. The so-called ministry of foreign affairs shows no inclination or capability to arrange a meeting. Time is running short.

The best approach is the simplest: go to the ministry of state security and ring the doorbell. It seems a trifle risky. This is not a building which outsiders normally enter willingly, especially Western journalists who write nasty things about gangster-ridden separatist enclaves. A friend cautions strongly against: it will be fruitless; the tea will contain polonium. A tiny spyhole opens, then a thick steel door; a soldier, in the uniform of the special forces, takes a proffered passport and visiting card.

Next stop is a long wait in a tiny, interrogation room, fitted out with a KGB version of IKEA furniture: flimsily constructed in pale plastic veneer. Steel shutters on the windows have slits for the muzzle of a gun, just in case.

Then General Antufeyev's secretary appears, ideally cast for the role in elegantly cut battledress, fishnet stockings, high heels and scarlet lipstick. She beams. The minister is available in 40 minutes. The soldiers beam. If they had been hoping to be told to take this unwelcome visitor on a one-way trip to the dungeons, they don't show it. Sadly, the canteen is not available, but an excellent café round the corner sells a passable borscht and dumplings. It is hard to imagine anywhere else in Europe where an impromptu request to meet the chief spook would meet such an accommodating and friendly reaction.

Face to face, General Antufeyev is charming and hospitable, not the sinister brute of popular repute. Coffee is offered at once; then a bottle of brandy as a souvenir, In crisp, vivid Russian he outlines his worldview: Transdniestria is a bastion against western hegemony; Moldova is intolerantly ethno-centric, pervaded by Romanian nationalism, and consequently unattractive to the liberty-loving Russophones of Transdniestria.

It is rather like talking to top *Stasi* people in East Berlin in 1988: the logic is fine and the brainpower impressive; but the assumptions are mistaken. In truth, Transdniestria is being squeezed: Russia is impatiently cutting back subsidies; Moldova, albeit slowly, is becoming more attractive. General Antufeyev will need all his wiles to survive.

Wednesday

In theory, Transdniestria is scary. It is the sort of place where thugs in leather jackets tote their guns in restaurants, a place where anything can be smuggled, laundered, bought or disposed of. Bad things can happen to the unwary or unlucky Westerner, and if they do, nobody will help you. Chisinau-based diplomats shun the illegal, unrecognised "Transdniestrian Moldovan Republic."

The place is run by a Ministry of State Security—"MGB," for its Russian initials—which has close and unexplained ties to powerful people in Moscow. That outfit is run by Vladimir Antufeyev, who—in the eyes of his enemies at least—is a villain straight out of a James Bond film. He is physically imposing, brainy, ruthless and has a suitably chequered background. He arrived here years ago under an assumed name, having staged an unsuccessful putsch in Latvia.

Driving from Chisinau to the Transdniestrian capital, Tiraspol, takes about 40 minutes. The "border" is the cease-fire line left over from a short, fierce, pointless war in 1992 between swaggering ethno-nationalist Moldovans and die-hard Soviet loyalists, which left the latter in charge of Transdniestria. You pass Moldovan customs officers, then Russian peace-keepers, then Transdniestrian customs, then the border control.

The checks are cursory. If your papers are in order, everything is OK. But what does it cost to get the right papers? One aim of this trip has been to see what the market is in forged or corruptly-obtained passports. A few years back your correspondent failed to take the chance, in the bazaar in Faizabad, of buying a spanking new Afghan passport for just $40, and has regretted it ever since.

So far, on this trip, the quest for a passport has been fruitless. In Bucharest they are said to be on sale in Moldova, no problem. Moldovans say that the trade there has dried up, but you can buy any documents you want in Transdniestria, no problem. But even the sleaziest looking taxi drivers and the barmen in the deepest dives of Tiraspol look shocked at the idea. Odessa in next-door Ukraine, they say, is the place for all kinds of papers, legal and illegal.

For a place that bangs on endlessly about its statehood, Transdniestria is pretty feeble when it comes to the details. The first stop is the foreign ministry. In most countries, the foreign ministry is a landmark. This one is tucked away in a backstreet. It lacks a national flag, a sign, and even a door bell.

Banging on what looks like a garage door produces reluctant admission that this is the foreign ministry, and eventually access to a dusty car park that leads on to a nondescript villa. Inside, it all seems unfinished. Bare wires dangle from the ceilings, some with lightbulbs, some without. It smells cold, damp and lifeless. "Have you just moved in?" "No, a couple of years ago" comes the answer.

The main road, October 25th St, tells you much of what you need to know. The most common shop signs are for money exchange—reflecting the inflow of remittances from émigrés that keep both Transdniestria and Moldova afloat. Transdniestrian roubles are dingy, scruffy scraps of paper, which manage to make even Moldova's tatty currency, the lei, look respectable. The coins are tiny discs of aluminium.

A hugely ornate new bank building shows the profits you can make from shovelling money into, and around, a place where international financial controls don't bite. A short way away is the headquarters of the MGB. Across the road is a fast-food joint that, oddly, is owned by the son of a senior Moldovan politician. The two sides may hate each other on one level, but on another their interests overlap in the most curious ways.

Tuesday

Moldova is run by Vladimir Voronin, the only serving head of state in the world to have won a contested election on a communist ticket. His views have changed a lot from 2001, when he said he would make his country the "Cuba of eastern Europe." Now he is pro-market and pro-European Union, He's pro-democracy too, in theory. But the justice system is dismal and the security services powerful. The authorities treat journalists they don't like with silly vindictiveness. The opposition finds it hard to get on telly: in short, it's a typical bureaucratic and fairly authoritarian presidential republic, a bit like Ukraine used to be before the "Orange Revolution."

The story the Moldovans want to tell is of their conversion to radical economic reform. It is certainly needed. Moldova is the poorest country in post-communist Europe; 47% of the population lives below the poverty line. At least 25% of the working age population has emigrated, Their remittances keep the place going.

Now Mr Voronin has announced an amnesty for illegal capital and unpaid taxes, and a sweeping tax cut for business. The idea, ministers and officials say with unconvincing confidence, is to make Moldova like Estonia.

That is a bit like announcing that Louisiana will in future be run like Switzerland. Estonia's post-communist trajectory is the most startling success story in the region. For much of the 1990s reform went at warp speed there, The civil service is hi-tech, anglophone, instinctively open in its approach, informal, liberal-minded and honest. The country also benefits from exceptionally close contacts with neighbouring Finland.

Dealing with the Moldovan government does not evoke memories of Estonia. Soldiers patrol the corridors in Soviet-style uniforms, saluting as minor bureaucrats go by. According to people who deal with it, the bureaucracy is old-fashioned and often corrupt. Ministries are run as Soviet-style hierarchies, where connections and status matter far more than good ideas, and everyone guards decision-making power and information jealously. No neighbouring country plays Finland's role. Most outsiders that come to Moldova from neighbouring countries offer bribes, not advice.

The economics minister wants to make the country a "logistics hub" for the Black sea region. Not a bad idea—but it will be hard to do that without allowing foreigners to buy land freely, or to compete with obese sacred cows such as the national airline.

Yet things are changing. People now move from Transdniestria to work in Chisinau. It used to be the other way round: in Soviet times Transdniestria was industrialised, whereas Moldova specialised in low-value-added agricultural produce. Moldova is even facing a huge influx of cash over the next few years: $1.2bn was pledged at a donor conference last year.

Every big international outfit seems to have an office in Chisinau. Some are run by inspirational people. Others seem to have been sent to Moldova as a punishment, or at the fag-end of their careers. Some foreign missions are run by locals of questionable outlook.

Given Moldova's exceptionally weak institutions, it is likely that some donated money will be stolen. Quite a lot will merely be wasted. Some will never be allocated at all, because Moldovan officialdom can't get its act together. But some may actually do good.

Monday

Moldova is not only the poorest ex-communist country in Europe; it is also last in the queue for love and attention, It lacks central Europe's glorious culture, the pungent romance of the Balkans, the charm and excitement of the Baltics, or the huge strategic importance of Central Asia and the Caucasus. Its main role is that of a country so obscure that it can safely be ridiculed, as it was in a book about a hapless British comedian's attempt to play tennis with the national football team.

Moldova is indeed flat, small, isolated, and ill-run, But it is not ridiculous, Its sadnesses spill over to other countries in the form of smuggling and prostitution. Bits of it—chiefly the breakaway puppet state of Transdniestria—are sinister. Its fate is tremendously important, As it wobbles between east and west, Moldova may be the first country that the Kremlin wins back from the west since the 1970s.

Simply getting there is quite hard. Deplorable state interference (to protect the national carrier) means that the low-cost European airlines that fly to the farthest corners of countries such as Poland don't serve the Moldovan capital, Chisinau. Municipal corruption means that no western hotel chains have opened. The best one is an ex-brothel, built for Turkish clients. After four lucrative years, the owners changed its name and went respectable, more or less.

The best way to get to Moldova is from Romania. Ties between these two countries ought to be the closest anywhere in eastern Europe. They share, broadly speaking, a common language and history, Moldova was part of Romania until 1940, when Stalin grabbed it as part of the Nazi-Soviet pact.

Now the mood is icy. Romanians mostly find it hard to think of Moldova as a separate country: rather the same way many English used to feel about Ireland, and still do about Scotland and Wales. Romania's beleaguered (and currently suspended) president, Traian Basescu, a cheerful former sea-captain, sees Moldova as a failed experiment that would be much better off rejoining Romania. His views would be fine in his old job, discussed in a lively harbourside bar, lubricated by a few glasses of Romania's national drink, *tuica*. Coming from a head of state, amid the delicate levantine gold filigrees and white plaster of the former royal palace, they sound crass.

Only the most flimsy euphemisms disguise his real views: Moldova is run by an incompetent provincial Soviet elite that has lost the confidence not only of the outside world, but also its own people. They are signing up for Romanian passports en masse—he reckons 800,000 out of a population of 4.5m. Romania's newly won membership of the European Union makes its citizenship—available to most Moldovans—irresistibly attractive, and the process of unification unstoppable.

Yet a few moments' thought show the difficulty with Mr Basescu's simplistic notions. Romania struggled to get into the EU and is now struggling to survive there. Moldova has far worse problems, and is not even in the waiting room for membership. The last thing the EU wants is another chunk of dirt-poor, ill-run, ex-communist nuisance. What would happen to Transdniestria, the mainly Russian-speaking territory that was stitched to Moldova in Soviet times, and now tries to be independent?

Crucially, reunification with Romania is not popular in Moldova. Mr Basescu's views may be coloured by the rapturous reception he received from his fans there during a recent visit. But less than a sixth of the population declare themselves as "Romanians." The majority have got used, over the past 50 years, to living in a separate country. They do not want to go back to being a neglected province of Greater Romania.

Counterterrorism Operations Continue in Tajikistan

ANDREW MCGREGOR

Having recently abandoned the pretense that they were running anti-narcotics sweeps rather than counter-terrorist operations in eastern Tajikistan, the Interior Ministry and the State Committee for National Security (GKNB) announced on July 8 that counterterrorist operations in Tavil-Dara district were completed (Interfax, August 5; see Terrorism Monitor, June 12 for the origin of the operations). Most operations have focused on the Rasht Valley Region, particularly the Tavil-Dara district. Tavil-Dara is part of the Region of Republican Subordination (RRS—formerly Karategin Province), one of the four administrative divisions of Tajikistan. Interior Ministry and GKNB troops returned to their bases in Dushanbe after "totally destroying" Shaykh Nemat's group of Islamist militants. Thirty members were arrested and 11 killed, including Shaykh Nemat and Lieutenant General Mirzo Ziyoev. The detainees included six Russian citizens. According to Tajik Interior Minister Abdurahim Qahhorov, military operations in Tavil-Dara have restored stability to the district (RFE/RL, July 23). Tavil-Dara was an Islamist stronghold during the 1992–1997 civil war in Tajikistan.

On July 8, a group of vehicles tried to force their way through a Tavil-Dara checkpoint but were repulsed by security forces in a prolonged exchange of gunfire. A group of "foreign militants" attacked a National Guard post at Havz-i Kabud in Tavil-Dara on July 16, but were driven back with losses. The Tajik Interior Ministry and the GKNB issued a joint statement on July 18 identifying the five Russian militants killed at Havz-i Kabud. They included two Daghestanis, an ethnic Tatar, an ethnic Kazakh from St. Petersburg and a native of the Siberian region of Tyumen. The militants were reported to have been armed with Kalashnikov assault rifles and grenades.

Dushanbe blames much of the increased violence on a resurgence of activity by the Islamic Movement of Uzbekistan (IMU). Three men arrested at a checkpoint in eastern Tajikistan on July 22 were reported by police to have been members of the IMU who fought alongside the Taliban in Afghanistan and Pakistan. Firearms, grenades, communications equipment and homemade bombs were seized in the investigation. Six Tajiks were sentenced on July 31 to eight years in prison for their membership in the IMU. The suspects were reported to have received training at a school in Afghanistan belonging to the IMU. Another alleged IMU operative was arrested in Khujand on August 4 in connection with the murder of two police officers in 2005. Two IMU militants were reportedly shot by security forces in separate incidents on August 9.

A police car was destroyed by a bomb in Dushanbe on July 30 during a summit meeting of the leaders of Tajikistan, Afghanistan, Pakistan and Russia. Two other bombs exploded several days earlier when the summit began. The bombings followed the arrest of three local men in early July on charges that they planned to commit terrorist acts in Dushanbe. All three suspects were reported to be members of the IMU and veterans of the fighting in Afghanistan. The Tajik press suggested that the bombings were connected to threats made by the Taliban regarding Tajikistan's cooperation in establishing new supply routes to NATO forces in Afghanistan.

Recent operations have focused on a number of different individuals and groups:

- Mirzo Ziyoev was the military commander of the Tajik Islamists during the civil war. Until recently he and his men were based in the Rasht Valley. In the reconciliation that followed the civil war, Ziyoev was made a Lieutenant General and Minister of Emergencies, a post that came with its own paramilitary. He was dismissed from the cabinet in 2006.

  According to the Interior Ministry, Lieutenant General Ziyoev joined an IMU unit led by Shaykh Nemat Azizov in June and was planning a series of attacks against state targets. An audiotape released by IMU leader Tahir Yuldash denied that Mirzo Ziyoev had ever been a member of the movement, suggesting instead that he had "fallen victim to intrigues of the government".

  He was captured by state security forces on July 11 in connection with the attack on a National Guard post. According to the Interior Ministry, Ziyoev was killed later that day when he led a security unit to convince an armed militant group to surrender. He was reported to have been killed in the resulting crossfire. Security officials reported that five Chechens were among the captured members of the militant gang. The five are alleged to have arrived in Tajikistan to transfer "a huge amount of money accumulated from the drug trade in order to fund terrorist organizations in Afghanistan and Pakistan" (Asia-Plus, July 28).

- Colonel Mahmud Khudoyberdiyev came to prominence in 1997, when he led an unsuccessful mutiny as commander of a Defense Ministry brigade. He escaped the rebellion's suppression only to return in 1998, briefly capturing the northern city of Khujand before fleeing for Uzbekistan. Tashkent has denied numerous requests for his extradition. Since the beginning of the year, Tajik security services have arrested seven former members of Khudoyberdiyev's group on charges of murder, kidnapping and terrorism, most dating back to the 1990s.

- Militant leader Shaykh Nemat Azizov was killed by members of the Tajik Interior Ministry's Special Forces in late July when he and a group of fighters were tracked down in a remote mountainous area in Tavil-Dara. Formerly a well-known guerrilla leader in the Tajik civil war, Azizov was made leader of the Tavil-Dara division of the Emergency Situations Ministry as part of the post-war reconciliation before he allegedly returned to armed opposition to the Tajik state. A Daghestani man alleged to be part of Shaykh Nemat's group was displayed on Tajik television in late July. Magomed Rukhullaevich Sabiullaev (also given as Satsiyullayev) claimed he flew from St. Petersburg to Dushanbe before participating in attacks on security forces in Nurabad and Tavil-Dara under Shaykh Nemat's command. The Daghestani's confession stated that he and two other Daghestanis met veteran Islamist commander Mullo Abdullo in Tavil-Dara. Mullo Abdullo blessed their jihad before Shaykh Nemat armed them.

  Shaykh Nemat was killed on July 29 when security forces tracked his group to a remote mountain village. Azizov refused to surrender and was killed in a firefight.

The Interior Ministry claimed Shaykh Nemat had entered Tajikistan to sell narcotics from Afghanistan in order to fund militant operations in Pakistan and Afghanistan. Security services also claimed Shaykh Nemat was a commander in the IMU.

- Forty-six members of the Tablighi Jamaat, an international Islamic revival group, were arrested in the Khatlon region in mid-July. The Tablighis insist they are strictly apolitical, though there are suspicions that Islamist militants elsewhere may have used the cover of the Tablighi movement to facilitate an international movement (see Terrorism Focus, February 13, 2008). The Tablighi Jamaat was banned in March 2006 in the belief that the movement aims to subvert constitutional order in Tajikistan and establish an Islamic Caliphate. The 46 Jamaat members will be tried for incitement to extremism and the creation of an extremist community. In a separate case, five Jamaat members of Tajik and Russian origin were sentenced to terms of three to six years each for calling for "forcible change" to the country's constitution.

It is still difficult to say with any accuracy exactly what groups Tajikistan's security forces have been battling in the mountains since May. While a mixture of native Tajiks and foreign militants appear ready to take up arms against the state, it remains to be seen if there is a taste in the Tajikistan public for a resumption of the disastrous and brutal civil war of the 1990s. Some of the violence appears to be generated by militants escaping the Pakistani military offensive on Pakistan's northwest frontier. But other incidents appear to be based in the personal grievances of Islamists who accepted lofty government posts as part of the national reconciliation after the civil war, and who are now being dropped from government. Many of the arrests for crimes committed in the 1990s indicate that the Islamists have lost the "immunity" they once enjoyed as part of the government's desire to restore stability to Tajikistan. As an Interior Ministry spokesman told a correspondent, "The overwhelming majority of people who remember the bitter lessons of the recent civil war well did not respond to [the militants'] provocative calls" (Itar-Tass, August 9).

Uzbekistan Challenges Regional Electricity Supplies Network

Erica Marat

Kyrgyzstan's growing list of troubles has recently been further complicated by yet another predicament. Tashkent has announced that Uzbekistan is likely to leave the Central Asian power supply cascade in the coming months. According to Tashkent's official interpretation, Uzbekistan can now provide its population with enough locally generated electricity and does not need to be part of the network created during the Soviet period. This means that Kyrgyzstan's south and parts of Tajikistan will experience severe electricity shortages due to the break in regional cycles.

The Toktogul hydro-power plant (HPP) cascade is part of a regional power supply network which runs through all five Central Asian countries. It delivers electricity produced by Toktogul HPP to southern areas of Kyrgyzstan through Uzbek territory. Electricity produced in Kyrgyzstan flows through 220 kV and 500 kV transmission lines that pass through Uzbekistan. Several years ago, Uzbekistan began the construction of the Novo-Angren thermal power plant, aimed at producing enough electricity for the residents of the Ferghana Valley. Tashkent's decision to break the old connections was announced abruptly and caught the Kyrgyz government unprepared.

It is unclear how Bishkek will provide electricity for its entire southern region should Uzbekistan implement its decision. Although some analysts have warned the Kyrgyz leadership about the probability of this scenario, no action was taken to avert the crisis. Instead, the Kyrgyz Minister of Energy Ilias Davydov rushed to visit Uzbekistan following the announcement of the decision.

The Kyrgyz government has already begun its winter electricity rationing throughout the country. The government promised that electricity cuts would be shorter and less frequent than last year. However, last winter was fairly mild and colder temperatures might hit the country this year.

Uzbekistan's decision in favor of greater energy independence reveals the inability of the Central Asian states to benefit from regional trade. The move appears more politically motivated than based on economic considerations. Uzbekistan is openly against upstream Kyrgyzstan and Tajikistan constructing new HPPs. New dams might leverage both countries with greater control over water resources. Yet, the Uzbek government

prefers to prevent these new projects in neighboring countries, rather than make an effort to negotiate new terms for inter-state collaboration.

Kazakhstan's government is also considering leaving the Soviet constructed regional electricity network to establish an autonomous domestic power supply. This will add an additional burden to Kyrgyzstan's north (www.ferghana.ru, September 28). Yet, unlike Tashkent, Kazakh government representatives insist that they will delay their decision to avoid destabilizing northern Kyrgyzstan.

Meanwhile, the Kyrgyz government's management of water resources remains plagued with corruption. The construction of the Kambarata HPPs on the Naryn River lacks transparency. Kyrgyz President Kurmanbek Bakiyev recently signed a bill allowing Kambarata to be privatized or to be used by the government as collateral for loans (www.akipress.kg, October 14). This means that the plants might be privatized by either local or foreign private companies. The bulk of the Russian $2 billion credit was allocated for the construction of the Kambarata-1 HPP last February. However, as to how these funds will be spent and when the HPP will be constructed remains undisclosed by the government.

One possible solution for Kyrgyzstan to meet the challenges it faces in the regional energy trade is the construction of the Central Asia-South Asia (CASA) 1,000 mega watt (MW) energy project that would connect Kyrgyzstan, Tajikistan, Afghanistan and Pakistan in one single network. The project would allow all countries to benefit from such cooperation. Even Russia, as the major investor in Tajikistan's Sangtuda-1 HPP and the Kambarata HPP, would gain from the project.

However, the participating countries still need to find investors for the project. Major international organizations, among them the World Bank and the Asian Development Bank, are ready to invest in the CASA 1,000 MW project. Given Bakiyev's corrupt management of the hydropower sector, he might see international investment flowing into neighboring countries bypassing Kyrgyzstan. According to one Bishkek-based representative of an international organization, since Bakiyev came to power in 2005, it has become virtually impossible to implement reform in the country. "The Kyrgyz government is deeply in debt, while most money is redistributed among state actors and state contractors" the expert told Jamestown.

According to regional experts, Uzbekistan might not be able to fully supply its population with only local electricity production. Furthermore, Tashkent might reverse its decision and remain part of the regional network in the short term. The ongoing tension between Uzbekistan and Kyrgyzstan has shown yet another side of the mutual antagonism between the leaders of both countries. It has also highlighted the weak national governance in both states. Unfortunately, the population suffers the most from this predicament.

From *Eurasia Daily Monitor,* vol. 6, Issue 189, October 15, 2009. Copyright © 2009 by Jamestown Foundation. Reprinted by permission. www.jamestown.org

Ukraine—From Orange Revolution to Failed State?

ANDREW WILSON

As the former Ukrainian ambassador to the U.S. Yurii Shcherbak has recently complained, the Russian press is currently full of *schadenfreude* at the prospect of Ukraine becoming a 'failed state'[1]—and not just from the usual suspects on the nationalist right.[2] For example, the March 2009 issue of Gleb Pavlovskii's *Russkii zhurnal* came out under the highly suggestive title—'Is the de-sovereignisation of Ukraine manageable?' Inside, Sergei Karaganov casually speculates on the difference between 'passive de-sovereignisation' resulting from local state incapacity to 'managed de-sovereignisation', i.e. a process that Russia actively encourages. The suggestion is that the only open question for Karaganov is whether the Ukrainians need pushing over the abyss, or whether they will jump off themselves.[3]

Ukraine's current travails are real enough. But what is the Russian definition of a 'failed state'? The Russian line, most notoriously expressed by Putin in his (originally) private remarks to Bush at the NATO Bucharest summit in 2008,[4] echoes the famous 1994 CIA report:

> Ukraine is a very complex state. Ukraine, in its current form, came to be in Soviet-era days. . . . From Russia the country obtained vast territories in what is now eastern and southern Ukraine. . . . Crimea was simply given to Ukraine by a CPSU Political Bureau's decision . . . one third of the population are ethnic Russians. According to official census statistics, there are 17 million ethnic Russians there, out of a population of 45 million. Some regions, such as Crimea, for example, are entirely populated by ethnic Russians. There are 90 percent of them there. . . . If the NATO issue is added there, along with other problems, this may bring Ukraine to the verge of existence as a sovereign state.

Apart from demonstrating Putin's casual approach to statistics, this statement says more about Russia than it does about Ukraine. Putin is really making an **'artificial state'** argument. Artificial states may be more likely to fail, but the two questions are logically distinct. As is the further question about how the Ukrainian public might react to growing state weakness, despite the prediction by at least one other Russian that the current 'systemic crisis' is a 'question of life or death', and will lead to 'a growth of lack of faith in the Ukrainian project as such'.[5] The logic of Putin's argument is presumably the following: first, Russia has more of a right to interfere in the affairs of 'artificial' states (with failed states, the 'right to interfere' might imply an 'obligation to clear up the mess'); second, Putin is trying to undermine Ukraine's credibility as a worthy partner for NATO, and increasingly for the EU—as the last sentence in the above quote clearly shows.

But the concept of a 'failed state' is notoriously elastic. Most of the phenomena often cited, such as corruption, a breakdown in the provision of public goods like law and order or welfare, and human flight, are symptoms not definitions of state failure, and are shared by many states to some degree. A fully failed state has lost the capacity for sovereign government: as evidenced by a loss of control over its territory or population or key private interests, or the interference of foreign powers.[6] The Crisis States Workshop defines "a 'failed state' as a condition of 'state collapse'—e.g., a state that can no longer perform its basic security, and development functions and that has no effective control over its territory and borders. A failed state is one that can no longer reproduce the conditions for its own existence."[7]

The 'Failed State Index' used by the Fund for Peace and Foreign Policy, looks mainly at symptoms—as it is looking for evidence—grouped in twelve 'Indicators of Insecurity'. But if we use its methodology, Ukraine actually improved its score from 2005 to 2006, though it has since stagnated.[8] Ukraine actually scores better on 'failed-state-ness' (a global 108th in 2008, first is worst) than it does on other measures, such as the Corruption Perceptions index (134th in 2008, first is best).[9] Ukraine is clearly not an archetypal failed state like Sierra Leone or Somalia. That said, the assessment for 2009 will be interesting. Ukraine's economic problems are well-known, although Anders Åslund has talked up the prospects for recovery.[10] Political crisis is endemic. Ukraine has also suffered a serious break-down in the judicial sphere.

Ukraine is not a failed state on the above definition. But it is not a successful state either. It could plausibly be described as a 'risk state' or a 'fragile state', or even a 'crisis state'. Various adjectives could be chosen, but I have chosen to label Ukraine an **'immobile state'**, in order to highlight what I see as the key problem of political stasis at an elite level. More exactly perhaps, Ukraine could be called a 'slow-moving state'. It has shown it can adapt under pressure, but it never quite adapts enough. It is a tortoise when it needs to be more of a hare in response to current crises.

Many of the reasons for this political gridlock are well-known. Ukraine's well-known regional, linguistic and other divisions will always be a factor, if not in the exaggerated form described by Putin. Others have talked about constitutional gridlock and the weaknesses of the party system. But, as Paul D'Anieri has shown,[11] none of these are in themselves a decisive factor. The same regional and ethno-linguistic divisions were also present under Kuchma in 1994–2004, when the political system was if anything too centralised and unbalanced towards autocracy. They are not the main reason why Ukraine should suddenly now be at risk of becoming a failed state. Ukraine did not fail in the 1990s. These underlying problems have, however, always made it difficult for Ukraine to become a successful state.

- They do, however, support the **balance of the clans.** Ukraine's well-known 'pluralism by default' has often been presented as a comparative advantage,[12] but in the years since the Orange Revolution it has meant an unhealthy gridlock between the half dozen or so business groups that finance the political system. One little-noticed aspect of the 2007 elections was the way in which several business groups simply switched sides. The major oligarchs played musical chairs: Renat Akhmetov's group in the Party of Regions actually went up from around sixty deputies to around ninety, but the party's main financier was now Dmytro Firtash. The Industrial Union of the Donbas (IUD) and Privat swapped places, with Privat now supporting Yushchenko and the IUD backing Tymoshenko, as her other sponsors like Kostiantyn Zhevaho's Ferrexpo took a big hit from the global economic crisis. Vasyl Khmelnytskyi, formerly close to ex-president Kuchma, invested an alleged $30 million in Volodymyr Lytvyn.[13]

This merry-go-round has allowed the oligarchs to protect their interests. But it may not last forever. Anders Åslund has suggested that the current economic crisis might be 'Ukraine's 1929'—i.e. a slump, but also the twilight of the old economic order, and the end of the 'robber baron' era.

Logically, this thought can be divided into three separate questions. First, have particular oligarchs been hit particularly badly by the current crisis, shifting the balance between the parties they support? (Though any given oligarch could decide to gamble his declining fortune by investing *more* in politics, which seems to be the best explanation by Firtash's rise and apparent fall in 2008–2009). Have particular sectors suffered more? Steel obviously, which should most affect those supported by the likes of Akhmetov and Zhevaho, but steel barons are everywhere in Ukrainian politics. Current politics also plays a part. *If* Rosukrenergo really is out of the gas transit market, both the president and the Party of Regions will lose massive campaigning resources. In the

spring it seemed that other business interests were lining up behind Tymoshenko and Yatseniuk, and migrating away from Yanukovych, but this changes every week.

Second, are 'bad' oligarchs likely to suffer more than 'good'? Possibly. There are some signs that the major source of rent-seeking by the likes of Rosukrenergo is in decline, as Central Asian gas prices go up, European prices go down, and Gazprom grows cash-desperate and is in no mood to share remaining rents.

Third, even if all 'oligarchs' are hit equally hard, their collective downfall could pave the way for a new type of politics. But Ukraine must survive the economic crisis first. Hryhorii Nemeriia's judgement that "No one has the money to pay for political technologists anymore" may therefore be premature.[14] In the boom years, the sheer amount of money to be made in the merry-go-round was one reason why it continued.[15] In the slump, it would seem to be irrational to opt out of the fight for declining resources.

- The way that **Russia** has played Ukrainian politics since 2004 also promotes political gridlock. Instead of backing one candidate onesidedly (Yanukovych obviously in 2004), Russia now plays across the political spectrum. It finds plenty of willing partners, but, as Dmitrii Trenin has said, 'there have never been and never can be pro-Russian politicians in Ukraine',[16] in the sense of slavish service at least. Ukrainian politics is too anarchic and too free. The current favourite can always take the money and run. Another will always be bidding to take his or her place. But then Russia will back his or her rivals in turn. Once again, it is difficult for a clear winner to emerge.

This brings me to another point about the global economic crisis. Ukraine is undoubtedly hit worst than most, but what matters most in a recession is relative strength—of who emerges relatively stronger. According to one Russian, 'Ukraine is cheap, we can buy it'.[17] Russia has clearly not retrenched yet. In the short-term it is upping its ambition. First, there are obvious economic moves. In January 2009 Vneshekonombank, where none other than Vladimir Putin is chair of the supervisory board, bought Ukraine's troubled Prominvestbank for a knock-down price of $150 million—and promptly injected $900 million.

But there is also a broader context. In times of recession, Western states revert to utilitarian cost-cutting; but Russia thinks geopolitically. It is prepared to invest to win friends. If we thought Russia got its fingers burnt at the last Ukrainian elections in 2004, then the extent to which Russia has been a factor in the recent Moldovan elections—and in supporting the authorities in the subsequent crack-down—shows that Russian ambition is back.

One might add the fact that, in broader foreign policy terms Ukraine's multi-vectoral equidistance between Russia and the West is both a strength and a weakness. It allows Ukraine to play like Tito, and build up some sovereign statehood. But it also allows Ukraine to play the balancing game or the chicken game (Ukraine is 'too big to fail') to extract resources, rather than embarking on serious reform.

The third reason for political gridlock is the persistence of '**political technology**', as shown in the recent dress-rehearsal vote in Ternopil—which is particularly depressing for a regime created by the protest against such fraud in the Orange Revolution.

The Ternopil vote was marred by outright vote-buying and the abuse of 'administrative resources'. But the grand-scale centralised theft that marred the 2004 campaign is unlikely in 2009–2010—in part precisely because Ukraine is too decentralised, and the central state is too weak. Most of the main candidates will be able to cheat to an extent, probably cancelling each other out—though the winner in a competitive fraud can gain an edge, as with Kennedy vs Nixon in the U.S. presidential election in 1960.[18]

Other types of 'political technology' that were supposed to have been marginalised by the Orange Revolution are back, as with the fake nationalist party 'Svoboda', covertly backed by the presidential administration. Suddenly, the air is full of renewed talk of 'technical parties' and 'technical candidates' in the upcoming presidential election. Much of the talk is conspiracy theory—such as whether Arsenii Yatseniuk is a 'life-raft' for Yushchenko forces or a genuine alternative. And whether he represents kosher oligarchs like Pinchuk and Taruta, or more controversial figures like Firtash and Khoroshkovskyi.

I can only suggest a tentative hypothesis as to why 'project' parties are making a comeback. The relatively free mass media that is one of the most enduring changes since the Orange Revolution has been the best antidote to such projects. In both the 2006 and 2007 elections a lot of money was wasted on such parties (Ne Tak!, Eko+25%, Viche in 2006; the 'renewed' Communists, Hromada and Regional Active in 2007—though the Lytvyn project was ultimately successful). Although you can fool some of the people some of the time, the kind of broad-based project equivalent to Russia's Unity in 1999 is now a difficult sell in Ukraine. The only answer I can suggest for their apparent partial rebirth is that people are (in the short-term at least) more receptive to quick-fix populist messages in a recession, and may filter out their scepticism to a greater extent.[19] Svoboda's crude economic nationalism fits this pattern, but so does the unproven idea of Yatseniuk as the 'Ukrainian Obama' (he has yet to show he can offer 'change': his current strategy has been described as sitting by the river Dnipro and watching the bodies float by).

If the return of 'project parties' and 'technical candidates' is surprising, 'black PR' and 'kompromat wars' were always going to be harder to eradicate. Accumulated kompromat would still be powerful, even if the underlying political and economic system had changed. But it hasn't. Economy and society are still dominated by 'circles of interest' (*krugovaia poruka*) that maintain a 'circle of kompromat' to check one another in a type of 'mutually assured destruction' or 'Mexican standoff'.[20]

Resurgent political technology is also contributing to political gridlock. It makes weak players competitive. Yushchenko's opinion poll ratings are rock-bottom, but he can still play a destructive role in the election—in fact possibly a decisive role in who reaches the second round. Negative political technology also prevents strong players, currently Tymoshenko and Yanukovych, from striking out in front.

In Russia, the Kremlin is the monopoly player, after it destroyed Yukos's attempt at political entrepreneurship in 2003. In Ukraine, the abuse of normal politics is competitive. All players check each other with 'spoiler parties'. If Arseniuk is a 'technical candidate' against Tymoshenko, then Serhii Tyhipko may well be a technical candidate against him in turn.

Lastly, of course, so much political manipulation rightly destroys public faith in all politicians.

So, to conclude, Ukraine is not a failed state in a conventional sense, but it is stuck in a twilight world, where it is difficult to fail, but difficult to be too successful either. And the longer-term structural problems I have briefly enumerated will still be around in a year or two, regardless of who is elected as president in January, and regardless of whether Ukraine recovers from the current economic crisis or not.

Notes

1. Yurii Shcherbak, 'Ukraïna yak failed state—mify ta real'nist", *Den*, 21 May 2009, www.day.kiev.ua/274251 and www.day.kiev.ua/274238/274238

2. Though those with strong stomachs can check out Aleksandr Dugin's send 'tanks to Kiev' speech at www.tap-the-talent.blogspot.com/search/label/Dugin

3. See the special issue of *Russkii zhurnal*, no. 6, 16 March 2008, "Opravlyaema li desuverenizatsiya Ukrainy?" ("Is the de-sovereignisation of Ukraine manageable?"). Karaganov's interview is at www.russ.ru/Mirovaya-povestka/Nikomu-ne-nuzhnye-chudischa

4. "What precisely Vladimir Putin said at Bucharest", www.mw.ua/1000/1600/62750/

5. Andrei Stavistskii, 'Ukraina i global'nyi krizis: prigovor uzhe vynesen?', www.odnarodyna.ru/articles/6/666.html, dated 21 May 2005

6. For a discussion, see Robert I. Rotberg (ed.), *Why States Fail: Causes and Consequences*, (Princeton, 2004).

7. See www.crisisstates.com/download/drc/FailedState.pdf

8. Ukraine rated at 88.8 in 2005 (39th), 72.9 in 2006 (86th), 71.4 in 2007 (106th) and 70.8 in 2008 (108th); see www.fundforpeace.org/web/index.php?option=com_content&task=view&id=17&Itemid=80

9. See www.transparency.org/policy_research/surveys_indices/cpi/2008

10. Anders Åslund, 'Ukraine Above the Rest in Crisis Management', *The Moscow Times,* 22 April 2009, www.themoscowtimes.ru/article/1028/42/376481.htm

11. Paul D'Anieri, *Understanding Ukrainian Politics: Power, Politics and Institutional Design,* (New York: M.E. Sharpe, November 2006).

12. Lucan D. Way, *Pluralism by Default: Challenges of Authoritarian State-Building in Belarus, Moldova and Ukraine,* (Glasgow: Centre for the Study of Public Policy, 2003).

13. Anders Åslund, *How Ukraine Became a Market Economy and Democracy,* (Washington, DC: Peterson Institute, 2009), p. 221.

14. Conversation with the author, 15 May 2009.

15. Edward Chow and Jonathan Elkind, "Where East Meets West: European Gas and Ukrainian Reality, *The Washington Quarterly,* vol. 32, no. 1 (January 2009), pp. 77–92; Margarita Balmaceda, *Energy Dependency, Politics, and Corruption in the Former Soviet Union: Russia's power, oligarchs' profits and Ukraine's missing energy policy, 1995–2006,* (London: Routledge, 2008).

16. Dmitrii Trenin, 'Russia-Ukraine: Problems will Remain', dated 29 March 2006, at www.carnegie.ru/en/pubs/media/73992.htm

17. Off the record interview with the author, 21 October 2008.

18. This suggestion was made by Paul D'Anieri.

19. Highly personalized presidential elections are also a factor.

20. Alena Ledeneva, *How Russia Really Works. The Informal Practices That Shaped Post-Soviet Politics and Business,* (Ithaca and London: Cornell University Press, 2006).

Speech by Andrew Wilson, *European Council on Foreign Relations,* May 2009. Copyright © 2009 by European Council on Foreign Relations. Reprinted by permission of Andrew Wilson, Senior Policy Fellow at the European Council on Foreign Relations. www.ecfr.eu

Glossary of Terms and Abbreviations

Bolshevik The left wing of the Russian Social Democratic Party (RSDP); Bolshevik members of the RSDP believed in the necessity of the violent overthrow of the czarist order to achieve change.

CIS (Commonwealth of Independent States) A loose alliance of eleven formerly inter-related parts of the Soviet Union. Three out of 15 Soviet republics—Latvia, Lithuania, and Estonia—never joined the CIS; Georgia quit the alliance in 2008; and Turkmenistan reduced its status to "associate member."

Civilization A term with multiple meanings. This book takes guidance from Samuel P. Huntington's definition, according to which a civilization is "the highest cultural grouping of people and the broadest level of cultural identity people have short of that which distinguishes humans from other species." Huntington distinguished such civilizations as Western, Orthodox, Islamic, African, Latin American, Sinic (China, Korea, and Vietnam), Hindu, Buddhist, and Japanese.

Cold War (from the late 1940s to the dissolution of the Soviet Union in 1991) was the continuing state of ideological and geopolitical conflict, military tension, and economic competition primarily between the U.S.S.R. and its satellite states, and the powers of the Western world, particularly the United States. Although the primary participants' military forces never officially clashed directly, they manifested the conflict through military and political alliances, conventional force deployments, a nuclear arms race, espionage, proxy wars, propaganda, and technological competition.

Collectivization The forced amalgamation under collective communal management of farms formerly owned privately.

Communism The ideal classless and stateless societies that workers in socialist countries strove to create in accordance with the teachings of Karl Marx and Vladimir Lenin.

CPSU Communist Party of the Soviet Union, the successor of the Bolshevik Party that seized power in Russia in November 1917.

Democratization A broadening of popular participation in electoral processes within politics, government, and the workplace.

Détente Relaxation of tensions between the Soviet Union and its allies and the Western countries.

Diaspora A dispersion of a people from their original homeland and the community formed by such a people. It has become habitual to talk about such diasporas as Jewish, Armenian, Greek, and Chinese. However, other ethnic groups form diasporas as well, including ethnic Russians residing in all former Soviet republics other than Russia and in countries outside the former U.S.S.R.

Duma The lower house or chamber of the Russian Parliament. Because the upper house (the Council of the Federation) consists of unelected or appointed officials and is not influential, the terms Duma or State Duma are frequently used as synonyms to Russian Parliament at large.

East Central Europe A term usually defining the countries located between German-speaking countries (Germany and Austria) and the former Soviet Union. Consists of formerly communist countries allied with the Soviet Union.

East Europe Formerly, this term encompassed the republics of the Soviet Union located west of the Urals and north of the Caucasus (Latvia, Estonia, Lithuania, Belarus, Ukraine, Moldova, and a part of Russia) as well as other communist countries located to the west of the U.S.S.R. Because the latter group is currently referred to as **East Central Europe,** East Europe is comprised of Ukraine, Belarus, and the European section of Russia (i.e., Russia west of the Urals).

Ethnic Homeland within the Russian Federation Areas whose indigenous populations are ethnically non-Russian. Such areas may have a status of a republic (e.g.; Tatar, Bashkir, Udmurt, Mari, Chechen, etc.) which is the highest ethnic autonomy status within the Russian Federation. Most of the other homelands have a status of an autonomous district or okrug (e.g.; Khanty-Mansi, Nenets, Evenk, Chukotka, etc.).

Ethnic Identity The extent to which one identifies with a particular ethnic group. Refers to one's sense of belonging to an ethnic group and the part of one's thinking, perceptions, feelings, and behavior that is due to ethnic group membership. The ethnic group tends to be one in which the individual claims heritage.

FSB A Russian acronym for *Federalnaya Sluzhba Bezopasnosti* or Federal Security Service, a successor to KGB.

Glasnost In the Soviet Union, referred in the Gorbachev phase to a new openness or candor of political and ordinary citizens on national issues and problems.

Great Purge The arrest, conviction, and punishment by death or imprisonment of high-ranking Soviet Communist Party and government officials by Joseph Stalin's regime in the mid-1930s.

G8 The Group of Eight (G8, and formerly the G6 or Group of Six and also the G7 or Group of Seven) is a forum, created in 1975, for governments of the six richest countries in the world: France, Germany, Italy, Japan, the United Kingdom, and the United States. In 1976, Canada joined the group (thus creating the G7). In becoming the G8, the group added Russia in 1997.

Genocide The deliberate and systematic destruction, in whole or in part, of an ethnic, racial, religious, or national group.

Geopolitics The term with multiples meanings, it may denote the study of the relationship among politics and geography, demography, and economics, especially with respect to the foreign policy of a nation. Geopolitics may also denote a combination of geographic and political factors relating to or influencing a nation or region. In the post-Soviet realm, this term often indicates foreign policy acts or initiatives which aim at expanding or sustaining the nation's sphere of influence.

KGB A Russian acronym for *Komitet Gosudarstvennoi Bezopasnosti* or Committee of State Security, a secret investigatory agency of the Soviet government responsible for, among other things, internal and external security and both strategic and counterintelligence work.

Korenizatsiya meaning "nativization" or "indigenization," was the early (1921–1929) Soviet ethnicity policy. The policy consisted of promoting representatives of titular ethnic groups of Soviet republics (e.g.; Ukrainians in Ukraine, Belarusians in Belarus, etc.) and national minorities on lower levels of administrative subdivision, into local government, management, and bureaucracy in the corresponding national entities. The term derives from the Russian term *korennoye naseleniye* for indigenous nationals.

National Delimitation refers to the policy and process of creating well-defined national territorial units within the Soviet Union (Soviet socialist republics—SSR, autonomous Soviet socialist republics—ASSR, autonomous provinces—oblasts, or autonomous national territories—okrugi) from the ethnic diversity of the Soviet Union's sub-regions. National delimitation is a translation of the Russian term *natsionalno-territorialnoye razmezhevanie,* which can also be rendered as national-territorial demarcation, partition, or disengagement. National delimitation was most consistently used in Central Asia and the Trans-Caucasus in the 1920s and 1930s. At least twelve out of fifteen former Soviet republics are products of national delimitation applied prior to World War II.

NEP The New Economic Policy under Vladimir Lenin. The development of the mixed Soviet economy in which some private ownership of the means of production continued, especially in agriculture, but other sectors of the economy, notably industry, were nationalized.

Oblast This is how the majority of Russia's civil subdivisions or constituent entities are called. The Russian Federation comprises 84 constituent entities, including 21 republics, 7 krais, 46 oblasts, 7 autonomous districts, and 1 autonomous oblast. Moscow and St. Petersburg are federal cities, each forming a separate constituent entity of Russia.

Oligarch One of business magnates possessing informal power over a certain region or country. The choice of the word oligarch, which means "one of the few rulers," denotes the significant influence such wealthy individuals may have on the life of a nation. Today, this term is used most often to describe the fast-increased wealth of some businessmen of Russia and Ukraine. In these countries, it is very common to apply the word *oligarch* to any business tycoon, regardless of whether or not he or she has real political power.

OSCE Organization for Security and Cooperation in Europe. It is the largest regional security organization in the world. Most of its 56 members are European countries. However, OSCE also includes the former Soviet Republics of the Trans-Caucasus (Georgia, Armenia, and Azerbaijan) and Central Asia (Kazakhstan, Kyrgyzstan, Tajikistan, Turkmenistan, and Uzbekistan) as well as Canada and the U.S.A. The OSCE Chairmanship is held for one calendar year by the OSCE participating state designated as such by a decision of the Ministerial Council. The function of the Chairperson-in-Office is exercised by the Minister of Foreign Affairs of that state. In 2010, Kazakhstan took over the chairmanship of the organization. The OSCE Secretariat is located in Vienna, Austria.

Patriarch The word has multiple meanings; in the Russian Orthodox Church, Patriarch is the highest-ranking bishop and the spiritual leader.

Perestroika (1985–1991) Means "restructuring" or "overhaul." This was a system of economic and political reforms associated with the leadership of Mikhail Gorbachev and designed to salvage the Soviet Union.

Personality Cult Popular worship of the individual, encouraged by a political leader to increase power.

Privatization The process of transferring ownership of a business, enterprise, agency, or public service from the public sector (government) to the private sector (business).

RSDRP Russian Social Democratic Worker's (*Rabochikh* in Russian) Party, a Marxist party established in 1898 to work for the improvement of the conditions of the Russian proletariat and divided between evolutionists (Mensheviks) and revolutionaries (Bolsheviks).

Separatism The advocacy of a state of cultural, ethnic, tribal, religious, or racial separation from the larger group, often with demands for greater political autonomy or for full political secession and the formation of a new state. In the post-Soviet context, mature separatist movements exist in Chechnya (aiming at separation from Russia), Abkhazia and South Ossetia (aiming at separation from Georgia), Transnistria (aiming at separation from Moldova), and Nagorno Karabakh (aiming at separation from Azerbaijan).

Serfdom The system of economic and political servitude prevalent in Czarist Russia until its abolition in 1861 by Czar Alexander II.

Shanghai Cooperation Organization An intergovernmental mutual-security organization which was founded in 2001 in Shanghai by the leaders of China, Kazakhstan, Kyrgyzstan, Russia, Tajikistan, and Uzbekistan. Except for Uzbekistan, the other countries had been members of the Shanghai Five, founded in 1996; after the inclusion of Uzbekistan in 2001, the members renamed the organization.

Shock Therapy In economics, shock therapy refers to the sudden release of price and currency controls, withdrawal of state subsidies, and immediate trade liberalization within a country, usually also including large scale privatization of previously publicly owned assets. Applied in Russia and Ukraine in the early 1990s, this policy led to the evaporation of people's lifelong savings and galloping inflation. At the same time shock therapy was instrumental in dismantling centrally planned economy and in replacing it with market economy.

Slavs or Slavic Peoples Groups speaking related languages belonging to the Slavic subfamily of the Indo-European language family and living mainly in East Central Europe and Eastern Europe. Slavic peoples are classified geographically and linguistically into West Slavic (including Czechs, Poles, and Slovaks), East Slavic (including Belarusians, Russians, and Ukrainians), and South Slavic (including Bosniaks, Bulgarians, Croats, Macedonians, Montenegrins, Serbs, and Slovenes).

Socialism A set of ideas and practices that call for community management and control of the scarce factors of production to achieve equality of well-being. Different kinds of socialism (Marxian socialism, Arab socialism, Indian socialism, etc.) in different national settings at different periods of national development determine the precise degree of political power assigned to the community to achieve social justice.

Soviet Federalism The federal type of intergovernmental relations established in the Soviet Union in 1922 wherein the bulk of administrative decision making was concentrated in the central government in Moscow and constituent units, in particular the largest and most populous, were based on the ethnic principle but enjoyed little autonomy.

Supreme Soviet The Parliament of the Soviet Union. Consisted of two houses, the Council of the Union, and the Council of

Nationalities. Alongside the Supreme Soviet of the U.S.S.R., each of its fifteen union republics had its own Supreme Soviet. From the dissolution of the Soviet Union (December 1991) to October 1993, Russia's Parliament was also called Supreme Soviet. According to the 1993 Constitution, it was replaced by the Federal Assembly consisting of the Council of the Federation and the State Duma (see Duma).

Turks This term may be used either in reference to people living in Turkey and speaking the Turkish language or in reference to Turkic peoples, the Eurasian peoples residing in northern, central, and western Eurasia who speak languages belonging to the Turkic subfamily of the Altaic language family. Turkish is just one member of that subfamily. In the former Soviet Union, the most numerous Turkic peoples are Uzbeks, Kazakhs, Azerbaijani, Tatars, Kyrgyz, and Turkmen.

U.S.S.R. The Union of the Soviet Socialist Republics, or Soviet Union, successor in 1922 of the Russian Soviet republic established in 1917. The Soviet Union broke up in December 1991, and its constituent 15 republics became sovereign countries.

War Communism Sometimes called "militant communism," this term refers to efforts of the Bolsheviks between 1917 and 1921 to respond to emergency conditions resulting from World War I and the Revolution of 1917, such as acute shortages of food, and simultaneously to inaugurate the Socialist transition to communism, for example, by nationalizing industries, redistributing land, and replacing money as a medium of exchange with a barter system. War communism, along with the Civil War and the allied blockade of Russia, caused the near total collapse of the Russian economy and was superseded by the New Economic Policy starting in 1921.

Warsaw Pact An alliance of six Central/Eastern European countries and the Soviet Union, linking them closely together in matters of defense. Established in 1955, ostensibly in response to West Germany's entry into NATO, the alliance, which was dominated by the Soviet Union, by far its most militarily powerful member, became another instrument of Soviet influence and control in Central/Eastern Europe. With the dissolution of the U.S.S.R., the Warsaw Pact was disbanded.

Bibliography

The following works are suggested for readers seeking more detailed coverage of a topic, country, or region.

SOURCES FOR STATISTICAL REPORTS

U.S. State Department, *Background Notes*
The World Factbook
World Statistics in Brief
World Almanac
The Statesman's Yearbook
Population Reference Bureau: *World Population Data Sheet*
World Bank, *World Development Report and Country Reports*

PUBLICATIONS OF GENERAL INTEREST ESSENTIAL FOR INTERPRETING POST-SOVIET DEVELOPMENTS

Amy Chua, *World on Fire: How Exporting Free Market Democracy Breeds Ethnic Hatred and Global Instability* (New York: Doubleday, 2003).

David Harvey, *A Brief History of Neoliberalism* (Oxford: Oxford University Press, 2005).

Samuel P. Huntington, *The Clash of Civilizations and the Remaking of World Order* (New York: Simon and Schuster, 1996).

"Interview Transcript: Mikhail Khodorkovsky," The Moscow Times, 11 September 2008.

Jim Jubak, *Why Russia's woes should worry you*, www.articles .moneycentral.msn.com/Investing/JubaksJournal/why-russias-woes-should-worry-you.aspx?page=2; published 01/06/2009.

Robert D. Kaplan, "The revenge of geography," *Foreign Policy*, May/June 2009; www.foreignpolicy.com/story/cms .php?story_id=4862

Robert D. Kaplan, "Was democracy just a moment?" *The Atlantic Monthly*, December 1997; www.theatlantic.com/doc/199712/democracy-kaplan

Paul Khlebnikov, *"Godfather of the Kremlin: The Decline of Russia in the Age of Gangster Capitalism"* (New York: A Harvest Book/Harcourt Inc.) Forbes, September 2000.

Henry Kissinger, *Does America Need a Foreign Policy? Toward a Diplomacy for the 21st Century* (New York: Simon and Schuster, 2001), 137–138.

Henry Kissinger, "Our nuclear nightmare," *Newsweek*, February 7, 2009; www.newsweek.com/id/183673

Naomi Klein, *The Shock Doctrine* (New York: Metropolitan Books, 2008).

Robert Putnam, Robert Leonardi, and Raffaella Y. Nanetti, *Making Democracy Work: Civic Traditions in Modern Italy* (Princeton: Princeton University Press, 1993).

Charles Tilly, *Democracy* (Cambridge: Cambridge University Press, 2007).

Don Van Atta, *The "Farmer Threat": The Political Economy of Agrarian Reform in Post-Soviet Russia*, (Boulder: Westview Press, 1993).

Fareed Zakaria, *The Future of Freedom: Illiberal Democracy at Home and Abroad* (New York: W. W. Norton, 2003).

Fareed Zakaria, "What the World Really Wants," *Newsweek*, May 29, 2006; www.newsweek.com/id/47895

Fareed Zakaria, "There is a silver lining," *Newsweek*, October 20, 2008; www.newsweek.com/id/163449

RUSSIAN DOMESTIC POLITICS, ECONOMY, SOCIETY, AND FOREIGN POLICY

A. S. Akhiezer, *Rossiya: Kritika Istoricheskogo Opyta* (Russia: A Critique of Historical Experience), (Moscow: FO, 1991).

A. S. Akhiezer, A. P. Davydov, et al, *Sotsio-Kul'turnye Osnovaniya i Smysl Bol'shevizma* (Socio-Cultural Foundations and Meaning of Bolshevism) (Novosibirsk: Sibirskii Khronograf, 2002).

Mark Almond, "In the shadow of the Bronze Soldier," August 19, 2008, www.markalmondoxford.blogspot.com/2008/08/in-shadow-of-bronze-soldier.html

Piotr Aven, "My zalozhili fundament dalneishei zhizni," (We have laid the foundations of further life) *PolitRu*, www.polit.ru/analytics/2006/12/20/aven.html

Nicolas Berdyaev, *The Origin of Russian Communism* (Ann Arbor: University of Michigan Press, 1960).

Pat Buchanan, "Why are we baiting Putin?" *The Conservative Voice*, May 8, 2006.

Pat Buchanan, "To die for Tallinn?" 2007; www.creators.com/opinion/pat-buchanan/to-die-for-tallinn.html

Mary Dejevsky, "The Big Question: Who is Mikhail Khodorkovsky, and why is he on trial again?" *The Independent*, March 6, 2009; www.independent.co.uk/news/world/europe/the-big-question-who-is-mikhail-khodorkovsky-and-why-is-he-on-trial-again-1638456.html

George Demko, Grigory Ioffe, and Zhanna Zayonchkovskaya, *Population under Duress: The Geodemography of Post-Soviet Russia* (Boulder, CO: Westview Press, 1999).

Donald E. Davis, and Eugene P. Trani, *Mirrors: Americans and Their Relations with Russia and China in the Twentieth Century* (Columbia, MO: University of Missouri Press, 2009).

Richard Dreyfuss, "The Rise and McFaul of Obama's Russia Policy," *The Nation*, 07/02/2008; www.thenation.com/blogs/dreyfuss/334120

"Enigma Variations: a Special Report on Russia," *The Economist*, November 29, 2008.

David S. Foglesong, *American Mission and "The Evil Empire": The crusade for a "Free Russia" since 1881* (Cambridge: Cambridge University Press, 2007).

Gosudarstvennyi Doklad o Sostoyanii i ob Okhrane Okruzhayushchei Sredy 2007 (The State Report on the Condition and Protection of Natural Environment); www.mnr.gov.ru/part/?act=more&id=2993&pid=1032

Grigory Ioffe and Tatyana Nefedova, *The Environs of Russian Cities* (Lewiston, NY: Edwin Mellen Press, 2000).

Grigory Ioffe and Tatyana Nefedova, *Continuity and Change in Rural Russia* (Boulder, CO: Westview Press, 1997).

Grigory Ioffe, Tatyana Nefedova, and Ilya Zaslavsky, *The End of Peasantry? The Disintegration of Rural Russia* (Pittsburgh: University of Pittsburgh Press, 2006).

Lucy Komisar, "Yukos Kingpin on Trial," May 10, 2005; www .corpwatch.org/article.php?id=12236

Veronika Krasheninnikova, *Rossiya—Amerika: Kholodnaya Voina Kultur (Russia—America: Cold War of Cultures)* (Moscow: Yevropa, 2007).

Anatol Lieven, "Russia's Limousine Liberals," *National Interest,* 06.10.2009, www.nationalinterest.org/Article.aspx?id=21586

Anatol Lieven, "A Tragic Feud: Alienation between the Russian State and the Liberal Intelligentsia, Past and Present," A meeting report at Kennan Institute 4 June 2007; www.wilsoncenter.org/index.cfm?topic_id=1424&fuseaction=topics.publications&doc_id=303419&group_id=7718

Yuri Mamchur, "Russia's economic crisis could have been avoided," *The Seattle Times,* December 30, 2008; www.seattletimes.nwsource.com/html/opinion/2008573115_opin31mamchur.html

Michael Mandelbaum, "Russia: ease Moscow's suspicions," *Newsweek,* December 8, 2008; www.newsweek.com/id/171258

Michael McFaul and Kathryn Stone-Weiss, "The Myth of the Authoritarian Model," *Foreign Affairs,* January/February 2008.

Dmitry A. Medvedev, "Building Russian-U.S. Bonds, *The Washington Post,* March 31, 2009.

Boris Nemtsov and Vladimir Milov, *Putin. Itogi* (Putin: the Balance Sheet) (Moscow: Novaya Gazeta, 2008) www.sps.ru/file/0034/6516/index.pdf

Viacheslav Nikonov, "Gordoye soznaniye mogushchestva" (Proud awareness of power), *Izvestia,* July 26, 2007, www.izvestia.ru/comment/article3106530/?print

V. V. Patsiorkovsky, *Sel'skaya Rossiya 1991–2001* (Rural Russia 1991–2001) (Moscow: Finansy i Statistika, 2003).

Andrei Piontkovsky, "Russians turning 'an Asian mug,'" *The Russia Journal,* 12.20.1999; www.russiajournal.com/node/2308

Anna Politkovskaya, *A Dirty War: A Russian Reporter in Chechnya* (London: Harvill Press, 2001).

Anna Politkovskaya, *A Small Corner of Hell: Dispatches from Chechnya* (Chicago: University of Chicago Press, 2003).

Vladimir Putin, *Speech at the 43rd Munich Conference on Security Policy,* 10/02/2007; www.securityconference.de/Putin-Rede-2007.381.0.html?&L=1&no_cache=1&sword_list[0]=putin

Leonid Radzikhovsky, Interview to the *Ekho Moskvy* devoted to Russian Anti-Americanism, 02.13.2009; www.echo.msk.ru/programs/personalno/572176-echo.phtml (in Russian)

Tanquary Robinson, *Rural Russia under the Old Regime* (London: Macmillan, 1932).

Natalya Ryzhova and Grigory Ioffe, "Trans-Border Exchange between Russia and China: The Case of Blagoveshchensk and Heihe," *Eurasian Geography and Economics,* vol. 50, no. 3: 348–364, 2009.

Georgiy Satarov, *Korruptsiya (Corruption)*; a series of twenty 2008–2009 articles at http://ej.ru

Victor Shenderovich, "Pochuvstvuite raznitsu" (Feel the difference), *Yezhednevny Zhurnal,* 6 January 2009; www.shender.ru/paper/text/?.file=204

Andrei Shleifer and Daniel Treisman, "A normal country," *Foreign Affairs* March/April 2004; www.foreignaffairs.com/articles/59707/andrei-shleifer-and-daniel-treisman/a-normal-country

Piotr Skwiecinski, "Kompleks Rosji" (The Russia complex), *Rzeczpospolita,* 13.09.2008.

Joseph Stiglitz, "The Ruin of Russia" *The Guardian,* April 9, 2003; www.guardian.co.uk/world/2003/apr/09/russia.artsandhumanities

Nickolas Vakar, *The Taproot of Soviet Society* (New York: Harper and Brothers, 1962).

Mikhail Veller, "Kriticheskiye zametki po natsionalnomu voprosu" (A critical essay on the national question), 2006; www.forum.canada.ru/index.php?action=printage;topic=11861.0

A. G. Vishvevsky, *Serp i Rubl* (The Scythe and the Ruble) (Moscow: OGI, 1999).

Tatyana Vorozheikina, *Samozashchita Kak Pervyi Shag k Solidarnosti* (Self-Defense as the First Step to Solidarity), August 18, 2008; www.polit.ru/research/2008/08/18/vorogejkina.html

Stephen White, Margot Light, and Ian McAllister, "Russia and the West: Is there a Values Gap?" *International Politics* 42(1), 2005: 314–33.

Tomasz Zarycki, "Uses of Russia: The role of Russia in the modern Polish national identity," *East European Politics and Societies,* vol. 18, no. 2, 2004: 1–33.

Zhanna Zayonchkovskaya, "Russia's Search for a New Migration Policy," *European View,* vol. 5, Spring 2007 (Europe and Immigration. Forum for European Studies): 137–145; www.migrocenter.ru/science/zayonchkovskaya.pdf

ADDITIONAL RECOMMENDED READING ON RUSSIA

Anders Aslund and Andrew Kuchins, *The Russia Balance Sheet* (Washington, D.C.: Peterson Institute for International Economics, 2009).

Timothy J. Colton, *Yeltsin: A Life* (New York: Basic Books, 2008).

Timothy J. Colton and Henry E. Hale, "The Putin Vote: Presidential electorates in a hybrid regime," *Slavic Review,* vol. 68, no. 3, 2009: 473–504.

Grigory Ioffe and Tatyana Nefedova, *The Environs of Russian Cities* (Lewiston, NY: Edwin Mellen Press, 2000).

Stephen Kotkin, "Comment: From overlooking to overestimating Russia's authoritarianism?," *Slavic Review,* vol. 68, no. 3, 2009: 548–552.

BIBLIOGRAPHY ON MULTIPLE SOVIET AND POST-SOVIET REPUBLICS

Andrei Danilov, Victor Kistanov, and Sergei Ledovskikh, (Editors), *Ekonomicheskaya Geografiya SSSR* (Economic Geography of the USSR) (Moscow: Vysshaya Shkola, 1983).

Piotr Eberhardt, "The concept of boundary between Latin and Byzantine civilizations in Europe," *Przeglad Geograficzny,* 76:2(2004): 169–208.

Modris Eksteins, *Walking Since Daybreak. A Story of Eastern Europe, World War II and the Heart of the Twentieth Century* (London: Papermac, 2000).

Jean Gottmann, *A Geography of Europe* (Austin: Holt, Rinehart and Winston, 1969).

Miroslav Hroch, *Social Preconditions of National Revival in Europe: A Comparative Analysis of Patriotic Groups among the Smaller European Nations,* 2nd ed. (New York: Columbia University Press, 2000).

Anatoly Khrushchev, Tatyana Kalashnikova, and Igor Nikolsky (eds.), *Ekonomicheskaya Geografiya SSSR* (Economic Geography of the USSR) (Moscow: MGU, 1989).

Roy Medvedev, "Ekonomika stran sodruzhestva. Na raznykh skorostiakh i po raznym dorogam" (The Economy of the CIS members: On different paths and at different speeds), *Ekonomichekiye Strategii,* 8-2005: 14–18; www.inesnet.ru/magazine/mag_archive/free/2005_08/medvedev.htm

Alexander B. Murphy, Terry Jordan-Bychkov, and Bella Bychkova Jordan, *The European Culture Area: A Systematic Geography,* 5th ed. (Lanham: Rowman and Littlefield, 2008).

Vitold Rom, (ed.), *Ekonomichskaya i Sotsialnaya Geografiya SSSR* (Economic and Social Geography of the USSR); Two volumes (Moscow: Prosveshcheniye, 1986).

Bibliography

Andrew Wilson, *Virtual Politics: Faking Democracy in the Post-Soviet World* (New Haven: Yale University Press, 2005).

Ukraine

Balasz Jarabik, *Belarus: Are the Scales Tipping? Madrid: FRIDE, a European Think Tank for Global Action,* 3 March 2009; www.fride.org/publication/576/belarus-are-the-scales-tipping

Maxim Kononenko, "Zapasayemsya popkornom" (Stashing popcorn), February 13, 2009, www.strana.ru/doc.html?id=124427&cid=8

Leonid Kravchuk, "Do chego dovoyevalas Ukraina?" (What has Ukraine achieved by fighting?) *Izvestia,* January 19, 2009; www.izvestia.ru/comment/article3124447/index.html

Leonid Kuchma, *Ukraina—Nie Rossiya* (Ukraine is No Russia) (Moscow: Vremya, 2003).

Roy, Medvedev, "Golodomor: natsionalizatsiya tragedii" (Holodomor: the nationalization of a tragedy), *Russky Zhurnal,* no. 11, November 2008; www.intelros.ru/readroom/2809-russkijj-zhurnal-1-2008g.html

William Green Miller, January 5, 2009, lecture at the Kennan Institute: "Priorities for U.S. Policy toward Ukraine in the Obama Administration," [report can be found at www.wilsoncenter.org/index.cfm?topic_id=1424&fuseaction=topics.publications&doc_id=506201&group_id=7718]

Mykola Ryabchuk, *Vid Malorossii do Ukraini* (From Little Russia to Ukraine) (Kyiv: Kritika, 2000).

Yuri Shevtsov, *Novaya Ideologiya: Holodomor* (A New Ideology: Holodomor) (Moscow: Yevropa, 2009).

Stephen Shulman, "National Identity and Public Support for Political and Economic Reform in Ukraine," *Slavic Review* 64:1(2005): 59–87.

Dmitry Tabachnik, "Ot Ribbentropa do Maidana" (From Ribbentrop to the Square), *Izvestia,* 09.23.2009; www.izvestia.ru/politic/article3133320/index.html

Oleg Tiagnibok, Answers to the Readers' Questions, www.lenta.ru/conf/tyagnibok

Yulia Tymoshenko, "Containing Russia," *Foreign Affairs,* (May/June 2007).

"Victor ludorum," *The Economist,* July 5, 2007; www.economist.com/world/europe/displaystory.cfm?story_id=E1_JQQPGTR

Andrew Wilson, "Cleaning up Ukraine's Pipes," *European Council on Foreign Relations,* January 13, 2009. www.ecfr.eu/content/entry/commentary_gas_crisis_ukraine_ecfr_andrew_wilson/

Andrew Wilson, "Ukraine—From Orange Revolution to Failed State?" a speech in Washington, D.C., May 29, 2009; www.ecfr.3cdn.net/6715d63db63156c9b8_9xm6bhd8j.pdf

Andrew Wilson, *Ukraine's Orange Revolution* (New Haven: Yale University Press, 2005).

Andrew Wilson, *The Ukrainians* (New Haven: Yale University Press, 2002).

Belarus

Igar Babkou, "Genealyogiya Belaruskai idei" (The genealogy of Belarusian idea), *Arche,* 3 (2005): 136–165.

Chernobyl's Legacy: Health, Environmental and Socio-Economic Impacts and Recommendations to the Governments of Belarus, the Russian Federation and Ukraine (New York: United Nations, 2005).

V. I. Golubovich (ed.), *Ekonomicheskaya Istoriya Belarusi* (Economic History of Belarus), 3rd ed. (Minsk: Up Ecoperspectiva, 2004).

Grigory Ioffe, "Belarus and Chernobyl: Separating Seeds from Chaff," *Post Soviet Affairs,* 2007, vol. 23, no. 4: 353–356.

Grigory Ioffe, *Understanding Belarus and How Western Foreign Policy Misses the Mark* (Lanham, MD: Rowman and Littlefield, 2008).

Vitaly Ivanov, "Ne nuzhno zabluzhdadtsia naschiot Belorussii" (One should not be misled on Belarus), *Izvestia,* January 18, 2007; www.izvestia.ru/comment/article3100239/index.html

John Loftus, *The Belarus Secret* (New York: Alfred Knopf, 1982).

David R. Marples, "Color Revolutions: The Belarus Case" *Communist and Post-Communist Studies* 39, no. 3 (2006): 351–64.

David R. Marples, *Belarus: A Denationalized Nation* (Amsterdam: Harwood, 1999).

David R. Marples, *Belarus from Soviet Rule to Nuclear Catastrophe* (New York: St. Martin's Press, 1996).

Svetlana Martovskaya, "Pochemu Belarus golosuyet za Lukashenko" (Why does Belarus vote for Lukashenka?), *Telegraf,* September 14, 2005; accessed at www.inosmi.ru/translation/222228.html.

Nina B. Mečkovskaya, *Belorusskii Yazyk: Sotsiolingvisticheskie Ocherki* (The Belarusian Language: Ethno-Linguistic Essays) (Munich: Verlag Otto Sagner, 2003).

Yelena Rakova, "Polucheniye kredita otlozhno," *Nashe Mneniye,* 25 May 2009; www.nmnby.eu/pub/0905/29j.html

Yury Shevtsov, *Obyedinionnaya Natsiya: Fenomen Belarusi* (A United Nation: the Phenomenon of Belarus) (Moscow: Yevropa, 2005).

Jerzy Turonek, *Bialorus pod Okupacja Niemiecka* (Belarus under German Occupation) (Warsaw: Ksiazka i Wiedza, 1993).

Nicholas Vakar, *Belorussia: The Making of A Nation* (Cambridge: Harvard University Press, 1956).

Stephen White, Elena Korosteleva, and John Loewenhardt (eds.), *Postcommunist Belarus* (Lanham, MD: Rowman & Littlefield Publishers, 2005).

Andrew Wilson, and Clelia Rontoyani, "Security or Prosperity: Belarusian and Ukrainian choices," Robert Legvold and Celleste Wallander (eds.), *Swords and Sustenance: The Economics of Security in Belarus and Ukraine* (Cambridge, MA: MIT Press, 2004).

Moldova

Xenia Fokina, "Budushcheye Moldavii—v sostave Rumynii, no bez nas" (The future of Moldova is within Romania but without us"), *Izvestia,* April 8, 2009.

Anatoly Maximov, "Pochemu molchit Yevropa?" (Why does Europe keep mum?), *Izvestia,* April 9, 2009.

"Moldova and Transdniestria: Another forgotten conflict," *The Economist,* November 13, 2008; www.economist.com/world/europe/displaystory.cfm?story_id=E1_TNGDJSVS

"Moldova burning," *The Economist,* April 8, 2009 www.economist.com/world/europe/displaystory.cfm?story_id=E1_TPQQSTTJ

Moldova, Country Profile 2009, The World Bank; www.enterprisesurveys.org/documents/EnterpriseSurveys/Reports/Moldova-2009.pdf

Moldova: The country that Europe forgot; *Economist.com,* May 18th 2007; www.economist.com/displaystory.cfm?story_id=E1_JTSPQNT&CFID=91132574&CFTOKEN=75531449

Moldova's elections: In the balance, *The Economist,* August 6, 2009; www.economist.com/world/europe/displaystory.cfm?story_id=14191327

Trafficking in Persons Report, June, 2009, Washington, D.C.: U.S. State Department; www.state.gov/g/tip/rls/tiprpt/2009/

Georgia

Lev Anninsky, Oni—gruziny (They are Georgians), *Druzhba Narodov,* 2004, no. 3; www.magazines.russ.ru/druzhba/2004/3/echo26.html

Ellen Barry, "Georgia Challenges Report That Says It Fired First Shot," *The New York Times,* September 30, 2009; www.nytimes.com/2009/10/01/world/europe/01russia.html?hpw

Archil Bezhanishvili, "Tbilisi: Krovavyi mart 1956 goda" (The treacherous March of 1956), March 9, 2008; www.apsny.ge/analytics/1205110952.php

Richard Carlson, "Georgia on his mind—Soros's Potemkin revolution," *The Weekly Standard,* May 23, 2004; www.defenddemocracy.org/index.php?option=com_content&task=view&id=11773241&Itemid=347

C. J. Chivers, and Thom Shanker, "Georgia Lags in Its Bid to Fix Army," *The New York Times,* December 17, 2008, www.nytimes.com/2008/12/18/world/europe/18georgia.html

Shalva K. Chktetia, *Tbilisi v XIX Stoletii* (Tbilisi in the 19th century) (Tbilisi: AN GSSR, 1942).

Robert English, "Georgia: The Ignored History," *NY Review of Books* no. 17, November 6, 2008; www.nybooks.com/articles/22011

Gareth Evans, "Putin twists UN Policy," *The Australian,* September 2, 2008.

Revaz Gachechiladze, *The New Georgia: Space, Society, Politics* (College Station: Texas A&M University Press, 1995).

Georgia Donors Conference, October 22, 2008; *European Commission External Relations;* www.ec.europa.eu/external_relations/georgia/conference/index_en.htm

"Georgia timeline: how crisis unfolded," *Times,* August 12, 2008; www.timesonline.co.uk/tol/news/world/europe/article4514939.ece

Inal-Ipa, S. D. , Zarubezhnye Abkhazy, (Sukhumi: Alashara, 1990).

Alexander Krylov, *Religiya i Traditsii Abkhazov* (Religion and Traditions of the Abkhazians) (Moscow: Institut Vostokovedeniya, 2001).

Georgy Nizharadze, "My—gruziny" (We are Georgians), *Druzhba Narodov,* 1999, no. 10; www.magazines.russ.ru/druzhba/1999/10/nizhar.html

Nicolai Petro, "The Legal Case for Russian Intervention in Georgia," *Fordham International Law Journal,* vol. 32, Issue 5 (May 2009), pp. 1524–1549.

Leonid Radzikhovsky, "Pervyi god uchioby" (The first anniversary of drawing lessons [about the Russian-Georgian war]), *Vzgliad,* August 10, 2009; www.vz.ru/columns/2009/8/10/315953.html

Wendell Steavenson, "Marching through Georgia, *The New Yorker,* December 15, 2008: 64–73.

The Circassian World: A collection of essays and linguistic maps devoted to Abkhaz and other related languages; www.circassianworld.com/northwest.html

Avtandil Tsuladze, "V gostiakh u skazki" (As fairy tale's guests), *Yezhednevnyi Zhurnal,* April 16, 2009; www.ej.ru/?a=note&id=8985

Stefan Wagstyl, and Isabel Gorst, "A harder power," *The Financial Times (London),* August 2, 2009; www.ft.com/cms/s/0/df6d3fca-7f85-11de-85dc-00144feabdc0.html

Yegorov i Kantariya byli ne pervymi (Yegorov and Kantaria were not first), *Argumenty i Fakty,* May 7, 2003; www.gazeta.aif.ru/online/aif/1176/06_01

Armenia

America and the Armenian genocide of 1915: Written statement of the USA, May 28, 1951; www.armenian-genocide.org/Affirmation.388/current_category.6/affirmation_detail.html

Armenian National Committee of America, Press Release: Monday, October 26, 2009, www.anca.org/press_releases/press_releases.php?prid=1584

Armenian National Institute in the USA [a major source of data on Armenian genocide]; www.armenian-genocide.org/

Alexander Arsenyev, "K voprosu o chislennosti naseleniya Armenii," (On Armenia population numbers), *Demoscope,* No. 131–132, October 2003; www.demoscope.ru/weekly/2003/0131/analit06.php

Background Notes: Armenia, October 2009, Washington, D.C.: Department of State, www.state.gov/r/pa/ei/bgn/5275.htm

Peter Balakian, The *Burning Tigris: The Armenian Genocide and America's Response* (New York: HarperCollins Publishers, 2003).

Bahar Baser, and Ashok Swain, "Diaspora design versus homeland realities: case study of Armenian diaspora," *Caucasian Review of International Affairs,* vol. 3(1) Winter 2009: 45–61; www.se2.isn.ch/serviceengine/Files/RESSpecNet/.../Chapter4.pdf

Paul De Bendern, and Hasmik Mkrtchyan, "Soccer diplomacy brings Turkey's Gul to Armenia," September 6, 2008; www.alertnet.org/thenews/newsdesk/L6345007.htm

Interview with Gary Kasparov, *Yerkramas* (The Newspaper of Armenians in Russia), December 6, 2008, www.yerkramas.org/news/2008-12-06-2489

Marc Landner, and Sebnem Arsu, "After Hitch, Turkey and Armenia Normalize Ties," *The New York Times,* October 10, 2009; www.nytimes.com/2009/10/11/world/europe/11armenia.html?_r=1&scp=2&sq=armenia&st=cse

Zvi Lerman, and A. Mirzakhanian, *Private Agriculture in Armenia* (Lanham, MD: Lexington Books, 2001).

Carol Migdalovitz, *Armenia-Azerbaijan Conflict,* Washington, D.C.: CRS Issue Brief for Congress, Updated August 8, 2003, www.au.af.mil/au/awc/awcgate/crs/ib92109.pdf

Naseleniye Armenii (Population of Armenia), 2008; www.am.spinform.ru/people.html

Yossi Shain, "Ethnic Diaspora and U.S. Foreign Policy," *Political Science Quarterly* (Winter 1994/95) 109(5): 811–842. "Troubling news from the Caucasus Bloodshed in Armenia worries both Russia and the West," *The Economist,* March 6, 2008; www.economist.com/world/europe/displaystory.cfm?story_id=E1_TDRDJDDG&source=login_payBarrier

Azerbaijan

Fuad Alekperi, "Kto my, ot kogo proizoshli, i kuda idiom?" (Who are we, who did we descend from, and where are we going?), Zerkalo, 08.08.2009, www.zerkalo.az/2009-08-08/politics/1630-read

Arsenyev, Alexander, "Legendy i mify demografii Zakavkazya" (Legends and myths of the Transcaucasian demography) Demoscope, No. 65–66, 2002; www.demoscope.ru/weekly/2002/065/analit01.php

Natig Aliyev, "The History of Oil in Azerbaijan" (1994) at Azerbaijan International, Summer 1994: 22–23; www.azer.com/aiweb/categories/magazine/22_folder/22_articles/22_historyofoil.html

Azerbaijan: Outcry at Commissars' Reburial, Institute for War and Peace Reporting, February 5, 2009, CRS No. 479, www.unhcr.org/refworld/docid/4991335819.html [accessed 27 October 2009]

Svante Cornell, "Iranian Azerbaijan: A brewing hotspot," Presentation to symposium "Human Rights and Ethnicity in Iran," November 22, 2004 organizd by the Moderate (conservative) Party, Swedish Parliament, Stockholm; www.docs.google.com/gview?a=v&q=cache:QYUGeVR0UyEJ:www.cornellcaspian.com/pub2/0411IRAN.pdf+Svante+Cornell+azerbaijan&hl=en&gl=us&sig=AFQjCNH3B0GKcUm8h70Uz5odcSi68z7ZaQ

Bibliography

Thomas De Waal, *Black Garden: Armenia and Azerbaijan through Peace and War* (New York: New York University Press, 2003).

Geidar Aliyev skonchalsia v Klivlende (Heydar Aliyev dies in Cleveland), NewsRu.com, December, 12 2003; www.newsru .com/arch/world/12dec2003/aliev.html

Karabakhsky Conflict: Azerbaijansky Vzgliad (The Karabakh Conflict: the Azerbaijani View) (Moscow: Yevropa, 2006).

Can Karpat, "Azerbaijani Vice-President: We are Turkey's Link to Russia," Axis Information and Analysis, November 14, 2005; www.axisglobe.com/article.asp?article=479

Karabakhskii Konflikt: Azerbaijanskii Vzglyad, (Moscow: Yevropa, 2006).

Carol Migdalovitz, *Armenia-Azerbaijan Conflict,* Washington, D.C.: CRS Issue Brief for Congress, Updated August 8, 2003, www.au.af.mil/au/awc/awcgate/crs/ib92109.pdf

Nagorno-Karabakh: The risk of a thaw, *The Economist,* November 29, 2007.

Soviet Census of 1989, Ethnic composition of population by republic, www.demoscope.ru/weekly/ssp/sng_nac_89.php?reg=7

Tadeusz Swietochowski, *Russia and Azerbaijan: a Borderland in Transition* (New York: Columbia University Press, 1995).

Kazakhstan

Abdizhapar Abdakayimov, *Istoriya Kazakhstana* (History of Kazakhstan) (Almaty: RIKUML, 1994; available at www .kazakhstan.awd.kz/

Alexander A. Akekseyenko, "Naseleniye Kazakhstana 1926–1939" (Population of Kazakhstan 1926–1939), *Komputer i Istoricheskaya Demografiya;* www.new.hist.asu.ru/biblio/histdem/1.html

Alexander A. Akekseyenko, Naseleniye Kazakhstana mezhdu proshlym i budushchim (Population of Kazakhstan between the past and future), *Demoscope Weekly,* No. 245–246, May 1–21, 2006; www.demoscope.ru/weekly/2006/0245/tema02 .php

Nurlan Aremkulov, "Zhuzy v sotsialno-politicheskoi zhizni Kazakhstana," (Jus'es in socio-political life of Kazakhstan), CA&CC Press, Sweden, September 2000; www.ca-c.org/journal/ cac-09-2000/16.Amrek.shtml

Marc E. Babej and Tim Pollak, "Unsolicited Advice: Kazakhstan versus Borat," *Forbes,* October 12, 2006: www.forbes .com/2006/10/11/unsolicited-advice-advertising-oped_meb_ 1012borat.html

Claire Bigg, "Solzhenitsyn Leaves Troubled Legacy Across Former Soviet Union," RFE/ RL, August 06, 2008; www.rferl .org/content/Solzhenitsyn_Leaves_Troubled_Legacy_Across_ Former_Soviet_Union/1188876.html

Beate Eshment, "Problemy russkikh Kazakhstana: etnichnost ili politika?" (Problems of Russians in Kazakhstan: Ethnicity or politics?), *Russky Mir;* www.archipelag.ru/ru_mir/rm-diaspor/ russ/russkie-kazahstana/

Phyllis Fletcher, "Kazakhstan Embassy Responds to Borat," October 16, 2006, *National Public Radio;* www.npr.org/ templates/story/story.php?storyId=6276973

Kazakh SSR KGB documents regarding December 1986 riots in Alma-Ata; (in Russian) www.libertykz.narod.ru/book/razdel9 .html

Kazakhstan Background, *Energy Information Administration, Official Energy Statistics from the US Government;* www.eia .doe.gov/emeu/cabs/Kazakhstan/Background.html

Kazakhstan Oil, *Energy Information Administration, Official Energy Statistics from the US Government,* www.eia.doe.gov/ emeu/cabs/Kazakhstan/Oil.html

Amanzhol Kuzembayuly, and Yerkin Amanzholluly, *Istoriya Respubliki Kazakhstan* (History of the Republic of Kazakhstan), (Astana: Foliant, 1999); available at www.kazakhstan.awd.kz/

Nancy Lubin, and Arustan Joldasov, "Central Asians take stock: Comparison of results from public opinion survey, Uzbekistan and Kazakhstan," 1993 & 2007; (Seattle: NCEEER, 2009). www.ucis.pitt.edu/nceeer/2009_822-08_Lubin.pdf

Roy Medvedev, "Stanovleniye kazakhstanskoi modeli ekonomiki," (The formation of the Kazakhstan economic model), October 6, 2006; *Central Asia,* www.centrasia.ru/newsA .php?st=1160088900

Nikolai Nasedkin, *Entsyklopediya Dostoyevsky* (Encyclopedia Dostoyevsky), "Valikhanov, Chokan Chingisovich," www.niknas .by.ru/dost_enz/dost_enz3-05.htm

Neftyanaya otrasl Kazakhstana (The Oil Industry of Kazakhstan), Kazmunaigaz, www.kmg.kz/page.php?page_id=294&lang=1

Neftyanaya promyshlennost Respubliki Kazakhstan (The Oil Industry of the Republic of Kazakhstan), Astana: Ministry of Energy and Mineral Resources, www.memr.gov.kz/?lng= rus&mod=oil

Dosym Satpayev, "Poyedut li russkiye v Rossiyu iz Kazakhstana?" (Will Russians leave Kazakhstan for Russia?) *APN Kazakhstan,* August 3, 2006; www.apn.kz/publications/article5135.htm

Alexander Solzhenitsyn, Kak Nam Obustroit Rossiyu (How We Ought to Upgrade Russia) (Moscow: *Komsomolskaya Pravda,* 1990). www.lib.ru/PROZA/SOLZHENICYN/s_kak_1990.txt

Yelena Tarasova, "Pochemu russkiye prodolzhayut uyezzhat iz Kazakhstana?" (Why do Russians continue to leave Kazakhstan?) *APN Kazakhstan,* March 21, 2006; www.apn.kz/publications/ article266.htm

Yevgeniya P. Zimovina, "Osnovnye etapy formirovaniya diasporalnoi struktury Kazakhstana I stran Tsentralnoi Azii," (The principal stages of the diasporas' formation in Kazakhstan and Central Asia), *Demoscope Weekly,* no. 377–378, May 18–31, 2009; www.demoscope.ru/weekly/2009/0377/analit04.php

Yevgeniya Sadovskaya, "Perenos stolitsy iz Alma-Aty v Astanu i yego vliyaniye na migratsionnye protsessy" (The transfer of the capital from Almaty to Astana and its influence upon migration processes), *Democsope Weekly,* No. 71–71, June 17–30, 2002; www.demoscope.ru/weekly/2002/071/analit03.php

Dickie Wallace, "Hyperrealizing "Borat" with the Map of the European "Other" *Slavic Review,* Vol. 67, No. 1 (Spring, 2008): 35–49.

Uzbekistan

Sergei Abashin, "The transformation of ethnic identity in Central Asia: a case study of the Uzbeks and Tajiks," (London: The International Institute for Strategic Studies, www.iiss.org/ programmes/russia-and-eurasia/copyof-russian-regional- perspectives-journal/copyof-rrp-volume-1-issue-2/the- transformation-of-ethnic-identity-in-central-asia/

Edward A. Allworth, *The Modern Uzbeks* (Stanford: Hoover Institution Press, 1990).

Central Asia and Uzbekistan History, The Twentieth Century; www.eastlinetour.com/uzbekistan/new_history.html

"Crisis resistant or crisis prone? Uzbekistan's economy is growing but so are the risks," *Economist Intelligence Unit ViewsWire,* July 21, 2009; www.economist.com/daily/news/displaystory.cfm? story_id=14070444

Integrated Water Management in Fergana Valley, IWRM 2006 (In Russian); www.cawater-info.net/library/rus/iwrm_fergana.pdf

L. Landa and H. Tursunov, "Natsionalno-gosudarstvennoye razmezhevaniye respublik Srednei Azii" (National delimitation of Central Asian republics) (Moscow: *Bolshaya Sovetskaya Entsyklopedia,* 1979).

Nancy Lubin and Arustan Joldasov, "Central Asians take stock: Comparison of results from public opinion survey, Uzbekistan and Kazakhstan," 1993 & 2007; Seattle: NCEEER 2009; www.ucis.pitt.edu/nceeer/2009_822-08_Lubin.pdf

Calcum Macleod and Bradley Mayhew, *Uzbekistan: The Golden Road to Samarkand* (Hong-Kong: Odyssey Publications, 6th ed., 2008).

Lyudmila Maksakova, "Demografichesky put Uzbekistana," (The demographic path of Uzbekistan), *Demoscope Weekly,* no. 235–236; February 20–March 5, 2006; www.demoscope.ru/weekly/2006/0235/tema01.php

Konstantin Smirnov, "Zalozhniki nischchety ili russkie v Uzbekistane" (Hostages of poverty or Russians in Uzbekistan), *Rossiiskaya Fedetratsiya Segodnya* October 23, 2006; Reprinted at www.demoscope.ru/weekly/2006/0263/gazeta030.php

Igor Torbakov, "Tajik-Uzbek relations: divergent national historiographies threaten to aggravate tensions," *EurasiaNet.org;* June 12, 2001; www.eurasianet.org/departments/culture/articles/eav061201.shtml

V odnikh stranakh snizhayetsya yestestvennyi prirost, v drugikh narastayet yestestvennaya ubyl (Whereas in some countries natural increase goes up, in other countries natural decrease does), *Demoscope Weekly* No. 197–198, April 4–17, 2005; www.demoscope.ru/weekly/2005/0197/barom03.php

Vasily Barthold o natsionalnom razmezhevanii v Srednei Azii (Vassily Barthold about the national delimitation in Central Asia), CA&CC Press, Sweden, www.ca-c.org/datarus/st_13_olimov.shtml; reprinted from the journal *Vostok,* No. 5, 1991.

Igor Zenkov, "Tashkentsky zavod budet rabotat na vooruzheniye Kitaya (The Tashkent aircraft factory will work to arm the Chinese army), February 21, 2006; *Ferghana.Ru,* www.ferghana.ru/article.php?id=4251

Kyrgyzstan

John Andersen, *Kyrgyzstan: Central Asian Island of Democracy?* (Abingdon: Routledge, 1999).

Abilabek Asankanov, "Ethnic conflict in the Osh region in summer 1990: Reasons and lessons," Rupesinghe, Kumar and Valery A. Tishkov (eds.), *Ethnicity and Power in the Contemporary World* (Tokyo: United Nations University Press, 1996); www.unu.edu/unupress/unupbooks/uu12ee/uu12ee0d.htm

Gold Mining in Kyrgyzstan, *MBendi Information Services,* www.mbendi.com/indy/ming/gold/as/kg/p0005.htm

Kyrgyzstan, *AdvanTour;* www.advantour.com/rus/kyrgyzstan/index.htm (in Russian)

"Kyrgyzstan's elections: Tulips Squashed," *The Economist,* July 30, 2009; www.economist.com/world/asia/displaystory.cfm?story_id=14140310

Michael Mainville, "U.S. Bases Overseas Show New Strategy," *Pittsburgh Post-Gazette* (Pennsylvania), July 26, 2004; www.globalsecurity.org/org/news/2004/040726-us-bases.htm

"Tripolar disorder: The new Great Game, from the ground up," *The Economist,* Oct 1st 2009; www.economist.com/displaystory.cfm?story_id=14530938

Tajikistan

Rasheed Abdullaev and Umed Babakhanov, "Thank the Taliban for the Tajik Peace Agreement," In: Rutland, Peter, (ed) *Annual Survey of Eastern Europe and the Former Soviet Union 1997: The Challenge of Integration* (Armonk, NY: M. E. Sharpe, 1998).

Christian Bleuer, "On Civil War in Tajikistan," April 2007; www.easterncampaign.wordpress.com/2007/04/10/afghanistans-nightmare-scenario-pakistani-civil-war-refugees/

Vladimir Dronov, "O vzaimootnosheniyakh mezhdu Tadzhikistanom i Uzbekistanom" (About the relationships between Tajikistan and Uzbekistan), *IA Center,* November 29, 2007; www.ia-centr.ru/archive/public_detailsa0f1.html?id=1022

Ekonomika Tajikistana n(Tajikistan's Economy), *EastTimeRu,* www.easttime.ru/countries/topics/1/10/75.html

Embassy of the United States of America in Tajikistan: Embassy History; www.dushanbe.usembassy.gov/history.html

Geografiya Tajikistana (The Geography of Tajikistan), *AdvanTour;* www.advantour.com/rus/tajikistan/geography.htm

Vladimir Georgiyev, "Optiko-elektronnyi kompleks Okno v Tadzhikistane de jure stanovitsya rossiiskim" (The optical-electronic complex Okno in Tajikistan becomes Russian de jure), FerghanaRu, April 17, 2006, www.ferghana.ru/article.php?id=4351

Yury Kazarin, "Tajiki ugrozhayut Rossii demograficheskim terrorom" (Tajiks threaten Russia by demographic terror), *Internet Protiv Teleekrana* 2009; www.contr-tv.ru/common/287

Artiom Kobzev, "Tajiki chasche drugikh migrantov narushayut zakon v Rossii" (Tajiks break law in Russia more often than other migrants), *GZT.ru,* August 21, 2009; www.gzt.ru/255948.html

Ann K. S. Lambton, *Landlord and Peasant in Persia* (London: I.B. Tauris, 1991).

Migratsiya iz Tajikistana glazami tajikov (Migration from Tajikistan through Tajik eyes), *Tochikiston Rossiya,* January 27, 2007; www.tajinfo.ru/tema/1170055755/pindex.shtml

Hairulloh Mirsaidov, "Aerodrom Aini ostayotsya poka za Tajikistanom" (The Aini airport is a Tajik property as of yet), *The Deutsche Welle,* July 10, 2009; www.dw-world.de/dw/article/0,4470841,00.html

Vladimir Nosov, "Grazhdanskaya voina v Tajikistane" (Civil war in Tajikistan), 2006; www.conflictologist.narod.ru/taj90.html

Operational Group of Russian Forces in Tajikistan; *Global Security,* www.globalsecurity.org/military/world/russia/ogrv-tajikistan.htm

Dilip Ratha, Sanket Mohapatra, K. M. Vijayalakshmi, and Zhimei Xu (2007-11-29), "Remittance Trends 2007. Migration and Development Brief 3; (Washington, D.C.: *World Bank,* 2008); www.siteresources.worldbank.org/EXTDECPROSPECTS/Resources/476882-1157133580628/BriefingNote3.pdf

Alexander Reutov, "Rogunskaya GES vydala pervoye napriazheniye" (The Rogoun Hydroelectric Station Has Started to Produce Electricity), *Kommersant,* January 29, 2009; www.kommersant.ru/doc.aspx?DocsID=1109698

James Risen, "A nation challenged: Qaeda diplomacy; Bin Laden Sought Iran as an Ally, U.S. Intelligence Documents Say," *The New York Times,* December 31, 2001; www.nytimes.com/2001/12/31/world/nation-challenged-qaeda-diplomacy-bin-laden-sought-iran-ally-us-intelligence.html

Rutland (ed.), *Annual Survey of Eastern Europe and the Former Soviet Union 1997: The Challenge of Integration,* M. E. Sharpe: 1998: 393.

Said Abdullo Nuri, Conciliation Resources, UK; www.c-r.org/our-work/accord/tajikistan/profiles.php

Bakhodir Sidikov, "Tursky uzel" (A turk knot), *Zavtra,* September 5, 2000; www.zavtra.ru/cgi/veil/data/zavtra/00/353/62.html

Tajik guests of Iran's Persian literature congress arrive, *Tehran Times,* January 11, 2009; www.irpedia.com/iran-news/1238/

Bibliography

Tajikistan: Industry, Encyclopedia of the Nations, www.nationsencyclopedia.com/Asia-and-Oceania/Tajikistan-INDUSTRY.html

Mikhail Tulsky, "Itogi perepisi naseleniya Tajikistana" (The results of Tajikistan's Census) *Demoscope Weekly,* no. 37–38, October 8–21, 2001; www.demoscope.ru/weekly/037/evro04.php

V Tajikistane poyavilas nasha voyennaya baza (In Tajikistan our military base has emerged), *Delovaya Pressa,* October 21, 2004; www.businesspress.ru/newspaper/article_mId_40_aId_319386.html

Sergei Zhigarev, "Stroitelstvo GES v Tajikistane i Kyrgyzstane—prestupnoye legkomysliye" (Building hydroelectric stations in Tajikistan and Kyrgyzstan is criminal carelessness), *Regnum,* September 8, 2009; www.regnum.ru/news/1203508.html

Turkmenistan

2007 Investment Climate Statement—Turkmenistan, Embassy of the USA to Turkmenistan; www.turkmenistan.usembassy.gov/ic_report07.html

A Short History of the Akhal-Teke Horse, Akhal-Teke Society of America; www.akhaltekesocietyofamerica.com/index_Page1157.htm

Rafis Abazov, *Historical Dictionary of Turkmenistan* (Lanham MO: The Scarecrow Press, 2005).

Ravil Atzhanov, "Turkmenistan zavoyiovyvayet mir" (Turkmenistan triumphs over the world), *Ogoniok* no. 52, 2008; www.ogoniok.com/archive/2001/4680-2/80-26-27/

Crude Accountability: Mitro International Limited www.crudeaccountability.org/ru/index.php?page=mitro

Glenn E. Curtis, ed. *Turkmenistan: A Country Study* (Washington DC: GPO for the Library of Congress, 1996); www.countrystudies.us/turkmenistan/

Encyclopedia Turkmenskoi SSR (Encyclopedia of Turkmen SSR), Ashkhabad: GRTSE 1984

Shohrat Kadyrov, *Rossiisko-Turkmensky Istorichesky Slovar* (Russian-Turkmen Historical Dictionary), Two volumes (Bergen, Norway: Bodoni Hus, 2001); www.turkmeny.h1.ru/

Oleg Kusov, "Mnogovektornaya gazovaya politika Turkmenistana: Pochemu opasno sidt u Rossii na privyazi" (Multi-vector gas policy of Turkmenistan: Why is it dangerous to be on Russia's leash?), *Radio Liberty,* September 8, 2009; www.svobodanews.ru/content/transcript/1819359.html

Merv (A description of the Merv oasis' glorious history), Owadan Tourism, www.owadan.net/turkmenistan/merv.htm

Niyazov, Saparmurat (a brief biography), LentaRu 2007, www.lenta.ru/lib/14160039/full.htm

Martha Brill Olcott, *International Gas Trade in Central Asia: Turkmenistan, Iran, Russia, and Afghanistan,* Working Paper No. 28, May 2004, James Baker III Institute for Public Policy, Program on Energy and Sustainable Development at the Center for Environmental Science and Policy Stanford Institute for International Studies, Stanford University; www.rice.edu/energy/publications/docs/GAS_InternationalGasTradeinCentralAsia.pdf

Priroda Turkmenistana (Nature of Turkmenistan), *AdvanTour,* www.advantour.com/rus/turkmenistan/nature.htm

Ruhnama, a book by Niyazov and a mandatory read in Turkmenistan; www.turkmenistan.gov.tm/ruhnama/ruhnama-eng.html

Spisok literatury po ashkhabadskomu zemletryaseniyu (Bibliography on the Ashgabat earthquake); www.scgis.ru/russian/cp1251/dgggms/2-98/reestr.htm

Sredneaziatskaya Zheleznaya Doroga: Istoriya (History of Central Asian Railway Lines); www.rrh.agava.ru/encyclopedia/railroads/sredasia.htm

Yevgenyi Trushin, "Drozhzhi dlya diktatora" (Nourishing the dictator), *Moskovsky Komsomolets,* July 12, 2005; www.compromat.ru/page_10213.htm

Turkmenia (Turkmenistan), www.militera.lib.ru/common/show/01_37.html

Turkmenistan: Velikoye Odinochestvo (Turkmenistan: A great solitude) (Saint Petersburg: Ekonomicheskaya Shkola, 2007); www.seinstitute.ru/Files/SSSR-ch12turk_p455-476.pdf

"Upper Jurassic Carbonates of Cheleken Structure, Turkmenistan." *Internet Geology News Letter* No. 64, September 25, 2000 www.geocities.com/internetgeology/L64a.html

Index

Index

Index

Index